CLASSICAL ELECTRODYNAMICS
SECOND EDITION

Contents

---

*Sections marked by an asterisk are optional. The material in these sections is not needed for later parts of the book.

Preface

Einstein's celebrated article "Zur Elektrodynamik bewegter Körper" is commonly regarded as laying the foundation of relativity. But it must also be regarded as placing the capstone on electrodynamics (see the back end papers of this book). In this article, Einstein achieved the complete unification of electricity and magnetism. He showed that electric and magnetic fields are merely different aspects of one single underlying entity, the four-dimensional field tensor. What is electric and what is magnetic depends on the observer's point of view—under Lorentz transformations the electric and magnetic fields metamorphose into one another. Paraphrasing Minkowski's well-known remark on the relativistic unification of space and time, we may say that "henceforth the electric field by itself, and magnetic field by itself, are doomed to fade away into mere shadows, and only a kind of union of the two will preserve an independent reality." Thus, Einstein established electrodynamics as the first example of a unified theory.

Regrettably, a century after Einstein, most textbooks on electrodynamics continue to treat electric and magnetic fields as independent entities. Of course, these fields are linked by Maxwell's equations; but in the three-dimensional formulation of the equations, this linkage is commonly misconstrued as a mere interaction between the fields. To gain a deeper insight into the essential unity of the electric and magnetic fields one must look at the four-dimensional formulation of Maxwell's equations. The first undergraduate textbook to approach electrodynamics from a relativistic point of view was, I believe, *Principles of Electrodynamics* by M. Schwartz. I taught this approach for several years in a senior-level course at Rensselaer Polytechnic Institute and gradually developed a set of lecture notes for the students. My book grew out of these lecture notes.

The first five chapters follow the traditional introductory road into electricity: vector calculus, electrostatic field and potential, boundary-value problems, dielectrics, and electric energy. The road then makes a sharp turn into the relativistic domain in Chapters 6 and 7; these chapters give an overview of the physical foundations of special relativity and of the four-dimensional tensor formalism. The road then takes us into the heartland of the book in Chapter 8; here, we see how the union of Coulomb's law with the laws of special relativity gives issue to the relativistic form of Maxwell's equations. From there, the road branches out into several applications of Maxwell's equations in Chapters 9 through 16: magnetostatics, induction, magnetic materials, electromagnetic waves, radiation by point charges and by a system of charges, resonant cavities and waveguides, and scattering and diffraction.

Throughout the chapters, I have paid careful attention to the experimental evidence that serves as the basis for the theoretical development. Thus, Chapters 1 and 2 include a discussion of the evidence for charge conservation and for the

$1/r^2$ force law; and Chapter 6 includes a thorough summary of the evidence for special relativity. However, I have not bothered to discuss the experimental evidence for the magnetic law of force or for the law of induction, because these laws are merely consequences of previously established laws. With this attitude, I do not mean to express contempt for the achievements of Jean Baptiste Biot, Félix Savart, or Michael Faraday—on the contrary, I have always stood in awe of Biot and Savart, being totally unable to comprehend how anybody could assemble anything as tangled as their law from bitty pieces of experimental data. By way of tribute to the incredible feats of the many generations of scientists who patiently assembled the great pyramid of electromagnetism, I have prepared a genealogical chart (see the back end papers). This chart displays only the main lines of relationships between discoveries and discoverers; many collateral liaisons have been discreetly omitted.

Wherever possible, I have explored areas of contact between classical electrodynamics and quantum physics, such as electrostatic energy of nuclei, image charges as confinement mechanism for free electrons, magnetic monopoles, angular momentum in electromagnetic waves, radiation by atoms and nuclei, and lasers. Although a complete treatment of these topics requires quantum mechanics, it is possible to gain considerable insight into the basic physics by extrapolating classical concepts.

Besides the problems at the ends of the chapters, the book contains exercises scattered throughout the chapters. These exercises fill in some details of the derivations, generalize discussions, or introduce some additional material. A reader traveling through the chapters will do well to stop occasionally at one of these exercises and work out the details

I employ CGS electrostatic units in this book (the appendix offers a list of equivalences between CGS and SI units and a dictionary for the translation of CGS equations into SI equations.) The reasons for giving preference to these units in a relativistic context are incontrovertible: since the electric and magnetic fields are merely different aspects of the same underlying entity, they should be measured in the same units. Of course, the same argument applies to space and time—they should also be measured in the same units. In a first draft of this book, I did just that; but in the final version, I felt compelled to revert to centimeters and seconds. The 1983 redefinition of the meter in terms of the distance traveled by light in a given interval of time is a commendable step toward a unified treatment of the units of length and time. The next logical step would be to regard the speed of light as a conversion factor between the units of length and time, as advocated by E. F. Taylor and J. A. Wheeler in *Spacetime Physics*. I am hopeful that after a few more 10^{18} centimeters have elapsed, time will finally be ripe for centimeters.

In this second edition, I have corrected a handful of errors and misprints. I am grateful to readers who brought some of these to my attention. Other changes in the new edition include updated experimental data (especially in the chapter dealing with relativity), expanded explanations, additional examples, additional problems, some reorganization (Sections 1.1 and 12.4 of the old edition have been split into two sections), discussion of the concept of negative oscillator strengths, and a new section dealing with lasers and the threshold condition for lasing.

I am indebted to John R. Albright (Florida State University), Robert Cordery (Northeastern University), John F. Dicello (Clarkson University), R. A. Guyer (University of Massachuesetts, Amherst), Kannan Jagannathan (Amherst College), John A. Jungerman (University of California, Davis), James A. McLennan (Lehigh University), William C. Oelfke (University of Central Florida), Gerhard L. Salinger (Rensselaer Polytechnic Institute), and Joseph E. Willett (University of Missouri-Columbia) for their helpful reviews of the manuscript and I am indebted to Farhang Amiri (Weber State University), Steven M. Grimes (Ohio University), C. A. Quarles (Texas Christian University), Mark D. Semon (Bates College), G. R. Storey (University of Auckland), George Vahala (College of William and Mary) for comments on the first edition and many suggestions for improvements. I am also indebted to the late Roland M. Lichtenstein (Rensselaer Polytechnic Institute) for his advice on the books in the bibliography and for bringing many of these books to my attention.

H. C. O. *Charlotte, Vermont, 2006*

NOTE: Although the book is intended for a two-semester course, it can be used for a one-semester course, with judicious deletion of some chapters and sections. The following list indicates which chapters and sections might be deleted if the book is used for a one-semester course.

Section 1.2
Section 2.5
Section 3.3, 2nd half (Bessel functions)
Section 4.4
Section 4.5
Section 5.4, 2nd half (coefficients of capacitance)
Section 6.3
Section 9.7
Section 12.5
Section 12.7
Section 12.8
Section 12.9
Chapter 13
Chapter 14
Chapter 15
Chapter 16

CHAPTER

Vector Calculus 1

Vector calculus deals with vectors and with their derivatives and integrals. Vector calculus plays a central role in the study of electromagnetic theory. The equations for the electric and magnetic fields (Maxwell's equations) link components of these fields in different directions; such equations can be expressed most concisely in vector notation. In fact, much of vector calculus was invented for the specific purpose of simplifying the equations of electromagnetic theory. To appreciate the advantages of vector notation, one need only compare the cumbersome formulas in Maxwell's old treatise *Electricity and Magnetism* with the much neater formulas in modern textbooks.

This first chapter is mostly mathematics. It introduces a concise superscript notation for vectors and tensors, and it establishes some valuable mathematical results, which will be much used in later chapters. Only in the last section of this chapter will we deal with an application of our mathematical formulas to a question in physics, that is, the conservation of electric charge.*

1·1 Scalars and Vectors

Roughly, a vector is a quantity that has magnitude and direction and that behaves like the position vector under addition and multiplication, and under rotations. Thus, the position vector is to be regarded as the prototype for all vectors. In this section, we begin with the precise definition of the position vector and afterward spell out in detail how any other vector "behaves like" the position vector.

We will describe points of space by means of their Cartesian coordinates x, y, and z relative to a rectangular coordinate system (see Fig. 1.1). We can then lay down the following definition:

• ***Definition:*** The position vector of a point with given coordinates x, y, z is the three-component object

$$(x, y, z) \qquad (1)$$

---

*Readers impatient with mathematics can begin with this last section and read the earlier sections as needed.

Figure 1·1

Point P with Cartesian coordinates x, y, z.

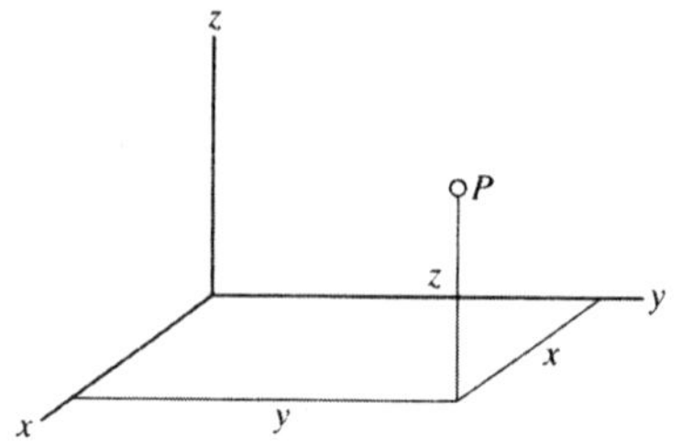

We will sometimes write the position vector in the form

$$\begin{pmatrix} x \\ y \\ z \end{pmatrix} \tag{2}$$

The objects in (1) and (2) are called *row* and *column* vectors, respectively. As an alternative to (x, y, z), we find it useful to introduce the notation

$$(x^1, x^2, x^3) \tag{3}$$

We will usually abbreviate this to

$$x^k \tag{4}$$

where it is understood that the superscript k has the values $k = 1, 2, 3$. Strictly speaking, x^k is the kth component of the vector. However, for the sake of brevity, we will often use the phrase "the vector x^k"; in this context, x^k then stands for the set of all three components, that is, it stands for the entire vector. The superscript notation for the vectors has the advantage that it easily generalizes to four-dimensional vectors, needed in the study of relativity (Chapter 7).

The expressions (1)–(4) are component notations for the position vector. We will also find it convenient to use the abstract (or geometrical) notation

$$\mathbf{x} \tag{5}$$

instead of (1)–(4). The vector $\mathbf{x}$ can be represented graphically as a directed line segment (see Fig. 1.2).

Figure 1·2

Position vector **x** of the point *P*.

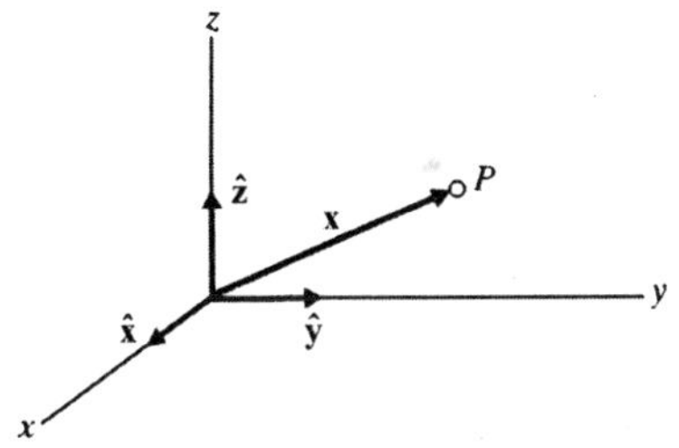

Addition of position vectors is defined as addition of their components—for example,

$$(5, 3, 0) + (-1, 0, 2) = (4, 3, 2)$$

Multiplication of vector by a number is defined as multiplication of each component—for example,

$$2(-1, 0, 2) = (-2, 0, 4)$$

The magnitude of the vector **x** is $|\mathbf{x}| = \sqrt{(x^1)^2 + (x^2)^2 + (x^3)^2}$. A vector is called a *unit vector* if its magnitude is 1.

The unit vectors along the rectangular axes are defined as

$$(1, 0, 0) \quad (0, 1, 0) \quad (0, 0, 1) \tag{6}$$

In abstract notation, we will write these unit vectors as*

$$\hat{\mathbf{x}},\ \hat{\mathbf{y}},\ \hat{\mathbf{z}} \tag{7}$$

We can express an arbitrary position vector **x** as a superposition of unit vectors:

$$\mathbf{x} = x(1, 0, 0) + y(0, 1, 0) + z(0, 0, 1)$$

or

$$\mathbf{x} = x\hat{\mathbf{x}} + y\hat{\mathbf{y}} + z\hat{\mathbf{z}} \tag{8}$$

---

*Note that the unit vector $\hat{\mathbf{x}}$ is *not* in the direction of the position vector **x**. The unit vector in the direction of the position vector **x** will be designated by $\hat{\mathbf{r}}$. An alternative notation, popular in introductory textbooks, employs **i, j, k** for the unit vectors. But this is awkward in electrodynamics because **j** and **k** are needed to represent the electric current density and the wave vector, respectively.

The *scalar product,* or *dot product,* of two position vectors **x** and **X** is

$$\mathbf{x}\bullet\mathbf{X} = xX + yY + zZ \qquad (9)$$

This definition is equivalent to the elementary definition of the dot product as the product of the magnitudes of the vectors and the cosine of the angle between them,

$$\mathbf{x}\bullet\mathbf{X} = |\mathbf{x}||\mathbf{X}|\cos\phi$$

This can also be interpreted as the product of the magnitude of one vector and the component of the other along the first.

■ ***Exercise 1.*** Verify the equivalence of these definitions [*Hint*: Suppose that, by choice of the coordinate axes, the vector **x** lies along the z direction; then Eq. (9) reduces to $\mathbf{x}\bullet\mathbf{X} = zZ$.] ■

The dot products of the unit vectors $\hat{\mathbf{x}}$, $\hat{\mathbf{y}}$, and $\hat{\mathbf{z}}$ are

$$\hat{\mathbf{x}}\bullet\hat{\mathbf{x}} = \hat{\mathbf{y}}\bullet\hat{\mathbf{y}} = \hat{\mathbf{z}}\bullet\hat{\mathbf{z}} = 1 \quad \text{and} \quad \hat{\mathbf{x}}\bullet\hat{\mathbf{y}} = \hat{\mathbf{x}}\bullet\hat{\mathbf{z}} = \hat{\mathbf{y}}\bullet\hat{\mathbf{z}} = 0$$

These dot products of the unit vectors can be regarded as consequences of Eq. (9); alternatively, Eq. (9) can be regarded as a consequence of these dot products of the unit vectors.

We can also write Eq. (9) as

$$\sum_{k=1}^{3} x^k X^k \qquad (10)$$

Here, it is convenient to introduce the *Einstein summation convention:* If a Latin vector index appears twice in a product, summation over this index is taken for granted. Thus, (10) is the same as

$$x^k X^k \qquad (11)$$

Note that k is a dummy index that is summed over and disappears when the summation is written out explicitly, as in Eq. (9); it therefore makes no difference whether we write $x^k X^k$, $x^n X^n$, or $x^l X^l$, or any other dummy index.

With the Einstein summation convention, we can write the magnitude of a vector concisely as

$$|\mathbf{x}| = \left(x^k x^k\right)^{1/2} \qquad (12)$$

The *vector product, or cross product*, of two position vectors $\mathbf{x}$ and $\mathbf{X}$ is*

$$\mathbf{x}\times\mathbf{X} = (yZ - zY)\hat{\mathbf{x}} + (zX - xZ)\hat{\mathbf{y}} + (xY - yX)\hat{\mathbf{z}} \qquad (13)$$

This definition is equivalent to the definition of the cross product as a vector with magnitude equal to the product of the magnitudes of the two vectors and the sine of the angle between them, and with a direction given by the right-hand rule. As a convenient mnemonic for the complicated expression on the right side of Eq. (13), we can use the determinant

$$\begin{vmatrix} \hat{\mathbf{x}} & \hat{\mathbf{y}} & \hat{\mathbf{z}} \\ x & y & z \\ X & Y & Z \end{vmatrix}$$

From Eq. (13) we can verify that the cross product is anticommutative,

$$\mathbf{x}\times\mathbf{X} = -\mathbf{X}\times\mathbf{x}$$

and that the cross products of the unit vectors $\hat{\mathbf{x}}, \hat{\mathbf{y}}$, and $\hat{\mathbf{z}}$ are

$$\hat{\mathbf{x}}\times\hat{\mathbf{y}} = \hat{\mathbf{z}} \qquad \hat{\mathbf{y}}\times\hat{\mathbf{z}} = \hat{\mathbf{x}} \qquad \hat{\mathbf{z}}\times\hat{\mathbf{x}} = \hat{\mathbf{y}}$$

To write Eq. (13) in component notation, we introduce the *alternating tensor* ε^{klm}, which is a $3\times3\times3$ matrix whose components are all zero except for the following:

$$\varepsilon^{123} = \varepsilon^{231} = \varepsilon^{312} = 1$$
$$\varepsilon^{213} = \varepsilon^{132} = \varepsilon^{321} = -1 \qquad (14)$$

Thus, the nonzero components of ε^{klm} equal +1 or −1, depending on whether *klm* is obtained from 123 by an even or an odd number of permutations of adjacent indices. In terms of ε^{klm}, we can write the cross product of x^l and X^m as

$$\varepsilon^{klm} x^l X^m \qquad (15)$$

According to the summation convention, Eq. (15) involves two (separate) summations over *m* and *l*. Written out explicitly, the expression (15) is

---

*This assumes that the coordinate system is "right-handed," with the axes oriented as shown in Fig. 1.2. We will always use right-handed coordinate systems.

$$\sum_{l=1}^{3}\sum_{m=1}^{3}\varepsilon^{klm}x^l X^m$$

■ ***Exercise 2.*** Verify that the components of Eq. (15) agree with those of Eq. (13). ■

■ ***Exercise 3.*** Verify that Eq. (15) implies that the cross product is anticommutative. ■

The definitions of the dot product and the cross product of two arbitrary vectors **A** and **B** (or A^k and B^k in component notation) are the same as for position vectors:

$$\mathbf{A}\cdot\mathbf{B} = A_x B_x + A_y B_y + A_z B_z$$

and

$$\mathbf{A}\times\mathbf{B} = (A_y B_z - A_z B_y)\hat{\mathbf{x}} + (A_z B_x - A_x B_z)\hat{\mathbf{y}} + (A_x B_y - A_y B_x)\hat{\mathbf{z}}$$

In component notation, these products can be simply expressed as $A^k B^k$ and $\varepsilon^{klm}A^k B^m$, respectively.

EXAMPLE: *dot cross identity*

As an example of the advantages of the component notation for vector products, let us prove the following "dot cross" identity for the multiplication of three vectors **A, B, C**:

$$\mathbf{A}\cdot(\mathbf{B}\times\mathbf{C}) = (\mathbf{A}\times\mathbf{B})\cdot\mathbf{C} \qquad (16)$$

In component notation, the cross product $\mathbf{B}\times\mathbf{C}$ is $\varepsilon^{klm}B^l C^m$, and therefore the left side of Eq. (16) can be expressed as $A^k\varepsilon^{klm}B^l C^m$, where summation over all the repeated indices *k*, *l*, *m* is understood. By two exchanges of adjacent indices, we see that this equals $A^k\varepsilon^{mkl}B^l C^m$ $=(\varepsilon^{mkl}A^k B^l)C^m$, which is exactly the right side of Eq. (16). This illustrates how the tensor ε^{klm} permits quick and efficient derivations of vector identities involving cross products. Another vector identity that can be derived by similar means (see Problem 10) is the "double cross" identity

$$\mathbf{A}\times(\mathbf{B}\times\mathbf{C}) = \mathbf{B}(\mathbf{A}\cdot\mathbf{C}) - \mathbf{C}(\mathbf{A}\cdot\mathbf{B}) \qquad (17)$$

1·2 Rotations of Coordinates; Tensors

The unit vectors $\hat{\mathbf{x}},\hat{\mathbf{y}},\hat{\mathbf{z}}$ belong to a particular set of rectangular axes. Each set of axes has its own unit vectors. Figure 1.3 shows the unit vectors belonging to the axes x, y and x', y', where the second set of axes is obtained from the first by a rotation through an angle θ in the *x*-*y* plane.

Figure 1·3

Unit vectors $\hat{\mathbf{x}},\hat{\mathbf{y}}$ and $\hat{\mathbf{x}}',\hat{\mathbf{y}}'$.

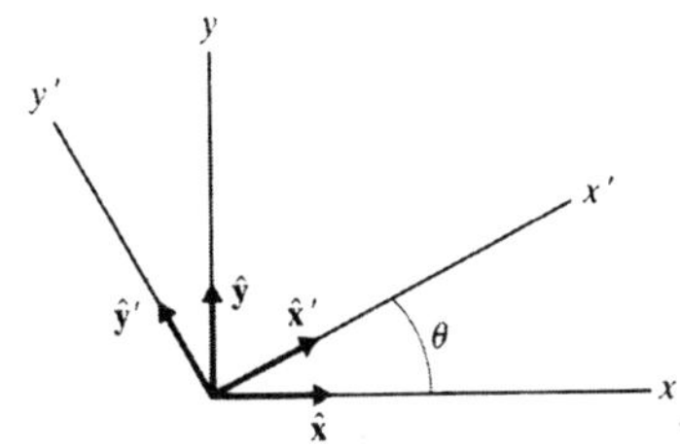

If the components of a vector **x** are x^k with respect to the first set of axes and x'^k with respect to the second set, then these components are related as follows:

$$\begin{aligned}x'^1 &= x^1\cos\theta + x^2\sin\theta\\ x'^2 &= -x^1\sin\theta + x^2\cos\theta\\ x'^3 &= x^3\end{aligned} \qquad (18)$$

■ ***Exercise 4.*** Derive the Eqs. (18). ■

The relation (18) is an instance of a linear transformation. We can write it as

$$x'^k = a^{kn}x^n \qquad (19)$$

where a^{kn} is a 3×3 matrix with components

$$\begin{aligned}&a^{11}=\cos\theta \quad && a^{12}=\sin\theta \quad && a^{13}=0\\ &a^{21}=-\sin\theta && a^{22}=\cos\theta && a^{23}=0\\ &a^{31}=0 && a^{32}=0 && a^{33}=1\end{aligned}$$

or, in compact notation,

$$a^{kn} = \begin{pmatrix}\cos\theta & \sin\theta & 0\\ -\sin\theta & \cos\theta & 0\\ 0 & 0 & 1\end{pmatrix} \qquad (20)$$

Here, the first superscript (k) refers to the rows of the matrix, and the second superscript (n) refers to the columns.

■ ***Exercise 5.*** Check that Eq. (19) agrees with Eqs. (18). ■

Note that the right side of Eq. (18) is equivalent to the product of the column vector by the matrix, according to the usual row-by-column multiplication rule:

$$\begin{pmatrix} x'^1 \\ x'^2 \\ x'^3 \end{pmatrix} = \begin{pmatrix} a^{11} & a^{12} & a^{13} \\ a^{21} & a^{22} & a^{23} \\ a^{31} & a^{32} & a^{33} \end{pmatrix} \begin{pmatrix} x^1 \\ x^2 \\ x^3 \end{pmatrix}$$

The matrix (20) describes a particularly simple case of rotation of axes, with a fixed z axis. In general, three independent angles are needed to describe the rotation of a set of axes. However, even in the general case, the relation between the old and the new components of the position vector is still a linear relation of the form

$$x'^k = a^{kn} x^n \tag{21}$$

where the coefficients a^{kn} are some functions of the angles between the axes. It is easy to understand why the components of a position vector must transform linearly: a given flat plane through the origin must have a linear equation, that is, an equation of the form $Ax^1 + Bx^2 + Cx^3 = 0$, no matter what rectangular coordinate system is used to describe it.

Some general remarks about rotations will be helpful. Suppose that we perform two successive rotations of our coordinate axes. If the first rotation changes the components of a given position vector from x^n to

$$x'^k = a^{kn} x^n \tag{22}$$

and the second rotation changes the components from x'^k to

$$x''^m = b^{mk} x'^k \tag{23}$$

then by substituting the second equation into the first, we find that x^n and x''^m are related by

$$x''^m = b^{mk} a^{kn} x^n \tag{24}$$

We can write this last relation as

$$x''^m = c^{mn} x^n \tag{25}$$

with

$$c^{mn} = b^{mk} a^{kn} \tag{26}$$

The rule (26) for the combination of two matrices is simply the familiar row-by-column multiplication rule.

EXAMPLE: *two successive rotations*

A simple example of the combination of two successive rotations about the same axis is contained in the following exercise.

■ ***Exercise 6.*** Show that if a^{kn} is as given by Eq. (20) and

$$b^{mk} = \begin{pmatrix} \cos\phi & \sin\phi & 0 \\ -\sin\phi & \cos\phi & 0 \\ 0 & 0 & 1 \end{pmatrix} \tag{27}$$

then matrix multiplication gives

$$c^{mn} = \begin{pmatrix} \cos(\phi+\theta) & \sin(\phi+\theta) & 0 \\ -\sin(\phi+\theta) & \cos(\phi+\theta) & 0 \\ 0 & 0 & 1 \end{pmatrix} \tag{28}$$

Explain why the sum of the angles ϕ and θ is expected to appear in Eq. (28). ■

If the angle ϕ in Eq. (27) equals $-\theta$, then

$$b^{mk} = \begin{pmatrix} \cos\theta & -\sin\theta & 0 \\ \sin\theta & \cos\theta & 0 \\ 0 & 0 & 1 \end{pmatrix}. \tag{29}$$

and

$$c^{mn} = b^{mk}a^{kn} = \delta^{mn} \tag{30}$$

where δ^{mn} is the *Kronecker delta* (or unit matrix)

$$\delta^{mn} = \begin{cases} 1 & \text{if } \ m = n \\ 0 & \text{if } \ m \neq n \end{cases} \tag{31}$$

The matrix b^{mk} of Eq. (29) is the *inverse* of the matrix a^{kn}; these two matrices represent opposite rotations that cancel one another, and according to Eq. (30), their product equals the unit matrix.

Comparison of Eqs. (29) and (20) shows that the matrix b^{mk} is the *transpose* of the matrix a^{kn}, that is, the matrix in Eq. (29) can be obtained from the matrix in Eq. (20) by exchanging rows and columns. Thus, the inverse of our rotation matrix equals the transpose of that matrix. A matrix that satisfies this condition is called *orthogonal*. In general, any matrix representing a rotation satisfies this condition, as we will now prove.

• ***Theorem:*** If a^{kn} is the matrix for a rotation and $(a^T)^{mk}$ is the transpose matrix, then their product equals the unit matrix,

$$(a^T)^{mk} a^{kn} = \delta^{mn} \tag{32}$$

The proof uses the fact that although the components of a vector change under rotation of the coordinates, the length of the vector remains the same:

$$x'^k x'^k = x^l x^l \tag{33}$$

With Eq. (21), this becomes

$$a^{km} x^m a^{kl} x^l = x^l x^l \tag{34}$$

We can use the identity

$$x^l = \delta^{ml} x^m \tag{35}$$

on the right side of Eq. (34) and obtain

$$a^{km} x^m a^{kl} x^l = \delta^{ml} x^m x^l \tag{36}$$

That is,

$$x^m x^l (a^{km} a^{kl} - \delta^{ml}) = 0 \tag{37}$$

But this cannot be true for all possible values of x^m unless

$$a^{km} a^{kl} = \delta^{ml} \tag{38}$$

■ ***Exercise 7.*** Deduce Eq. (38) from Eq. (37) by adopting diverse values of x^m, x^l. [*Hint*: To show that Eq. (38) is valid for $m = 1$, $l = 1$, adopt x^m and x^l = (1,0,0), and substitute into Eq. (37); for $m = 1$, $l = 3$, adopt x^m and x^l = (1,0,1), etc. Alternatively, take the partial derivative $\partial^2 / \partial x^1 \partial x^3$ of Eq. (37).] ■

Since the *km* component of any matrix is equal to the *mk* component of the transpose matrix, Eq. (38) can be rewritten as

$$(a^T)^{mk} a^{kl} = \delta^{ml} \tag{39}$$

which establishes Eq. (32).

The inverse matrix permits us to express x^m in terms of x'^k. If we multiply

$$x'^k = a^{kn} x^n \tag{40}$$

by a^{km} and sum over *k*, we obtain

$$a^{km} x'^k = a^{km} a^{kn} x^n \tag{41}$$

According to Eq. (38), the right side of Eq. (41) equals $\delta^{mn}x^n = x^m$. Hence, we can write x^m in terms of x'^k as follows:

$$x^m = a^{km}x'^k \tag{42}$$

Equation (40) expresses the transformation of the old components of the position vector into the new components; Eq. (42) expresses the inverse transformation of new components into old components.

We are now ready to give the general definition of a vector. We will use the notation (B_x, B_y, B_z), or (B^1, B^2, B^3), or **B** for a vector.

• ***Definition:*** A vector is an object with three components that transforms as follows under rotations:

$$B'^k = a^{kn}B^n \tag{43}$$

The matrix a^{kn} is, of course, exactly the same as that appearing in Eq. (21). What our definition says is that any vector must transform under rotations in the same way as the position vector, and thus the position vector serves as prototype for all vectors. With this definition, we can readily show that whenever A^k and B^k are vectors, any linear combination $\alpha A^k + \beta B^k$ (with arbitrary numerical coefficients α and β) is also a vector. Thus, our definition gives vectors precisely those properties that mathematicians demand of vectors in linear algebra.

■ ***Exercise 8.*** Show that if A^k and B^k are vectors, then $\alpha A^k + \beta B^k$ is also a vector. ■

The definition of a scalar also hinges on the behavior under rotations, according to the following definition.

• ***Definition:*** A scalar is a one-component object that remains unchanged under rotations.

The length of the position vector and the square of this length obviously are scalars [see Eq. (33)]. More generally, we can show that if A^k and B^k are vectors, then the dot product A^kB^k is necessarily a scalar. What we must show is that

EXAMPLE: *dot products as scalars*

$$A'^kB'^k = A^kB^k \tag{44}$$

Consider the left side of this equation. By means of the transformation law (43) for vectors and the orthogonality of the matrix a^{kn}, we find

$$A'^kB'^k = a^{km}A^m a^{kn}B^n = (a^T)^{mk}a^{kn}A^mB^n = \delta^{mn}A^mB^n = A^nB^n \tag{45}$$

The last expression coincides with the right side of Eq. (44).

EXAMPLE: *scalars and vectors in physics*

In physics, we encounter many examples of scalars and vectors. Mass and time (in nonrelativistic physics) are scalars because regardless of what rotation we perform on the coordinate axes, the mass of a particle and the time registered by a clock will be unchanged. Kinetic energy is a scalar because it depends only on the mass and on the *magnitude* of the velocity. Velocity is a vector because it is the product of a vector $d\mathbf{x}$ by a scalar $1/dt$. Acceleration, momentum, and force are vectors for similar reasons.

Finally, we give the definition of a *tensor*. Suppose we have a set of nine numbers arranged as a 3×3 matrix. We will use the notation T^{kl} for this matrix, where, as before, the first superscript refers to the rows of the matrix and the second to the columns. Under what conditions is this matrix a tensor?

• ***Definition:*** A tensor is an object with nine components that transform as follows under rotations:

$$T'^{kl} = a^{kn}a^{lm}T^{nm} \tag{46}$$

This simply says that T^{nm} must transform as a vector in each superscript [compare Eq. (43)]. Keep in mind that not every matrix is a tensor—it all depends on whether the components of the matrix obey the correct transformation law under rotations.

■ ***Exercise 9.*** Show that if T^{kl} and Q^{kl} are tensors, then $\alpha T^{kl} + \beta Q^{kl}$ is a tensor. ■

■ ***Exercise 10.*** Show that the Kronecker delta is a tensor; that is, show that

$$\delta'^{kl} = a^{kn}a^{lm}\delta^{nm} \tag{47}$$

where $\delta'^{kl} = 1$ if $k = l$, and $\delta'^{kl} = 0$ if $k \neq l$. ■

■ ***Exercise 11.*** Show that the transpose of any tensor is a tensor. ■

EXAMPLE: *construction of a tensor from two vectors*

One simple procedure for constructing a tensor is to begin with two vectors A^k and B^l; then define the kl component of the tensor as the product of the k component of the first vector by the l component of the second. This procedure yields the nine-component object

$$Q^{kl} = A^k B^l \tag{48}$$

By making use of the transformation law of the vectors, we can check that this nine-component object satisfies the transformation law of a tensor:

$$Q'^{kl} = A'^k B'^l = a^{kn} A^n a^{lm} B^m = a^{kn} a^{lm} A^n B^m = a^{kn} a^{lm} Q^{nm} \tag{49}$$

An example of such a tensor is the moment-of-inertia tensor used in mechanics. For a particle of mass m and position x^k, this tensor is

$$I^{kl} = m(-x^k x^l + \delta^{kl} x^n x^n) \tag{50}$$

Note that in this expression, the second term in parentheses is the product of a tensor δ^{kl} by a scalar $x^n x^n$.

If T^{kl} is a tensor and B^l is a vector, then the product $T^{kl} B^l$ is a vector. To show this, we start with the transformation laws for the tensor and the vector, and we carry out a calculation quite similar to that involved in Eq. (45). We leave this calculation as an exercise.

■ ***Exercise 12.*** Show that if T^{kl} is a tensor and B^l is a vector, then $T^{kl} B^l$ is a vector. ■

■ ***Exercise 13.*** Show that if T^{kl} and Q^{ln} are tensors, then $T^{kl} Q^{ln}$ is a tensor. ■

■ ***Exercise 14.*** Show that if T^{kl} is a tensor, then T^{kk} is a scalar. This scalar is called the *trace* of the tensor. ■

A tensor is said to be *symmetric* if

$$T^{kl} = T^{lk} \tag{51}$$

and it is said to be *antisymmetric* if

$$T^{kl} = -T^{lk} \tag{52}$$

Any arbitrary tensor can be written as a sum of a symmetric plus an antisymmetric tensor. This can be seen from the trivial identity

$$T^{kl} = \tfrac{1}{2}(T^{kl} + T^{lk}) + \tfrac{1}{2}(T^{kl} - T^{lk}) \tag{53}$$

Obviously, the first set of parentheses contains a symmetric tensor, and the second contains an antisymmetric tensor.

The 3×3 tensors we have considered so far are called tensors of the second rank. Tensors of higher rank can be defined in a similar way. Thus, a tensor of the third rank is an object T^{klm} with $3 \times 3 \times 3$ components that transforms as follows under rotations:

$$T'^{klm} = a^{kn} a^{lr} a^{ms} T^{nrs} \tag{54}$$

The alternating tensor ε^{klm} defined in Eq. (14) is a tensor of the third rank.

■ ***Exercise 15.*** Show that if A^k, B^l, and C^m are vectors, then $A^k B^l C^m$ is a tensor of the third rank. ■

■ ***Exercise 16.*** Show that the alternating tensor ε^{klm} defined in Eq. (14) is a tensor of the third rank; that is, show that

$$\varepsilon'^{klm} = a^{kn} a^{lr} a^{ms} \varepsilon^{nrs} \tag{55}$$

where $\varepsilon'^{123} = \varepsilon'^{231} = \varepsilon'^{312} = 1$, $\varepsilon'^{213} = \varepsilon'^{132} = \varepsilon'^{321} = -1$, and all other components of ε'^{klm} are zero. [*Hint:* Evaluate the right side of Eq. (55) explicitly, using the expression (20) for the rotation matrix.] ■

■ ***Exercise 17.*** Show that if A^l and B^m are vectors, then $\varepsilon^{klm} A^l B^m$ is a vector. Hence the cross product of vectors does indeed give a vector. ■

1·3 Fields and Differential Operators

By a *field* is meant a function of the position vector $\mathbf{x}$. Fields can be scalars, vectors, or tensors. In physics, we use fields to describe the state of continuous systems. Consider, for instance, the gas that makes up the body of a star. This gas has a (mass) density ρ, which is a function of position. Clearly, the density at a given point does not depend on how the coordinates are chosen; hence, the density is a scalar and $\rho(\mathbf{x})$ is a scalar field. If the star is *not* in static equilibrium, then the gas at each point has some velocity $\mathbf{v}$, which depends on position; the function $\mathbf{v}(\mathbf{x})$ is a vector field. Graphically, we can represent the scalar field $\rho(\mathbf{x})$ by drawing the surfaces $\rho(\mathbf{x})$ = constant. For example, Fig. 1.4 shows the surfaces of constant density inside a gravitationally collapsing star at one instant of time. We can represent the vector field $\mathbf{v}(\mathbf{x})$ by drawing the velocity vector associated with each point, as in Fig. 1.5. We can sometimes represent vector fields by drawing field lines, but this method works only for some very special vector fields, such as the velocity field for an incompressible fluid or the electric field;* we will discuss field lines later on.

---

* Field lines can be drawn only for vector fields that have zero divergence [see Eq. (67)].

Figure 1·4

Contours of equal density in the *x*-*z* plane for a star undergoing gravitational collapse. The numbers give the density relative to the central density. [After T. Nakamura, *Ann. N.Y. Acad. Sci.* **433**, 56 (1984).]

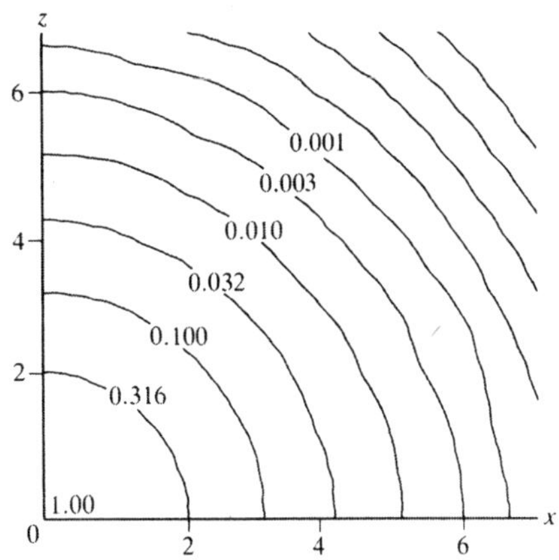

Figure 1·5

Velocity vectors for a collapsing star. (After T. Nakamura, ibid.)

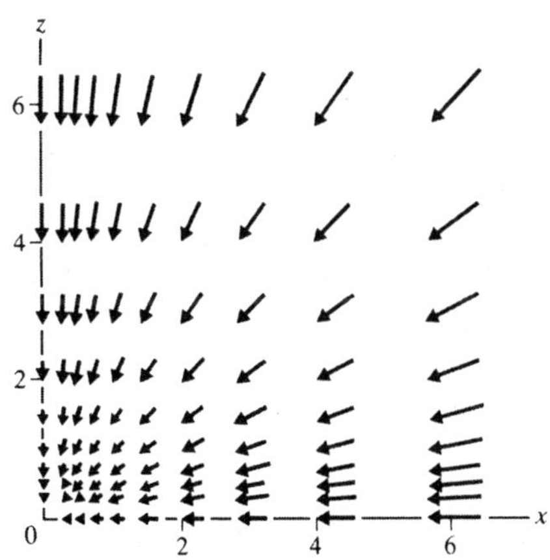

Given an arbitrary scalar field $\Phi(\mathbf{x})$, we can construct the vector field A^k whose components are the derivatives of $\Phi(\mathbf{x})$ with respect to x, y, and z:

$$(A^1, A^2, A^3) \equiv \left(\frac{\partial \Phi}{\partial x}, \frac{\partial \Phi}{\partial y}, \frac{\partial \Phi}{\partial z} \right) \tag{56}$$

or, in compact notation,

$$A^k = \frac{\partial \Phi}{\partial x^k} \tag{57}$$

This vector field is called the *gradient* of $\Phi(\mathbf{x})$. To show that A^k is a vector, consider a rotation of coordinates described by a rotation matrix a^{kl}. For such a rotation, the transformation of the position vector and the corresponding inverse transformation are

$$x'^k = a^{kl} x^l \quad \text{and} \quad x^l = a^{nl} x'^n \tag{58}$$

(here, as in the preceding sections, summation over the repeated index l in the first equation and over the repeated index n in the second equation is understood). Accordingly,

$$\frac{\partial x^l}{\partial x'^k} = a^{kl} \tag{59}$$

The chain rule for derivatives then tells us that

$$\frac{\partial \Phi}{\partial x'^k} = \frac{\partial \Phi}{\partial x^l}\frac{\partial x^l}{\partial x'^k} = \frac{\partial \Phi}{\partial x^l} a^{kl} \tag{60}$$

that is,

$$A'^k = a^{kl} A^l \tag{61}$$

This is exactly the transformation law of a vector and establishes that A^k is, indeed, a vector.

The chain rule for derivatives used in the preceding paragraph also tells us that the differential operator $\partial / \partial x^k$ is a vector operator—it obeys the transformation law of a vector,

$$\frac{\partial}{\partial x'^k} = a^{kl} \frac{\partial}{\partial x^l} \tag{62}$$

The operator $\partial / \partial x^k$ is called the *gradient* operator or *del* operator. It is often written as ∇,

$$\nabla = \hat{\mathbf{x}}\frac{\partial}{\partial x} + \hat{\mathbf{y}}\frac{\partial}{\partial y} + \hat{\mathbf{z}}\frac{\partial}{\partial z} \tag{63}$$

and Eq. (56) is then written as

$$\mathbf{A} = \nabla \Phi \tag{64}$$

We can use the gradient to find small changes in a function produced by small displacements. Consider two points separated by a small displacement $d\mathbf{x} = \hat{\mathbf{x}}dx + \hat{\mathbf{y}}dy + \hat{\mathbf{z}}dz$ (see Fig. 1.6). Then the difference between the values of the function $\Phi(\mathbf{x})$ at these two points is

$$d\Phi = \Phi(\mathbf{x} + d\mathbf{x}) - \Phi(\mathbf{x}) = dx\frac{\partial\Phi}{\partial x} + dy\frac{\partial\Phi}{\partial y} + dz\frac{\partial\Phi}{\partial z} \quad (65)$$

With the formula (64) for the gradient, this becomes

$$d\Phi = d\mathbf{x}\bullet\nabla\Phi \quad (66)$$

Thus, the change of the function in the direction of $d\mathbf{x}$ equals the component of $\nabla\Phi$ in that direction multiplied by $|d\mathbf{x}|$.

EXAMPLE: *gradient of a function*

For example, consider the function $\Phi(\mathbf{x}) = x^2 + y^2 + z^2$. Its gradient is

$$\begin{aligned}\nabla\Phi &= \hat{\mathbf{x}}\frac{\partial}{\partial x}(x^2 + y^2 + z^2) + \hat{\mathbf{y}}\frac{\partial}{\partial y}(x^2 + y^2 + z^2) + \hat{\mathbf{z}}\frac{\partial}{\partial z}(x^2 + y^2 + z^2) \\ &= 2(x\hat{\mathbf{x}} + y\hat{\mathbf{y}} + z\hat{\mathbf{z}})\end{aligned}$$

Accordingly, the change in the function produced by a displacement $d\mathbf{x}$ is $d\Phi = 2d\mathbf{x}\bullet(x\hat{\mathbf{x}} + x\hat{\mathbf{y}} + z\hat{\mathbf{z}})$. Evidently, $d\Phi$ is maximum when $d\mathbf{x}$ is in the direction of the gradient vector. Note that the gradient vector $2(x\hat{\mathbf{x}} + x\hat{\mathbf{y}} + z\hat{\mathbf{z}})$ is in the radial direction, and it is perpendicular to the spherical surfaces defined by $\Phi(\mathbf{x}) = x^2 + y^2 + z^2 =$ constant.

This is a general result: the direction of $\nabla\Phi$ is always perpendicular to the surfaces $\Phi(\mathbf{x})$ = constant. To deduce this from Eq. (66),

Figure 1·6

A small displacement $d\mathbf{x}$ takes us from a surface $\Phi(\mathbf{x}) = \text{constant}$ to a surface $\Phi + d\Phi = \text{constant}$. The gradient of Φ is perpendicular to the surface $\Phi(\mathbf{x}) = \text{constant}$.

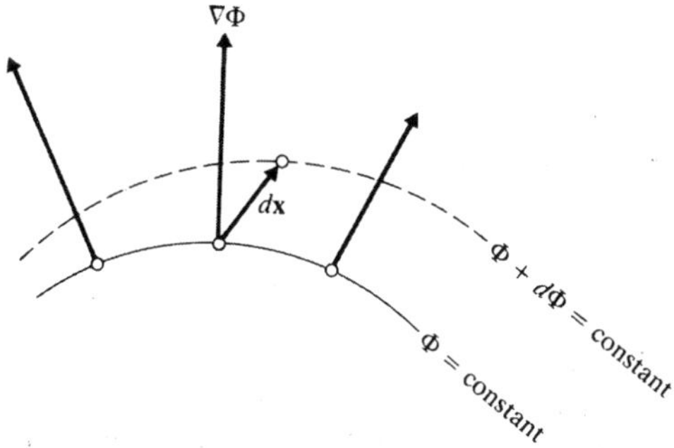

we select a displacement $d\mathbf{x}$ parallel to one of the surfaces $\Phi(\mathbf{x})$ = constant; then Eq. (66) tells us $0 = d\mathbf{x}\cdot\nabla\Phi$, which means that this displacement and $\nabla\Phi$ are perpendicular. Figure 1.6 shows the direction of the gradient at several points on some general curved surface $\Phi(\mathbf{x})$ = constant.

Since $\partial/\partial x^k$ behaves like a vector, we can take the scalar product of $\partial/\partial x^k$ with a vector field B^k and obtain a scalar field*

$$\frac{\partial B^k}{\partial x^k} \tag{67}$$

This is called the *divergence* of the vector field B^k. One often uses the notation $\nabla\cdot\mathbf{B}$ for (67). Written out in full, the divergence is

$$\nabla\cdot\mathbf{B} = \frac{\partial}{\partial x}B_x + \frac{\partial}{\partial y}B_y + \frac{\partial}{\partial z}B_z \tag{68}$$

Furthermore, we can take the cross product of $\partial/\partial x^l$ with B^m to obtain the vector

$$\varepsilon^{klm}\frac{\partial}{\partial x^l}B^m \tag{69}$$

This is called the *curl* of the vector field B^m. Instead of the component notation, one often uses the notation $\nabla\times\mathbf{B}$ for Eq. (69). Written out in full, the curl is

$$\nabla\times\mathbf{B} = \hat{\mathbf{x}}\left(\frac{\partial}{\partial y}B_z - \frac{\partial}{\partial z}B_y\right) + \hat{\mathbf{y}}\left(\frac{\partial}{\partial z}B_x - \frac{\partial}{\partial x}B_z\right) + \hat{\mathbf{z}}\left(\frac{\partial}{\partial x}B_y - \frac{\partial}{\partial y}B_x\right) \tag{70}$$

■ ***Exercise 18.*** Show that if B^m is the gradient of a scalar field, then its curl vanishes identically,

$$\varepsilon^{klm}\frac{\partial}{\partial x^l}\frac{\partial\Phi}{\partial x^m} = 0 \tag{71}$$

or

$$\nabla\times\nabla\Phi = 0 \quad ■$$

■ ***Exercise 19.*** Show that, for any vector B^m, the divergence of the curl vanishes identically:

$$\frac{\partial}{\partial x^k}\varepsilon^{klm}\frac{\partial}{\partial x^l}B^m = 0$$

---

*Note that we could take the scalar product in the opposite order and get $B^k\partial/\partial x^k$. This is not a function but a differential operator. An instance of such a differential operator appears in Eq. (66).

or

$$\nabla \cdot (\nabla \times \mathbf{B}) = 0 \tag{72}$$

■

We have seen that if $\Phi(\mathbf{x})$ is a scalar, then $\partial\Phi/\partial x^k$ is a vector. By taking the divergence of this vector, we obtain the scalar

$$\frac{\partial}{\partial x^k}\left(\frac{\partial}{\partial x^k}\Phi\right) \tag{73}$$

The differential operator $\partial^2/\partial x^k \partial x^k$, for which one also uses the notation $\nabla \cdot \nabla$ or ∇^2, is called the *Laplacian.* Explicitly,

$$\nabla^2 = \frac{\partial^2}{\partial x^2} + \frac{\partial^2}{\partial y^2} + \frac{\partial^2}{\partial z^2} \tag{74}$$

This differential operator plays an important role in potential theory (see Chapter 4).

In Eq. (73), the Laplacian operator acts on a scalar, and the result is a scalar. But we can also let this operator act on a vector; the result is then a vector:

$$\frac{\partial^2}{\partial x^k \partial x^k} A^l \tag{75}$$

or, in vector notation,

$$\nabla^2 \mathbf{A} \tag{76}$$

Written out in full, this becomes

$$\nabla^2 \mathbf{A} = \hat{\mathbf{x}} \nabla^2 A_x + \hat{\mathbf{y}} \nabla^2 A_y + \hat{\mathbf{z}} \nabla^2 A_z \tag{77}$$

In a later chapter, we will have occasion to use the identity

$$\nabla \times (\nabla \times \mathbf{A}) = \nabla (\nabla \cdot \mathbf{A}) - \nabla^2 \mathbf{A} \tag{78}$$

This is a consequence of the double cross identity

$$\mathbf{C} \times (\mathbf{B} \times \mathbf{A}) = \mathbf{B}\,\mathbf{C} \cdot \mathbf{A} - \mathbf{C} \cdot \mathbf{B}\,\mathbf{A} \tag{79}$$

if we make the substitutions $\mathbf{C} = \nabla$ and $\mathbf{B} = \nabla$. Since the components of ∇ appear in the dot product and in the cross product just as though they were ordinary vector components [see Eqs. (68) and (70)], the identity (79) remains valid under this substitution. Note that we have taken the precaution of writing **A** to the right of each of the terms in Eq. (79); this is important, because when we change **B** and **C** into operators, both must act on **A**.

1·4 Cylindrical and Spherical Coordinates

Equation (63) expresses the gradient operator ∇ in rectangular coordinates. For our later work, we will need an expression for ∇ in cylindrical and in spherical coordinates. Figure 1.7 shows the cylindrical coordinates $\rho,\ \phi,\ z$ of a point and the unit vectors $\hat{\boldsymbol{\rho}}$, $\hat{\boldsymbol{\phi}}$, $\hat{\mathbf{z}}$. These vectors are mutually perpendicular, and they point in the direction of increasing $\rho,\ \phi,\ z$, respectively. Note that both $\hat{\boldsymbol{\rho}}$ and $\hat{\boldsymbol{\phi}}$ depend on position.

The relationship between the cylindrical and Cartesian coordinates and unit vectors is as follows:

$$x = \rho\cos\phi \qquad y = \rho\sin\phi \qquad z = z \tag{80}$$

$$\hat{\boldsymbol{\rho}} = \hat{\mathbf{x}}\cos\phi + \hat{\mathbf{y}}\sin\phi \qquad \hat{\boldsymbol{\phi}} = -\hat{\mathbf{x}}\sin\phi + \hat{\mathbf{y}}\cos\phi \qquad \hat{\mathbf{z}} = \hat{\mathbf{z}} \tag{81}$$

■ ***Exercise 20.*** Show that

$$\frac{\partial\hat{\boldsymbol{\rho}}}{\partial\phi} = \hat{\boldsymbol{\phi}} \qquad \frac{\partial\hat{\boldsymbol{\phi}}}{\partial\phi} = -\hat{\boldsymbol{\rho}} \tag{82}$$

and that all other derivatives of the unit vectors are zero. ■

An arbitrary vector **A** can be written as a superposition of the unit vectors:

$$\mathbf{A} = A_\rho\hat{\boldsymbol{\rho}} + A_\phi\hat{\boldsymbol{\phi}} + A_z\hat{\mathbf{z}} \tag{83}$$

Figure 1·7

Cylindrical coordinates and the unit vectors $\hat{\boldsymbol{\rho}}$, $\hat{\boldsymbol{\phi}}$, $\hat{\mathbf{z}}$.

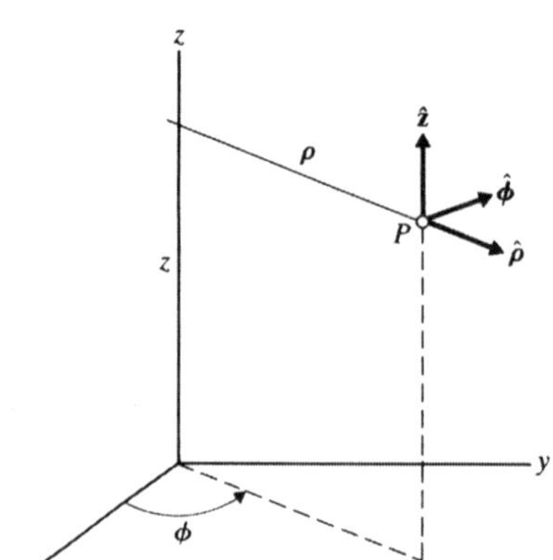

where A_ρ, A_ϕ, A_z are the cylindrical components of the vector along the $\hat{\boldsymbol{\rho}}$, $\hat{\boldsymbol{\phi}}$, and $\hat{\mathbf{z}}$ directions, respectively. These components can be expressed as

$$A_\rho = \hat{\boldsymbol{\rho}}\cdot\mathbf{A} \qquad A_\phi = \hat{\boldsymbol{\phi}}\cdot\mathbf{A} \qquad A_z = \hat{\mathbf{z}}\cdot\mathbf{A}$$

The gradient operator in cylindrical coordinates takes the form

$$\nabla = \hat{\boldsymbol{\rho}}\frac{\partial}{\partial\rho} + \hat{\boldsymbol{\phi}}\frac{1}{\rho}\frac{\partial}{\partial\phi} + \hat{\mathbf{z}}\frac{\partial}{\partial z} \tag{84}$$

We can verify this by using the chain rule for partial derivatives to express $\partial/\partial\rho$, $\partial/\partial\phi$ in terms of $\partial/\partial x$, $\partial/\partial y$ and by using Eq. (81) to express $\hat{\boldsymbol{\rho}}$, $\hat{\boldsymbol{\phi}}$ in terms of $\hat{\mathbf{x}}$, $\hat{\mathbf{y}}$; the result of this calculation is Eq. (63).

■ ***Exercise 21.*** Carry out the calculation leading from Eq. (84) to Eq. (63). ■

A more elegant way to arrive at Eq. (84) is as follows. If $f(\mathbf{x})$ is an arbitrary function, then

$$df = d\mathbf{x}\cdot\nabla f(\mathbf{x}) \tag{85}$$

is the change in the function produced by a displacement $d\mathbf{x}$. The expression (85) is also valid in cylindrical coordinates, but, of course, both $d\mathbf{x}$ and ∇f must then be expressed in these coordinates. Since, in cylindrical coordinates,

$$d\mathbf{x} = \hat{\boldsymbol{\rho}}\,d\rho + \hat{\boldsymbol{\phi}}\,\rho\,d\phi + \hat{\mathbf{z}}\,dz \tag{86}$$

we obtain

$$df = d\rho\,\hat{\boldsymbol{\rho}}\cdot\nabla f + \rho\,d\phi\,\hat{\boldsymbol{\phi}}\cdot\nabla f + dz\,\hat{\mathbf{z}}\cdot\nabla f \tag{87}$$

Comparing this with

$$df = d\rho\frac{\partial f}{\partial\rho} + d\phi\frac{\partial f}{\partial\phi} + dz\frac{\partial f}{\partial z} \tag{88}$$

we see that the components of ∇ must be

$$\hat{\boldsymbol{\rho}}\cdot\nabla = \frac{\partial}{\partial\rho} \qquad \hat{\boldsymbol{\phi}}\cdot\nabla = \frac{1}{\rho}\frac{\partial}{\partial\phi} \qquad \hat{\mathbf{z}}\cdot\nabla = \frac{\partial}{\partial z} \tag{89}$$

and hence

$$\nabla = \hat{\boldsymbol{\rho}}\frac{\partial}{\partial\rho} + \hat{\boldsymbol{\phi}}\frac{1}{\rho}\frac{\partial}{\partial\phi} + \hat{\mathbf{z}}\frac{\partial}{\partial z} \tag{90}$$

The divergence of a vector in cylindrical coordinates is

$$\nabla \bullet \mathbf{A} = \left(\hat{\boldsymbol{\rho}} \frac{\partial}{\partial \rho} + \hat{\boldsymbol{\phi}} \frac{1}{\rho} \frac{\partial}{\partial \phi} + \hat{\mathbf{z}} \frac{\partial}{\partial z} \right) \bullet \left(\hat{\boldsymbol{\rho}} A_\rho + \hat{\boldsymbol{\phi}} A_\phi + \hat{\mathbf{z}} A_z \right) \tag{91}$$

In working out the right side, we must remember to differentiate the unit vectors in the second set of parentheses. For instance,

$$\left(\hat{\boldsymbol{\phi}} \frac{1}{\rho} \frac{\partial}{\partial \phi} \right) \bullet \left(\hat{\boldsymbol{\rho}} A_\rho \right) = \hat{\boldsymbol{\phi}} \bullet \hat{\boldsymbol{\rho}} \frac{1}{\rho} \frac{\partial A_\rho}{\partial \phi} + \frac{1}{\rho} A_\rho \hat{\boldsymbol{\phi}} \bullet \frac{\partial \hat{\boldsymbol{\rho}}}{\partial \phi} \tag{92}$$

In the first term on the right side, $\hat{\boldsymbol{\phi}} \bullet \hat{\boldsymbol{\rho}} = 0$; but in the second term,

$$\hat{\boldsymbol{\phi}} \bullet \frac{\partial \hat{\boldsymbol{\rho}}}{\partial \phi} = \hat{\boldsymbol{\phi}} \bullet \hat{\boldsymbol{\phi}} = 1 \tag{93}$$

and hence the expression (92) reduces to $(1/\rho)A_\rho$. Carrying out these and all other differentiations in Eq. (91), we finally find

$$\nabla \bullet \mathbf{A} = \frac{\partial A_\rho}{\partial \rho} + \frac{1}{\rho} A_\rho + \frac{1}{\rho} \frac{\partial A_\phi}{\partial \phi} + \frac{\partial A_z}{\partial z} \tag{94}$$

■ ***Exercise 22.*** Supply the steps leading to Eq. (94). ■

■ ***Exercise 23.*** Show that Eq. (94) can also be written as

$$\nabla \bullet \mathbf{A} = \frac{1}{\rho} \frac{\partial}{\partial \rho} \left(\rho A_\rho \right) + \frac{1}{\rho} \frac{\partial A_\phi}{\partial \phi} + \frac{\partial A_z}{\partial z} \tag{95}$$

■

Note that the dot product of two vectors **B** and **A** in cylindrical coordinates is

$$\mathbf{B} \bullet \mathbf{A} = B_\rho A_\rho + B_\phi A_\phi + B_z A_z \tag{96}$$

Compared with this, the dot product (94) of the vector operator ∇ and the vector **A** has an extra term $(1/\rho)A_\rho$. Such extra terms are a characteristic feature of curvilinear coordinates; as we found in connection with Eq. (92), the extra term arises from the differentiation of the unit vectors. Only in rectangular coordinates are the unit vectors constant, and products involving vector operators look the same as ordinary products [see Eq. (68)].

The Laplacian operator can be obtained from Eq. (90) by a straightforward calculation:

$$\nabla^2 = \nabla\bullet\nabla = \left(\hat{\boldsymbol{\rho}}\frac{\partial}{\partial\rho} + \hat{\boldsymbol{\phi}}\frac{1}{\rho}\frac{\partial}{\partial\phi} + \hat{\mathbf{z}}\frac{\partial}{\partial z}\right)\bullet\left(\hat{\boldsymbol{\rho}}\frac{\partial}{\partial\rho} + \hat{\boldsymbol{\phi}}\frac{1}{\rho}\frac{\partial}{\partial\phi} + \hat{\mathbf{z}}\frac{\partial}{\partial z}\right)$$

In working out the right side, we must again remember to differentiate the unit vectors, as in Eqs. (91)–(94). The final result is

$$\nabla^2 = \frac{\partial^2}{\partial\rho^2} + \frac{1}{\rho}\frac{\partial}{\partial\rho} + \frac{1}{\rho^2}\frac{\partial^2}{\partial\phi^2} + \frac{\partial^2}{\partial z^2} \qquad (97)$$

■ ***Exercise 24.*** Supply the steps leading to Eq. (97). ■

■ ***Exercise 25.*** Show that Eq. (97) can also be written as

$$\nabla^2 = \frac{1}{\rho}\frac{\partial}{\partial\rho}\left(\rho\frac{\partial}{\partial\rho}\right) + \frac{1}{\rho^2}\frac{\partial^2}{\partial\phi^2} + \frac{\partial^2}{\partial z^2} \qquad (98)$$

■

Next, we must deal with spherical coordinates. Figure 1.8 shows the spherical coordinates r, θ, ϕ and the corresponding unit vectors $\hat{\mathbf{r}}$, $\hat{\boldsymbol{\theta}}$, $\hat{\boldsymbol{\phi}}$. These unit vectors are mutually perpendicular, and they point in the direction of increasing r, θ, ϕ, respectively. All these vectors depend on position.

The relationship between spherical and Cartesian coordinates and unit vectors is

$$x = r\sin\theta\cos\phi \qquad y = r\sin\theta\sin\phi \qquad z = r\cos\theta \qquad (99)$$

Figure 1·8

Spherical coordinates and the unit vectors $\hat{\mathbf{r}}$, $\hat{\boldsymbol{\theta}}$, $\hat{\boldsymbol{\phi}}$.

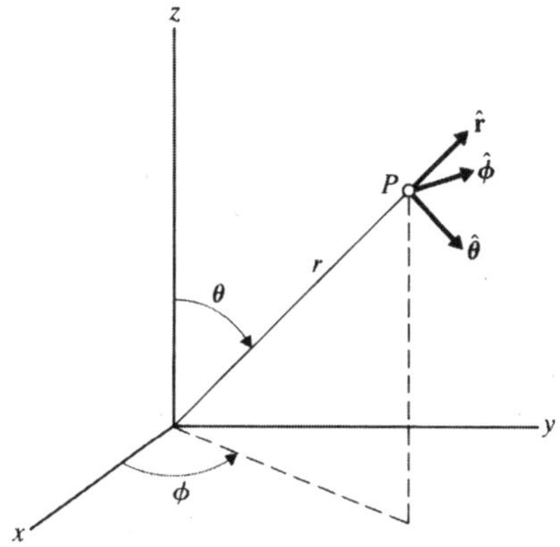

and

$$\hat{\mathbf{r}} = \hat{\mathbf{x}}\sin\theta\cos\phi + \hat{\mathbf{y}}\sin\theta\sin\phi + \hat{\mathbf{z}}\cos\theta$$

$$\hat{\boldsymbol{\theta}} = \hat{\mathbf{x}}\cos\theta\cos\phi + \hat{\mathbf{y}}\cos\theta\sin\phi - \hat{\mathbf{z}}\sin\theta$$

$$\hat{\boldsymbol{\phi}} = -\hat{\mathbf{x}}\sin\phi + \hat{\mathbf{y}}\cos\phi \tag{100}$$

■ ***Exercise 26.*** Show that

$$\frac{\partial\hat{\mathbf{r}}}{\partial\theta} = \hat{\boldsymbol{\theta}} \qquad \frac{\partial\hat{\mathbf{r}}}{\partial\phi} = \hat{\boldsymbol{\phi}}\sin\theta$$

$$\frac{\partial\hat{\boldsymbol{\theta}}}{\partial\theta} = -\hat{\mathbf{r}} \qquad \frac{\partial\hat{\boldsymbol{\theta}}}{\partial\phi} = \hat{\boldsymbol{\phi}}\cos\theta$$

$$\frac{\partial\hat{\boldsymbol{\phi}}}{\partial\phi} = -\hat{\mathbf{r}}\sin\theta - \hat{\boldsymbol{\theta}}\cos\theta \tag{101}$$

and that all other derivatives are zero. ■

An arbitrary vector can be expressed as

$$\mathbf{A} = A_r\hat{\mathbf{r}} + A_\theta\hat{\boldsymbol{\theta}} + A_\phi\hat{\boldsymbol{\phi}} \tag{102}$$

where A_r, A_θ, and A_ϕ are its components along the $\hat{\mathbf{r}}$, $\hat{\boldsymbol{\theta}}$, and $\hat{\boldsymbol{\phi}}$ directions, that is,

$$A_r = \hat{\mathbf{r}}\bullet\mathbf{A} \qquad A_\theta = \hat{\boldsymbol{\theta}}\bullet\mathbf{A} \qquad A_\phi = \hat{\boldsymbol{\phi}}\bullet\mathbf{A}$$

The gradient operator in spherical coordinates is

$$\nabla = \hat{\mathbf{r}}\frac{\partial}{\partial r} + \hat{\boldsymbol{\theta}}\frac{1}{r}\frac{\partial}{\partial\theta} + \hat{\boldsymbol{\phi}}\frac{1}{r\sin\theta}\frac{\partial}{\partial\phi} \tag{103}$$

The divergence of a vector is

$$\nabla\bullet\mathbf{A} = \frac{1}{r^2}\frac{\partial}{\partial r}\left(r^2 A_r\right) + \frac{1}{r\sin\theta}\frac{\partial}{\partial\theta}\left(\sin\theta A_\theta\right) + \frac{1}{r\sin\theta}\frac{\partial A_\phi}{\partial\phi} \tag{104}$$

The Laplacian operator is

$$\nabla^2 = \frac{1}{r^2}\frac{\partial}{\partial r}\left(r^2\frac{\partial}{\partial r}\right) + \frac{1}{r^2\sin\theta}\frac{\partial}{\partial\theta}\left(\sin\theta\frac{\partial}{\partial\theta}\right) + \frac{1}{r^2\sin^2\theta}\frac{\partial^2}{\partial\phi^2} \tag{105}$$

■ ***Exercise 27.*** Derive Eqs. (103), (104), and (105) by the same methods as were used for the cylindrical case. ■

■ ***Exercise 28.*** Show that the Laplacian operator can be put in the alternative form

$$\nabla^2 = \frac{1}{r}\frac{\partial^2}{\partial r^2} r + \frac{1}{r^2 \sin\theta}\frac{\partial}{\partial\theta}\left(\sin\theta\frac{\partial}{\partial\theta}\right) + \frac{1}{r^2\sin^2\theta}\frac{\partial^2}{\partial\phi^2} \qquad (106)$$

■

Incidentally: the identity (78) *is not* valid in cylindrical or in spherical coordinates. The trouble is that, although the vector identity (79) is valid in these coordinates, we cannot convert it into an operator identity by the simple substitutions $\mathbf{C} = \nabla$ and $\mathbf{B} = \nabla$ because, as we saw in the discussion following Eq. (94), products involving the vector operator ∇ in curvilinear coordinates have extra terms.*

The table on the front endpapers summarizes the expressions for the gradient, the Laplacian, the divergence, and the curl in rectangular, cylindrical, and spherical coordinates (the expressions for the curl can be obtained by arguments similar to those employed to obtain the divergence; see Problems 28 and 29).

1·5 Gauss' and Stokes' Theorems

The divergence of a vector field satisfies an important theorem known as *Gauss' theorem*. Figure 1.9 shows a closed surface; in the following discussion, we will use the notation $\hat{\mathbf{n}}$ for a unit vector perpendicular to the surface and pointing in the *outward* direction. Suppose that **A** is a vector field defined throughout the volume enclosed by this surface. We can then state the theorem as follows:

• ***Gauss' Theorem:*** If S is a closed surface and V is the volume enclosed by S, then

$$\int_V \nabla\cdot\mathbf{A}\,dV = \int_S \mathbf{A}\cdot d\mathbf{S} \qquad (107)$$

where

$$d\mathbf{S} = \hat{\mathbf{n}}\,dS$$

is the vector representing a surface element. Note that the vector $d\mathbf{S}$ has a magnitude equal to the area dS and points in the outward direction.

---

*In an attempt to preserve the validity of (78) in cylindrical and spherical coordinates, some authors use (78) as a *definition* of $\nabla^2\mathbf{A}$. But this means that $\nabla^2\mathbf{A}$ becomes no more than an abbreviation for $\nabla(\nabla\cdot\mathbf{A}) - \nabla\times(\nabla\times\mathbf{A})$. It is then better to write this expression out in full whenever it is needed and avoid any confusing abbreviation.

Figure 1·9

An arbitrary closed surface. The unit vector $\hat{\mathbf{n}}$ is perpendicular to the surface, in the outward direction

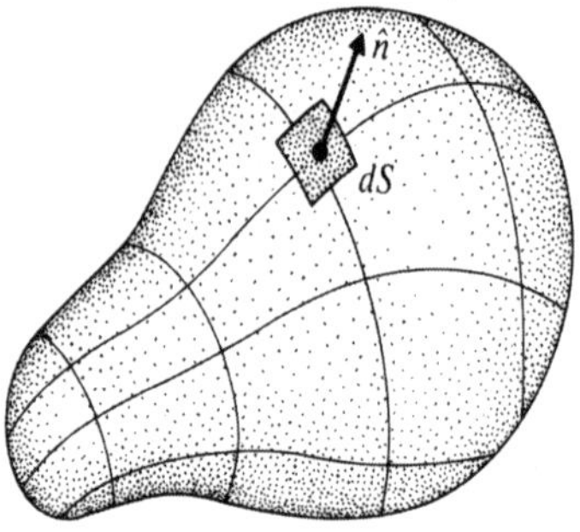

We can prove the theorem by first taking a volume in the shape of a cube; any other volume may be thought of as built from a large number of small cubes. Suppose that the cube occupies the region $x = x_0$ to $x_0 + L$, $y = y_0$ to $y_0 + L$, $z = z_0$ to $z_0 + L$ (see Fig 1.10). Then the volume integral $\int \nabla \cdot \mathbf{A}\, dV$ becomes

$$\iiint \left(\frac{\partial}{\partial x} A_x + \frac{\partial}{\partial y} A_y + \frac{\partial}{\partial z} A_z \right) dxdydz$$

$$= \iint \left(\int \frac{\partial}{\partial x} A_x dx \right) dydz + \iint \left(\int \frac{\partial}{\partial y} A_y dy \right) dxdz + \iint \left(\int \frac{\partial}{\partial z} A_z dz \right) dxdy$$

$$= \iint [A_x(x_0 + L, y, z) - A_x(x_0, y, z)] dydz$$

$$+ \iint \left[A_y(x, y_0 + L, z) - A_y(x, y_0, z) \right] dxdz$$

$$+ \iint [A_z(x, y, z_0 + L) - A_z(x, y, z_0)] dxdy \qquad \textbf{(108)}$$

Figure 1·10

A closed surface in the shape of a cube.

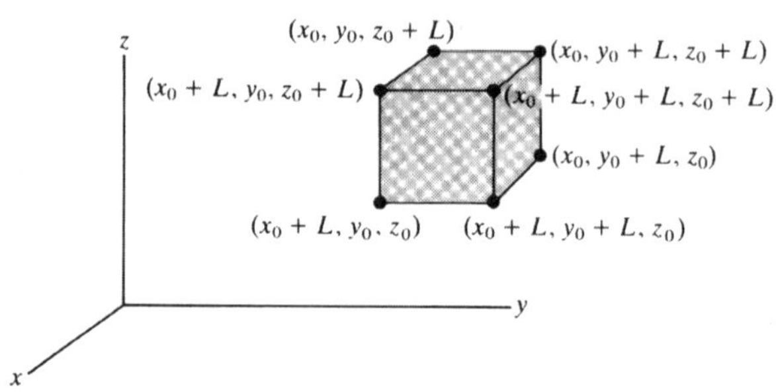

On the other hand, if we perform the surface integral $\int \mathbf{A}\cdot d\mathbf{S}$ over the face of the cube at $x = x_0 + L$ and over the face at $x = x_0$, we obtain

$$\iint A_x(x_0 + L, y, z)dydz - \iint A_x(x_0, y, z)dydz \qquad (109)$$

where the minus sign arises because, at $x = x_0$, the direction of $d\mathbf{S}$ is opposite to that of $\hat{\mathbf{x}}$. Expression (109) agrees with the two terms in the first line of (108). The remaining terms in (108) are exactly what we get from the surface integrals over the remaining faces of the cube.

■ ***Exercise 29.*** Check the remaining terms. ■

This establishes Gauss' theorem for a cube. An arbitrary volume can be approximated by a large number of adjacent small cubes (see Fig. 1.11). The sum of the integrals $\int \mathbf{A}\cdot d\mathbf{S}$ over the surface of all the cubes receives a net contribution only from those faces that are on the exterior; the contributions from adjacent interior faces cancel in pairs, because the unit vectors are opposite (see Fig 1.11). Hence, summation over all the cubes gives us the final result (107).*

If a vector field has zero divergence, then Gauss' theorem tells us that

$$\int_S \mathbf{A}\cdot d\mathbf{S} = 0 \qquad (110)$$

for any arbitrary closed surface. A vector field that satisfies this condition can be graphically represented by *field lines* whose direction at any

Figure 1·11

An arbitrary volume is approximated by a collection of small cubes. The unit vectors of adjacent interior faces are opposite.

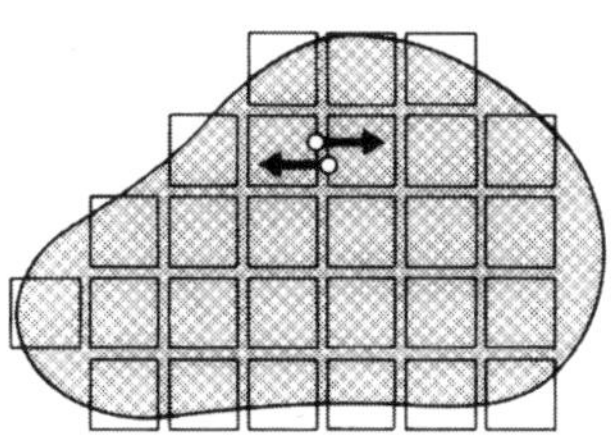

*The normal to the cube face does not converge to the normal to the curved surface; even in the limit of infinitesimal cubes, the normals to the cube faces can point in only three mutually perpendicular directions. This does not matter, because the normals will agree *on the average* if the cubes are taken sufficiently small.

point gives the direction of the vector and whose density—or number of lines crossing a unit area perpendicular to the lines—gives the magnitude of the vector. Of course, such a graphical representation is always possible at one point in any arbitrary vector field, but we cannot prolong the field lines continuously over all space unless the number of lines entering any given volume equals the number of lines leaving the volume, that is, unless the *flux* of lines is zero. Equation (110) expresses this condition of zero flux. Thus, a vector field can be represented by field lines whenever its divergence is zero. Figure 1.12 shows some examples of such vector fields represented by field lines.

The curl of a vector field also satisfies another important theorem known as *Stokes' theorem*. Figure 1.13 shows an open surface; as

Figure 1·12

(*a*) Streamlines representing the velocity of an incompressible fluid flowing out of a pipe immersed in a large pool. The streamlines in this diagram were calculated for a two-dimensional arrangement. (*b*) Streamlines representing the velocity of an incompressible fluid flowing in a pipe of nonuniform cross section.

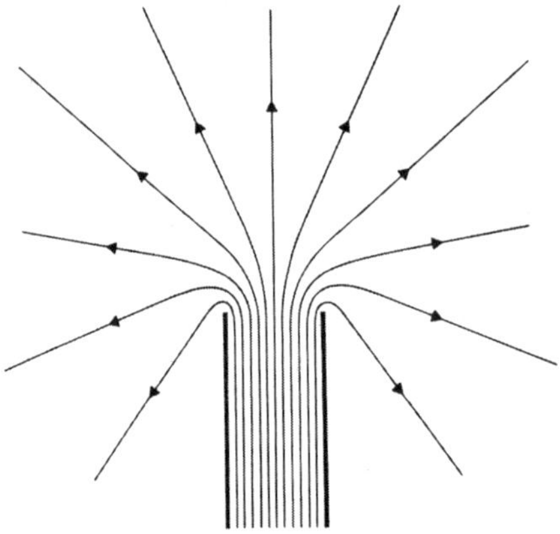

(a)

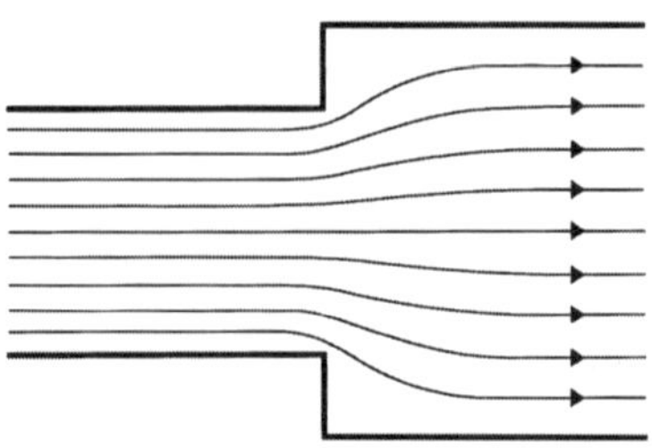

(b)

Figure 1·13

An arbitrary open surface. The unit vector is perpendicular to the surface. If the fingers of your right hand lie along $d\mathbf{l}$, your thumb will lie along $\hat{\mathbf{n}}$.

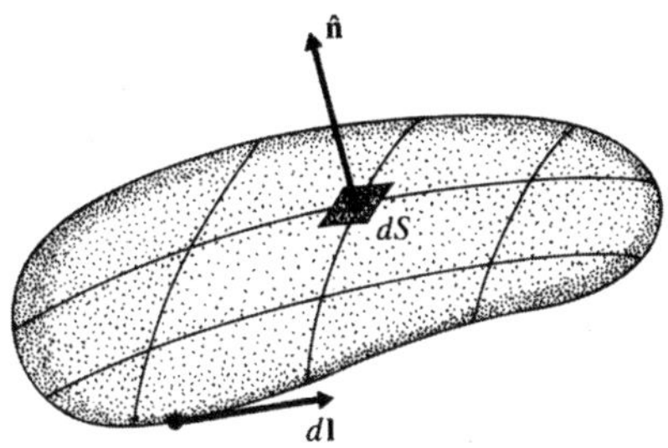

before, $\hat{\mathbf{n}}$ is a unit vector perpendicular to the surface, and $d\mathbf{l}$ is a small displacement along the boundary of the surface. Note that the direction of $d\mathbf{l}$ is assumed to be related to the direction of $\hat{\mathbf{n}}$ by the right-hand rule (see Fig. 1.13; the displacement $d\mathbf{l}$ is counterclockwise if $\hat{\mathbf{n}}$ points out of the plane of the page). We can then prove the following theorem:

• ***Stokes' Theorem:*** Suppose S is an open surface with a boundary C. Then

$$\int_S \nabla \times \mathbf{A} \cdot d\mathbf{S} = \oint_C \mathbf{A} \cdot d\mathbf{l} \qquad (111)$$

We first give the proof for the case of a flat surface in the shape of a square lying in the x-y plane. If the square consists of the region $x = x_0$ to $x_0 + L$, $y = y_0$ to $y_0 + L$ (see Fig. 1.14), then

Figure 1·14

A square surface.

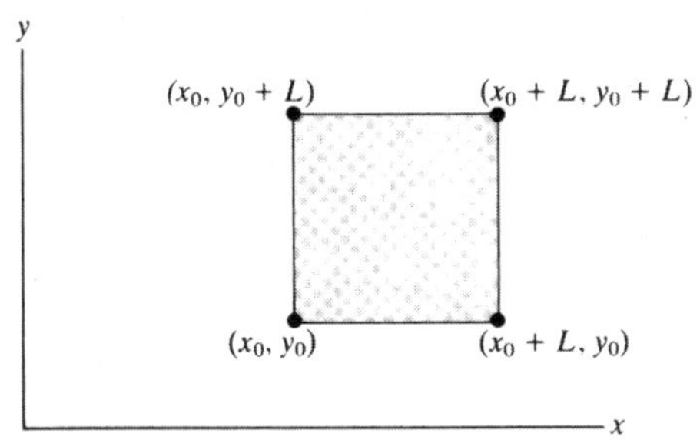

$$\begin{aligned}\int \nabla\times\mathbf{A}\bullet d\mathbf{S} &= \iint \nabla\times\mathbf{A}\bullet\hat{\mathbf{z}}\,dxdy \\ &= \iint\left(\frac{\partial}{\partial x}A_y-\frac{\partial}{\partial y}A_x\right)dxdy \\ &= \int\left[A_y(x_0+L,y)-A_y(x_0,y)\right]dy-\int\left[A_x(x,y_0+L)-A_x(x,y_0)\right]dx\end{aligned}\tag{112}$$

But this is the same as $\int \mathbf{A}\bullet d\mathbf{l}$ over the four sides of the square; hence, Eq. (111) holds.

This establishes Stokes' theorem for a square. A *flat* surface of arbitrary shape can be approximated by a large number of adjacent small squares. The sum of the integrals $\int \mathbf{A}\bullet d\mathbf{S}$ over the sides of all these squares receives a contribution only from those sides at the outer boundary of the surface; the contributions from adjacent interior sides cancel in pairs, because adjacent squares have opposite displacements $d\mathbf{l}$ (see Fig. 1.15). Therefore, Eq. (111) holds for such an arbitrary flat surface. A surface that is not flat can be approximated by polygonal flat surfaces; hence, Eq. (111) holds in general.

From Gauss' and Stokes' theorem we can gain some insight into the geometrical meaning of the divergence and the curl. Consider a vector field $\mathbf{A}(\mathbf{x})$ that has a nonzero divergence—say, a positive divergence—in all or in some part of a spherical region. Then Gauss' theorem tells us that the integral $\int \mathbf{A}\bullet d\mathbf{S}$ is positive over the surface of the sphere. This means that the vectors $\mathbf{A}$ at different locations on the surface of the sphere are predominantly radially outward; that is, these vectors "diverge" in the outward direction, like the radially outward electric field vectors of a positive point charge or the radially outward velocity field vectors of the flow of water from a source placed within a large pool [see Fig. 1.16(*a*)].*

Now consider a vector field $\mathbf{A}(\mathbf{x})$ that has a nonzero curl in some region surrounding a point. Stokes' theorem tells us that for a circular path perpendicular to the direction of the curl, $\oint \mathbf{A}\bullet d\mathbf{l}$ is nonzero. This means that the vectors $\mathbf{A}$ at different locations along the circular path are predominantly tangential to the circle; that is, these vectors "curl" along the circle, like the tangential magnetic field vectors of the magnetic field of a current on a straight wire or the tangential velocity field vectors of the flow of water in a circular pool, when the water has been given a rotational motion by a paddle or by a tangential stream from a recirculating

---

*For the purposes of this example, we have to pretend that the water is actually created at the source—say, by synthesis from oxygen and hydrogen. If the water were supplied by a pipe, then the velocity vectors in the pipe would be radially *inward* [see Fig. 1.12(*a*)], and this would compensate the outward velocity vectors in the pool, so the integral of these vectors over the spherical surface would be zero, in accord with zero divergence for the flow of water.

Figure 1·15

An arbitrary surface is approximated by small squares. The vectors $d\mathbf{l}$ of adjacent sides are opposite.

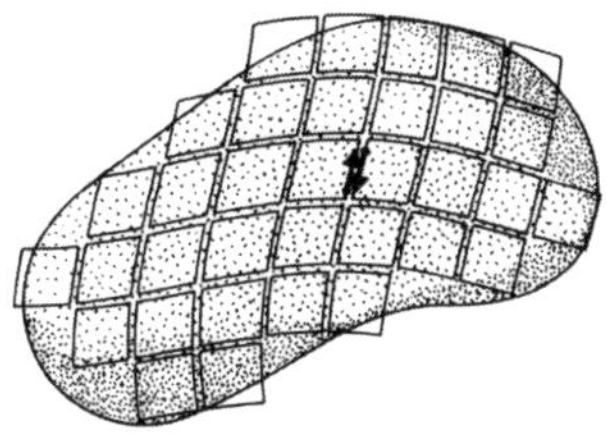

Figure 1·16

(*a*) Velocity vectors of water flowing outward from a source at the center (each vector indicates the velocity for the point located at its tail). These velocity vectors "diverge" from the center. (*b*) Velocity vectors of water flowing in circles around a central axis. These vectors "curl" along circles.

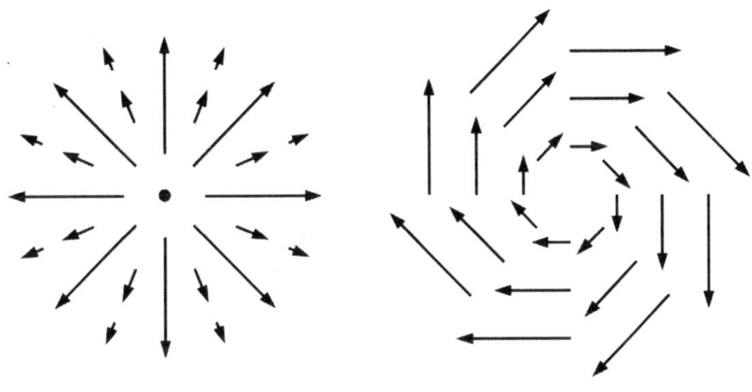

pump [see Fig. 1.16(*b*)]. Of course, it is also possible for a vector field to have both a nonzero divergence and a nonzero curl. An example of this is the field of velocity vectors for the flow of water in a pool, when the water has been given an initial rotational motion and a source has been placed in the center, so the water spirals outward, with concurrent rotational and divergent flows. Note an important difference between the two examples presented in Fig. 1.16: in (*b*) the curl is nonzero throughout the entire velocity field, whereas in (*a*) the divergence is nonzero only at one point, at the source of water at the center (for a surface that does not include the center, the integral of the velocity vectors over the surface is zero).

A vector field $\mathbf{A}(\mathbf{x})$ is said to be *conservative* if the path integral of $\mathbf{A}$ around any arbitrary closed path is zero:

Figure 1·17

Two distinct paths from $\mathbf{x}_0$ to $\mathbf{x}$.

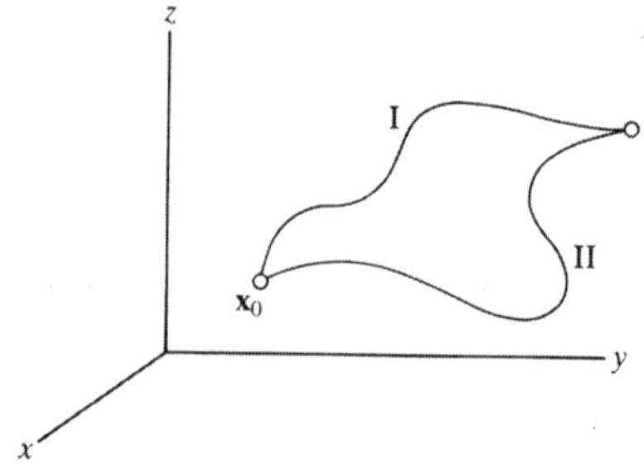

$$\oint_C \mathbf{A} \cdot d\mathbf{l} = 0 \qquad (113)$$

Such a vector field can be written as the gradient of a scalar function, or potential function:

$$\mathbf{A}(\mathbf{x}) = \nabla\Phi(\mathbf{x}) \qquad (114)$$

To prove that such a scalar exists, we construct the scalar function explicitly by means of the formula

$$\Phi(\mathbf{x}) = \int_{\mathbf{x}_0}^{\mathbf{x}} \mathbf{A} \cdot d\mathbf{l} \qquad (115)$$

where $\mathbf{x}_0$ is some reference point and the integral is evaluated along any path from $\mathbf{x}_0$ to $\mathbf{x}$. The rules for differentiation of an integral then immediately yield Eq. (114).

■ ***Exercise 30.*** With $d\mathbf{l} = \hat{\mathbf{x}}dx' + \hat{\mathbf{y}}dy' + \hat{\mathbf{z}}dz'$, Eq. (115) becomes

$$\Phi(\mathbf{x}) = \int_{x_0}^{x} A_x dx' + \int_{y_0}^{y} A_y dy' + \int_{z_0}^{z} A_z dz'$$

Show that the gradient of this function equals $\mathbf{A}(\mathbf{x})$. ■

Note that a crucial ingredient in our construction is the path independence of the integral in Eq. (115)—if this integral had any dependence on the shape of the path from $\mathbf{x}_0$ to $\mathbf{x}$, then $\mathbf{\Phi}$ would be a function of this path, rather than a function of the endpoint. This independence follows from Eq. (113). Consider two distinct paths leading from $\mathbf{x}_0$ to $\mathbf{x}$ (see Fig. 1.17). We then have

$$\int_{\mathrm{I}\ \mathbf{x}_0}^{\mathbf{x}} \mathbf{A}\cdot d\mathbf{l} + \int_{\mathrm{II}\ \mathbf{x}}^{\mathbf{x}_0} \mathbf{A}\cdot d\mathbf{l} = 0 \tag{116}$$

or

$$\int_{\mathrm{I}\ \mathbf{x}_0}^{\mathbf{x}} \mathbf{A}\cdot d\mathbf{l} = -\int_{\mathrm{II}\ \mathbf{x}}^{\mathbf{x}_0} \mathbf{A}\cdot d\mathbf{l} = \int_{\mathrm{II}\ \mathbf{x}_0}^{\mathbf{x}} \mathbf{A}\cdot d\mathbf{l} \tag{117}$$

so the integrals for both paths are equal. With this, we have established that any conservative vector is the gradient of a scalar (conversely, the gradient of any scalar is a conservative vector).

The following theorem relies on Stokes' theorem to establish that if a vector field has zero curl in a (suitably restricted) region, then the vector field is conservative in this region and, consequently, can be expressed as the gradient of a scalar function.

• ***Theorem:*** Suppose that a vector field **A** has zero curl in a simply connected region V. Then, within this region, the vector field is conservative, and there exists a scalar function Φ such that

$$\mathbf{A}(\mathbf{x}) = \nabla\Phi(\mathbf{x})$$

A region of space is said to be *simply connected* if any closed path in this region is shrinkable to a point. The interior of a sphere, the exterior of a sphere, and the spherical shell bounded by the two concentric spheres are examples of simply connected regions. The interior of a torus, the exterior of a torus, and the exterior of an infinitely long cylinder are examples of multiply connected regions (see Fig. 1.18).

To prove the theorem, we consider any closed path C in the region V. Since V is simply connected, we can find a surface S that has this path as boundary and lies entirely with V. By Stokes' theorem, the integral of **A** around the path is

Figure 1·18

(*a*) The interior of a sphere is a simply connected region. Any path can be shrunk to a point while remaining within the region. (*b*) The interior of a torus is a multiply connected region. The path shown cannot be shrunk to a point while remaining within the region.

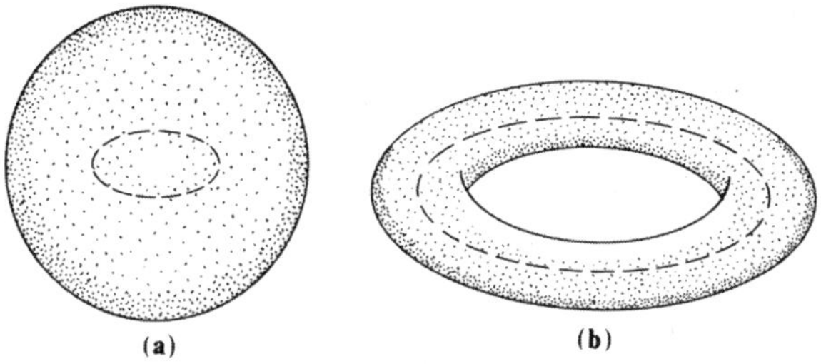

$$\oint_C \mathbf{A} \cdot d\mathbf{l} = \int_S \nabla \times \mathbf{A} \cdot d\mathbf{S}$$

Since $\nabla \times \mathbf{A} = 0$, this tells us that

$$\oint_C \mathbf{A} \cdot d\mathbf{l} = 0$$

The rest of the proof follows exactly the same steps as in Eqs. (115) — (117); we will not bother to repeat these steps.

1·6 The Continuity Equation; Conservation of Charge

Ordinary matter is made of electrons, protons, and neutrons. Only the first two kinds of particles carry an electric charge: negative electric charge on electrons and positive electric charge on protons. In the following discussion, we will concentrate on electrons because electric currents are usually produced by the motion of the electrons rather than of the protons or ions, but we could easily generalize the discussion to include protons.

Consider the free (conduction) electrons in a metal, the electrons in a plasma, or the electrons in the beam of an electron gun. Instead of regarding these electrons as individual particles with individual positions and velocities, we can regard them as a *fluid*, which has a time-dependent density and velocity at each point of space. The density of this fluid, or the number of electrons per unit volume, will be some function $n(t,\mathbf{x})$ of time and of position. Likewise, the flow velocity of this fluid will be some function $\mathbf{v}(t,\mathbf{x})$. The *electric charge density* is then simply the electron density multiplied by the charge per electron. We will write this charge density as

$$\rho(t,\mathbf{x}) = -en(t,\mathbf{x}) \qquad \textit{(118)}$$

where e is the magnitude of the charge per electron. The value of e is

$$e = 4.80 \times 10^{-10} \text{ esu} \qquad \textit{(119)}$$

in CGS electrostatic units, or

$$e = 1.60 \times 10^{-19} \text{ coulomb} \qquad \textit{(120)}$$

in SI units. Here, we have anticipated the definition of the unit of charge given in Chapter 2.

The *electric current density* is

$$\mathbf{j}(t,\mathbf{x}) = \rho(t,\mathbf{x})\mathbf{v}(t,\mathbf{x}) = -en(t,\mathbf{x})\mathbf{v}(t,\mathbf{x}) \tag{121}$$

This current density gives the rate of flow of electric charge per unit area; for instance, the x component j_x gives the amount or charge that crosses a unit area perpendicular to the x axis in unit time. In the CGS electrostatic system of units, the unit of current is the statamp,

$$1\,\text{statamp} = 1\,\text{esu/s}$$

and the unit of current density is the statamp/cm^2.

There is excellent experimental evidence that electric charge is conserved. Thus, the electric charge in a given volume can change only if charge moves across the boundary of the volume; the charge cannot be created or destroyed. Leaving the discussion of this evidence for later, let us assume charge conservation holds. We can then prove that $\mathbf{j}$ and ρ must satisfy the relation

$$\nabla\boldsymbol{\cdot}\mathbf{j} + \frac{\partial\rho}{\partial t} = 0 \tag{122}$$

To see this, we begin with Gauss' theorem applied to a small volume V (see Fig. 1.9):

$$\int \nabla\boldsymbol{\cdot}\mathbf{j}\,dV = \int \mathbf{j}\boldsymbol{\cdot}d\mathbf{S} \tag{123}$$

But

$$\begin{aligned}\mathbf{j}\boldsymbol{\cdot}d\mathbf{S} &= dS\hat{\mathbf{n}}\boldsymbol{\cdot}\mathbf{j} = dS \times (\text{component of current density normal to } dS\,)\\ &= (\text{amount of charge that flows } \textit{out} \text{ through the area } dS \text{ in unit time})\end{aligned} \tag{124}$$

and hence

$$\int \nabla\boldsymbol{\cdot}\mathbf{j}\,dV = (\text{rate of outflow of charge from } V\,) \tag{125}$$

The loss of charge from the volume V can also be expressed as

$$-\frac{d}{dt}\int \rho\,dV = -\int \frac{\partial\rho}{\partial t}dV \tag{126}$$

The conservation of charge tells us that this rate of loss must match the rate of outflow of charge from the volume V:

$$\int \nabla\boldsymbol{\cdot}\mathbf{j}\,dV = -\int \frac{\partial\rho}{\partial t}dV \tag{127}$$

If the volume is small, the integrands can be approximated as constants; this implies

$$\nabla\cdot\mathbf{j}\,\Delta x\Delta y\Delta z = -\frac{\partial\rho}{\partial t}\Delta x\Delta y\Delta z \qquad \textit{(128)}$$

or

$$\nabla\cdot\mathbf{j}+\frac{\partial\rho}{\partial t}=0 \qquad \textit{(129)}$$

The relation (129) is called the *continuity equation;* it is the differential form of the law of conservation of charge.

The most precise empirical evidence in favor of charge conservation comes from an inquiry into the stability of the electron. Why is the electron stable? The answer is that charge conservation forbids the decay of the electron.[1] A particle can decay only into less massive particles; the only particles less massive than an electron are neutrinos, photons, and gravitons;* since these particle are all neutral, the decay of the electron into them would necessarily involve the disappearance of one electronic charge. Thus, we see that charge conservation implies the stability of the electron. But we can also argue for the converse: the stability of the electron implies charge conservation. The reason is that a reaction such as

$$\text{electron} \rightarrow \quad \text{neutrino} + \text{photon} \qquad \textit{(130)}$$

is forbidden *only* by charge conservation. This reaction violates no other conservation law and would proceed if charge conservation did not hold.

If the reaction (130) were to proceed, the photon would be emitted as a gamma ray of one-half of the rest-mass energy of the electron, and experimenters have attempted to detect such gamma-ray emission from samples of ordinary materials. No gamma rays of the characteristic energy were detected, and from the best available experimental data it can be concluded that the lifetime of the electron with respect to the reaction (130) is 5×10^{26} years.[2]

There are, of course, other conceivable reactions that violate charge conservation, such as decay of an electron into several neutrinos and several photons. A limit can be placed on all such decay processes of the electron by looking for the spontaneous emission of X rays by atoms. If an electron in an inner shell of an atom were to disappear suddenly, one of the outer electrons would jump into the "hole" left by the disappearance, and an X ray would be emitted during this jumping process. In careful experiments with isotopes of iodine and germanium,[3] no such X rays were observed, and it was concluded that the electron lifetime is certainly greater than 4×10^{24} years. This means that, on the average, the

---

*Gravitons are the quanta of gravity, just as photons are the quanta of light.

electric charge disappears at a rate no greater than 10^{-34} esu per year per electron.

Similar arguments can be made in connection with the decay of the neutron. The normal decay of the neutron—into a proton, an electron, and an antineutrino—obeys charge conservation. However, we can contemplate the possibility of an exotic decay, such as

$$\text{neutron} \rightarrow \text{proton} + \text{neutrino} + \text{antineutrino} \qquad (131)$$

which is forbidden *only* by charge conservation. The absence of this reaction therefore constitutes evidence for charge conservation. Precise limits on the rate of this reaction have been obtained by experiments with various isotopes. For instance, if the neutrons in the nucleus of gallium could decay by the reaction (131), then the gallium would transmute into germanium, with the ejection of neutrinos. In observations on a sample of gallium, no such transmutation was seen, and it was established that the lifetime of the neutron with respect to the reaction (131) is in excess of 3×10^{21} years.[4]

According to this evidence, we can feel quite confident that the charges on charged particles do not suddenly disappear. However, we might contemplate an alternative mechanism for the violation of charge conservation: Could the "constant" *e,* which represents the fundamental unit of charge in atomic physics, gradually change with time? Gamow,[5] following some earlier speculations of Dirac about changes in fundamental "constants," suggested that e^2 might be proportional to the time that has elapsed since the beginning of the universe. We can obtain evidence against such a hypothesis by a careful study of the abundances and decay rates of radioactive isotopes—such as U^{238}, K^{40}, Re^{187}—found in rocks on Earth. The decay rates of these isotopes are sensitive to the value of *e*, and from this we can set an upper limit of 10^{-23} esu per year on the rate of change of *e*.[6] In view of this tight limit, we will assume from now on that charge is exactly conserved.*

References

1. G. Feinberg and M. Goldhaber, *Proc. Natl. Acad. Sci. USA* **45**, 1201 (1959).
2. H. O. Back et al., *Phys. Lett.* B **525**, 29 (2002).

---

* It is worth noting that since *e* enters the equations of atomic physics as a parameter determining the strength of the electric force, a change in *e* could be reinterpreted as a change in the dielectric constant of the vacuum, which would have the same effect on the strength of the force. If there were any evidence for a change in *e*, such a reinterpretation would permit us to avoid any conflict with charge conservation or with Maxwell's equations.

3. P. Belli et al., *Phys. Lett.* B **460**, 236 (1999).
4. E. B. Norman, J. N. Bahcall, and M. Goldhaber, *Phys. Rev.* D **53**, 4086 (1996).
5. G. Gamow, *Phys. Rev. Lett.* **19**, 759 (1967).
6. F. J. Dyson, *Phys. Rev. Lett.* **19**, 1291, (1967); R. Gold, *Phys. Rev. Lett.* **20**, 219 (1968).

Problems

1. From the definition (13), show that if **A** and **B** are arbitrary vectors, then $|\mathbf{A}\times\mathbf{B}| = AB\sin\phi$, where ϕ is the angle between the vectors. [*Hint:* Orient the coordinate axes so that **A** is along the x axis and **B** is in the x-y plane.]

2. Find a unit vector perpendicular to both $3\hat{\mathbf{x}}+\hat{\mathbf{y}}+\hat{\mathbf{z}}$ and $\hat{\mathbf{x}}+2\hat{\mathbf{y}}-\hat{\mathbf{z}}$.

3. Use vector methods to find the angle between the diagonal of a cube and one of the edges it intersects.

4. Find a unit vector whose direction bisects the angle between $2\hat{\mathbf{x}}+\hat{\mathbf{y}}-2\hat{\mathbf{z}}$ and $3\hat{\mathbf{x}}-4\hat{\mathbf{z}}$.

5. Show that $(\mathbf{x}-\mathbf{A})\cdot\mathbf{B}=0$ is the equation of a plane that includes the point **A** and is perpendicular to the vector **B**.

6. Given the vectors $\mathbf{A}=\hat{\mathbf{x}}-\hat{\mathbf{y}}+3\hat{\mathbf{z}}$ and $\mathbf{B}=\hat{\mathbf{x}}-2\hat{\mathbf{y}}-2\hat{\mathbf{z}}$, find the component of **B** perpendicular to the direction of **A** in the plane of **A** and **B.**

7. If **A** is an arbitrary vector and $\hat{\mathbf{n}}$ a unit vector, show that

$$\mathbf{A}=\hat{\mathbf{n}}(\mathbf{A}\cdot\hat{\mathbf{n}})+\hat{\mathbf{n}}\times(\mathbf{A}\times\hat{\mathbf{n}})$$

This expresses **A** as a sum of a vector parallel to $\hat{\mathbf{n}}$ and a vector perpendicular to $\hat{\mathbf{n}}$.

8. Starting with the definitions (81), evaluate $\hat{\boldsymbol{\rho}}\times\hat{\boldsymbol{\phi}}$, $\hat{\boldsymbol{\phi}}\times\hat{\mathbf{z}}$, and $\hat{\boldsymbol{\rho}}\times\hat{\mathbf{z}}$.

9. Starting with the definitions (100), evaluate $\hat{\mathbf{r}}\times\hat{\boldsymbol{\theta}}$, $\hat{\boldsymbol{\theta}}\times\hat{\boldsymbol{\phi}}$, and $\hat{\mathbf{r}}\times\hat{\boldsymbol{\phi}}$.

10. (a) Show that

$$\varepsilon^{klm}\varepsilon^{mrs}=\delta^{ls}\delta^{kr}-\delta^{ks}\delta^{lr}$$

[*Hint*: Consider $k = 1$, $l = 2$. The sum over m then collapses to $\varepsilon^{123}\varepsilon^{3rs}$. But this is zero unless $r = 1, s = 2$ or $r = 2, s = 1$. The first alternative yields the first term in the above equation, the second yields the second.]

(b) Use this result to obtain the double cross identity

$$\mathbf{A}\times(\mathbf{B}\times\mathbf{C})=\mathbf{B}(\mathbf{A}\cdot\mathbf{C})-\mathbf{C}(\mathbf{A}\cdot\mathbf{B})$$

11. Show that $\varepsilon^{klm}\varepsilon^{mrk}=-2\delta^{lr}$ and that $\varepsilon^{klm}\varepsilon^{mlk}=-6$.

12. Prove the dot cross identity (16) and the double cross identity (17) by direct substitution of the explicit formula (13) for the cross products. [*Hint*: This is tedious, but it can be simplified somewhat by assuming that one of the vectors has only one component, by choice of coordinate axes.]

13. Prove the identity

$$(\mathbf{A}\times\mathbf{B})\cdot(\mathbf{C}\times\mathbf{D}) = (\mathbf{A}\cdot\mathbf{C})(\mathbf{B}\cdot\mathbf{D}) - (\mathbf{A}\cdot\mathbf{D})(\mathbf{B}\cdot\mathbf{C})$$

14. Suppose that we attempt to define a new vector product, called the "star product," as follows: The star product **A*B** is a vector of length $|\mathbf{A}||\mathbf{B}|$ in the direction of **A** + **B**.

(a) Is this a vector? That is, does it behave correctly under coordinate rotations? What are the star products of the unit vectors?

(b) Is the star product distributive, so **A*(B** + **C)** = **A*B** + **A*C**? What can you conclude about the star product?

15. Show that if **A**, **B**, and **C** are three arbitrary vectors, then

$$\mathbf{A}\cdot(\mathbf{B}\times\mathbf{C}) = \varepsilon^{klm}A^k B^l C^m = \begin{vmatrix} A^1 & A^2 & A^3 \\ B^1 & B^2 & B^3 \\ C^1 & C^2 & C^3 \end{vmatrix}$$

and show that this dot cross product is the volume in the parallelepiped with edges along **A**, **B**, **C**.

16. Prove that $\varepsilon^{klm}Q^{k1}Q^{l2}Q^{m3}$ is the determinant of the matrix Q^{kl}.

17. Suppose that the three-component object $(-2,1,4)$ becomes $(-2,1,0)$ when the coordinates are rotated by 45° around the x axis. Is the object a vector?

18. Suppose we rotate the axes x,y,z by ϕ around the z axis and obtain new axes x',y',z'. Then we rotate the axes x',y',z' by θ around the x' axis and obtain new axes x'',y'',z''. What is the matrix that represents the rotation from x,y,z axes to x'',y'',z'' axes? What is the inverse of this matrix?

19. Show that $T^{kl}Q^{lk}$ is a scalar whenever T^{kl} and Q^{lm} are tensors.

20. A vector defined at the point $(2,-1,0)$ in rectangular coordinates has components $(-2,-1,3)$. What are the components of this vector in cylindrical coordinates? In spherical coordinates? Why is it necessary to specify the point at which the vector is defined?

21. (a) Express $\hat{\mathbf{x}}$ and $\hat{\mathbf{y}}$ in terms of the cylindrical unit vectors $\hat{\boldsymbol{\rho}}$ and $\hat{\boldsymbol{\phi}}$.

(b) Express $\hat{\mathbf{x}}$, $\hat{\mathbf{y}}$, and $\hat{\mathbf{z}}$ in terms of the spherical unit vectors $\hat{\mathbf{r}}$, $\hat{\boldsymbol{\theta}}$, and $\hat{\boldsymbol{\phi}}$.

22. Show that the following equations are true:

(a) $\nabla\cdot\mathbf{x} = 3$

(b) $\nabla\dfrac{1}{|\mathbf{x}|} = -\dfrac{\mathbf{x}}{|\mathbf{x}|^3}$

(c) $\nabla^2\dfrac{1}{|\mathbf{x}|} = 0$

(d) $\nabla\cdot\dfrac{\mathbf{x}}{|\mathbf{x}|} = \dfrac{2}{|\mathbf{x}|}$

(e) $(\mathbf{B}\cdot\nabla)\mathbf{x} = \mathbf{B}$

(f) $\nabla\times\mathbf{x} = 0$

(g) $\nabla\times\dfrac{\mathbf{x}}{|\mathbf{x}|} = 0$

23. Prove the following identities:
(a) $\nabla(\Phi\Psi) = \Phi\nabla\Psi + \Psi\nabla\Phi$
(b) $\nabla\cdot(\Phi\mathbf{A}) = \Phi\nabla\cdot\mathbf{A} + \mathbf{A}\cdot\nabla\Phi$
(c) $\nabla\times(\Phi\mathbf{A}) = \Phi\nabla\times\mathbf{A} - \mathbf{A}\times\nabla\Phi$

24. Prove the following identities:
(a) $\nabla\cdot(\mathbf{A}\times\mathbf{B}) = \mathbf{B}\cdot(\nabla\times\mathbf{A}) - \mathbf{A}\cdot(\nabla\times\mathbf{B})$
(b) $\nabla\times(\mathbf{A}\times\mathbf{B}) = (\mathbf{B}\cdot\nabla)\mathbf{A} - (\mathbf{A}\cdot\nabla)\mathbf{B} + \mathbf{A}(\nabla\cdot\mathbf{B}) - \mathbf{B}(\nabla\cdot\mathbf{A})$
(c) $\nabla(\mathbf{A}\cdot\mathbf{B}) = (\mathbf{B}\cdot\nabla)\mathbf{A} + (\mathbf{A}\cdot\nabla)\mathbf{B} + \mathbf{B}\times(\nabla\times\mathbf{A}) + \mathbf{A}\times(\nabla\times\mathbf{B})$

25. Given $\Phi(\mathbf{x}) = (2z^2 - x^2 - y^2)/(x^2 + y^2 + z^2)^{5/2}$, evaluate $\nabla\Phi$ and evaluate $\nabla^2\Phi$.

26. Suppose that $\mathbf{A}(\mathbf{x}) = r^n\mathbf{x}$, where $r = (x^2 + y^2 + z^2)^{1/2}$.
(a) Show that $\nabla\times\mathbf{A} = 0$, and find a scalar field Φ such that $\mathbf{A} = \nabla\Phi$.
(b) For what value of n is $\nabla\cdot\mathbf{A} = 0$?

27. (a) Given that $\mathbf{A}(\mathbf{x}) = 4xy\hat{\mathbf{x}} + 2x^2\hat{\mathbf{y}} + 3z^2\hat{\mathbf{z}}$, find a scalar field such that $\mathbf{A} = \nabla\Phi$.
(b) Given that $\mathbf{A}(\mathbf{x}) = 4xy\hat{\mathbf{x}} + 3x^2\hat{\mathbf{y}} + 3z^2\hat{\mathbf{z}}$, show that there exists no scalar field such that $\mathbf{A} = \nabla\Phi$.

28. Show that the components of $\nabla\times\mathbf{A}$ in cylindrical coordinates are

$$(\nabla\times\mathbf{A})_\rho = \frac{1}{\rho}\frac{\partial A_z}{\partial\phi} - \frac{\partial A_\phi}{\partial z}$$

$$(\nabla\times\mathbf{A})_\phi = \frac{\partial A_\rho}{\partial z} - \frac{\partial A_z}{\partial\rho}$$

$$(\nabla\times\mathbf{A})_z = \frac{1}{\rho}\left[\frac{\partial}{\partial\rho}(\rho A_\phi) - \frac{\partial A_\rho}{\partial\phi}\right]$$

29. Show that the components of $\nabla\times\mathbf{A}$ in spherical coordinates are

$$(\nabla\times\mathbf{A})_r = \frac{1}{r\sin\theta}\left[\frac{\partial}{\partial\theta}(A_\phi\sin\theta) - \frac{\partial A_\theta}{\partial\phi}\right]$$

$$(\nabla\times\mathbf{A})_\theta = \frac{1}{r\sin\theta}\frac{\partial A_r}{\partial\phi} - \frac{1}{r}\frac{\partial}{\partial r}(rA_\phi)$$

$$(\nabla\times\mathbf{A})_\phi = \frac{1}{r}\left[\frac{\partial}{\partial r}(rA_\theta) - \frac{\partial A_r}{\partial\theta}\right]$$

30. Prove that if S is a closed surface, then for any vector field $\mathbf{A}$,

$$\int_S \nabla\times\mathbf{A}\cdot d\mathbf{S} = 0$$

31. Prove that if S is a closed surface and V is the volume contained in it, then for any scalar field Φ,

$$\int_V \nabla\Phi\, dV = \int_S \Phi\, d\mathbf{S}$$

32. Prove that if S is an open surface and C is its boundary, then for any scalar field Φ,

$$-\int_S \nabla\Phi \times d\mathbf{S} = \int_C \Phi\, d\mathbf{l}$$

33. (a) Show that for any point except the origin, $\nabla^2(1/|\mathbf{x}|) = 0$.

(b) Show that for any volume V that includes the origin,

$$\int_V \nabla^2 \frac{1}{|\mathbf{x}|} dV = -4\pi$$

This means that $-(1/4\pi)\nabla^2(1/|\mathbf{x}|)$ is a delta function. (*Hint:* Since $\nabla^2 = \nabla\cdot\nabla$, the integrand is a divergence. Shrink the volume of integration to a small sphere centered on the origin, and apply Gauss' theorem.)

34. Show that if S is a closed surface and V is the volume contained in it, then for any vector field $\mathbf{A}$,

$$\int_V \nabla\times\mathbf{A}\, dV = -\int_S \mathbf{A}\times d\mathbf{S}$$

35. The flow of an incompressible fluid around a cylinder of radius R is described by the velocity field (for $\rho^2 = x^2 + y^2 \geq R^2$)

$$v_x = v_0\left(1 - \frac{R^2}{\rho^2}\right) + 2v_0\frac{y^2R^2}{\rho^4} \qquad v_y = -2v_0\frac{xyR^2}{\rho^4} \qquad v_z = 0$$

(a) Show that $\nabla\cdot\mathbf{v} = 0$. What is the physical significance of this result?
(b) Find $\nabla\times\mathbf{v}$. Where in the x-y plane is $\nabla\times\mathbf{v} = 0$?

36. Suppose that the electric current density in a region filled with plasma is $\mathbf{j} = C(xr\hat{\mathbf{x}} + yr\hat{\mathbf{y}})$, where C is a constant. What is the instantaneous rate of change of the electric charge in the spherical region bounded by $r = R$?

37. Consider the flow of water in a circular pool, when the water has been given a rotational motion. If the water rotates rigidly about the center, what is the velocity field $\mathbf{v}(\mathbf{x})$? Verify that $\nabla\times\mathbf{v}(\mathbf{x}) \neq 0$ and that $\nabla\cdot\mathbf{v}(\mathbf{x}) = 0$.

38. Prove the following two-dimensional version of Gauss' theorem: If S is an area in, say, the x-y plane and C is its boundary, then for any vector function $\mathbf{A}(x, y)$,

$$\int_S \left(\hat{\mathbf{x}}\frac{\partial}{\partial x} + \hat{\mathbf{y}}\frac{\partial}{\partial y}\right)\cdot\mathbf{A}\, dS = \oint_C \mathbf{A}\cdot\hat{\mathbf{n}}\, dl$$

where $\hat{\mathbf{n}}$ is a unit vector perpendicular to the boundary in the outward direction.

CHAPTER

Electrostatics

2

Electrons, and protons, and some other particles ("charged" particles) exert a long-range force on one another. Like gravitation, this force is inversely proportional to the square of the distance between the particles; but unlike gravitation, this force is attractive or repulsive, depending on what particles are involved. Furthermore, this force is much stronger than the force of gravitation—between an electron and a proton, it is about 2×10^{39} times stronger. This force is called the *electric force*, and particles that exert these forces are said to carry *electric charge.* By definition, we will regard the electric charge of an electron as negative, and we will regard the charge of any other particle as positive or as negative, depending on whether it attracts or repels electrons. Macroscopic electric forces, such as the electric forces between two capacitor plates, are the result of the superposition of the attractive and repulsive electric forces of a large number of electrons and protons within these plates.

In this chapter, we will study the description of the electric force in terms of the electric field and potential. Throughout this chapter and the next three, we will assume that the charges exerting these forces are either at rest or moving very slowly; the electric forces produced under these conditions are called *electrostatic* forces. In later chapters, we will see how these forces are modified when the particles move at high speed or with acceleration.

2·1 Coulomb's Law

The electric force that a point charge q' exerts on a point charge q at a distance r has a magnitude

$$F_{\text{on } q \text{ by } q'} = \frac{|qq'|}{r^2} \qquad (1)$$

This force is directed along the line joining the charges (see Fig. 2.1). The force is repulsive if the charges have the same sign, and it is attractive if the charges have opposite signs. If the charge q' is at the origin of our coordinate system, then we can write the force in vector notation as

Figure 2·1

Electric force that a point charge q' exerts on a point charge q. The force is repulsive if both charges have the same sign.

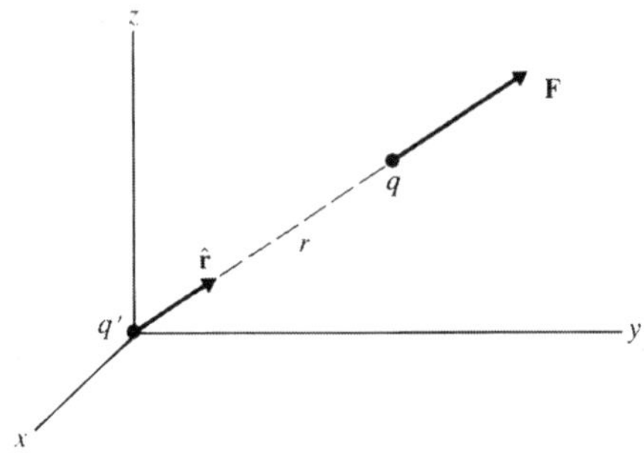

$$\mathbf{F}_{\text{on } q \text{ by } q'} = qq'\frac{\hat{\mathbf{r}}}{r^2} \tag{2}$$

This law of force is *Coulomb's law.*

In Eq. (2), all quantities are expressed in the CGS *electrostatic system* of units, also called the *Gaussian system.** In this CGS system, the units of length, mass, and time are the centimeter, the gram, and the second, respectively. The unit of force is the dyne, where 1 dyne $= 1\,\text{g}\cdot\text{cm/s}^2 = 10^{-5}$ newton; the unit of energy is the erg, where 1 erg $= 1\,\text{g}\cdot\text{cm}^2/\text{s}^2 = 10^{-7}$ joule. The unit of charge is the *electrostatic unit* (esu), defined as follows: Take two identical charges and place them 1 cm apart; the magnitude of each charge is said to be 1 esu if the force between them is exactly 1 dyne. This means that the unit of charge is determined by

$$1 \text{ dyne} = \frac{(1 \text{ esu})^2}{(1 \text{ cm})^2}$$

that is,

$$1 \text{ esu} = 1 \text{ cm} \sqrt{1 \text{ dyne}} = 1\,(\text{cm})^{3/2}(\text{g})^{1/2}/\text{s} \tag{3}$$

Since the right side of Eq. (3) is such an unwieldy concoction, we will always call this unit 1 esu, but we ought to keep in mind that this is a combination of centimeters, grams, and seconds.

*Confusingly, this name is also sometimes applied to the CGS *electromagnetic* system, which differs from the CGS electrostatic system by the inclusion of sundry factors of c in the definitions of the electric units. In this book, we will use only the CGS electrostatic system.

The charge of an electron is $-e$, that of a proton is $+e$, and that of a neutron is zero. According to the best available measurements, the value of e in electrostatic units is*

$$e = 4.8032042 \times 10^{-10} \text{ esu} \tag{4}$$

The equality of the magnitudes of the electron and proton charges has been verified with high precision. The difference between them is no more than one part in 10^{21}. The charge on the neutron is no more than 10^{-21} times the charge of a proton. These precise limits were set by experiments that tested the neutrality of molecules in a gas and by experiments that tested the neutrality of small, magnetically levitated steel spheres.[1]

All elementary particles discovered so far have electric charges that are integral multiples of e; that is, the electric charge is quantized:

$$q = ne \quad \text{with} \quad n = 0, \pm 1, \pm 2, \ldots \tag{5}$$

The reason is not well understood. However, the theories that unify the fundamental forces of physics—treating the weak, electromagnetic, and strong forces as different aspects of a single underlying force—are not viable unless the charge is quantized. Hence, such a unification of all forces demands the quantization of charge.

The statement of Coulomb's law given above applies to *point charges.* The force between charged bodies of appreciable size will usually deviate from the simple inverse-square law; we will discuss such deviations later on. Bodies can be approximated as point charges, provided that the distance between them is very large compared with their size. Since electrons and protons are so very small, they can always be regarded as point charges for the purpose of calculating their orbits in the electric or magnetic fields of accelerators, TV tubes, or other macroscopic devices.† However, in high-energy collisions between, say, electrons and protons, the distance between the particles can be small enough so the internal structure of the particles comes into play. The resulting deviations from Coulomb's law permit us to deduce some features of the internal structure. From the study of such high-energy collisions between

---

*In the SI system of units, the charge is measured in coulombs, and $e = 1.6021765 \times 10^{-19}$ C. The relation between CGS and SI units is 2.99792458×10^9 esu $\Leftrightarrow$ 1C (the number arises from the value of the speed of light). Note that we have written this relation as an equivalence and not as an equality. In SI units, Coulomb's law contains an extra, vexatious factor $1/4\pi\varepsilon_0$ and the coulomb and the esu are therefore dimensionally incompatible; they cannot be equated. For purposes of comparison, the numerical relationships among the units of the CGS electrostatic system and the SI system are listed in the table in Appendix 2.

†For calculating orbits in strongly inhomogeneous magnetic fields, it is necessary to take into account that both particles have not only an electric charge but also a magnetic dipole moment.

electrons and protons, it is known that protons are not exact point charges; rather, they resemble charged balls with a complicated charge distribution over their volume (see Fig. 2.2). Neutrons are neutral; but they do have a charge distribution in their interior, with equal amounts of positive and negative charge (see Fig. 2.2) Electrons, in contrast, seem to be (quantum-mechanical) point charges; if there is any structure within an electron, it is certainly smaller than 10^{-16} cm.

How do we know that the electric force of a point charge varies exactly as $1/r^2$? It is convenient to rephrase this question in terms of the potential energy. We will give a precise definition of the potential energy later on. For now, suffice it to remark that the potential energy must vary as $1/r$ if the force varies as $1/r^2$. The potential energy associated with the Coulomb force is

$$U(r) = \frac{qq'}{r} \tag{6}$$

Our question can then be rephrased: How do we know that the potential energy varies as $1/r$? Before we turn to the experimental evidence, we note that, according to general theoretical arguments, the potential energy must be of the form

$$U(r) = qq'\frac{e^{-\kappa r}}{r} \tag{7}$$

where the constant κ is positive or zero. This means that potentials of the form $1/r^{1.0002}$, $1/(r+10^{-5}r^2)$, or $1/r+10^{-5}/r^2$ are excluded. The arguments that lead to the expression (7) rest on some basic ideas of relativity and field theory, and we cannot go into these details here. The function appearing in Eq. (7) is called a *Yukawa potential.* If κ equals zero, then Eq. (7) coincides with Eq. (6). So our question can be rephrased as follows: Is κ equal to zero?

Figure 2·2

Charge distributions of the proton and the neutron deduced from scattering experiments.

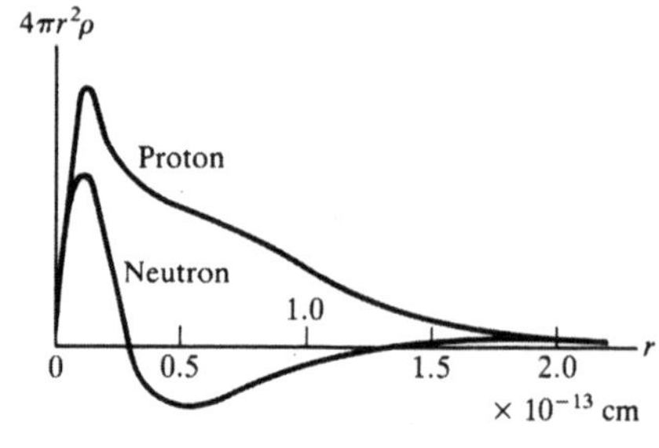

Cavendish, Maxwell, and, more recently, Plimpton and Lawton and Williams, Faller, and Hill[2] carried out tests of Coulomb's law. Their experiments relied on the following peculiarity of the Coulomb law: Under static equilibrium conditions, the charge of a conductor resides on its surface; furthermore, any closed empty cavity in the conductor will not have any charge on its surface. For example, if the conductor is in the shape of two concentric spherical shells connected by a wire (see Fig. 2.3), then charge can reside only on the outermost surface; all other surfaces face empty cavities and must be free of charge. This theorem, whose proof we will leave for a later section, depends crucially on the inverse-square law. A test of this law can therefore be carried out by placing charge on the outer shell in Fig. 2.3 and checking whether any of this charge moves to the inner shell. The most precise experiment of this kind was performed by Williams, Faller, and Hill, who found that the potential is proportional to $1/r^{\alpha}$, where $\alpha = 1 \pm 1.5\times10^{-16}$. According to our discussion, the deviation of the potential from $1/r$ should be expressed as an exponential function [see Eq. (7)] rather than as a change in the power of r. From the analysis of their data, Williams et al. showed that the implied upper limit on κ is

$$\kappa < 3\times10^{-10}\,\text{cm}^{-1} \tag{8}$$

If $\kappa > 0$, then Coulomb's law is a good approximation for small distances between the charges (where $e^{-\kappa r} \simeq 1$), but it fails at large distances (where $e^{-\kappa r} \ll 1$). Roughly, we may say that the failure in the Coulomb law becomes severe when

$$e^{-\kappa r} \simeq e^{-1}$$

that is, when

$$r \simeq \frac{1}{\kappa} \tag{9}$$

Figure 2·3

Two concentric metallic shells connected by a wire. Coulomb's law requires that under static equilibrium conditions, all the electric charge resides on the outer surface of the outer conductor.

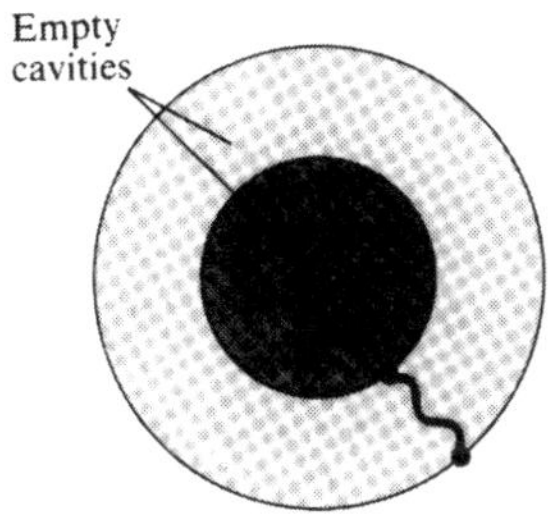

Beyond this distance, the exponential factor in Eq. (7) produces a strong decrease in the potential and soon reduces it to an insignificant level (see Fig. 2.4). The quantity $1/\kappa$ is called the *range* of the potential; note that the Coulomb potential, with $\kappa = 0$, has infinite range. The result (8) shows that the Coulomb law is a good approximation if the distance is much less than

$$\frac{1}{3\times10^{-10}\ \text{cm}^{-1}} \simeq 10^{9}\ \text{cm}$$

A somewhat better limit on κ has been set by a method originally proposed by Schrödinger. If the equations for the electrostatic field are modified so as to agree with Eq. (7), then it is also necessary to introduce some corresponding modifications in the equations for the magnetic field.[3] For instance, the factor $e^{-\kappa r}$ will have to appear in the magnetic field of a magnetic dipole and control the decrease of the field at large distances. We can then test whether the magnetic field of the Earth, measured by artificial satellites, is in agreement with these modifications. Such measurements set the limit[4]

$$\kappa < 4\times10^{-11}\ \text{cm}^{-1} \tag{10}$$

This is a significant improvement over Eq. (8). An even better limit can be set by examining the magnetic field of our Galaxy; this magnetic field is known to extend over thousands of light-years, which permits us to conclude that[5]

$$\kappa < 1\times10^{-22}\ \text{cm}^{-1} \tag{11}$$

Figure 2·4

Coulomb potential and Yukawa potential.

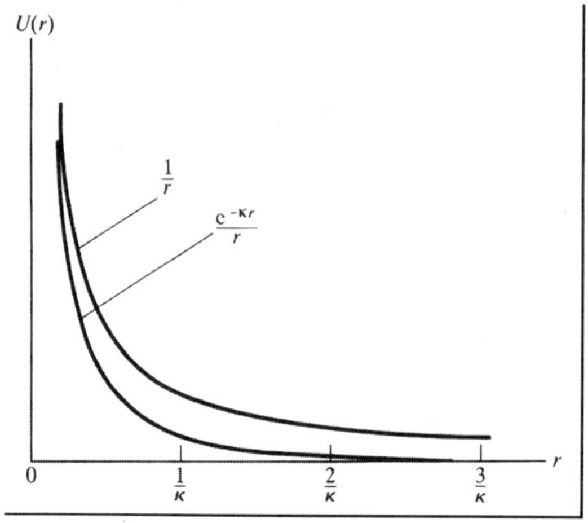

From this observational evidence, it is clear that the value of κ certainly is very small and probably is zero.

Incidentally: The value of κ is related to the mass of the photon. A nonzero value of κ corresponds to a photon mass

$$m = \frac{h\kappa}{2\pi c} \tag{12}$$

where c is the speed of light and h is Planck's constant. The limit (11) set on κ implies that the photon mass either is zero or else is very small,

$$m < 5\times 10^{-60}\text{ g} \tag{13}$$

2·2 The Electric Field; Gauss' Law

For a point charge q' placed at the origin of coordinates, we define the *electric field* that this charge produces at some distance r as

$$\mathbf{E}(r) = q'\frac{\hat{\mathbf{r}}}{r^2} \tag{14}$$

According to Eq. (2), we can then express the force on a second point charge as

$$\mathbf{F} = q\mathbf{E} \tag{15}$$

This equation tells us that the electric field equals the force per 1 esu of (positive) charge. The CGS electrostatic unit of electric field is the dyne per esu (this unit is the same as statvolt per centimeter; see Section 2.3).

■ ***Exercise 1.*** Express dyne per esu in terms of centimeters, grams, seconds. ■

It is often claimed that **E** should be defined by taking the limit $q \to 0$, but because Eq. (14) is completely independent of q, this limit is clearly not needed. The reason why the limiting procedure is not needed here is that in Eq. (15), we assumed that the position of charge q' is held precisely fixed. In practice, the positions of the charges that produce the field will often not be rigidly fixed; if we insert a test charge q at some point, then the electric force exerted by this charge will usually displace the other charges somewhat and produce a change in the electric field acting on q. These disturbances can be eliminated by proceeding to the limit $q \to 0$.

Graphically, we can represent the electric field of Eq. (14) by *field lines* whose direction is that of **E** and whose density is such that the number of lines piercing a unit area oriented perpendicular to the lines is proportional to the magnitude $\mathbf{E}$. Figure 2.5(*a*) shows these field lines for a positive charge q'. Note that this figure shows the field in three dimensions. Since it is difficult to make such a drawing, one often shows the field lines in just two dimensions, as in Fig. 2.5(*b*). But this is somewhat misleading—in two dimensions, the density of field lines decreases only as $1/r$ and therefore does not give a good feeling for how strongly the field decreases in three dimensions.

The electric field is a useful concept because it lets us think of electric effects as being due to local action, or action-by-contact. This means that we can think of the force on a charge as due to the altered state of the space surrounding the charge, not as due to action-at-a-distance of the other charge. Action-at-a-distance is unacceptable because it does not agree with relativity: if a charge could act directly on a second, distant charge, then any change in the position of the former would *immediately* affect the latter, and this would make it possible to propagate signals with a speed exceeding that of light. If the action between charges is mediated by fields, then there is no such instantaneous propagation. Any change in the position of the first charge produces a disturbance in its nearby field; and this disturbance spreads out, as a wave moving at the speed of light, from the first charge to the second.

As we will see later, fields have energy. They therefore are a form of matter, and they can be regarded as the fifth state of matter (solid, liquid, gas, and plasma are the other four states of matter).

If the charge q' that produces the electric field is located at $\mathbf{x}'$, then the electric field at a point $\mathbf{x}$ is given by the equation

$$\mathbf{E}(\mathbf{x}) = q' \frac{\mathbf{x} - \mathbf{x}'}{|\mathbf{x} - \mathbf{x}'|^3} \tag{16}$$

Figure 2·5

(*a*) Field lines of a positive charge as seen in three dimensions. (*b*) Field lines of a positive charge as seen in two dimensions.

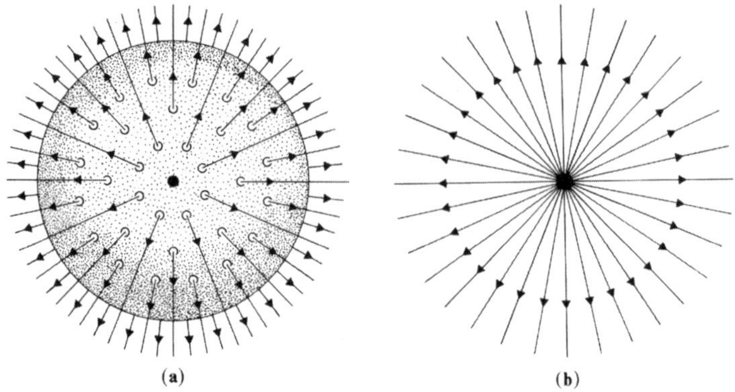

Note that the magnitude of the vector expression on the right side of this equation is $q'/|\mathbf{x}-\mathbf{x}'|^2$, and the direction is from $\mathbf{x}'$ to $\mathbf{x}$, as it should be (see Fig. 2.6). In Eq. (16), we regard $\mathbf{E}$ as a function of $\mathbf{x}$, because $\mathbf{x}'$ is fixed.

The electric field obeys a principle of *linear superposition:* The net electric field of several charges acting simultaneously is the vector sum of the individual electric fields. Thus, if the charges are $q_1, q_2, q_3, \ldots$ with positions $\mathbf{x}_1, \mathbf{x}_2, \mathbf{x}_3, \ldots$, then the net electric field at $\mathbf{x}$ is

$$\mathbf{E}(\mathbf{x}) = \sum_j q_j \frac{\mathbf{x}-\mathbf{x}_j}{|\mathbf{x}-\mathbf{x}_j|^3} \tag{17}$$

This principle of linear superposition of electric fields, and the corresponding principle for magnetic fields, constitute the basis of most of the calculations of electrodynamics. The excellent agreement of these calculations with experiment bears witness that deviations from linear superposition, if any, are certainly very small. The phenomena of "nonlinear" optics observed with laser light do not contradict linear superposition. What is nonlinear in this case is not the superposition of electric fields but only the response (polarization) of a dielectric material to an intense electric field. In brief, we can say that the vacuum is linear, but dielectric materials are (sometimes) nonlinear.

Although the available experimental evidence is in agreement with the linearity of electric and magnetic fields in a vacuum, theoretical calculations predict deviations from linearity under extreme conditions. One such deviation involves a quantum-mechanical effect, the scattering of light by light. If the behavior of electromagnetic fields were exactly linear, then two light beams would be able to pass through each other and

Figure 2·6

A charge q' at the point $\mathbf{x}'$ produces an electric field $\mathbf{E}(\mathbf{x})$ at the point $\mathbf{x}$.

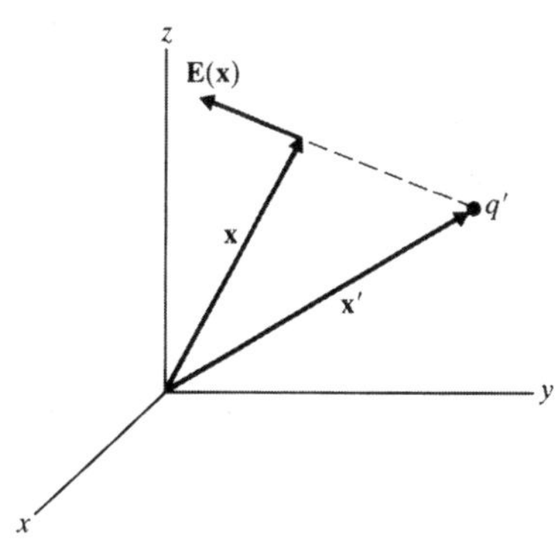

emerge unchanged. But relativistic quantum theory predicts that if two very intense light beams intersect, they will interact. This effect becomes appreciable only if the electric field is of a strength of about 10^{16} statvolt/cm (about 10^{18} volt/cm), which is far beyond what the lasers of today can produce. For practical purposes, such deviations from linearity can therefore safely be neglected.

In the calculation of electric fields, it is often a good enough approximation to ignore the discrete character of the fundamental charges and to describe the charges by a continuous charge density $\rho(\mathbf{x}')$. Then $\rho(\mathbf{x}')dx'dy'dz'$ is the amount of charge in the volume $dV' = dx'dy'dz'$ near the point $\mathbf{x}'$, and Eq. (17) becomes

$$\mathbf{E}(\mathbf{x}) = \iiint \rho(\mathbf{x}') \frac{\mathbf{x}-\mathbf{x}'}{|\mathbf{x}-\mathbf{x}'|^3} dx'dy'dz' \tag{18}$$

or, in a more convenient notation,

$$\mathbf{E}(\mathbf{x}) = \int \rho(\mathbf{x}') \frac{\mathbf{x}-\mathbf{x}'}{|\mathbf{x}-\mathbf{x}'|^3} dV' \tag{19}$$

This equation is valid regardless of whether the point $\mathbf{x}$ is outside or inside the charge distribution. In the latter case, cursory inspection of Eq. (19) suggests a singularity in the integrand when $|\mathbf{x}-\mathbf{x}'|$ vanishes. But careful examination shows that this is an illusion, because if we assume that the origin coincides with the $\mathbf{x}$, then $dV' = r'^2 dr' \times (\text{angular factors})$, and if we let $\mathbf{x}'$ approach $\mathbf{x}$, the quantity $[(\mathbf{x}-\mathbf{x}')/|\mathbf{x}-\mathbf{x}'|^3]dV'$ behaves like $(1/r'^2)r'^2 dr' \times (\text{angular factors})$, which is free of singularity.

EXAMPLE: *electric field of charged disk*

As an example of the calculation of the electric field of a continuous charge distribution, we will evaluate the electric field on the axis of a thin disk carrying a charge Q distributed uniformly over its surface. Figure 2.7 shows the disk in the x-y plane; the radius of the disk is R, and hence the charge per unit area is $Q/\pi R^2$. A small piece of the disk, of area $r'd\phi'dr'$, carries a charge $(Q/\pi R^2)r'd\phi'dr'$; at a height z above the disk, this small charge contributes an electric field of magnitude

$$dE = \frac{(Q/\pi R^2)r'd\phi'dr'}{z^2 + r'^2} \tag{20}$$

This electric field has both a vertical and a horizontal component. In view of the symmetry of the charge distribution, the horizontal components cancel upon integration. Hence we need consider only the vertical component:

Figure 2·7

A thin charged disk in the x-y plane.

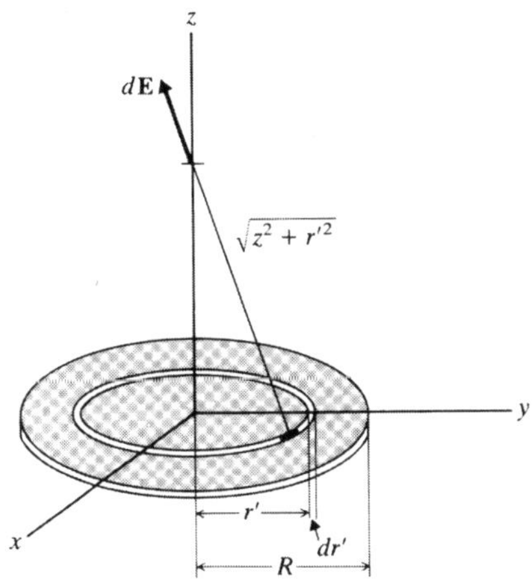

$$dE_z = \cos\theta \frac{(Q/\pi R^2) r' d\phi' dr'}{z^2 + r'^2}$$

$$= \frac{z}{(z^2 + r'^2)^{1/2}} \frac{(Q/\pi R^2) r' d\phi' dr'}{z^2 + r'^2} \qquad (21)$$

Integration gives

$$E_z = \frac{Q}{\pi R^2} \int_0^R \int_0^{2\pi} \frac{z r'}{(z^2 + r'^2)^{3/2}} d\phi' dr'$$

$$= \frac{2Q}{R^2}\left[1 - \frac{z}{(z^2 + R^2)^{1/2}}\right] \qquad (22)$$

EXAMPLE: *electric field of charged sheet (direct integration)*

From this formula for the electric field of a disk, we can obtain the electric field of a very large flat sheet of charge by taking the limit $R \to \infty$. Before we proceed to this limit, we have to express Eq. (22) in terms of the charge per unit area, $\sigma = Q/\pi R^2$. Then

$$E_z = \lim_{R\to\infty} 2\pi\sigma\left[1 - \frac{z}{(z^2 + R^2)^{1/2}}\right] = 2\pi\sigma \qquad (23)$$

For the case of the disk, the direct calculation of the electric field by means of Eq. (19) is fairly simple. But for many other charge distributions, such a direct calculation leads to unmanageable integrals, and we will find it worthwhile to develop alternative methods of calculation, taking advantage of some general mathematical properties of electric fields.

The electric field of any given charge distribution $\rho(\mathbf{x})$ satisfies the following equation:

$$\nabla \cdot \mathbf{E}(\mathbf{x}) = 4\pi\rho(\mathbf{x}) \tag{24}$$

This is known as *Gauss' law,* in differential form. For a proof of this equation, it is best to begin with the electric field of a single point charge q. Consider an arbitrary mathematical surface enclosing the point charge (see Fig. 2.8), and evaluate $\int \mathbf{E} \cdot d\mathbf{S}$ over this surface. If $\hat{\mathbf{n}}$ is the unit vector perpendicular to the surface and $\hat{\mathbf{r}}$ is the unit vector in the radial direction shown in Fig. 2.8, then

$$\begin{aligned} \mathbf{E} \cdot d\mathbf{S} = \mathbf{E} \cdot \hat{\mathbf{n}} dS &= q' \frac{\hat{\mathbf{r}}}{r^2} \cdot \hat{\mathbf{n}} dS \\ &= q' \frac{\cos\theta \, dS}{r^2} \end{aligned} \tag{25}$$

But $\cos\theta \, dS$ is the projection of the area dS on a plane perpendicular to $\hat{\mathbf{r}}$; hence $\cos\theta \, dS / r^2$ is the amount of solid angle $d\Omega$ subtended by dS, as seen from the position of the point charge.* Since the total solid angle subtended by any closed surface is 4π, the integration of Eq. (25) over the entire closed surface gives

Figure 2·8

A point charge q' is at the origin of coordinates. The unit vector $\hat{\mathbf{n}}$ is perpendicular to the surface, and the unit vector $\hat{\mathbf{r}}$ is in the radial direction.

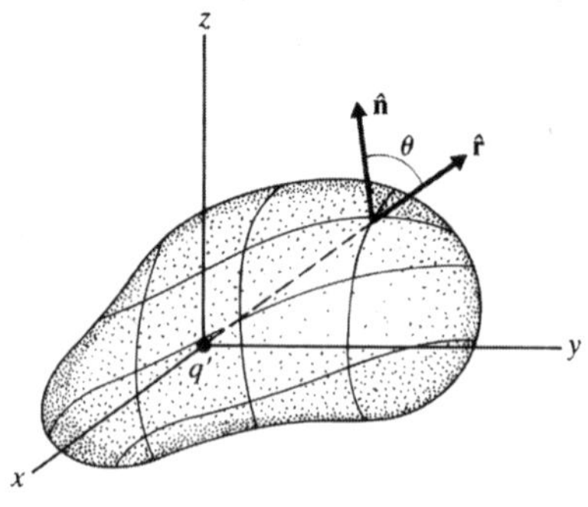

*The *solid angle* subtended by a small rectangular area is the product of angular width and breadth.

$$\int_S \mathbf{E}\cdot d\mathbf{S} = \int q' d\Omega = 4\pi q' \tag{26}$$

If, instead of one charge q', the surface encloses a distribution of many charges, then we can perform such a calculation for the individual electric field of each charge, and the sum of these electric fields gives

$$\int_S \mathbf{E}\cdot d\mathbf{S} = 4\pi Q \tag{27}$$

where, of course, **E** is now the total electric field produced by all the charges enclosed by the surface, and Q is the total charge.

Equation (27) is Gauss' law in integral form. To convert this into the differential form, we note that, by Gauss' theorem, $\int \mathbf{E}\cdot d\mathbf{S}$ equals $\int \nabla\cdot\mathbf{E}dV$, where the volume integral is over the region enclosed by the surface. Hence

$$\int \nabla\cdot\mathbf{E}dV = 4\pi Q \tag{28}$$

For a continuous charge distribution $\rho(\mathbf{x})$, we can write Eq. (28) as

$$\int \nabla\cdot\mathbf{E}(\mathbf{x})dV = 4\pi \int \rho(\mathbf{x})dV \tag{29}$$

Since this equation is true for *any* volume, let us apply it to a very small volume $\Delta x\Delta y\Delta z$ near the point **x**. Then Eq. (29) becomes

$$\nabla\cdot\mathbf{E}(\mathbf{x})\Delta x\Delta y\Delta z = 4\pi\rho(\mathbf{x})\Delta x\Delta y\Delta z \tag{30}$$

or, canceling the volume,

$$\nabla\cdot\mathbf{E}(\mathbf{x}) = 4\pi\rho(\mathbf{x}) \tag{31}$$

This completes the proof of Eq. (24).

Note that in any region free of charge, the electric field has zero divergence, and hence, according to the arguments of Section 1.4, it can be represented by field lines. However, in any region where there is some charge density $\rho(\mathbf{x})$, the electric field has a nonzero divergence, and it cannot be represented by field lines.

EXAMPLE: *electric field of charged sheet, by Gauss' law*

Although Eqs. (24) and (27) are mathematically equivalent, some calculations of electric fields, involving charge distributions with a high degree of symmetry, can be carried out most conveniently with the integral form of Gauss' law. For example, let us consider again the infinite sheet of charge whose electric field we obtained by direct integration [see Eq. (23)]. To calculate this electric field by means of Gauss' law, we begin with the observation that the symmetry of the charge distribution requires the electric field lines to be perpendicular to the sheet (see Fig. 2.9). If we then evaluate $\int \mathbf{E}\cdot d\mathbf{S}$ for the small cylindrical surface of base area A shown in Fig. 2.9, we find that only the top and bottom of the cylinder contribute to the integral, so $\int \mathbf{E}\cdot d\mathbf{S} = 2E_z A$. By Gauss' law, this integral must be proportional to the charge within the cylinder, that is

$$2E_z A = 4\pi(\sigma A)$$

or

$$E_z = 2\pi\sigma \tag{32}$$

which is the same result we obtained before.

■ ***Exercise 2.*** A sphere of radius R has a charge Q uniformly distributed over its volume. Apply the integral form of Gauss' law to a spherical surface of radius r concentric with the charge distribution. Show that the electric field is

EXAMPLE: *electric fields of charged sphere and of charged line*

$$\mathbf{E}(r) = Q\frac{\hat{\mathbf{r}}}{r^2} \quad \text{for } r \geq R$$

$$\mathbf{E}(r) = Q\frac{r\hat{\mathbf{r}}}{R^3} \quad \text{for } r \leq R \tag{33}$$

■

■ ***Exercise 3.*** A long straight line carries a uniform distribution of charge, with λ units of charge per unit length. Apply the integral form of Gauss' law to a cylindrical surface of radius r concentric with this line of charge. Show that the electric field is radial, of magnitude

$$E(r) = \frac{2\lambda}{r} \tag{34}$$

■

The electric field of a static charge distribution also satisfies the equation

$$\nabla \times \mathbf{E} = 0 \tag{35}$$

Figure 2·9

Field lines of a very large, uniformly charged sheet.

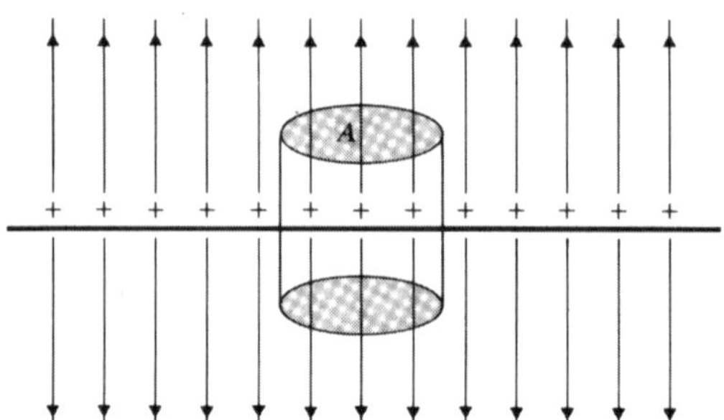

As in the proof of Gauss' law, it is best to begin with the electric field of a single point charge q'. If this charge is at the point $\mathbf{x}'$, its electric field is

$$E(\mathbf{x}) = q' \frac{\mathbf{x}-\mathbf{x}'}{|\mathbf{x}-\mathbf{x}'|^3}$$

The x component of $\nabla \times \mathbf{E}$ is then [see Eq. (1.70)]

$$\frac{\partial}{\partial y}E_z - \frac{\partial}{\partial z}E_y = \frac{\partial}{\partial y}\frac{q'(z-z')}{[(x-x')^2+(y-y')^2+(z-z')^2]^{3/2}} - \frac{\partial}{\partial z}\frac{q'(y-y')}{[(x-x')^2+(y-y')^2+(z-z')^2]^{3/2}}$$

When we carry out the differentiations, we immediately find that the two terms on the right side cancel; that is, the x component of $\nabla \times \mathbf{E}$ is zero. Likewise, the y and z components are zero. This establishes Eq. (35) for the case of a point charge. Since the electric field of an arbitrary charge distribution is the sum of the individual electric fields of point charges, it follows that Eq. (35) must be true in general.

Equations (24) and (35) for the divergence and the curl of **E** are two of Maxwell's equations, in a form valid for the special case of static fields. As we will see in Chapter 8, Eq. (24) remains valid even if the fields are not static, but Eq. (35) will have to be modified by a term involving the time derivative of the magnetic field [see Eq. (8.64)].

2·3 The Electrostatic Potential; Poisson's Equation

We know from Section 1.4 that a vector field of zero curl is conservative and can be expressed as the gradient of scalar function. The electrostatic field satisfies this condition of zero curl. It is therefore conservative; that is, for any arbitrary closed path,

$$\oint \mathbf{E} \cdot d\mathbf{l} = 0$$

According to Eq. (1.114), the function $\mathbf{E}(\mathbf{x})$ can then be expressed as the gradient of a scalar function,

$$\mathbf{E}(\mathbf{x}) = -\nabla\Phi(\mathbf{x}) \qquad (36)$$

The function $\Phi(\mathbf{x})$ is called the *electrostatic potential.* Note that Eq. (36), in contrast to Eq. (1.114), has a minus sign on the right side.

We have inserted this extra minus sign so that the potential has the same sign as the potential energy of a positive test charge (see below).

According to Eq. (1.115), we can express the electrostatic potential in terms of the electric field as follows:

$$\Phi(\mathbf{x}) = -\int_{\mathbf{x}_0}^{\mathbf{x}} \mathbf{E}(\mathbf{x}')\cdot d\mathbf{x}' \tag{37}$$

where $\mathbf{x}_0$ is some reference point at which the potential is zero by definition, $\Phi(\mathbf{x}_0) = 0$. If the electric field is that of a point charge q' at the origin of coordinates, then it is convenient to choose a reference point at infinity. All points at infinity (that is, all points on the surface of an infinite sphere) are at the same potential, because the electric field has zero component along the surface. In Eq. (37), the direction of $\mathbf{x}_0$ can therefore be assumed to coincide with the direction of $\mathbf{x}$. Any path of integration from $\mathbf{x}_0$ to $\mathbf{x}$ will give the same result for the integral in Eq. (37). For the sake of simplicity, we take a straight radial path along which $d\mathbf{x}' = \hat{\mathbf{r}}\,dr'$ (see Fig. 2.10). The integral in Eq. (37) then becomes

$$\Phi(r) = -\int_{\infty}^{r} q' \frac{\hat{\mathbf{r}}}{r'^2}\cdot\hat{\mathbf{r}}\,dr' = -\int_{\infty}^{r} \frac{q'}{r'^2}\,dr'$$

that is,

$$\Phi(r) = \frac{q'}{r} \tag{38}$$

This formula for the potential of a point charge at the origin of coordinates shows that the potential is inversely proportional to the distance. For a point charge at some arbitrary position $\mathbf{x}'$, the potential is then

Figure 2·10

Path of integration for the potential of a point charge q'.

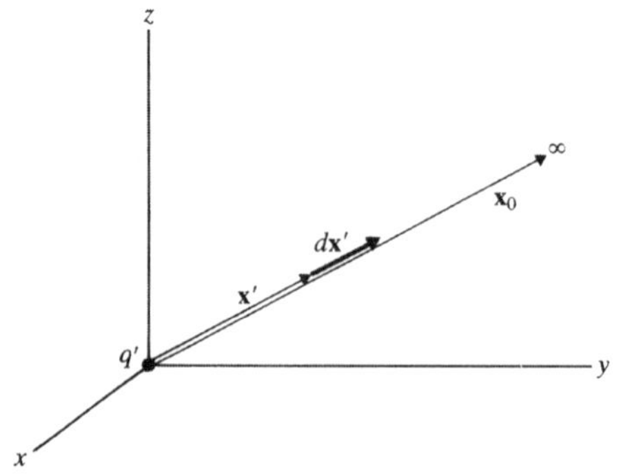

$$\Phi(\mathbf{x}) = \frac{q'}{|\mathbf{x} - \mathbf{x}'|} \tag{39}$$

Furthermore, the potential of an arbitrary distribution of charges with a charge density $\rho(\mathbf{x}')$ is

$$\Phi(\mathbf{x}) = \int \frac{\rho(\mathbf{x}')}{|\mathbf{x} - \mathbf{x}'|} dV' \tag{40}$$

A warning: Equation (40) is valid for any charge distribution of finite extent. But if the charge distribution extends to infinity (for example an infinite charged plane), then Eq. (40) will fail, because for such a charge distribution, the points at infinity cannot be used as reference points with zero potential. We must then return to Eq. (37) and make some other choice of reference point.

■ ***Exercise 4.*** By direct differentiation, calculate the gradient of Eq. (39), and show that the result leads to Eq. (16). Likewise, calculate the gradient of Eq. (40), and show that the result leads to Eq. (19). ■

EXAMPLE: *potential of charged disk*

As an example of the use of Eq. (40) in the calculation of the potential, let us again consider the thin charged disk discussed in the preceding section (see Fig. 2.7). A small piece of the disk carries charge $(Q/\pi R^2)r'd\phi'dr'$ and hence gives a contribution

$$d\Phi = \frac{(Q/\pi R^2)r'd\phi'dr'}{(z^2 + r'^2)^{1/2}}$$

to the potential. Upon integration, we obtain the potential generated by the entire disk:

$$\begin{aligned}\Phi(z) &= \frac{Q}{\pi R^2}\int_0^R \int_0^{2\pi} \frac{r'}{(z^2 + r'^2)^{1/2}} d\phi' dr' \\ &= \frac{2Q}{R^2}[(z^2 + R^2)^{1/2} - z]\end{aligned}$$

If we calculate the electric field from this potential, using $E_z = -\partial\Phi/\partial z$, we again obtain the result (22). This method of calculating the electric field via the potential is often more convenient than the direct calculation via Eq. (19), which involves the integration of a vector, with three separate components. However, if the charge distribution has enough symmetry to permit the calculation of the electric field via the integral form of Gauss' law, then it may be most convenient to begin with such a calculation of the electric field and to obtain the potential (if needed) by means of Eq. (37).

■ ***Exercise 5.*** Starting with Eq. (33), show that the potential generated by a sphere with a uniform distribution of charge is

$$\Phi(r) = \frac{Q}{r} \qquad \text{for } r \geq R$$
$$\Phi(r) = Q\left(-\frac{r^2}{2R^3} + \frac{3}{2R}\right) \qquad \text{for } r \leq R \tag{41}$$

■

The physical interpretation of the potential is as follows: Suppose that we place a point charge q in a given electric field (produced by some other charges). Then $q\mathbf{E}$ is the electric force on the charge q, and according to Eq. (37), $q\Phi$ is the potential energy of the charge q:

$$U(\mathbf{x}) = q\Phi(\mathbf{x}) \tag{42}$$

Thus the electric potential equals the potential energy per 1 esu of (positive) charge.

The CGS electrostatic unit of potential is the statvolt,

$$1\,\text{statvolt} = 1\,\text{erg}/1\,\text{esu} \tag{43}$$

For practical purposes, it is good to know the conversion factor from statvolt to volt:*

$$1\text{ statvolt} \Leftrightarrow 299.792458\text{ volts} \simeq 300\text{ volts} \tag{44}$$

Electric fields can be measured in statvolt/cm; this is the same thing as dyne/esu.

■ ***Exercise 6.*** Express the statvolt in terms of centimeters, grams, and seconds. ■

If we combine the differential form (24) of Gauss' law with $\mathbf{E} = -\nabla\Phi$, we obtain

$$\nabla\cdot\nabla\Phi(\mathbf{x}) = -4\pi\rho(\mathbf{x})$$

that is,

$$\nabla^2\Phi(\mathbf{x}) = -4\pi\rho(\mathbf{x}) \tag{45}$$

This is known as *Poisson's equation.* In any charge-free region, where $\rho(\mathbf{x}) = 0$, the equation reduces to

---

*The numerical factor in this equation arises from the speed of light. As with the units of electric charge (see Section 2.1), we write this equation as an equivalence rather than as an equality because the CGS and SI units of electric potential do not have the same dimensions.

$$\nabla^2\Phi(\mathbf{x}) = 0 \tag{46}$$

This is called *Laplace's equation.*

Equations (45) and (46) are second-order partial differential equations. In the next chapter, we will examine some methods for the solution of these equations.

The following theorem is an immediate corollary of Laplace's equation:

• ***Earnshaw's Theorem:*** The potential cannot have an extremum in a charge-free region.

To prove this, we note that if the function Φ has a maximum at some point, then the first derivatives $\partial\Phi/\partial x, \partial\Phi/\partial y$, and $\partial\Phi/\partial z$ must be zero, and the second derivatives $\partial^2\Phi/\partial x^2, \partial^2\Phi/\partial y^2$, and $\partial^2\Phi/\partial z^2$ must be negative. But if each of these second derivatives is negative, then their sum cannot be zero, as required by Laplace's equation. This argument rules out a maximum; a similar argument rules out a minimum.

According to this theorem, a system of pointlike charges cannot remain in stable, static equilibrium under the influence of their mutual electric forces. Each of the charges experiences the electric field and the electric potential of the *other* charges; thus each of the charges can be regarded as being located in a charge-free region; and since the potential in such a region cannot have an extremum, stable equilibrium is impossible. This instability led physicists to the recognition that the structure of atoms and molecules cannot be described by classical physics. As Jeans wrote in 1920:

> One physical application of Earnshaw's Theorem is of extreme importance. The theorem shows that an electron cannot rest in stable equilibrium under the forces of attraction and repulsion from other charges, so long as these forces are supposed to obey the law of the inverse square of the distance. Thus, if a molecule is to be regarded as a cluster of electrons and positive charges . . . then the law of force must be something different from that of the inverse square . . .
>
> Earnshaw's Theorem accordingly limits us to two alternatives. Either the molecule does not consist of a cluster of electrons in relative rest, or else the law of the inverse square fails at molecular distances.
>
> Recent experimental investigations decide very definitely against the second alternative and in favour of the first. Recent experiments on the deflection of the positively charged □-particles by matter indicate that the law of the inverse square holds down to distances of the order of 10^{-11} cms, a distance which is less than a thousandth part of the radius of the hydrogen atom, and a large mass of other evidence suggests, with a probability approximating to certainty, that the electrons in an atom or molecule must be in rapid orbital motion. Thus the problem of the structure of the molecule is removed from the province of Earnshaw's Theorem.[6]

Unfortunately, the orbital motion of the electrons gives rise to another conflict with classical physics. According to classical electromagnetic theory, electrons in rapid orbital motion must radiate. The atom should therefore lose energy by radiation, and the electrons should spiral inward until they collide with the nucleus and the atom ceases to exist. Thus the stability of atoms is in glaring contradiction to classical physics, and atomic physics must be quantum physics.

Finally, we note that Earnshaw's theorem implies the following: Given a charge-free region with some boundary, then the maximum value and the minimum value of the potential necessarily occur on the boundary.

2·4 The Electric Dipole

An electric dipole consists of two pointlike charges of equal magnitudes separated by a small distance (see Fig. 2.11). In this context, "small distance" means that the distance between the charges is much smaller than the distance from the charges to the point P. We can calculate the electric field of this dipole by direct addition of the Coulomb fields of the charges, but it is more convenient to calculate the field via the potential.

With the charges $\pm q$ located at $z' = \pm b/2$ on the z axis [see Fig. 2.11(a)], the potential at the point P is

$$\Phi(\mathbf{x}) = \frac{q}{r_1} - \frac{q}{r_2} \tag{47}$$

Figure 2·11

(a) An electric dipole consisting of two pointlike charges $\pm q$ at $z = \pm b/2$. (b) The approximation (49) is equivalent to the assumption that r_1 and r_2 are nearly parallel, which implies $r_1 \simeq r - \frac{1}{2} b\cos\theta$.

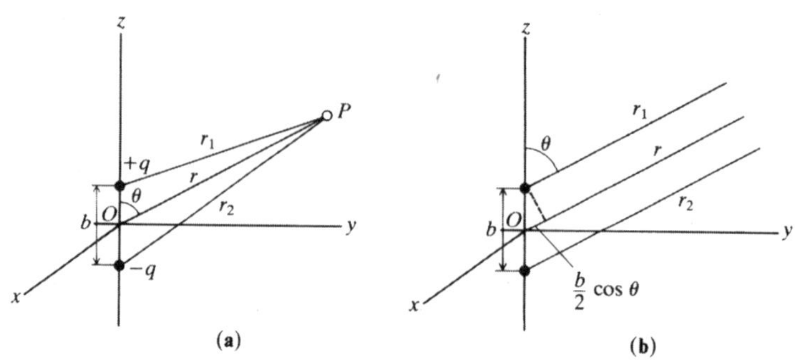

To obtain an approximate expression for r_1, we apply the law of cosines to the triangle $+qOP$, which gives

$$\frac{1}{r_1} = \frac{1}{\sqrt{r^2 - 2(b/2)r\cos\theta + (b/2)^2}}$$

$$= \frac{1}{r\sqrt{1-(b/r)\cos\theta + (b/2r)^2}} \tag{48}$$

If $r \gg b$, we can approximate the square root by a Taylor-series expansion, keeping only the first two terms:

$$\frac{1}{r_1} \simeq \frac{1}{r} + \frac{b}{2r^2}\cos\theta \tag{49}$$

This approximation is equivalent to the assumption that r_1 and r are nearly parallel, so $r_1 \simeq r - \frac{1}{2}b\cos\theta$, which gives $1/r_1 \simeq 1/(r - \frac{1}{2}b\cos\theta)$ $\simeq 1/r + \frac{1}{2}b\cos\theta/r^2$ [see Fig. 2.119(*b*)]. Likewise,

$$\frac{1}{r_2} \simeq \frac{1}{r} - \frac{b}{2r^2}\cos\theta$$

The expression (47) for the potential then reduces to

$$\Phi(r,\theta) \simeq qb\frac{\cos\theta}{r^2} \tag{50}$$

The product qb is called the *dipole moment*, usually designated by the symbol p,

$$p = qb \tag{51}$$

Hence

$$\Phi(r,\theta) \simeq p\frac{\cos\theta}{r^2} \tag{52}$$

Since this equation expresses the potential as a function of the spherical coordinates r and θ, we find it convenient to calculate the components of the electric field in these coordinates. Evaluating the gradient of Φ in spherical coordinates, we obtain

$$E_r = -\frac{\partial\Phi}{\partial r} \simeq 2p\frac{\cos\theta}{r^3} \tag{53}$$

$$E_\theta = -\frac{1}{r}\frac{\partial\Phi}{\partial\theta} \simeq p\frac{\sin\theta}{r^3} \tag{54}$$

$$E_\phi = -\frac{1}{r\sin\theta}\frac{\partial\Phi}{\partial\phi} = 0 \tag{55}$$

Note that the strength of the electric field of the dipole decreases with the cube of the distance. This decrease comes about by the (partial) cancellation of the opposite electric fields of the two opposite charges. Figure 2.12 shows the field lines for the dipole.

Although we here introduced the concept of dipole moment by examining the case of two opposite point charges, a dipole moment can also exist in systems of more than two point charges and in systems with a continuous distribution of positive and negative charge. Such systems will have dipole moments whenever there is a displacement between the average positions of the positive and the negative parts of their charge distributions (we will give a general definition of the dipole moment for arbitrary systems of charges in Section 3.5). Many molecules have dipole moments that arise from such a displacement between the average positions of their positive charges (nuclei) and their negative charges (electrons). For instance, the water molecule (H_2O) has a dipole moment of 1.85×10^{-18} esu · cm. This is a rather large dipole moment for a small molecule—it is equivalent to displacing one electron over a distance of about one atomic radius relative to the average position of the positive nuclear charges. This large dipole moment permits the water molecule to interact with other molecules by means of the electric dipole fields (53)–(55), which explains why water is such a good solvent.

From Eqs. (53)–(55), we can calculate the force that the dipole exerts on a charge or on a distribution of charges immersed in the electric

Figure 2·12

Field lines of a dipole as seen in two dimensions.

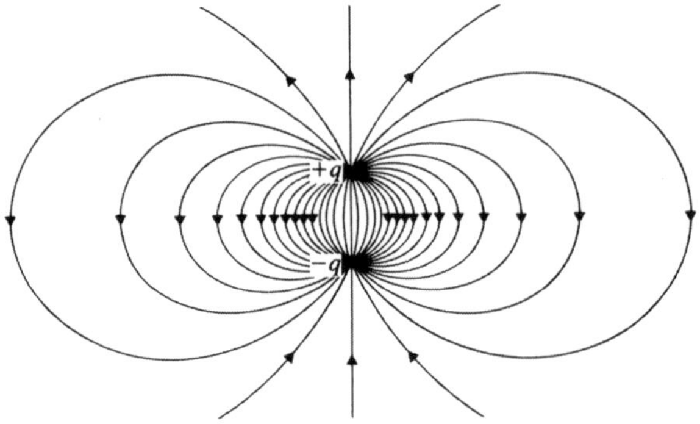

field of the dipole. Conversely, a charge or a distribution of charges will generate an electric field, and the dipole will experience a force when it is immersed in this electric field. Let us calculate this "external" force on the dipole exerted by some given "external" electric field. We will assume that the charges $\pm q$ of our dipole are held in a fixed position relative to one another by some mechanical restraint, so the dipole can be treated as a rigid body with a constant separation between the charges $\pm q$. If the external electric field is uniform, the force on the charge $+q$ exactly cancels the force on the charge $-q$, and the dipole experiences no net external force. But if the electric field is not uniform, there will be a net force.

Figure 2.13 shows the dipole immersed in a nonuniform electric field $\mathbf{E}(\mathbf{x})$. The center of the dipole is at the point $\mathbf{x}$. The x component of the force on the charge $+q$ is

$$qE_x(\mathbf{x}+\mathbf{b}/2) \simeq qE_x(\mathbf{x}) + q\frac{\mathbf{b}}{2}\bullet\nabla E_x(\mathbf{x}) \qquad (56)$$

Likewise, the x component of the force on the charge $-q$ is

$$-qE_x(\mathbf{x}-\mathbf{b}/2) \simeq -qE_x(\mathbf{x}) + q\frac{\mathbf{b}}{2}\bullet\nabla E_x(\mathbf{x}) \qquad (57)$$

The net x component of the force on the dipole is the sum of these components:

$$F_x = q(\mathbf{b}\bullet\nabla)E_x(\mathbf{x}) \qquad (58)$$

The corresponding expressions for the other components of the force lead to

$$\mathbf{F} = q(\mathbf{b}\bullet\nabla)\mathbf{E}(\mathbf{x}) \qquad (59)$$

Figure 2·13

Dipole immersed in a nonuniform external electric field. The displacement from $-q$ to $+q$ is $\mathbf{b}$.

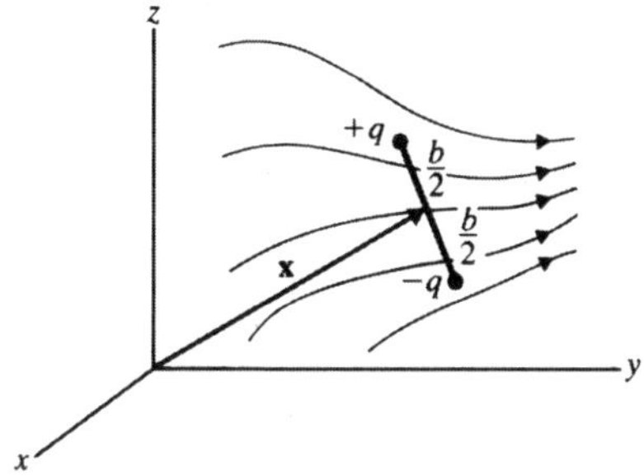

If we regard the dipole moment as a vector

$$\mathbf{p} = q\mathbf{b} \tag{60}$$

pointing from the negative charge to the positive charge, then we can write the force as

$$\mathbf{F} = (\mathbf{p}\cdot\nabla)\mathbf{E}(\mathbf{x}) \tag{61}$$

According to this equation, a dipole aligned with the field is pulled in the direction of increasing field strength, whereas a dipole aligned against the field is pushed in the opposite direction. We can easily understand this in terms of the forces on the two charges: if the electric field is not constant, then the force on one of the charges is larger than that on the other, and there is a net force on the dipole.

In the same way, we can calculate the potential energy of the dipole due to its position in the external electric field. If the potential at the center of the dipole is $\Phi(\mathbf{x})$, then the potential energies of the charges $\pm q$ are $\pm q\Phi(\mathbf{x} \pm \mathbf{b}/2) \simeq \pm q[\Phi(\mathbf{x}) \pm \frac{1}{2}\mathbf{b}\cdot\nabla\Phi]$ and

$$\begin{aligned} U &= q\Phi(\mathbf{x}+\mathbf{b}/2) - q\Phi(\mathbf{x}-\mathbf{b}/2) \\ &\simeq q\Phi(\mathbf{x}) + q\frac{\mathbf{b}}{2}\cdot\nabla\Phi - q\Phi(\mathbf{x}) + q\frac{\mathbf{b}}{2}\cdot\nabla\Phi \end{aligned} \tag{62}$$

Since $\nabla\Phi = -\mathbf{E}$, this becomes

$$U = -\mathbf{p}\cdot\mathbf{E}(\mathbf{x}) \tag{63}$$

Note that this potential energy does not include the "internal" energy of the dipole, that is, the energy that the two charges have by virtue of their position relative to each other. This internal energy is irrelevant for the calculation of the external force on the dipole.

We can, of course, calculate this external force on the dipole by taking the gradient of the potential in Eq. (63):

$$\mathbf{F} = -\nabla U = \nabla[\mathbf{p}\cdot\mathbf{E}(\mathbf{x})] \tag{64}$$

Although this new formula for the force seems to differ from the previous formula (61), both formulas give the same result for the force.

■ ***Exercise 7.*** Prove the vector identity

$$\mathbf{p}\times(\nabla\times\mathbf{E}) = \nabla(\mathbf{p}\cdot\mathbf{E}) - (\mathbf{p}\cdot\nabla)\mathbf{E}$$

and then use $\nabla\times\mathbf{E} = 0$ to establish that Eqs. (61) and (64) coincide. ■

Besides this translational force, the external electric field also exerts a rotational torque on the dipole. It is easy to show that this torque is

$$\boldsymbol{\tau} = \mathbf{p} \times \mathbf{E} \tag{65}$$

(where the nonuniformity of **E**, if any, has been neglected). This result for the torque can also be obtained by differentiating the potential energy (63) with respect to the angle of orientation of the dipole relative to the electric field.

■ ***Exercise 8.*** Verify that differentiation of Eq. (63) gives a torque with the magnitude and direction specified by Eq. (65). [*Hint:* Write Eq. (63) as $U = -pE\cos\alpha$.] ■

2·5 The Mean-Value Theorem

We learned in the preceding section that in any charge-free region, the electrostatic potential obeys the Laplace equation $\nabla^2\Phi = 0$. Any function obeying this equation satisfies the following theorem:

• ***Mean-Value Theorem:*** If S is the surface of a (mathematical) sphere whose interior contains no charge, then the potential at the center is equal to the average potential over the surface S.

To prove this, we consider a sphere of radius r. The average of Φ over S is

$$\bar{\Phi} = \frac{1}{4\pi r^2}\int_S \Phi dS = \frac{1}{4\pi r^2}\int_S \Phi r^2 d\Omega = \frac{1}{4\pi}\int_S \Phi d\Omega \tag{66}$$

Here $d\Omega$ is the solid angle that corresponds to dS. Thus

$$\frac{\partial\bar{\Phi}}{\partial r} = \frac{1}{4\pi}\int \frac{\partial\Phi}{\partial r} d\Omega = \frac{1}{4\pi}\int \nabla\Phi\cdot\hat{\mathbf{n}}d\Omega = \frac{1}{4\pi r^2}\int \nabla\Phi\cdot d\mathbf{S} \tag{67}$$

By Gauss' theorem, the last expression is equal to

$$\frac{1}{4\pi r^2}\int \nabla\cdot\nabla\Phi dV \tag{68}$$

But $\nabla\cdot\nabla\Phi = \nabla^2\Phi = 0$ in a charge-free region; hence $\partial\bar{\Phi}/\partial r = 0$. This means that $\bar{\Phi}$ is independent of r. So let us take a very small sphere of radius $r \to 0$. In this case, clearly, $\bar{\Phi} = \Phi_{\text{center}}$, which establishes the theorem.

Figure 2·14

Two large, parallel conducting plates, one with a step. The potential of the lower plate is 5 statvolts; that of the upper plate is 0 statvolt. The coordinate grid consists of squares of equal size.

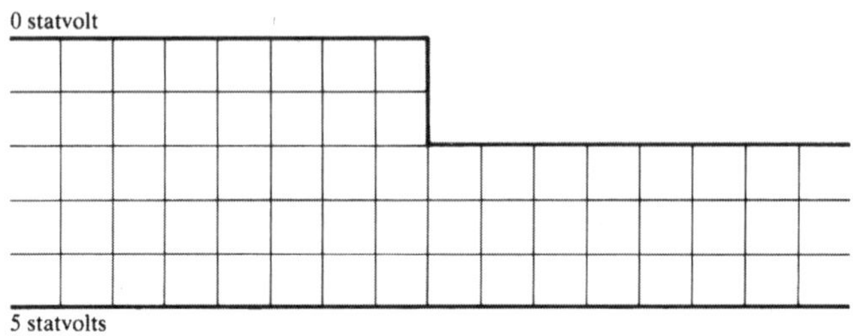

This theorem finds an important practical application in a relaxation method for the numerical solution of Laplace's equation. This relaxation method relies on the mean-value theorem to construct a solution by successive approximations, starting with a crude guess as a first approximation. Instead of presenting a general recipe, we will explain this method by means of a concrete example. Figure 2.14 shows a pair of large, parallel conducting plates, one of which has a step. Suppose that the plates are equipotential surfaces; the potential of the lower plate is 5 statvolts and that of the upper plate is 0 statvolt. What is the potential in the space between the plates?

EXAMPLE: *potential of parallel plates with step*

To apply numerical methods to this problem, we must represent the space between the plates by a discrete set of points. Figure 2.14 shows a coordinate grid superimposed on the plates. We will try to find approximate values of the potential at the intersection points of this grid. Note that the grid is two-dimensional. The third dimension (out of the plane of the page in Fig. 2.14) can be ignored because the potential at all points above or below the plane of the page is the same as that at the points shown in Fig. 2.14.

To obtain the values of the potential at the grid points, we proceed by successive approximations: We start with some reasonable first guess for the potential. Since, for flat plates, the potential in the space between would increase linearly as a function of position from 0 statvolt to 5 statvolts, the values shown in Fig 2.15(a) are a reasonable first approximation. To find a second approximation to the potential, we rely on the mean-value theorem. According to this theorem, the potential at any point should equal the average of the potentials of all the points that are at equal distances from the given point. In our grid, this average can be approximated by an average over the four nearest-neighbor points.* We

*In three dimensions, with a cubical grid, each point has six nearest-neighbor points, and we actually should average over these six. But it is easy to verify that, since the potential at the points above or below the plane of the page is the same as that in the plane, the average over these six points gives the same result as the average over the

next obtain a third approximation by repeating this procedure again, and so on.

Figure 2·15

(*a*) First approximation for the potential in the space between the plates.

(*b*) Fifteenth approximation. These values are self-consistent to two decimals.

(*c*) Equipotentials, obtained by interpolation on the values given in (*b*).

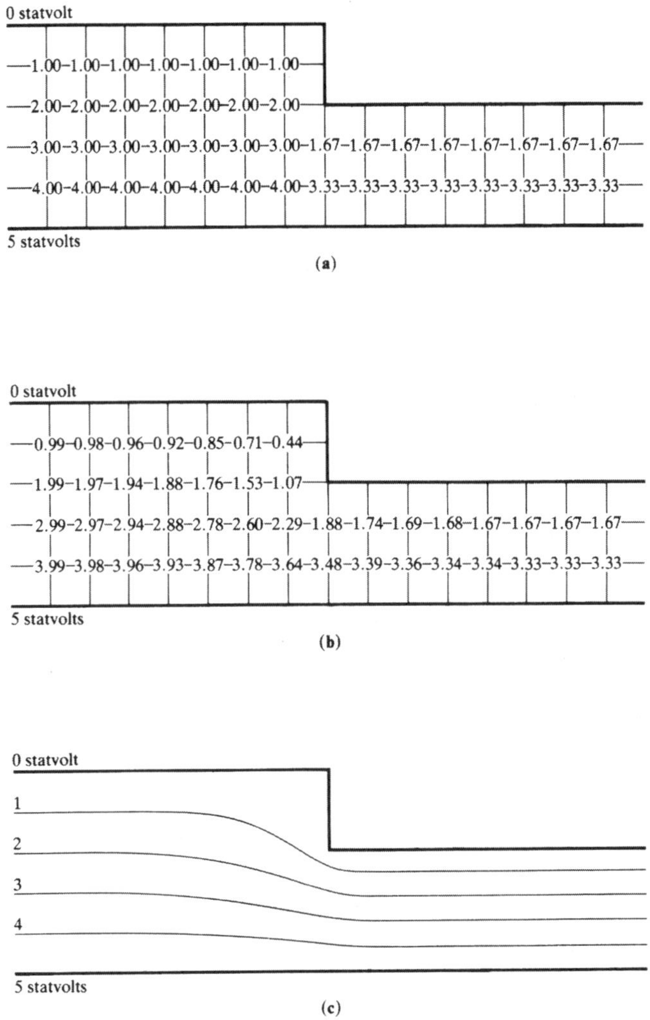

four points in the plane of the page. Alternatively, averaging over the four points in the plane of the page can be justified by formulating a mean-value theorem for two dimensions (see Problem 36).

After a few successive steps, this iteration procedure yields values that do not change appreciably from one step to the next [see Fig. 2.15(*b*)]. These values are an approximate answer to the problem. The equipotential surfaces can be obtained by interpolating on the values of the potential [see Fig. 2.15(*c*)]. And the components of the electric field can be obtained from the values of the potential by numerical calculation of the derivatives of the potential in the horizontal and vertical directions.

As a check on the approximation, we can repeat the calculation with a finer grid size. This tends to get tedious, but the relaxation method lends itself very well to programming on a digital computer. The method can also be adapted to cylindrical or spherical coordinate grids if the geometry of the problem makes these more suitable.[7]

References

1. H. F. Dylla and J. G. King, *Phys. Rev. A* **7**, 1224 (1973); M. Marinelli and G. Mopurgo, *Phys. Lett. B* **137**, 439 (1984).
2. S. J. Plimpton and W. E. Lawton, *Phys. Rev.* **50**, 1066 (1936); E. R. Williams, J. E. Faller, and H. A. Hill, *Phys. Rev. Lett.* **26**, 721 (1971).
3. For a general review of limits that can be set on κ by magnetic and electromagnetic methods, see A. S. Goldhaber and M. M. Nieto, *Rev. Mod. Phys.* **43**, 277 (1971); see also Williams et al., ref. 2.
4. E. Fishbach et al. *Phys. Rev. Lett.* **73**, 514 (1994).
5. G. V. Chibisov, *Sov. Phys. Usp.* **19**, 624 (1976).
6. J. H. Jeans, *The Mathematical Theory of Electricity and Magnetism* (University Press, Cambridge, 1920), p. 168.
7. M. DiStasio and W. C. McHarris, *Am. J. Phys.* **47**, 5 (1979).

Problems

1. Calculate the ratio of the strengths of the electric and gravitational forces between an electron and proton placed some distance apart.

2. Suppose that the potential of a point charge is not the Coulomb potential but the Yukawa potential,

$$\Phi = \frac{q\mathrm{e}^{-\kappa r}}{r}$$

Find the electric field $\mathbf{E} = -\nabla\Phi$. Is Gauss' law valid for this electric field? Is $\nabla \times \mathbf{E} = 0$?

3. The nucleus of uranium has a radius of 7.4×10^{-13} cm, and it has a charge of $92e$ uniformly distributed over its volume.
   (a) What is the electric field at the surface of the nucleus? At the center? Give your answer in statvolts/cm and in volts/m.
   (b) What is the potential at the surface of the nucleus? At the center? Give your answer in statvolts and in volts.

4. Two identical thin rods of plastic of length L carry equal charges Q distributed uniformly along their lengths. The rods are aligned, with a distance d between their nearest ends. Calculate the repulsive electric force that each rod exerts on the other.

5. A thick ellipsoidal shell is bounded by an outer surface consisting of an ellipsoid and an inner surface consisting of a similar but smaller ellipsoid with the same center (see Fig 2.16). The shell carries a charge distribution of uniform density. Prove *Newton's theorem:* The force on a point charge placed anywhere in the empty space within the shell is zero. [*Hint:* Take a thin ellipsoidal shell and consider the forces exerted by the charges intercepted by a thin double cone with apex at the point charge. Show that these forces cancel.]

6. A circular annulus of inner radius R_1 and outer radius R_2 has a uniform surface charge density σ. What is the electric field on the axis of the annulus at a distance z from the center?

7. A very long, thin rod lies along the x axis from $x = 0$ to $x = \infty$. Another very long thin rod lies along the y axis from $y = 0$ to $y = \infty$. Both rods carry a uniform charge λ per unit length. Find the electric field in the first quadrant in the x-y plane.

8. According to the classical "plum pudding" model of J. J. Thomson, the atom consists of a diffuse cloud of positive charge within which sit the negative electrons held in their equilibrium positions by the electric forces. Suppose that the hydrogen atom consists of such a cloud with a positive charge e distributed uniformly over a sphere of radius 0.05 nm. The equilibrium position of the electron is at the center. What is the frequency of small oscillations of the electron about this equilibrium position?

9. According to the classical model of J. J. Thomson (see also Problem 8), the atom of helium consists of a cloud with a positive charge $2e$ uniformly distributed over a spherical volume. Suppose that the radius of the cloud is 0.05 nm and that the two electrons are symmetrically placed with respect to the center.
   (a) What is the equilibrium separation of the electrons?

Figure 2·16

An ellipsoidal shell. This shell is obtained from a spherical shell by expanding the horizontal dimensions.

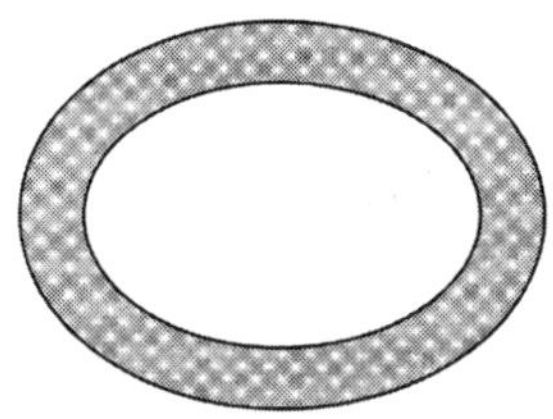

(b) What is the frequency of small oscillations of the electrons along the radial direction? Assume that the electrons move in phase (with equal displacements).

(c) Repeat part (b), but assume that the electrons move out of phase (with opposite displacements).

10. If a point charge q of mass m moving at a high speed v passes by a stationary point charge q', the effect of the electric force can be described as a transverse impulse, which produces a (small) transverse deflection of the direction of motion. To a first approximation, we can ignore this deflection, and we can pretend that the point charge q moves approximately along a straight line, say the x axis. Assume that the stationary point charge q' is on the y axis at $y=-b$. Then the transverse impulse given to q is

$$\int_{-\infty}^{\infty} F_y dt = \frac{q}{v}\int_{-\infty}^{\infty} E_y dx$$

where E_y is the electric field produced by q'. Use Gauss' law to evaluate the integral $\int E_y dx$, and show that the transverse impulse is $2qq'/(vb)$. From this, deduce that the transverse angular deflection is approximately $2qq'/(mv^2b)$. Evaluate this angular deflection for an alpha particle of 6.0 MeV passing by a lead nucleus with an impact parameter 1.0×10^{-10} cm.

11. In a region of space, the electric field is parallel to the x axis; that is, $E_x \neq 0, E_y = 0$, $E_z = 0$. Show that in this region, the electric field is independent of y and z. Show that if there is no charge density in this region, then the electric field is also independent of x.

12. Assume that the (infinite) volume between the planes $x=L$ and $x=-L$ is filled with a uniform distribution of charge of density ρ.

(a) Calculate the electric field in the regions $-L<x<L$, $x\leq -L$, and $x \geq L$.

(b) Calculate the potential difference between the points $x=-L$ and $x=L$.

13. A sphere of radius R has a charge Q uniformly distributed over its volume. A spherical cavity of radius R' is cut out of this sphere, and the charge in the cavity is discarded. Assume that the cavity is *not* concentric with the sphere. Show that the electric field in the cavity is constant, and find the magnitude of this electric field.

14. A very long cylinder of radius R has charge uniformly distributed over its volume. The amount of charge per unit length of the cylinder is λ. A cylindrical cavity of radius R', whose axis is parallel to the axis of the cylinder, is cut out of the cylinder and the charge in the cavity is discarded. Find the electric field in the cavity.

15. A very long rod of radius R_1 is surrounded by a long, thin, concentric cylindrical shell of radius R_2. Positive charge is uniformly distributed over the volume of the rod and over the shell; the amounts of charge per unit length on the rod and on the shell are $\pm\lambda$. Find the potential difference between the center of the rod and the shell.

16. Charge Q is uniformly distributed along a thin rod of length L. The rod lies on the x axis, from $x=-L/2$ to $x=L/2$.

(a) Find the potential in the space surrounding the rod.

(b) Show that the equipotential surfaces are ellipsoids of revolution whose foci coincide with the ends of the rod.

17. Three thin rods of length L form an equilateral triangle. Each rod carries a charge Q uniformly distributed along its length. Find the electric field and the potential as a function of distance along a line perpendicular to the triangle and passing through its center.

18. A cardboard tube of radius R and length L has a charge Q uniformly distributed over its surface. Find the electric field and the potential at a point on the axis of the tube at a distance d from the center.

19. An amount of charge Q is uniformly distributed over a cylindrical volume of radius R and length L. Find the electric field and the potential at a point on the axis of the cylinder at a distance d from the center.

20. A spherically symmetric charge distribution has a charge density $\rho = kr^n$, where k is a constant and $n > -3$.
(a) Find the electric field as a function of r.
(b) Find the potential difference between the points $r = a$ and $r = b$.

21. Suppose that the potential in some region is given by

$$\Phi(x, y, z) = ax^2 + by^3$$

Find the charge density in this region.

22. Consider the dipole described in Fig. 2.11(a). Without making the assumption $r \gg b$, calculate the exact potential and the exact electric field at an arbitrary point. Express your answer in spherical coordinates.

23. Find the equation for the field lines shown in Fig. 2.12. [*Hint:* Along each field line, $E_r / E_\theta = dr / rd\theta$.]

24. A dipole consists of (massless) charges $+q$ and $-q$ attached to the ends of a thin uniform rod of length b and mass m. This dipole is placed in a uniform electric field **E**. What is the frequency of small rotational oscillations of this dipole about its center of mass?

25. (a) Use Eq. (61) to calculate the force that a point charge q' exerts on a dipole of moment p. Assume that the point charge is at the position r, θ and that the dipole is at the origin, aligned with the z axis.
(b) Use Eqs. (53) and (54) to calculate the force that the dipole exerts on the point charge. Are your results in agreement with Newton's Third Law?

26. Calculate the torque that a point charge q' exerts on a dipole of moment p. Assume that the point charge is at the position r, θ and that the dipole is at the origin, aligned with the z axis. First calculate the torque about the center of the dipole. Then calculate the torque that the dipole exerts on the point charge (about the center of the dipole). Is the result as expected?

27. An electric quadrupole consists of three pointlike charges arranged on the z axis: a charge $+q$ at $z = b$, a charge $-2q$ at $z = 0$, and a charge $+q$ at $z = -b$. Show that for $r \gg b$, the potential is approximately

$$\Phi(r, \theta) = \frac{2qb^2}{r^3} \frac{3\cos^2\theta - 1}{2}$$

and find the electric field as a function of the spherical coordinates r, θ, and ϕ. [*Hint:* Take into account the additional terms of order $1/r^3$ in the Taylor-series expansion of Eq. (48).]

28. A spherical charge distribution is placed in a given electric field produced by charges external to the spherical distribution. Prove that the net force experienced by the spherical distribution is as though all of the charge were concentrated at the center of the distribution. [*Hint:* Use the mean-value theorem.]

29. (a) Sketch the field lines in Fig. 2.15(*c*)
(b) By numerical differentiation of the values of the potential given in Fig. 2.15(*b*), find the electric field components E_x and E_y at the grid points. Assume that the grid squares are 1 cm × 1 cm.

30. Two long, concentric tubes of sheet metal have a rectangular cross section (see Fig. 2.17). The outer rectangle is at a potential of 9 statvolts; the inner rectangle is at a potential of 0 statvolt. Use the numerical relaxation method for a calculation (to three decimals) of the potential at the grid points shown in Fig. 2.17.

Figure 2·17

Two concentric, rectangular tubes seen in cross section.

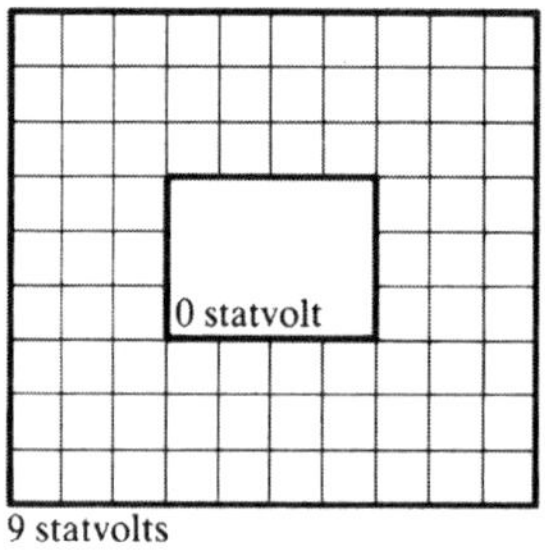

31. A large conducting plate is at zero potential. Another large, parallel conducting plate is at a potential of 8 statvolts. This second plate has a thin vertical ridge (see Fig. 2.18). Use the numerical relaxation method to find the potential at the grid points to within at least two decimals. Roughly sketch the equipotentials. Roughly sketch the field lines by drawing orthogonal trajectories to the equipotentials.

Figure 2·18

Two large, parallel conducting plates, the lower with a ridge.

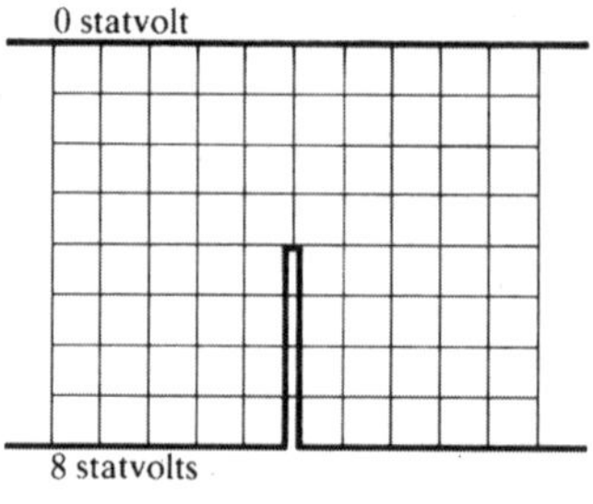

32. Repeat the numerical calculation of the potential for the configuration described in Fig. 2.14, but use a finer coordinate grid, with half the spacing shown in Fig. 2.14. Compare your new results with the old.

33. Two large, parallel conducting plates are at potentials of 0 statvolt and 4 statvolts, respectively. One of the plates has a triangular kink. Find the values of the potential at the grid points shown in Fig. 2.19, to two decimals.

Figure 2·19

Two large, parallel conducting plates, one with a kink.

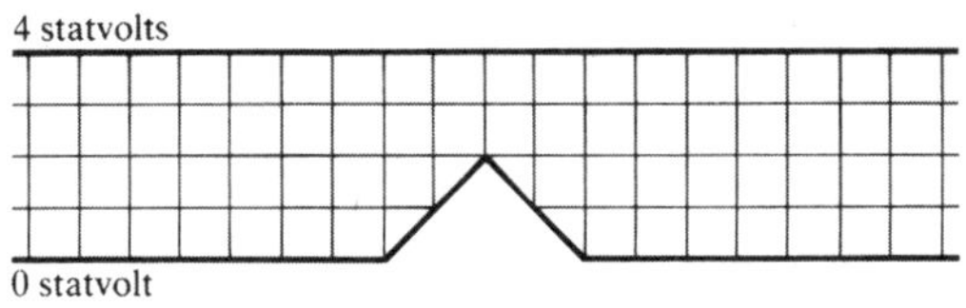

34. A long, thin conducting strip in empty space is at a potential of 8 statvolts. Figure 2.20 shows the strip in cross section and shows grid points in the vicinity of the strip. Find the values of the potential at the grid points to within two decimals.

Figure 2·20

A long conducting strip, seen in cross section.

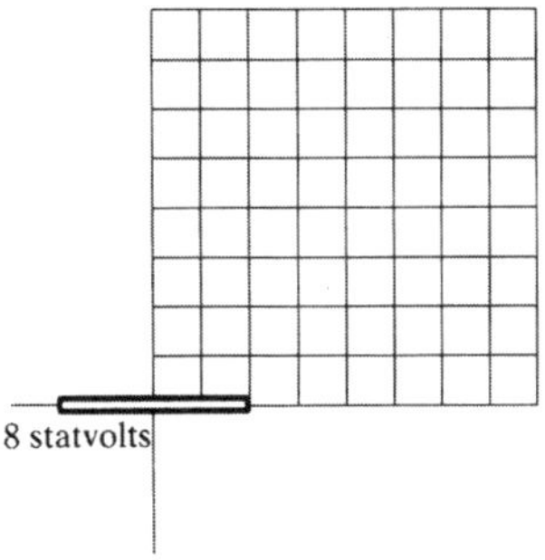

35. A cubical box measuring $6\ \text{cm} \times 6\ \text{cm} \times 6\ \text{cm}$ is made of sheet metal. The top and bottom faces are insulated from the side faces; these top and bottom faces are at a potential of 10 statvolt, whereas the side faces are at 0 statvolt. Adopt a cubical lattice of grid points with a 1-cm spacing for the interior of the box. Find the potential at the grid points in one octant of the box to within two decimal places. Display your results on three diagrams, showing three vertical sections through the box at 1-cm intervals. Where is the potential maximum? What is the maximum value of the potential?

36. Consider a potential $\Phi(x, y)$ that obeys the Laplace equation in two dimensions,

$$\left(\frac{\partial^2}{\partial x^2} + \frac{\partial^2}{\partial y^2}\right)\Phi(x, y) = 0$$

By means of the two-dimensional version of Gauss' theorem (see Problem 1.38), show that the potential obeys a two-dimensional version of the mean-value theorem: If a circle in the x-y plane contains no charge, then the potential at the center of the circle is equal to the average potential over the circumference of the circle.

CHAPTER 3

The Boundary-Value Problem

The solution of a problem asking for the electric field of a given charge distribution involves no more than a sum or an integral of the electric fields of all the individual charges [see Eqs. (2.17) and (2.18)]. The solution of this kind of problem is therefore a trivial—but possibly tiresome—exercise in summation or integration. Unfortunately, many of the problems in electrostatics do not completely specify the charge distribution in advance. Instead, these problems specify some given potentials on some conducting surfaces and then ask for the resulting potentials and electric fields in the space surrounding these surfaces.

In practice, this kind of problem arises whenever we connect conducting bodies to batteries or other sources of electromotive force (emf) that maintain given potential differences between these bodies. For example, we might connect the plates of a capacitor to a battery and then ask: What are the potential and the electric field in the space between the plates? If the plates have some complicated shape, the solution of this problem will be quite difficult. Furthermore, there might be some extra electric charges placed in the space between the plates, adding to the difficulty of the solution.

This kind of problem is a *boundary-value problem*. It involves the calculation of the potential and electric field within some given region of space, starting from the known values of the potential (or the electric field) at the boundaries of the region and from the known values of the charge within the region.

3·1 The Uniqueness Theorem

In mathematical language, we can state the boundary-value problem as follows: Suppose we have some region V with a given charge density $\rho(\mathbf{x})$. Suppose that the region has as boundary some surface S (this surface need *not* be a connected surface; see Fig. 3.1). We seek to solve the Poisson equation

$$\nabla^2\Phi = -4\pi\rho \qquad (1)$$

Figure 3·1

An example of a region V (shown shaded) that has two disconnected surfaces S_1 and S_2 as boundary.

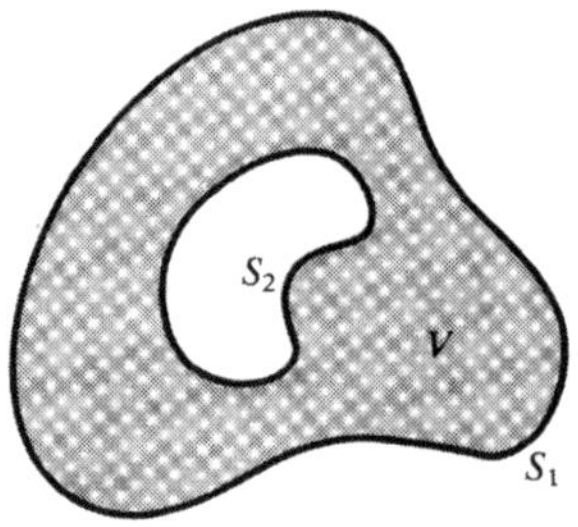

in the region V. This differential equation has many solutions. To single out the solution appropriate to our problem, we must take into account the *boundary conditions* that $\mathbf{\Phi}$ satisfies on the surface S. One possible boundary condition is the *Dirichlet condition:*

$$\Phi(\mathbf{x}) \text{ prescribed on } S \tag{2}$$

Another possible condition is the *Neumann condition:*

$$\hat{\mathbf{n}}\cdot\mathbf{E} \text{ prescribed on } S \tag{3}$$

where $\hat{\mathbf{n}}\cdot\mathbf{E} = -\hat{\mathbf{n}}\cdot\nabla\Phi$ is the component of $\mathbf{E}$ normal to S.

We will now prove a theorem that guarantees that either of these boundary conditions determines a unique solution for the electric field.

• ***Uniqueness Theorem:*** If Φ_1 and Φ_2 are two solutions of Poisson's equation with the same charge density and the same boundary conditions, then they differ at most by an additive constant.

To prove this theorem, we define ψ as the difference between Φ_1 and Φ_2,

$$\psi \equiv \Phi_2 - \Phi_1 \tag{4}$$

and note that

$$\nabla^2\psi = \nabla^2\Phi_2 - \nabla^2\Phi_1 = -4\pi\rho + 4\pi\rho = 0 \tag{5}$$

Hence

$$\nabla\cdot(\psi\,\nabla\psi) = (\nabla\psi)^2 + \psi\,\nabla^2\psi = (\nabla\psi)^2 \tag{6}$$

If we integrate both sides of Eq. (6) over the volume V and apply Gauss' theorem, we obtain

$$\int_V (\nabla\psi)^2\, dV = \int_V \nabla\bullet(\psi\,\nabla\psi)\, dV = \int_S \hat{\mathbf{n}}\bullet(\psi\,\nabla\psi)\, dS = \int_S \psi\hat{\mathbf{n}}\bullet\nabla\psi\, dS \qquad (7)$$

For the Dirichlet boundary condition, $\psi = 0$ on S; for the Neumann boundary condition, $\hat{\mathbf{n}}\bullet\nabla\psi = 0$ on S. In either case, the surface integral in Eq. (7) vanishes, and therefore,

$$\int_V (\nabla\psi)^2\, dV = 0 \qquad (8)$$

But this is impossible unless $\nabla\psi = 0$ everywhere. It follows that ψ is a constant throughout the region V, as was to be shown.

In the case of Dirichlet boundary conditions, we can go one step further and conclude that, since the difference between Φ_1 and Φ_2 vanishes on the boundary, it must vanish everywhere; therefore, the potentials Φ_1 and Φ_2 will be exactly the same. In the case of Neumann boundary conditions, we cannot rule out a constant difference between Φ_1 and Φ_2, but this does not affect the *physical* uniqueness of the solution because the electric fields $-\nabla\Phi_1$ and $-\nabla\Phi_2$ will be exactly the same.

In practice, the most convenient way to set up boundary conditions is to take conducting surfaces of suitable shape and connect them to batteries that generate the desired potentials. Metals and other homogeneous conducting materials have the characteristic property that under conditions of static equilibrium, all points in the material are at the same potential—that is, the electric field in the interior is everywhere zero. It is easy to understand why this must be so: In a metal, a fair fraction of the electrons are *free*; they are not held to any particular atom and can move throughout the entire volume of metal. If the potential is not uniform, the free electrons will move toward those regions where the potential is high, and the accumulation of the negatively charged electrons will decrease the potential. Equilibrium is attained when all regions are at the same potential and the electric field is everywhere zero.*

We can also say something about the electric field just *outside* a conductor: At the surface of a conductor, the electric field must be perpendicular to the surface. This is obvious because the surface is an equi-

---

*We should keep in mind that when two conductors made of two *different* metals are brought into contact, the potential will not be exactly the same on both; there will exist a *contact potential* difference, typically a fraction of 1 volt. This potential difference arises as follows: The "free" electrons in a metal are still bound to the metal; they are free only in the volume of the metal. If two different metals are brought into contact, the one with the stronger binding force for electrons will take away some electrons from the other; the accumulation of charge results in a potential difference. We will usually ignore these small contact potentials.

potential, and therefore $\nabla\Phi$ is necessarily perpendicular to the surfaces $\Phi =$ constant.

Another way to reach this conclusion is by comparison of the electric fields inside and outside the conductor. Since the electric field has zero curl, the path integral of **E** around any closed path is zero,

$$\oint \mathbf{E} \cdot d\mathbf{l} = 0 \tag{9}$$

Let us evaluate the left side of this around a small rectangular path of dimensions $l \times l'$ that is half inside the conductor and half outside (see Fig. 3.2). In the limit $l' \to 0$, we can ignore the contribution from the short sides. Taking into account that the interior electric field is zero, we obtain

$$\oint \mathbf{E} \cdot d\mathbf{l} = lE_{\|\text{ext}} - lE_{\|\text{int}} = lE_{\|\text{ext}} \tag{10}$$

where $E_{\|}$ designates the component of **E** parallel to the long side. Since the path integral is zero, it follows from Eq. (10) that the component of $\mathbf{E}_{\text{ext}}$ parallel to the surface vanishes—that is, there can exist only a perpendicular component of the electric field.

The magnitude of the perpendicular component of the electric field just outside a conductor can be related to the surface charge density. We note that there can be no volume charge density in a conductor. This follows from Eq. (2.24): if $\mathbf{E} = 0$ in some region, then necessarily $\rho = 0$ in this region. Thus, if a conductor is charged, the charge must then reside only on its surface. We define the *surface charge density* σ as the amount of charge per unit area; this surface charge density is some function of position along the surface. Now calculate the surface integral

$$\int_S \mathbf{E} \cdot d\mathbf{S}$$

Figure 3·2

Path of integration for Eq. (9). The path is half in the conductor, half outside.

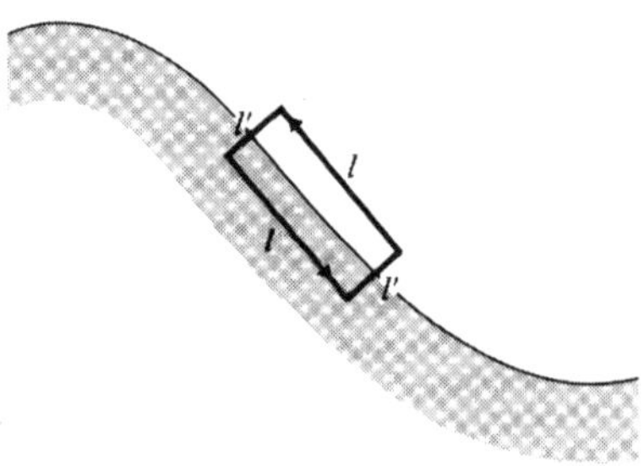

over the surface of a small cylindrical box ("pillbox") of base area A, height h that is half inside the conductor and half outside (see Fig. 3.3.). In the limit $h \to 0$, we can ignore the contributions from the sides. Taking into account that the interior electric field is zero, we obtain

$$\int_S \mathbf{E} \cdot d\mathbf{S} = AE_{\perp \text{ext}} - AE_{\perp \text{int}} = AE_{\perp \text{ext}} \tag{11}$$

where $E_\perp$ designates the component of $\mathbf{E}$ perpendicular to the surface ($E_\perp$ is positive if it is in the direction of the *outward* normal). But by Gauss' law,

$$\int_S \mathbf{E} \cdot d\mathbf{S} = 4\pi Q = 4\pi\sigma A \tag{12}$$

Comparison of Eqs. (11) and (12) shows

$$E_{\perp \text{ext}} = 4\pi\sigma \tag{13}$$

Thus the magnitude of the electric field just outside the conductor is proportional to the local surface charge density on the conductor.

EXAMPLE: *force exerted on surface charge*

In connection with Eq. (13), there arises an interesting question: What is the force that the electric field exerts on the surface charge? At first sight, one might think that the force per unit area is $\sigma E_{\perp \text{ext}} = 4\pi\sigma^2$. But this is wrong, because $E_{\perp \text{ext}}$ includes a contribution produced by the local surface charge σ, and self-forces (exerted by a charge on itself) must be excluded. To see how much of $E_{\perp \text{ext}}$ is due to the local surface charge, let us approximate a piece of the surface as a plane. If we are sufficiently near the surface, edge effects are unimportant, and we may regard the plane as infinite. The electric field generated by such an infinite plane has a magnitude $2\pi\sigma$ [see Eq. (2.23)]. Accordingly, exactly *one-half* of (13) is a self-field produced by the local surface charge:

Figure 3·3

Surface of integration for Eq. (11). The surface is half in the conductor, half outside.

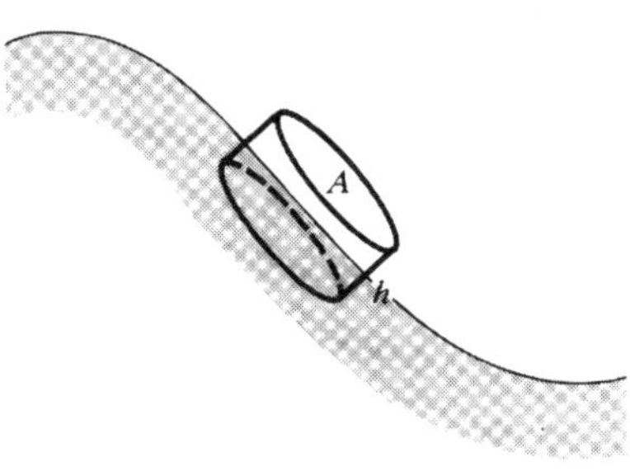

$$E_{\text{self}} = 2\pi\sigma \quad (14)$$

The remainder must be produced by charges in other places (for example, charges on the rest of the surface of the conductor, charges on other conductors, and so on):

$$E_{\text{other}} = 2\pi\sigma \quad (15)$$

Hence the force per unit area that the electric field exerts on the surface of a conductor is $\sigma E_{\text{other}} = 2\pi\sigma^2$.

Just outside the conductor, the "self" and "other" contributions add to give $E_{\perp\text{ext}} = 4\pi\sigma$. Inside the conductor, the "self" and "other" contributions cancel to give $\mathbf{E} = 0$. The absence of an electric field in a conductor is therefore not due to an inability of the field of a charge to penetrate a conductor [the field of a charge is always given by Eq. (2.14)]; rather, it is due to a cancellation against the field of the surface charges.

If a conductor has a closed, empty cavity in its interior (see Fig. 3.4), then the electric field will be zero in this cavity. This conclusion follows directly from the uniqueness theorem: We can regard the empty cavity as a region of space in which the potential obeys the Laplace equation $\nabla^2\Phi = 0$. The boundary condition imposed on the solution of this equation is that $\Phi =$ constant along the walls of the cavity (Dirichlet condition). Obviously, $\Phi(\mathbf{x}) =$ constant is then a possible solution of the Laplace equation in the cavity. By the uniqueness theorem, it is the *only* solution. Thus the potential throughout the cavity is constant, and the electric field is zero. From this vanishing of the electric field, we can draw a further conclusion: the absence of electric field in the cavity implies the absence of charge on the walls of the cavity [see Eq. (13)]; thus the electric charge on the conductor must reside exclusively on the *outside* surface of the conductor.

Figure 3·4

Conductor with a closed, empty cavity.

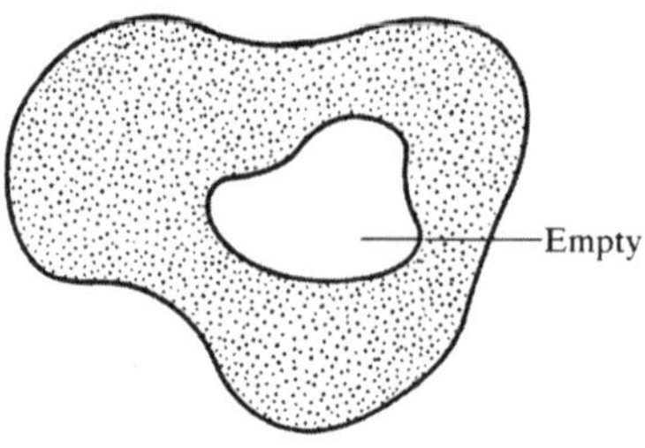

Of course, the preceding argument relies on the vanishing of the electric field in the material of a conductor, and thereby hangs a tale. Analysis of experiments on the free fall of electrons[1] in the Earth's gravitational field raised the question of whether the electric field inside a (metallic) vacuum chamber is *exactly* zero. The following qualitative argument suggests that the action of the gravitational field of the Earth on the metal will generate an electric field in the metal. Consider the "free" electrons in a metal. A gravitational force $m_e g$ pulls the electrons downward, and they begin to accumulate on the lower surface of the conductor, generating a downward electric field. For equilibrium, the upward force exerted by this electric field on the remaining electrons must compensate the gravitational force:

$$E = -\frac{m_e g}{e} \qquad (16)$$

But before we draw the hasty conclusion that (16) represents the typical electric field within the volume of a metal, let us see what happens when we consider the nucleus of an atom (with atomic number Z). The gravitational force on this nucleus is Mg; for equilibrium, this force must be compensated by an electric field

$$E = +\frac{Mg}{Ze} \qquad (17)$$

pointing *upward*.

Since the electrons occupy most of the volume of the metal, we might expect that the dominant contribution to the average electric field in the metal would be given by Eq. (16) rather than by Eq. (17). However, careful quantum-mechanical calculations, taking into account lattice-compression effects, show that the field should be of the order of magnitude given by Eq. (17). The field in a cavity in the conductor will then also be of this order of magnitude.

■ ***Exercise 1.*** Show that, for a typical metal, the field given by Eq. (17) is about 10^{-11} statvolt/cm, or about 10^{-7} volt/m. ■

Incidentally: The free-fall experiments on electrons dropped within metallic cavities seem to disagree with Eq. (17). It is believed that this disagreement can be blamed on some exotic surface effects at the walls of the cavity.[2]

With this discussion of boundary conditions at conductors, we are ready to solve some problems. We will begin a systematic investigation of analytic solutions of Laplace's equation in the next section. However, it is often possible to solve electrostatic problems by some clever trick. One such trick is the *method of images,* which relies on replacing a problem whose solution we do not know (but want to know) by a problem whose solution we do know and that has the same charge distribution and satisfies the *same boundary conditions*. The uniqueness theorem then tells us that the solution of the first problem is the same as that of the second.

EXAMPLE: *image charge in a conducting plate*

The simplest example of this method is as follows: Suppose we have a large (infinite) conducting plate at zero potential ("grounded") and a point charge q placed at some distance above this plate. What is the potential $\Phi(\mathbf{x})$? Figure 3.5 shows the plate (x-y plane) and the charge at $z = b$. We begin by looking at a different problem: two charges $\pm q$ separated by a distance $2b$ with *no* conducting plate (see Fig. 3.6).

■ ***Exercise 2.*** Show that the x-y plane in Fig. 3.6 is an equipotential surface. ■

The preceding exercise shows that at the boundaries of the region $0 \le z \le \infty$, the first and second problems satisfy exactly the same boundary condition; also, the charge densities in this region are the same. Since the potential for the second problem is

$$\Phi(\mathbf{x}) = \frac{q}{\sqrt{x^2 + y^2 + (z-b)^2}} - \frac{q}{\sqrt{x^2 + y^2 + (z+b)^2}} \tag{18}$$

this must also be the potential for the first problem (to be used for $z \ge 0$ only). Insofar as the region $z \ge 0$ is concerned, the effect of the conducting plate can be replaced by the effect of the charge $-q$. The virtual charge $-q$ is called the *image charge.* Note that since the image charge simulates the effect of the conducting plate, we can immediately conclude that the force that the plate exerts on the charge q is simply $q^2/4b^2$.

The method of images also provides the solution for the problem of a point charge placed near a conducting sphere (see Problem 4), the problem of a point charge placed near the plane interface between two different dielectrics, and a few other select problems.

Figure 3·5

A point charge q placed above a large conducting plate.

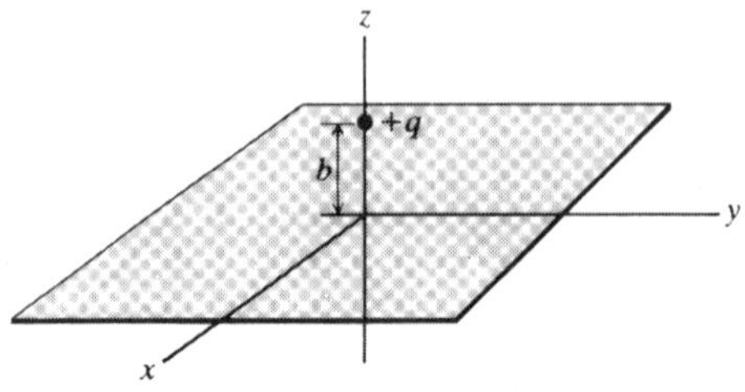

Figure 3·6

Two point charges $\pm q$ without the conducting plate.

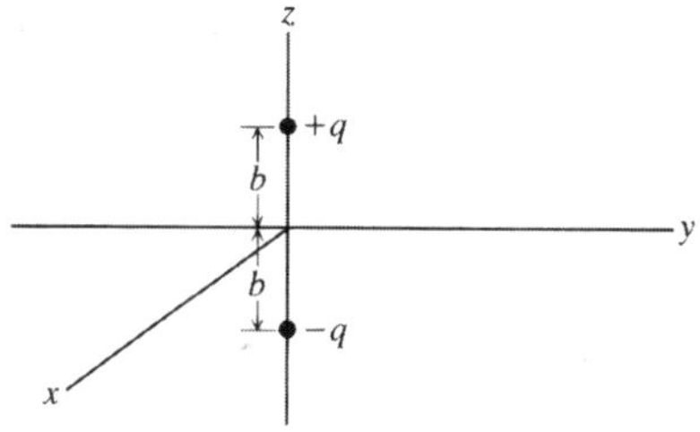

EXAMPLE: *conduction electrons confined by image charges*

Image charges play an important role in the confinement of the free conduction electrons in a metal. These conduction electrons are free to move within the volume of the metal, but when they reach the surface of the metal, they experience a binding force that pulls them back and prevents their escape. This binding force is generated by an image-charge effect. To understand this, think of an electron that is approaching the surface of the metal from outside. The negative electric charge $-e$ of the electron induces a positive charge distribution on the surface, which, by the method of images, is equivalent to a positive charge $+e$ symmetrically placed with respect to the surface. If the electron is at a distance d outside the surface, the image charge is at a distance d inside, and the potential energy of these two charges is $-e^2/(2d)$. The lowest (most negative) value of this potential energy is attained when $d \simeq$ atomic dimension $\simeq 10^{-8}$ cm, at which point the image-charge method fails because the electron begins to interact with individual atoms, and the surface can no longer be treated as a uniform, smooth conductor. This permits us to estimate the depth of the potential well confining the electrons in a metal as $e^2/(2d) \simeq e^2/(2\times10^{-8}\text{ cm}) \simeq 10^{-11}$ erg = 6 eV, a value that is of the same order of magnitude as the depth measured by electron diffraction experiments in metals, typically 10 eV.*

3·2 Laplace's Equation in Rectangular Coordinates

Whether rectangular, cylindrical, or spherical coordinates are most convenient in the solution of a given problem depends on the symmetry of the problem. In the case of rectangular boundaries, rectangular coordi-

*This depth of the potential well must not be confused with the "work function" of the electrons, which is the difference between the depth of the potential well and the kinetic energy of the electrons at the top of the Fermi distribution of free electrons. Typically, the work function is a few electron-volts.

nates will usually be convenient. The Laplace equation $\nabla^2\Phi = 0$ is then written as

$$\left(\frac{\partial^2}{\partial x^2}+\frac{\partial^2}{\partial y^2}+\frac{\partial^2}{\partial z^2}\right)\Phi(x,y,z)=0 \tag{19}$$

This is a second-order partial differential equation in *x, y, z*.

We will look for special solutions of the type

$$\Phi(x,y,z)=X(x)Y(y)Z(z) \tag{20}$$

where Φ is the product of a function of x, a function of y, and a function of z. Obviously, the function $1/\sqrt{x^2+y^2+z^2}$, which is a solution of (19) and represents the potential of a point charge at the origin, is *not* of the form of Eq. (20); we are therefore not going to obtain all possible solutions of the Laplace equation. However, it turns out that if we take *all* possible solutions of the type (20) and combine them cleverly, then we can construct the most general solution. Functions of the type $X(x)Y(y)Z(z)$ form a *complete set:* any solution of the Laplace equation can be written as superposition of these functions.

Substitution of Eq. (20) into Eq. (19) yields

$$YZ\frac{\partial^2}{\partial x^2}X+XZ\frac{\partial^2}{\partial y^2}Y+XY\frac{\partial^2}{\partial z^2}Z=0$$

or

$$\frac{1}{X}\frac{d^2X}{dx^2}+\frac{1}{Y}\frac{d^2Y}{dy^2}+\frac{1}{Z}\frac{d^2Z}{dz^2}=0 \tag{21}$$

The first term is a function of x only, the second of y only, and the third of z only. Each of these terms must equal a *constant.* To see this, suppose that $(1/X)d^2X/dx^2$ is *not* constant; then, if we change x while keeping y and z fixed, the first term in Eq. (21) changes while the other two do not; this is impossible because the sum of the three terms must remain zero. We therefore have

$$\frac{1}{X}\frac{d^2X}{dx^2}-\alpha^2 \tag{22}$$

$$\frac{1}{Y}\frac{d^2Y}{dy^2}=\beta^2 \tag{23}$$

$$\frac{1}{Z}\frac{d^2Z}{dz^2}=\gamma^2 \tag{24}$$

where α, β, γ are three constants. Instead of dealing with one *partial* differential equation, we now deal with three, separate *ordinary* differential equations for the *x, y,* and *z* dependence. This method for handling partial differential equations is called *separation of variables.*

Equation (21) will be satisfied if

$$\alpha^2 + \beta^2 + \gamma^2 = 0 \tag{25}$$

This shows that one or more of these constants must be a complex number.

■ ***Exercise 3.*** Show that the solutions of Eqs. (22)–(24) are, to within a multiplicative constant,

$$X = e^{\alpha x},\ e^{-\alpha x} \tag{26}$$

$$Y = e^{\beta y},\ e^{-\beta y} \tag{27}$$

$$Z = e^{\gamma z},\ e^{-\gamma z} \tag{28}$$

■

Our solution for $\Phi(x, y, z)$ is therefore

$$\Phi(x, y, z) = e^{\pm\alpha x} e^{\pm\beta y} e^{\pm\gamma z} \tag{29}$$

multiplied by some arbitrary constant.

EXAMPLE: *parallel plates with strips of alternating potential*

The following example illustrates how the superposition of solutions of the type (29) may be used in the solution of a problem. Figure 3.7 shows two large (infinite) parallel plates. The plate $z = h$ is at a potential $\Phi = 0$. The plate $z = 0$ is divided into strips of width $b/2$; these strips are, respectively, at potentials of alternating signs, $+V_0, -V_0, +V_0, -V_0, \ldots$. What is the potential in the region between the

Figure 3·7

Two large parallel plates. The lower plate is divided into strips at alternating potentials $\pm V_0$. The upper plate is at zero potential.

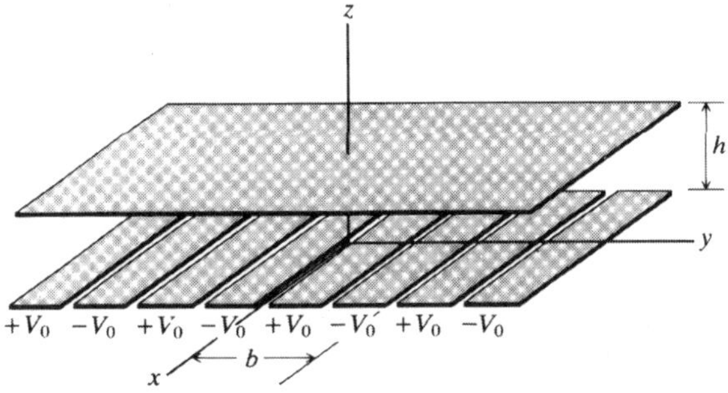

plates?

Since there is no x dependence in the problem, we must take $\alpha = 0$. Since the arrangement is periodic in y, with a period b, we must have a y dependence with the same period; this is achieved by taking β imaginary, $\beta = 2\pi in/b$, where n is a positive or negative integer. The functions (29) are then

$$\Phi \propto e^{2\pi iny/b} e^{\pm\gamma z} \tag{30}$$

According to Eq. (25), γ must have the value $2\pi n/b$. It remains to find the coefficients that are to be inserted in front of the solutions (30). The boundary condition $\Phi = 0$ at $z = h$ requires that we use a combination of $e^{+\gamma z}$ and $e^{-\gamma z}$ that vanishes at $z = h$. The following combination satisfies this requirement:

$$\frac{e^{-\gamma h} e^{\gamma z} - e^{\gamma h} e^{-\gamma z}}{e^{-\gamma h} - e^{\gamma h}} \tag{31}$$

Here the denominator is a normalization factor, inserted for convenience. If we now write Φ as a superposition of such functions

$$\Phi(y,z) = \sum_{n=-\infty}^{n=+\infty} A_n e^{2\pi iny/b} \left(\frac{e^{-2\pi nh/b} e^{2\pi nz/b} - e^{2\pi nh/b} e^{-2\pi nz/b}}{e^{-2\pi nh/b} - e^{2\pi nh/b}} \right) \tag{32}$$

then the only remaining boundary condition we have to worry about is the one at $z = 0$. We can fit this boundary condition by adjusting the coefficients A_n. At $z = 0$, Eq. (32) gives

$$\Phi(y,0) = \sum_{n=-\infty}^{n=+\infty} A_n e^{2\pi iny/b} \tag{33}$$

To find A_m, multiply both sides of Eq. (33) by $e^{-2\pi imy/b}$ and integrate over y, from $y = 0$ to $y = b$:

$$\int_0^b \Phi(y,0) e^{-2\pi imy/b}\, dy = \sum_{n=-\infty}^{n=+\infty} A_n \int_0^b e^{2\pi i(n-m)y/b}\, dy \tag{34}$$

■ ***Exercise 4.*** Show that

$$\int_0^b e^{2\pi i(n-m)y/b}\, dy = \begin{cases} 0 & \text{for } n \neq m \\ b & \text{for } n = m \end{cases} \tag{35}$$

■

With Eq. (35), the right side of Eq. (34) reduces to bA_m, and we find

$$A_m = \frac{1}{b}\int_0^b \Phi(y,0)e^{-2\pi imy/b}\,dy \tag{36}$$

The function $\Phi(y,0)$ for our problem is plotted in Fig. 3.8. Substituting this function into Eq. (36), we obtain

$$A_m = \frac{-V_0}{\pi im}(e^{-\pi im}-1) \tag{37}$$

■ ***Exercise 5.*** Integrate Eq. (36) and obtain Eq. (37). ■

Note that the argument leading to Eq. (36) is valid for any arbitrary function $\Phi(y)$. Equation (33) is the *Fourier series* for an arbitrary periodic function of y (with period b); it expresses the function as a superposition of complex exponentials or, equivalently, sines and cosines. Equation (36) gives the Fourier coefficient that must be used in this series. Our argument contains the implicit assumption that the series converges toward the function $\Phi(y)$. Concerning this, suffice it to say that for the case of the function plotted in Fig. 3.8, the series converges toward the function except at the points of discontinuity; the deviation at this set of isolated points is of no physical importance

If we substitute Eq. (37) into Eq. (32), we obtain the potential function that solves our problem:

$$\Phi(y,z) = \sum_{n=-\infty}^{n=+\infty} \frac{-V_0}{\pi in}(e^{-\pi in}-1)e^{2\pi iny/b}\left(\frac{e^{-2\pi nh/b}e^{2\pi nz/b} - e^{2\pi nh/b}e^{-2\pi nz/b}}{e^{-2\pi nh/b} - e^{2\pi nh/b}}\right) \tag{38}$$

■ ***Exercise 6.*** Show that the expression on the right side of Eq. (38) is a real function, and express the Fourier series in terms of sines and cosines. ■

It is of some interest to examine a special case of Eq. (38), with $h \gg b$.

■ ***Exercise 7.*** Show that in the limit $h/b \to \infty$,

Figure 3·8

Potential of the lower plate as a function of y.

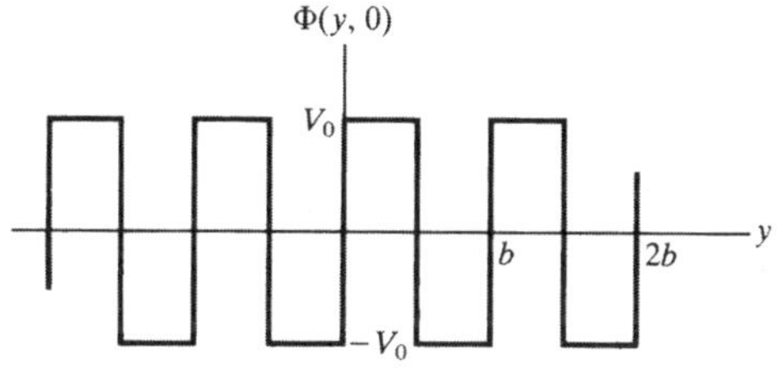

$$\frac{e^{2\pi n(z-h)/b} - e^{-2\pi n(z-h)/b}}{e^{-2\pi nh/b} - e^{2\pi nh/b}} \to e^{-2\pi|n|z/b}$$

and that

$$\Phi(y,z) \to \sum_{n=-\infty}^{n=+\infty} \frac{-V_0}{\pi i n}(e^{-\pi i n} - 1)e^{2\pi i n y/b} e^{-2\pi|n|z/b} \tag{39}$$

■

Unless z is very small, the main contribution to the sum in Eq. (39) comes from terms of *low* n; terms of high n fall off quickly as z increases. If we keep only $n = \pm 1$ in Eq. (39), we find

$$\Phi(y,z) \simeq \frac{4V_0}{\pi} \sin\frac{2\pi y}{b} e^{-2\pi z/b} \tag{40}$$

Equation (40) shows that the potential goes to zero exponentially as z increases. The electric field also falls off exponentially—there is very little field left if $z > b$. At a distance large compared with b, the "strip" plane at $z = 0$ looks very much like a grounded plane.

■ ***Exercise 8.*** Calculate the components E_z and E_y from Eq. (40), and verify that both of these components fall off exponentially with z. ■

EXAMPLE: *Faraday cage*

The exponential decrease in the field illustrates the principle of the *Faraday cage,* which is a box made of fine wire mesh. Roughly, the wires and the spaces between them are analogous to the strips of our problem. The external electric field is attenuated exponentially as it tries to penetrate the mesh, and the interior of the cage is (nearly) free of fields. To reduce the field *exactly* to zero, we must use a closed box made of solid sheet metal. As we have seen in the preceding section, no electric field can exist in an empty (that is, charge-free) cavity in a conductor.

3·3 Laplace's Equation in Cylindrical Coordinates

The differential operator ∇^2 in cylindrical coordinates ρ, ϕ, z was given in Chapter 1. In these coordinates, Laplace's equation takes the form

$$\frac{1}{\rho}\frac{\partial}{\partial\rho}\left(\rho\frac{\partial\Phi}{\partial\rho}\right) + \frac{1}{\rho^2}\frac{\partial^2\Phi}{\partial\phi^2} + \frac{\partial^2\Phi}{\partial z^2} = 0 \tag{41}$$

(Here the radial cylindrical coordinate ρ must not be confused with the charge density.)

Let us begin by assuming that we are dealing with a problem in which the z dependence is absent, as will be the case if the charges are parallel line charges and the conducting surfaces are parallel cylinders (of arbitrary cross section). Then we must solve

$$\frac{1}{\rho}\frac{\partial}{\partial\rho}\left(\rho\frac{\partial\Phi}{\partial\rho}\right)+\frac{1}{\rho^2}\frac{\partial^2\Phi}{\partial\phi^2}=0 \tag{42}$$

Again, we try separation of variables by assuming

$$\Phi(\rho,\phi)=R(\rho)Q(\phi) \tag{43}$$

■ ***Exercise 9.*** Show that this assumption results in the ordinary differential equations

$$\frac{\rho}{R}\frac{d}{d\rho}\left(\rho\frac{dR}{d\rho}\right)=k^2 \tag{44}$$

and

$$\frac{1}{Q}\frac{d^2Q}{d\phi^2}=-k^2 \tag{45}$$

where k is a constant. ■

■ ***Exercise 10.*** Show that Eq. (45) has the solutions

$$Q(\phi)=\begin{cases}\cos k\phi, \sin k\phi & \text{if } k\neq 0\\ a_0+a_1\phi & \text{if } k=0\end{cases} \tag{46}$$

where a_0, a_1 are constants. ■

As we increase ϕ by 2π, the potential must remain unchanged (the potential is single-valued); hence k must be an integer, and the value of a_1 must be zero. The possible functions $Q(\phi)$ are then

$$Q(\phi)=\cos n\phi,\ \sin n\phi \qquad n=0,1,2,... \tag{47}$$

The radial equation that remains to be solved is

$$\rho\frac{d}{d\rho}\left(\rho\frac{dR}{d\rho}\right)=n^2R \tag{48}$$

■ ***Exercise 11.*** Show that this equation has the solutions

$$R(\rho)=\begin{cases}\rho^n, \rho^{-n} & \text{if } n\neq 0\\ b_0+b_1\ln\rho & \text{if } n=0\end{cases} \tag{49}$$

■

Putting all of this together, we find that the solutions are

$$\Phi(\rho,\phi) = \ln\rho,\ \rho^{\pm n}\cos n\phi,\ \rho^{\pm n}\sin n\phi \qquad n = 0,1,2,\ldots \tag{50}$$

EXAMPLE: *conducting cylinder in uniform electric field*

As an example, we will consider the following problem: A very long conducting cylinder of radius R is placed in an initially uniform electric field (that is, the field *was* uniform before the cylinder was placed in it), with the axis of the cylinder at right angles to the electric field. What is the potential everywhere?

We take the initially uniform field in the x direction (see Fig. 3.9). The potential that corresponds to such a uniform field of strength E_0 is $-E_0 x$, or $-E_0\rho\cos\phi$ in cylindrical coordinates. The cylinder can change the potential only nearby; at large distances, the potential will remain unchanged. This gives us the boundary condition

$$\Phi \to -E_0\rho\cos\phi \quad \text{as } \rho\to\infty \tag{51}$$

Another boundary condition is that the electric field at the cylinder must be normal to the surface:

$$\frac{\partial\Phi}{\partial\phi} = 0 \quad \text{at } \rho = R \tag{52}$$

According to Eq. (51), the potential anywhere in space consists of $-E_0\rho\cos\phi$ plus extra terms that vanish as $\rho\to\infty$. Of the set of solutions given by Eq. (50), only $\rho^{-n}\cos n\phi$ and $\rho^{-n}\sin n\phi$ have the property of vanishing at infinity; the potential must therefore be

Figure 3·9

Conducting cylinder placed in an (initially) uniform electric field.

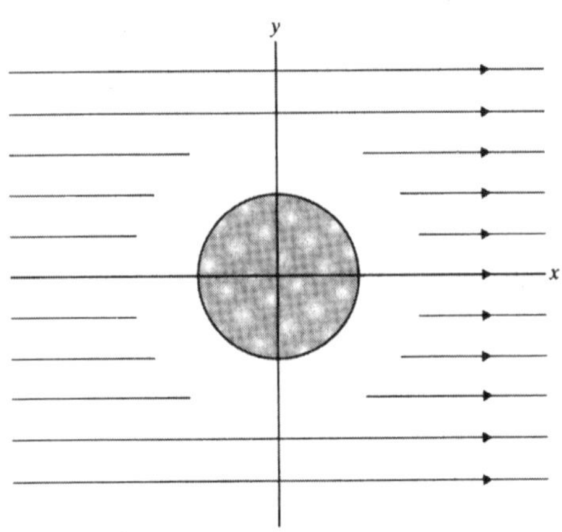

$$\Phi(\rho,\phi) = -E_0\rho\cos\phi + \sum_{n=1}^{\infty} A_n\rho^{-n}\cos n\phi + \sum_{n=1}^{\infty} B_n\rho^{-n}\sin n\phi \qquad (53)$$

The coefficients A_n, B_n are determined by the boundary condition at $\rho = R$:

$$0 = \left.\frac{\partial\Phi}{\partial\phi}\right|_R = E_0 R\sin\phi - \sum_{n=1}^{\infty} A_n R^{-n} n\sin n\phi + \sum_{n=1}^{\infty} B_n R^{-n} n\cos n\phi \qquad (54)$$

Equation (54) cannot be true for all angles ϕ unless

$$\begin{aligned} A_1 &= E_0 R^2 \\ A_2 &= A_3 = A_4 = \ldots = 0 \\ B_1 &= B_2 = B_3 = \ldots = 0 \end{aligned} \qquad (55)$$

The final result is then

$$\Phi(\rho,\phi) = -E_0\rho\cos\phi + E_0\frac{R^2}{\rho}\cos\phi \qquad (56)$$

■ ***Exercise 12.*** Calculate the components $E_\rho = -\partial\Phi/\partial\rho$ and $E_\phi = -(1/\rho)\partial\Phi/\partial\phi$ of the electric field. ■

The field lines for this electric field are shown in Fig. 3.10. The surface charge density on the cylinder can be calculated from the value of the electric field (see Problem 12).

The method of separation of variables can also be applied if the z dependence is *retained* in Eq. (41). If we take

Figure 3·10

Field lines around the conducting cylinder.

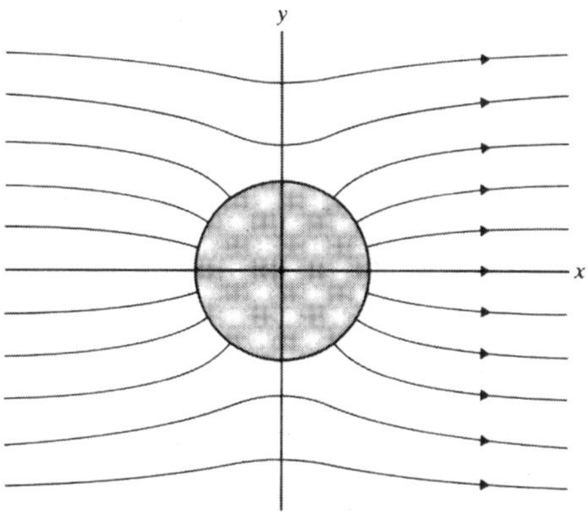

$$\Phi(\rho,\phi,z) = R(\rho)Q(\phi)Z(z) \tag{57}$$

then the Laplace equation becomes

$$\frac{\rho}{R}\frac{d}{d\rho}\left(\rho\frac{dR}{d\rho}\right) + \frac{1}{Q}\frac{d^2Q}{d\phi^2} + \frac{\rho^2}{Z}\frac{d^2Z}{dz^2} = 0 \tag{58}$$

The second term is a function of ϕ; the other terms are not. Hence the second term must be a constant, and we again obtain Eq. (45); the solution for Q is then the same as before. Dividing Eq. (58) by ρ^2, we are left with

$$\frac{1}{\rho R}\frac{d}{d\rho}\left(\rho\frac{dR}{d\rho}\right) - \frac{n^2}{\rho^2} + \frac{1}{Z}\frac{d^2Z}{dz^2} = 0 \tag{59}$$

The last term is a function of z; the other terms are not. Hence

$$\frac{1}{Z}\frac{d^2Z}{dz^2} = \gamma^2 \tag{60}$$

and

$$\frac{1}{\rho R}\frac{d}{d\rho}\left(\rho\frac{dR}{d\rho}\right) - \frac{n^2}{\rho^2} = -\gamma^2 \tag{61}$$

where γ is a constant.

The solutions of Eq. (60) are

$$Z(z) = \mathrm{e}^{\pm\gamma z}$$

The solutions of Eq. (61) are somewhat more difficult to find. It is convenient to introduce a new variable

$$\xi = \gamma\rho \tag{62}$$

In terms of this variable, the differential equation (61) can be written as

$$\frac{1}{\xi}\frac{d}{d\xi}\left(\xi\frac{dR}{d\xi}\right) + \left(1 - \frac{n^2}{\xi^2}\right)R = 0 \tag{63}$$

Figure 3·11

Bessel functions $J_n(\xi)$ and $N_n(\xi)$.

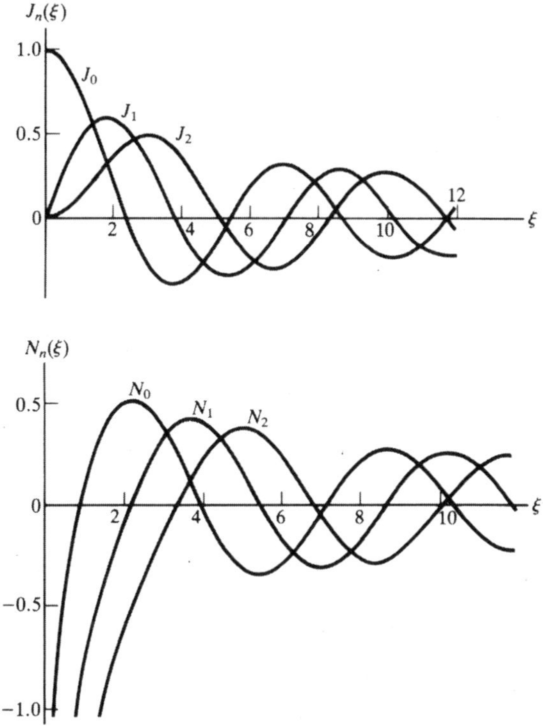

This is called *Bessel's equation,* and its solutions are called *Bessel functions.* These functions are usually designated by J_n and by N_n, and they can be written as infinite series in ξ :*

$$J_n(\xi) = \left(\frac{\xi}{2}\right)^n \sum_{j=0}^{\infty} \frac{(-1)^j}{j!(j+n)!} \left(\frac{\xi}{2}\right)^{2j} \tag{64}$$

$$N_n(\xi) = \frac{1}{\pi}\left[\frac{dJ_n(\xi)}{dn} - (-1)^n \frac{d}{dn} J_{-n}(\xi)\right] \tag{65}$$

*The following definition for N_n is valid for integer values of n; there exists a separate definition for noninteger values of n (which is of no interest to us). The derivative of the factorial function involved in Eq. (65) is defined by

$$\frac{dn!}{dn} = n!\left[-0.5772\ldots + \sum_{k=1}^{\infty} \frac{n}{k(n+k)}\right]$$

We will not give the arguments leading to Eqs. (64) and (65). Figure 3.11 displays plots of some of the functions J_n and N_n; all the functions N_n are singular at $\xi = 0$.

■ ***Exercise 13.*** Substitute expression (64) into Eq. (63) and verify that it solves this equation. ■

In terms of the functions J_n and N_n, the solutions of Laplace's equation are then

$$\Phi(\rho,\phi,z) = \begin{cases} J_n(\gamma\rho)\sin n\phi\, e^{\pm\gamma z}, & J_n(\gamma\rho)\cos n\phi\, e^{\pm\gamma z} \\ N_n(\gamma\rho)\sin n\phi\, e^{\pm\gamma z}, & N_n(\gamma\rho)\cos n\phi\, e^{\pm\gamma z} \end{cases} \tag{66}$$

This holds for $\gamma \neq 0$. In the case $\gamma = 0$ (no z dependence), we must, of course, return to Eq. (50).

3·4 Laplace's Equation in Spherical Coordinates

In spherical coordinates r,θ,ϕ, Laplace's equation reads

$$\frac{1}{r}\frac{\partial^2}{\partial r^2}(r\Phi) + \frac{1}{r^2 \sin\theta}\frac{\partial}{\partial\theta}\left(\sin\theta\frac{\partial\Phi}{\partial\theta}\right) + \frac{1}{r^2\sin^2\theta}\frac{\partial^2\Phi}{\partial\phi^2} = 0 \tag{67}$$

It is convenient to replace θ by the variable

$$\mu = \cos\theta \tag{68}$$

■ ***Exercise 14.*** Show that in terms of μ, Eq. (67) becomes

$$\frac{1}{r}\frac{\partial^2}{\partial r^2}(r\Phi) + \frac{1}{r^2}\frac{\partial}{\partial\mu}\left[(1-\mu^2)\frac{\partial\Phi}{\partial\mu}\right] + \frac{1}{r^2(1-\mu^2)}\frac{\partial^2\Phi}{\partial\phi^2} = 0 \tag{69}$$

■

For the sake of simplicity, we will assume that the problem to be solved has axial symmetry so that there is no ϕ dependence. Our equation then reduces to

$$\frac{1}{r}\frac{\partial^2}{\partial r^2}(r\Phi) + \frac{1}{r^2}\frac{\partial}{\partial\mu}\left[(1-\mu^2)\frac{\partial\Phi}{\partial\mu}\right] = 0 \tag{70}$$

This equation can be solved by separation of variables.

■ ***Exercise 15.*** Show that with

$$\Phi(r,\mu) = F(r)P(\mu) \tag{71}$$

we obtain the ordinary differential equations

$$\frac{r}{F}\frac{d^2}{dr^2}(rF) = l(l+1) \tag{72}$$

and

$$\frac{1}{P}\frac{d}{d\mu}\left[(1-\mu^2)\frac{dP}{d\mu}\right] = -l(l+1) \tag{73}$$

In Eqs. (72) and (73), the separation constant has been written in the strange form $l(l+1)$ because this will be convenient later on. Note that at this stage, l is completely arbitrary. ■

■ ***Exercise 16.*** Show that the functions

$$F(r) = r^l, r^{-l-1} \tag{74}$$

are solutions of Eq. (72). ■

The solution of Eq. (73) is not as obvious. But it so happens that if we assume that l is an integer ($l \geq 0$), then this equation can be integrated by means of a trick or two. If we restrict l to integral values, we will not obtain the most general solution of Eq. (73); but this need not worry us too much because we have already given up our claim to generality by assuming the special form (71) for $\Phi(r,\theta)$. All that matters is that the functions we finally obtain form a sufficiently large set so that any (axially symmetric) solution can be expressed as a superposition of our functions; this will indeed be the case.

If we indicate derivatives by a prime, then Eq. (73) reads

$$[(\mu^2-1)P']' = l(l+1)P \tag{75}$$

Let us write $P(\mu)$ as the lth derivative of some auxiliary function $u(\mu)$:

$$P(\mu) = \frac{d^l u(\mu)}{d\mu} \equiv u^{(l)}(\mu) \tag{76}$$

where the notation $u^{(l)}$ indicates l primes, that is, l derivatives. Then Eq. (75) becomes

$$[(\mu^2-1)u^{(l+1)}]' = l(l+1)u^{(l)}$$

If we integrate this once, we obtain

$$(\mu^2-1)u^{(l+1)} = l(l+1)u^{(l-1)} \tag{77}$$

Here we have ignored any constant of integration, because we are still looking only for a *particular* solution of Eq. (73). Next, observe that, by adding $l(l-1)u^{(l-1)}+2\mu l u^{(l)}$ to each side, we can write Eq. (77) as

$$(\mu^2-1)u^{(l+1)} + l(l-1)u^{(l-1)} + 2\mu l u^{(l)} = 2l^2 u^{(l-1)} + 2\mu l u^{(l)} \tag{78}$$

which has the nice property that each side is a perfect derivative, as follows:

$$(\mu^2-1)u^{(l+1)} + l(l-1)u^{(l-1)} + 2\mu l u^{(l)} = [(\mu^2-1)u']^{(l)} \tag{79}$$

and

$$2l^2 u^{(l-1)} + 2\mu l u^{(l)} = (2l\mu u)^{(l)} \tag{80}$$

■ ***Exercise 17.*** Check this. ■

Our Eq. (78) then reduces to

$$[(\mu^2-1)u']^{(l)} = (2l\mu u)^{(l)} \tag{81}$$

and if we integrate this l times, we find

$$[(\mu^2-1)u'] = 2l\mu u \tag{82}$$

This equation has the solution

$$u = (\mu^2-1)^l$$

According to Eq. (76), the function $P(\mu)$ is then

$$P(\mu) = \frac{d^l}{d\mu^l}(\mu^2-1)^l \tag{83}$$

It is customary to normalize these functions by a factor $1/(2^l l!)$ and to designate them by the symbol $P_l(\mu)$:

$$P_l(\mu) = \frac{1}{2^l l!}\frac{d^l}{d\mu^l}(\mu^2-1)^l \tag{84}$$

These functions are called *Legendre polynomials,* and Eq. (84) is called the *Rodrigues formula.* Since $(\mu^2-1)^l$ is a polynomial of order $2l$, it follows from Eq. (84) that $P_l(\mu)$ is a polynomial of order l.

■ **Exercise 18.** Show that Eq. (84) gives

$$P_l(1)=1 \tag{85}$$

$$P_l(-1)=(-1)^l \tag{86}$$

[*Hint:* Write Eq. (84) as $(1/2^l l!)(d^l/d\mu^l)(\mu-1)^l(\mu+1)^l$.] ■

■ **Exercise 19.** Show that

$$P_l(-\mu)=(-1)^l P_l(\mu) \tag{87}$$

■

■ **Exercise 20.** Show that

$$\begin{aligned} P_0(\mu)&=1 & P_1(\mu)&=\mu \\ P_2(\mu)&=½(3\mu^2-1) & P_3(\mu)&=½(5\mu^3-3\mu) \end{aligned} \tag{88}$$

■

Figure 3.12 gives plots of some of the functions $P_l(\mu)$, or $P_l(\cos\theta)$.

Our solutions of the Laplace equation then take the form

$$\Phi(r,\theta)=r^l P_l(\cos\theta),\ r^{-l-1}P_l(\cos\theta) \qquad l=0,1,2,\ldots \tag{89}$$

These solutions are sufficiently general for our purposes because the functions $P_l(\cos\theta)$ form a complete set: any function of θ can be expressed as a superposition of Legendre polynomials.

Figure 3·12

Legendre polynomials $P_l(\mu)$.

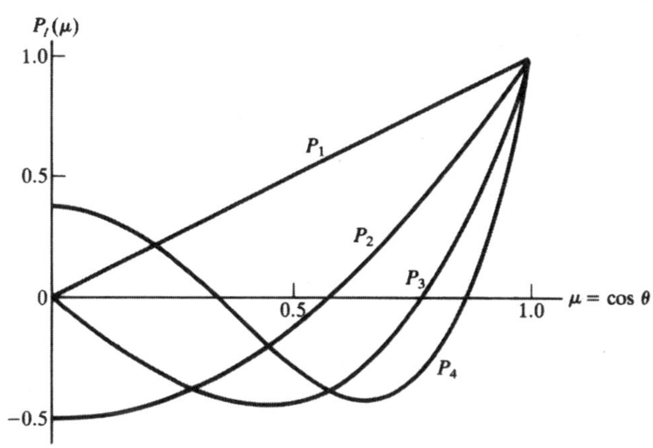

Before we look at an example of the use of these functions, we need to learn something about the integration of these functions. The following exercises contain two important results.

■ ***Exercise 21.*** Show that if $l \neq n$,

$$\int_{-1}^{1} P_l(\mu)P_n(\mu)\,d\mu = 0 \tag{90}$$

[*Hint:* If $l > n$, integration by parts repeated l times gives

$$\int_{-1}^{1} \frac{d^l}{d\mu^l}(\mu^2-1)^l \frac{d^n}{d\mu^n}(\mu^2-1)^n\,d\mu = (-1)^l \int_{-1}^{1} (\mu^2-1)^l \frac{d^{l+n}}{d\mu^{l+n}}(\mu^2-1)^n\,d\mu$$

This expression is zero, because for $l > n$, the $(l+n)$th derivative of a polynomial of order $2n$ vanishes.] ■

■ ***Exercise 22.*** Show that

$$\int_{-1}^{1} P_l(\mu)P_l(\mu)\,d\mu = \frac{2}{2l+1} \tag{91}$$

[*Hint:* Begin by integrating by parts; this simplifies the integrand.] ■

EXAMPLE: *conducting shell surrounded by two hemispheres at different potentials*

As one example of the solution of Laplace's equation in spherical coordinates, let us find the potential in the space between the two concentric spherical shells shown in Fig. 3.13. The inner shell, of radius R, is at zero potential; the outer shell, of radius R', is split into two hemispheres at potentials V_0 and $-V_0$, respectively. The potential in the region $R \le r \le R'$ will then be of the form

Figure 3·13

Two concentric spherical shells. The outer shell is split into two hemispheres at potentials $\pm V_0$. The inner shell is at zero potential.

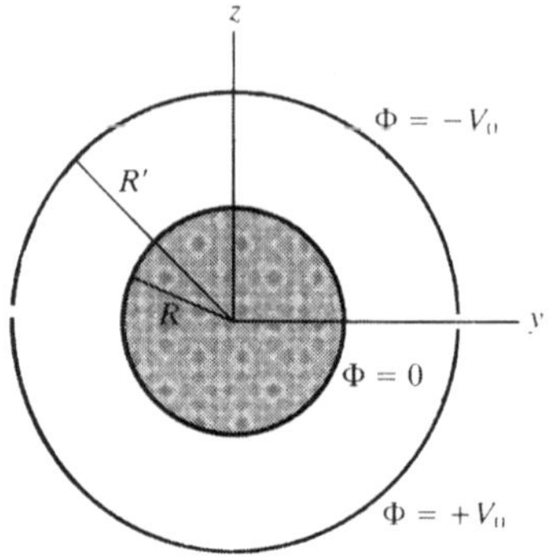

$$\Phi(r,\theta)=\sum_{l=0}^{\infty}[A_l r^l P_l(\cos\theta)+B_l r^{-l-1}P_l(\cos\theta)] \tag{92}$$

The boundary condition at $r=R$ tells us that

$$0=\Phi(R,\theta)=\sum_{l=0}^{\infty}[A_l R^l P_l(\cos\theta)+B_l R^{-l-1}P_l(\cos\theta)] \tag{93}$$

This requires

$$A_l R^l + B_l R^{-l-1}=0 \tag{94}$$

or

$$B_l=-A_l R^{2l+1} \tag{95}$$

The boundary condition at $r=R'$ then becomes

$$\Phi(R',\theta)=\sum_{l=0}^{\infty}\left(A_l R'^l - A_l\frac{R^{2l+1}}{R'^{l+1}}\right)P_l(\cos\theta) \tag{96}$$

To find A_m, multiply both sides of Eq. (96) by $P_m(\cos\theta)\sin\theta$ and integrate over θ, from $\theta=0$ to $\theta=\pi$:

$$\int_0^\pi \Phi(R',\theta)P_m(\cos\theta)\sin\theta\,d\theta$$

$$=\sum_{l=0}^{\infty}A_l\left(R'^l-\frac{R^{2l+1}}{R'^{l+1}}\right)\int_0^\pi P_m(\cos\theta)P_l(\cos\theta)\sin\theta\,d\theta \tag{97}$$

Upon changing the variable of integration to $\mu=\cos\theta$, we see that, according to Eq. (90), the integral on the right side is 0 unless $l=m$; and that, according to Eq. (91), this integral is $2/(2m+1)$ if $l=m$. The summation on the right side therefore collapses to a single term involving only the coefficient A_m. Solving Eq. (97) for this coefficient, we find

$$A_m=\frac{(2m+1)/2}{R'^m-R^{2m+1}/R'^{m+1}}\int_0^\pi \Phi(R',\theta)P_m(\cos\theta)\sin\theta\,d\theta \tag{98}$$

In the evaluation of this integral over θ, it is again best to change the variable to $\mu=\cos\theta$, so

$$\int_0^\pi \Phi(R',\theta)P_m(\cos\theta)\sin\theta\,d\theta=\int_{-1}^{1}\Phi(R',\mu)P_m(\mu)\,d\mu \tag{99}$$

The function $\Phi(R',\mu)$ is an odd function of μ (see Fig. 3.14); the function $P_m(\mu)$ is an even function if m is even and an odd function if m is odd [see Eq. (87)]. Hence, for even values of m, the integral in Eq. (99) vanishes, that is, $A_m = 0$. For odd values of m, the integral in Eq. (99) may be written as

$$2\int_{-1}^{0} V_0 P_m(\mu)\,d\mu = 2V_0\int_{-1}^{0}\frac{1}{2^m m!}\frac{d^m}{d\mu^m}(\mu^2-1)^m\,d\mu$$
$$= 2V_0\left[\frac{1}{2^m m!}\frac{d^{m-1}}{d\mu^{m-1}}(\mu^2-1)^m\right]_{-1}^{0} \tag{100}$$

■ ***Exercise 23.*** Show that

$$\left[\frac{1}{2^m m!}\frac{d^{m-1}}{d\mu^{m-1}}(\mu^2-1)^m\right]_{-1}^{0} = \frac{1}{2^{m+1}}\frac{(-1)^{m/2+1/2}}{\left[(m/2+1/2)!\right]^2}\frac{(m+1)!}{m} \tag{101}$$

[*Hint:* The contribution from the lower limit $(\mu=-1)$ is zero; the contribution from the upper limit ($\mu = 0$) depends only on the coefficient of μ^{m-1} in the polynomial $(\mu^2-1)^m$. Find this coefficient by means of the binomial theorem,

$$(x-1)^m = \sum_{k=0}^{m}\frac{m!}{(m-k)!k!}(-1)^k x^{m-k}$$

with $x = \mu^2$.] ■

Our result for A_m is then

$$A_m = \frac{(2m+1)V_0}{R'^m - R^{2m+1}/R'^{m+1}}\frac{1}{2^{m+1}}\frac{(-1)^{m/2+1/2}}{\left[(m/2+1/2)!\right]^2}\frac{(m+1)!}{m} \tag{102}$$

Figure 3·14

Potential of the outer spherical shell as a function of $\mu = \cos\theta$.

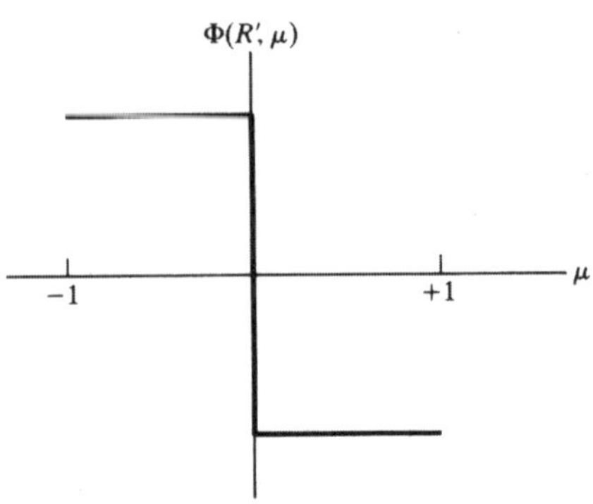

and our result for the potential (92) is

$$\Phi(r,\theta)=\sum_{\substack{l\\ \text{odd}}}\left[\frac{(2l+1)V_0}{R'^l - R^{2l+1}/R'^{l+1}}\frac{1}{2^{l+1}}\frac{(-1)^{l/2+1/2}}{\left[(l/2+1/2)!\right]^2}\frac{(l+1)!}{l}\right]$$
$$\times\left[r^l P_l(\cos\theta)-\frac{R^{2l+1}}{r^{l+1}}P_l(\cos\theta)\right] \quad \textbf{(103)}$$

An interesting special case emerges from this expression if we take the limit $R'\to\infty$ and simultaneously the limit $V_0\to\infty$ in such a way that $3V_0/2R'=E_0=\text{constant}$. Then, only the terms with $l=1$ survive in the summation, and we are left with

$$\Phi(r,\theta)=-E_0 r\cos\theta+E_0\frac{R^3}{r^2}\cos\theta \quad \textbf{(104)}$$

EXAMPLE: *conducting sphere in uniform electric field*

For large values of r, this potential approaches $\Phi(r,\theta)\to -E_0 r\cos\theta = -E_0 z$, which is the potential corresponding to a uniform electric field E_0 in the z direction. Thus Eq. (104) represents the potential for a conducting sphere placed in a uniform electric field. [Of course, we could also have obtained Eq. (104) by proceeding as we did in the calculation of the potential of a conducting cylinder placed in a uniform field; see Eqs. (51)–(56)]. Figure 3.15 shows the field lines for the electric field surrounding this conducting sphere.

Figure 3·15

Field lines around a conducting sphere placed in an (initially) uniform electric field.

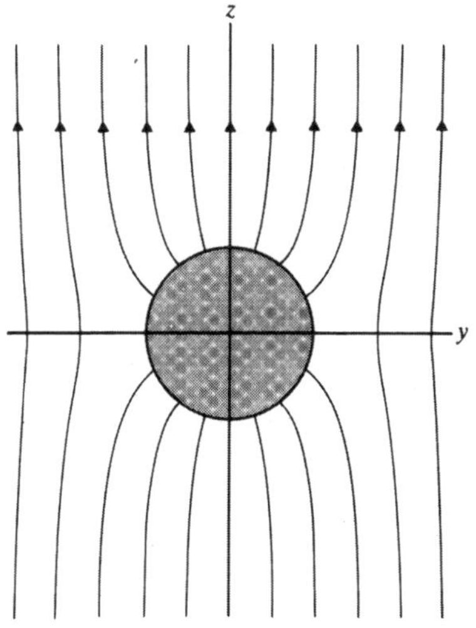

3·5 The Multipole Expansion

Consider a point charge q placed on the z axis at a distance r' from the origin (see Fig. 3.16). The potential of this point charge is spherically symmetric about the point charge. However, about the origin, the potential is *not* spherically symmetric—it is only axially symmetric. In terms of spherical coordinates,

$$|\mathbf{x}-\mathbf{x}'| = \sqrt{r^2 + r'^2 - 2rr'\cos\theta} \tag{105}$$

and hence

$$\Phi(r,\theta) = \frac{q}{\sqrt{r^2 + r'^2 - 2rr'\cos\theta}} \tag{106}$$

According to the claims of the preceding section, it should be possible to express the function of r on the right side of Eq. (106) as a superposition of the set of functions given by Eq. (89). Let us try to find this superposition. What we want is an expansion of the following form:

$$\frac{1}{\sqrt{r^2 + r'^2 - 2rr'\cos\theta}} = \sum_{l=0}^{\infty}(A_l r^l + B_l r^{-l-1})P_l(\cos\theta) \tag{107}$$

The regions $r > r'$ and $r < r'$ must be treated separately; we will concentrate on $r > r'$. Obviously, in this case $A_l = 0$, because otherwise the right side would not approach zero at infinity. To discover the values of B_l, consider the special case $\cos\theta = 1$. Since $P_l(1) = 1$,

$$\frac{1}{r - r'} = \sum_{l=0}^{\infty}\frac{B_l}{r^{l+1}} \tag{108}$$

But the series expansion for $1/(r - r')$ is well known; if $r > r'$, it is

Figure 3·16

A point charge q at a point $\mathbf{x}'$ on the z axis.

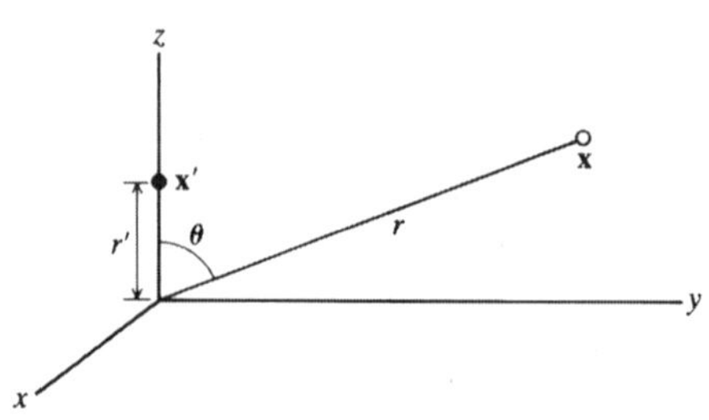

$$\frac{1}{r-r'} = \frac{1}{r}\frac{1}{1-r'/r} = \frac{1}{r}\left[1+\frac{r'}{r}+\left(\frac{r'}{r}\right)^2+\ldots\right] \qquad (109)$$

Comparison of Eqs. (108) and (109) shows that $B_l = (r')^l$, and hence

$$\frac{1}{\sqrt{r^2+r'^2-2rr'\cos\theta}} = \sum_{l=0}^{\infty}\frac{r'^l}{r^{l+1}}P_l(\cos\theta) \quad \text{for } r > r' \qquad (110)$$

Equation (110) is a very useful identity, which is sometimes taken as a *definition* of $P_l(\cos\theta)$. A similar identity holds for the case $r < r'$:

$$\frac{1}{\sqrt{r^2+r'^2-2rr'\cos\theta}} = \sum_{l=0}^{\infty}\frac{r^l}{(r')^{l+1}}P_l(\cos\theta) \quad \text{for } r < r' \qquad (111)$$

■ ***Exercise 24.*** Derive this identity. ■

Let us examine the first few terms of the potential obtained from Eq. (110):

$$\Phi(r,\theta) = \frac{q}{r}+\frac{qr'}{r^2}\cos\theta+\frac{qr'^2}{r^3}\frac{3\cos^2\theta-1}{2}+\ldots \qquad (112)$$

The coefficients qr' and qr'^2 in front of the functions of r, θ in Eq. (112) are called the *dipole moment* and the *quadrupole moment* of the charge, respectively. Note that a single point charge can have a dipole moment; the dipole moment vanishes if the charge is at the origin ($r' = 0$), but it is nonzero if the charge is at some distance from the origin. This makes it clear that the value of the dipole moment—and also of the quadrupole moment—depends on the position of the charge with respect to the origin of coordinates.

Next, consider a point charge at some point $\mathbf{x}'$ *not* on the z axis. We want to find an expansion in powers of $1/r$ analogous to Eq. (112). A convenient procedure is as follows: Introduce an auxiliary set of Cartesian coordinates x^*, y^*, z^* such that the z^* axis passes through the point charge (see Fig. 3.17). If θ^* is the polar angle of the field point $\mathbf{x}$ as measured in this new coordinate system, then formula (112) applies:

$$\Phi(r,\theta^*) = \frac{q}{r}+\frac{qr'}{r^2}\cos\theta^*+\frac{qr'^2}{r^3}\frac{3\cos^2\theta^*-1}{2}+\ldots \qquad (113)$$

But $\cos\theta^* = \mathbf{x}'\cdot\mathbf{x}/rr'$, and therefore Eq. (113) becomes

Figure 3·17

A point charge q at a point $\mathbf{x}'$ not on the z axis. The z^* axis passes through the point charge.

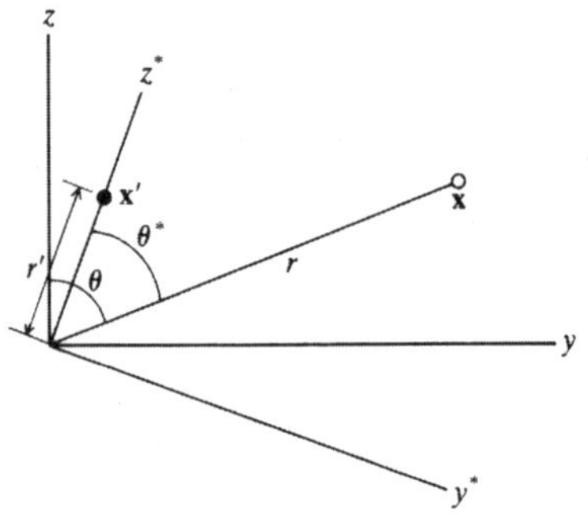

$$\Phi(\mathbf{x}) = \frac{q}{r} + \frac{q}{r^3}\mathbf{x}'\cdot\mathbf{x} + \frac{q}{r^5}\frac{3(\mathbf{x}'\cdot\mathbf{x})^2 - r'^2 r^2}{2} + \dots \qquad (114)$$

This last expression does not contain any explicit reference to the x^*, y^*, z^* coordinates; we can now forget about these coordinates and evaluate the dot product $\mathbf{x}'\cdot\mathbf{x}$ in our original coordinates, if we want to.

Our result (114) can be used to find an expansion in powers of $1/r$ for the potential of an arbitrary charge distribution. If the charge density is $\rho(\mathbf{x}')$, then

$$\Phi(\mathbf{x}) = \frac{1}{r}\int \rho\, dV' + \frac{1}{r^3}\int \mathbf{x}\cdot\mathbf{x}'\rho\, dV'$$

$$+ \frac{1}{2r^5}\int\left[3(\mathbf{x}\cdot\mathbf{x}')^2 - r^2 r'^2\right]\rho\, dV' + \dots \qquad (115)$$

This can also be written as

$$\Phi(\mathbf{x}) = \frac{Q}{r} + \frac{1}{r^3}\mathbf{x}\cdot\mathbf{p} + \frac{1}{2r^5}\sum_{i=1}^{3}\sum_{j=1}^{3} x^i x^j Q^{ij} + \dots \qquad (116)$$

where

$$Q \equiv \int \rho(\mathbf{x}')\, dV' \qquad (117)$$

$$\mathbf{p} \equiv \int \mathbf{x}'\rho(\mathbf{x}')\, dV' \qquad (118)$$

$$Q^{ij} \equiv \int (3x'^i x'^j - \delta^{ij} r'^2)\rho(\mathbf{x}')\, dV' \qquad (119)$$

The quantity Q is, of course, the total charge of the distribution; **p** is the dipole-moment vector; and Q^{ij} is the *quadrupole-moment tensor.*

The expression (116) is called the *multipole expansion* for the potential; it is valid only for points **x** outside the charge distribution (see Fig. 3.18). In the limit $r \to \infty$, the dominant term in Eq. (116) is Q/r; hence, seen from a very large distance, any charge distribution looks like a point charge. As we approach nearer the charge distribution, the dipole term becomes important, and then the quadrupole term, and so on.

EXAMPLE: *potential of pure dipole*

Finally, let us look at some special cases of Eq. (116). Figure 3.19 shows two point charges $\pm q$ at $z' = \pm b/2$ on the z axis. These charges form a dipole. We already obtained the (approximate) potential for this charge distribution in Section 2.4; now we will obtain the potential from the multipole expansion. Obviously, the total charge is $Q = 0$. The dipole moment is

$$p_x = p_y = 0$$

$$p_z = \int z' \rho \, dV' = q\frac{b}{2} + (-q)\frac{(-b)}{2} = qb \qquad (120)$$

and the quadrupole-moment tensor is zero.

■ ***Exercise 25.*** Check that Q^{ij} is zero for this charge distribution. ■

The only terms that remain in Eq. (116) are then the second term $\mathbf{x} \cdot \mathbf{p}/r^3$ and the higher-order terms indicated by dots. If we proceed to the limit $q \to \infty$, $b \to 0$ in such a way that qb remains constant, then these higher-order terms will drop out of Eq. (116) (the extra terms are proportional to qb^3, qb^4, and so on). We are then left with the potential of an *ideal dipole*, or *pure dipole*,

Figure 3·18

A charge distribution (shown shaded). The point **x** is outside the charge distribution.

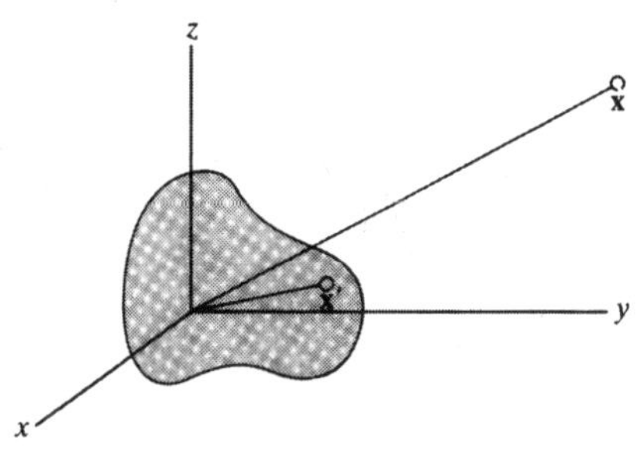

Figure 3·19

A charge distribution consisting of a pair of charges $\pm q$ on the z axis.

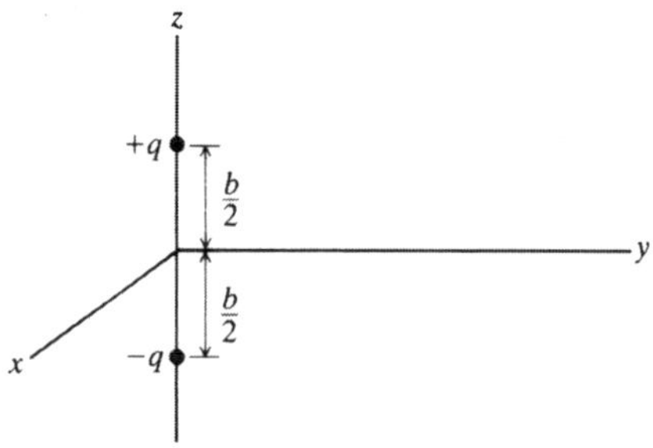

$$\Phi(\mathbf{x}) = \frac{1}{r^3}\mathbf{x}\cdot\mathbf{p} = p_z\frac{\cos\theta}{r^2} \tag{121}$$

We examined this potential in Section 2.4. The electric field of such a dipole is given in Eqs. (2.53)–(2.55). This electric field falls off as $1/r^3$ with distance. The field lines for the pure dipole are shown in Fig. 3.20.

■ ***Exercise 26.*** Show that the dipole moment of the charges $\pm q$ is independent of the position of the center of the dipole relative to the origin of coordinates. In general, show that for any charge distribution of zero total charge, the dipole moment (118) is proportional to the displacement between the average position vectors of the positive and the negative parts of the charge distribution, which is independent of the overall position of the charge distribution relative to the origin of coordinates [for this purpose, write the right side of Eq. (118) as a sum of two separate integrals, one for the positive part $(\rho > 0)$ and one for the negative part $(\rho < 0)$]. ■

■ ***Exercise 27.*** Show that a pure dipole (with, of course, zero net charge) located at the point $\mathbf{x}'$ produces a potential

Figure 3·20

Field lines for the pure dipole.

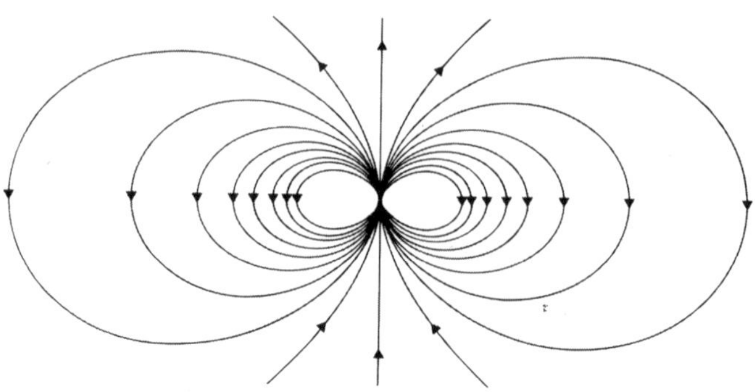

$$\Phi(\mathbf{x}) = \frac{1}{|\mathbf{x}-\mathbf{x}'|^3}(\mathbf{x}-\mathbf{x}')\cdot\mathbf{p} \tag{122}$$

Show that this can also be written as

$$\Phi(\mathbf{x}) = -\mathbf{p}\cdot\nabla\frac{1}{|\mathbf{x}-\mathbf{x}'|} \tag{123}$$

and as

$$\Phi(\mathbf{x}) = +\mathbf{p}\cdot\nabla'\frac{1}{|\mathbf{x}-\mathbf{x}'|} \tag{124}$$

■

EXAMPLE: *potential of quadrupole*

Figure 3.21 shows three charges on the z axis: a charge $+q$ at $z'=b$, a charge $-2q$ at $z'=0$, and a charge $+q$ at $z'=-b$. This charge distribution is called a *linear quadrupole.* This charge distribution has zero net charge and zero dipole moment. The quadrupole-moment tensor is

$$Q^{ij} = \int(3x'^i x'^j - \delta^{ij}r'^2)\rho(\mathbf{x}')dV' = \begin{pmatrix} -2qb^2 & 0 & 0 \\ 0 & -2qb^2 & 0 \\ 0 & 0 & 4qb^2 \end{pmatrix} \tag{125}$$

Thus the only terms that appear in the multipole expansion (116) are the quadrupole term and high-order terms. If we take the limit $q\rightarrow\infty$ and $b\rightarrow 0$ in such a way that qb^2 remains constant, all the higher-order terms will drop out. What remains is the potential of a *pure quadrupole*:

$$\Phi(\mathbf{x}) = \frac{1}{2r^5}2qb^2(2z^2-x^2-y^2)$$

$$= \frac{2qb^2}{r^3}\frac{3\cos^2\theta-1}{2} \tag{126}$$

Figure 3·21

Charges $+q, -2q$, and $+q$ on the z axis.

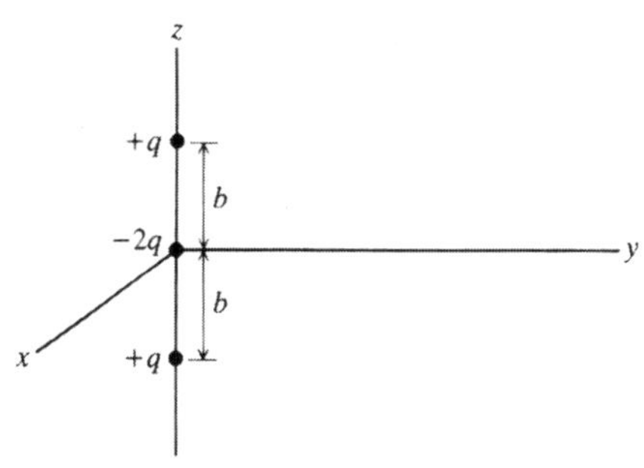

The electric field calculated from this potential falls off as $1/r^4$ with distance.

References

1. F. C. Witteborn and W. M. Fairbanks, *Phys. Rev. Lett.* **19**, 1049 (1967).
2. L. I. Schiff, *Phys. Rev.* B **1**, 4649 (1970).

Problems

1. If the olive oil inside a closed tin can has a charge Q, prove that an equal charge Q will be induced on the outside surface of the can.

2. Consider a charged cylindrical conducting surface of arbitrary (not necessarily circular) cross section. Use Gauss' law to show that, at the surface of the cylinder, the derivative of the electric field in the normal direction is

$$\frac{\partial E}{\partial l} = -\frac{E}{R}$$

where R is the (local) radius of curvature of the surface.

3. Two infinite, grounded conducting planes (x-z plane and y-z plane) intersect perpendicularly. A point charge q is placed at $(b,b,0)$. Use the method of images to find the potential for the region $x>0, y>0$. Find the force that the conducting planes exert on the point charge.

4. A grounded conducting sphere of radius a is placed at the origin. A point charge q is placed on the x axis at $x=3a$. Find the potential everywhere. Find the surface charge density σ induced on the sphere. [*Hint:* Try an image charge q' at $x=b$; adjust q' and b so as to make the surface of the sphere into an equipotential.]

5. Two infinite conducting plates are parallel, separated by a distance $2d$; the plates are grounded. A point charge q is located midway between the plates. Find the potential in the region between the plates. [*Hint:* Try an infinite sequence of image charges. The potential is an infinite series; do not bother to sum it.]

6. A very long line with a charge λ per unit length lies along the x axis. A very large, grounded conducting plane is at $z=-b$. What is the potential in the region $z \geq -b$?

7. In Section 3.1, we found the electrostatic potential generated by a point charge placed at a distance b from a very large conducting plate.
(a) Calculate the surface charge density on the plate.
(b) Calculate the electric field at the plate.
(c) By integration of the force per unit area [see Eq. (15)], find the net force on the plate, and compare it with the force $q^2/4b^2$ that the plate exerts on the point charge.

8. A point charge is placed at a distance b from a large, flat conducting plate.
   (a) What is the work needed to remove the point charge from the distance b to infinity? Does this work agree with the mutual potential energy between charge and the (virtual) image charge?
   (b) Assume that an electron is released at a large distance from a conducting plate. What will be the kinetic energy of the electron when it has accelerated to within a distance of 1 nm from the plate?

9. Find the potential in the (empty) region between two infinite parallel planes given the following boundary conditions: The lower plane coincides with the x-y plane, and the potential on it is

$$\Phi(x,y,0) = 5\sin(4x)\cos(3y)$$

The upper plane is at $z = 2/5$, and the potential on it is

$$\Phi(x,y,2/5) = 2\sin(4x)\cos(3y)$$

10. Find the potential in the interior of an (empty) cube of edge L, given that the potential is V_0 on the top face of the cube and zero on all the other faces.

11. Two long, concentric conducting cylinders of sheet metal have radii a and b and are at potentials V_a and V_b, respectively. The region between the cylinders is filled with a uniform charge density $\rho_0 = \text{constant}$. Find the potential in the region between the cylinders.

12. Consider the long conducting cylinder in an (initially) uniform electric field described in Section 3.3. Calculate the surface charge density on this cylinder.

13. A long, horizontal conducting cylinder of radius 1 cm is placed in an initially uniform, vertical electric field $E_0 = 20\,\text{statvolts/cm}$. What is the force per unit area of the electric field on the cylinder at its uppermost point?

14. A very long conducting cylinder of radius R is divided into four 90ϒ sectors by thin insulating plates passing through the axis. The sectors are at potentials $+V_0$, $-V_0$, $+V_0$, and $-V_0$, respectively. What is the potential in the region outside the cylinder?

15. On a cylindrical surface of radius R whose axis is along the z axis, the potential satisfies the boundary condition

$$\Phi(R,\phi) = \begin{cases} 0 & \text{for } -\pi/4 < \phi < \pi/4 \\ V_0 & \text{for } \pi/4 < \phi < 7\pi/4 \end{cases}$$

Find the potential in the region within the cylinder.

16. An empty can has the shape of a sector of a cylinder. The volume of the can is defined by the cylindrical coordinates $0 \le \rho \le R, 0 \le \phi \le \pi/2$, and $-\infty < z < \infty$. The surface $\rho = R$ is at potential V_0. The other surfaces are at zero potential. Find the potential everywhere within the can.

17. Prove that each of the Legendre polynomials, except P_0, has an average value of zero, that is,

$$\int_{-1}^{1} P_l(\mu)\,d\mu = 0$$

18. Consider the split spherical shell of radius R' described in Section 3.4. What is the potential in the region $r > R'$ *outside* this shell?

19. Equation (104) gives the electrostatic potential in the region surrounding a conducting sphere placed in an initially uniform electric field.

(a) Calculate the electric field at the surface of the sphere, and calculate the surface charge density on the sphere.

(b) Calculate the electric force on the upper half $(0 < \theta < \pi/2)$ of the sphere, and calculate the electric force on the lower half $(\pi/2 < \theta < \pi)$ of the sphere.

20. An amount of charge Q has been deposited on a conducting sphere of radius R, and the sphere has been placed in a uniform electric field $\mathbf{E}_0$ in the z direction. What is the potential in the region surrounding the sphere?

21. A very large, horizontal conducting plate has a convex hemispherical boss of radius R at its center. The plate is placed in an electric field that, at a large distance above the plate, is vertical and of strength E_0. Find the potential in the region above the plate, and find the surface charge density on the boss and on the plate.

22. A thin spherical shell made of insulator is centered on the origin. The shell has a radius R and carries a surface charge $\sigma = \sigma_0 \cos\theta$. Prove that this charge distribution generates a uniform electric field in the spherical region inside the shell, and find the magnitude of this electric field. Prove that the electric field outside the shell is that of a dipole, and find the dipole moment. [*Hint:* Assume that the electric field inside is constant, and take advantage of the uniqueness theorem.]

23. Find the electrostatic potential inside a spherical volume of radius R, given that the volume contains a uniform charge density ρ_0 and that the potential at the surface $r = R$ is

$$\Phi(R,\theta) = V_0(2 + 5\cos\theta)$$

24. The region bounded by two concentric spherical surfaces is filled with a uniform charge density $\rho_0 = \text{constant}$. On the inner boundary $(r = a)$ of this region, the potential is

$$\Phi(a,\theta,\phi) = V_0 \cos\theta$$

On the outer boundary $(r = b)$ of this region, the potential is

$$\Phi(b,\theta,\phi) = 2V_0$$

Find the solution of Poisson's equation in the region $a < r < b$.

25. Derive Eq. (112) by a Taylor-series expansion of $1/\sqrt{r^2 + r'^2 - 2rr'\cos\theta}$ about $r' = 0$.

26. A total charge Q is distributed uniformly along a ring of radius b. Find the quadrupole-moment tensor for this charge distribution and write down the first three terms of the multipole expansion for the potential. Assume that the ring is in the x-y plane centered on the origin.

27. For the case of the ring of charge described in Problem 26, the multipole expansion is *not* valid for $r < b$. Find an expansion for the potential valid in this region.

28. Equation (116) gives the first three terms in the multipole expansion. Find the next term in this expansion.

29. An ellipsoid defined by the equation

$$\frac{x^2}{a^2}+\frac{y^2}{b^2}+\frac{z^2}{c^2}\le 1$$

has a charge Q uniformly distributed over its volume. What is the quadrupole-moment tensor of this ellipsoid?

30. A small charge distribution with a quadrupole-moment tensor Q^{ij} and with zero charge $(Q=0)$ and zero dipole moment $(p=0)$ is placed in an external electric field $\mathbf{E}(\mathbf{x})$. Show that the force of this electric field on the charge distribution is approximately

$$F^k=\frac{1}{6}Q^{ij}\frac{\partial^2 E^j}{\partial x^k \partial x^i}+\ldots$$

31. A cylinder of radius b and height h carries a charge Q uniformly distributed over its volume. Assuming that the axis of the cylinder coincides with the z axis and that the center of the cylinder is at the origin, calculate the potential outside the cylinder by means of the multipole expansion, Eq. (116).

32. (a) Show that Q^{ij} has zero trace.

(b) For a body with a uniform distribution of charge and mass, express Q^{ij} in terms of the moment of inertia tensor $I^{ij}=\int(\delta^{ij}r'^2-x'^i x'^j)\rho_{\text{mass}}(\mathbf{x}')dV'$. Is it possible to express I^{ij} in terms of Q^{ij}?

(c) Use the result of part (b) to deduce Q^{ij} for a uniformly charged square plate of dimensions $L\times L$ lying in the x-y plane, with center at the origin and sides parallel to the axes. (*Note*: The moment of inertia of the plate about the x or the y axis is that of a rod, $mL^2/12$; the moment of inertia about the z axis is $mL^2/6$, by the perpendicular-axis theorem.)

33. A large sheet of paper lying in the x-y plane carries a uniform surface charge density σ. A narrow annulus of exterior radius $R+\Delta R/2$ and interior radius $R-\Delta R/2$ centered on the origin has been cut out of this sheet. Use the multipole expansion to find the potential at a distance $r\gg R$.

CHAPTER

Dielectrics 4

Conductors have free electrons or other mobile charges in their interior. When a conductor is placed in an electric field, the free electrons rearrange themselves, settling into a new equilibrium configuration such that the electric field is totally canceled everywhere within the conductor. Insulators have no free electrons or other free charges. When an insulator is placed in an electric field, the field will not be canceled, at least not totally. Insulators have *bound charges.* In response to the electric field, some limited motion of these bound charges occurs, and this leads to a partial cancellation of the electric field. The bound charges rearrange themselves in such a way that there is a slight separation between the average positions of the positive and the negative charges; such a separation is called a *polarization* of the charge distribution. Materials that acquire a polarization when placed in an electric field are called *dielectrics.*

4·1 The Polarization

We can summarize the effect of the electric field on a dielectric material by saying that dipoles are produced in the material. Of course, since what really is involved here is a charge separation, not only dipoles but also quadrupoles and higher multipoles will be produced; but it usually is a good approximation to ignore everything except dipoles. One mechanism for the production of dipoles in a dielectric material is the distortion of the charge distribution within each atom. If we place an atom in an electric field, the forces exerted by this electric field on the electrons and on the nucleus are in opposite directions; this causes a small separation between the centroids of the negative and the positive charge distributions of the atom and gives the atom an induced dipole moment (see Fig. 4.1).

Another mechanism relies on the existence of molecules with permanent dipole moments. In liquid or gaseous materials, the molecules are usually oriented at random by the thermal motions, and the component of the dipole moment along any direction is zero on the average. But if we place the material in an electric field, the forces on the charges within the molecules tend to align the dipole moments with the electric field. The alignment will not be total because the random thermal motions contribute some disorder, but the average component of the dipole

Figure 4·1

In these atoms, the electric forces have displaced the centroids of the negative charge distributions from the nuclei (dots). The vectors indicate the dipole moments.

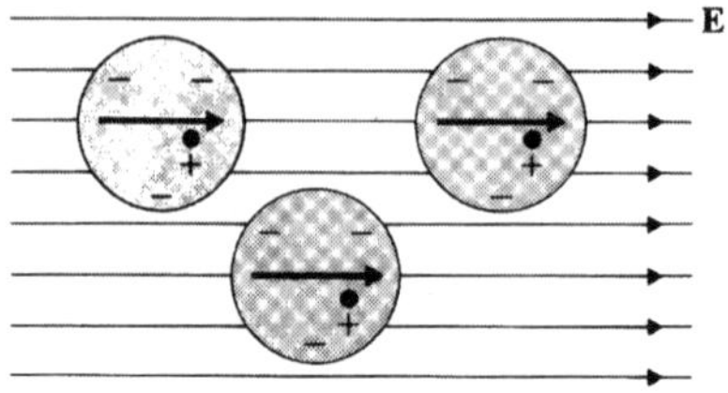

moment along the direction of the electric field will be positive (see Fig. 4.2).

Suppose that, in the presence of an external electric field, the average (induced or permanent) dipole moment of one atom or molecule along the direction of the electric field is **p**. We can then define the dipole moment per unit volume as

$$\mathbf{P} = n\mathbf{p} \tag{1}$$

where n is the number of atoms or molecules per unit volume. The vector **P** is also called the polarization vector or, simply, the *polarization*. In general, **P** will be a function of position because **p** varies with the strength of the electric field and because n is not necessarily uniform throughout the material. In terms of $\mathbf{P}(\mathbf{x}')$, the potential contributed by the dipoles in a small volume dV' is, according to Eq. (3.124),

$$d\Phi(\mathbf{x}) = \mathbf{P}(\mathbf{x}')\cdot\nabla'\frac{1}{|\mathbf{x}-\mathbf{x}'|}dV' \tag{2}$$

Figure 4·2

In this sample of molecules with permanent dipole moments, the electric forces have produced a partial alignment—more point toward the right than toward the left. The net dipole moment is toward the right.

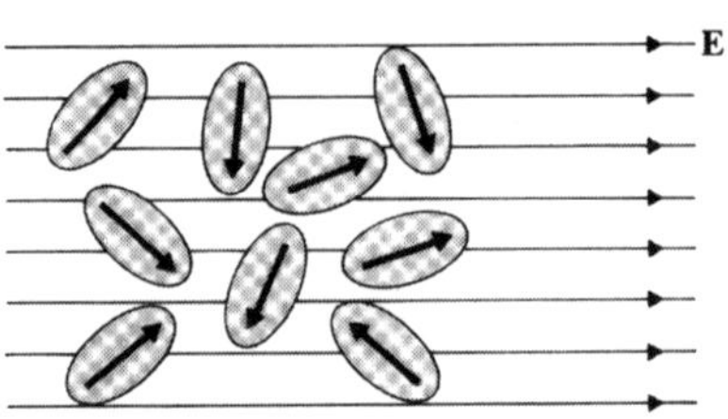

If, furthermore, a certain amount of free charge of density $\rho_F(\mathbf{x}')$ is distributed over the dielectric, then this charge will contribute an extra potential

$$d\Phi(\mathbf{x}) = \frac{\rho_F(\mathbf{x}')}{|\mathbf{x}-\mathbf{x}'|}dV' \qquad (3)$$

Hence, the total potential produced by both the bound and free charges in the dielectric is

$$\Phi(\mathbf{x}) = \int \frac{\rho_F(\mathbf{x}')}{|\mathbf{x}-\mathbf{x}'|}dV' + \int \mathbf{P}(\mathbf{x}')\boldsymbol{\cdot}\nabla'\frac{1}{|\mathbf{x}-\mathbf{x}'|}dV' \qquad (4)$$

This equation is valid for any point $\mathbf{x}$, outside or inside the dielectric. If the point $\mathbf{x}$ is inside the dielectric, then Eq. (2)—which is based on the dipole approximation—breaks down when $\mathbf{x}'$ approaches $\mathbf{x}$. But the points $\mathbf{x}'$ that are in the immediate vicinity of $\mathbf{x}$ make a negligible contribution to the second integral in Eq. (4) and do not affect the validity of the equation.

The second integral in Eq. (4) is identical to

$$\int \nabla'\boldsymbol{\cdot}\left(\frac{\mathbf{P}}{|\mathbf{x}-\mathbf{x}'|}\right)dV' - \int \frac{1}{|\mathbf{x}-\mathbf{x}'|}\nabla'\boldsymbol{\cdot}\mathbf{P}\,dV' \qquad (5)$$

Let us apply Gauss' theorem to the first integral in the expression (5). We take a Gaussian surface that lies just inside the surface of the dielectric* (see Fig. 4.3). Then

$$\int \nabla'\boldsymbol{\cdot}\left(\frac{\mathbf{P}}{|\mathbf{x}-\mathbf{x}'|}\right)dV' = \int \frac{\mathbf{P}\boldsymbol{\cdot}\hat{\mathbf{n}}}{|\mathbf{x}-\mathbf{x}'|}dS' \qquad (6)$$

Combining all this, we obtain

$$\Phi(\mathbf{x}) = \int \frac{\rho_F}{|\mathbf{x}-\mathbf{x}'|}dV' + \int \frac{\mathbf{P}\boldsymbol{\cdot}\hat{\mathbf{n}}}{|\mathbf{x}-\mathbf{x}'|}dS' + \int \frac{-\nabla'\boldsymbol{\cdot}\mathbf{P}}{|\mathbf{x}-\mathbf{x}'|}dV' \qquad (7)$$

Equation (7) has an interesting interpretation. The potential $\Phi(\mathbf{x})$ is produced by three separate charge distributions: a charge density ρ_F in

---

*This is the best we can do. It is tempting to take a Gaussian surface *outside* the dielectric where the surface integral in (6) would vanish. But we cannot do that because $\mathbf{P}$ is discontinuous *at* the surface of the dielectric and $\nabla'\boldsymbol{\cdot}\mathbf{P}$ does not exist there.

Figure 4·3

A dielectric (shown shaded) and a Gaussian surface just inside the dielectric.

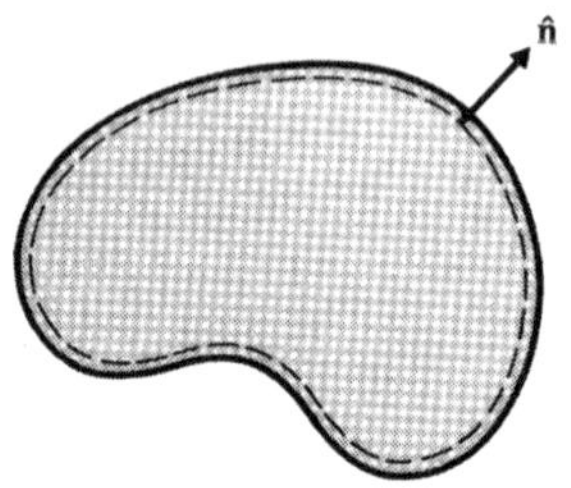

the volume of the dielectric, a surface charge density $\mathbf{P}\cdot\hat{\mathbf{n}}$ on the surface, and another charge density $-\nabla'\cdot\mathbf{P}$ in the volume. The quantity

$$\rho_P(\mathbf{x}') = -\nabla'\cdot\mathbf{P}(\mathbf{x}') \tag{8}$$

is called the *polarization charge density*, and

$$\sigma_P(\mathbf{x}') = \mathbf{P}(\mathbf{x}')\cdot\hat{\mathbf{n}} \tag{9}$$

is called the *polarization surface charge density*.

The charge densities (8) and (9) are due to bound charges; they result from the displacement of the bound charges from their equilibrium positions. To gain a better understanding of how (9) comes about, consider a piece of dielectric with a uniform polarization **P**. Such a polarization means that the positive and negative charges are displaced by a constant amount (see Fig. 4.4). Obviously, the left end of the dielectric will then carry an excess negative charge, and the right end will carry an excess positive charge; this is in agreement with Eq. (9).

If the polarization in the dielectric is not uniform, a volume charge density will usually exist in the dielectric. This can be understood from Fig. 4.5, which shows a volume element inside a dielectric with a

Figure 4·4

The distribution of negative charges (shaded) is displaced to the left of the distribution of positive charges (hatched).

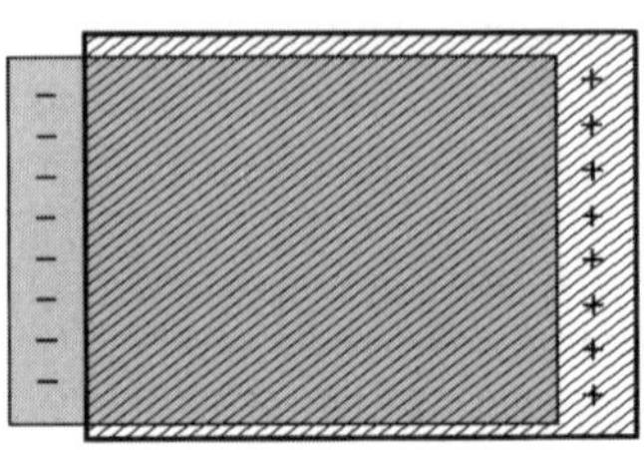

Figure 4·5

The heavy outline shows a volume element inside a dielectric. The polarization of the dielectric moves some negative charges (shaded) out of the volume element and some into the volume element.

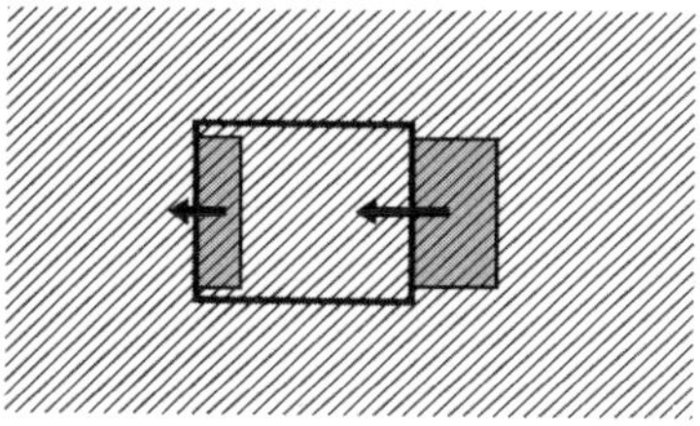

polarization **P** that increases from left to right. For simplicity, let us suppose that the polarization is entirely due to displacement of negative charges that are initially uniformly distributed. The displacement of these charges moves all the negative charges in the shaded region on the left out of the volume element, and it moves all the negative charges in the shaded region on the right *into* the volume element. Since the displacement is larger on the right than it is on the left, more negative charge moves into the volume element than moves out, and the result is a net negative charge density.

4·2 The Electric Displacement Field

In order to write Gauss' law for the interior of a dielectric, we must use the total volume charge density that includes both the free and the bound charge density:

$$\rho = \rho_F + \rho_P \tag{10}$$

or

$$\rho = \rho_F - \nabla \cdot \mathbf{P} \tag{11}$$

The electric field then satisfies the equation

$$\nabla \cdot \mathbf{E} = 4\pi(\rho_F + \rho_P) = 4\pi(\rho_F - \nabla \cdot \mathbf{P}) \tag{12}$$

Hence

$$\nabla \cdot (\mathbf{E} + 4\pi \mathbf{P}) = 4\pi\rho_F \tag{13}$$

It is customary to define the *electric displacement field* **D** as

$$\mathbf{D} = \mathbf{E} + 4\pi\mathbf{P} \tag{14}$$

so that

$$\nabla \cdot \mathbf{D} = 4\pi\rho_F \tag{15}$$

This is Gauss' Law in a dielectric. We can, of course, also express this law in integral form, by means of the usual argument:

$$\int \mathbf{D} \cdot d\mathbf{S} = Q_F \qquad (16)$$

Although the differential equation (15) for **D** contains only the density of free charges ρ_F, we must not draw the hasty conclusion that **D** is independent of the density of bound charges. Equation (15) is, in general, *not sufficient* to determine **D** (it would be sufficient if **D** were the gradient of a potential, but that is not the case). Hence, when we seek to determine **D** for some given distribution of charges and dielectrics, we must use the additional Eq. (14) or its equivalent; this brings the polarization and the distribution of bound charges into the picture. We will see more of this in the next section.

It is important to keep in mind that the polarization $\mathbf{P}(\mathbf{x}')$ is defined by an averaging process: we must take a small volume of material surrounding the point $\mathbf{x}'$, say a cube $10^{-4}\,\text{cm} \times 10^{-4}\,\text{cm} \times 10^{-4}\,\text{cm}$, evaluate the sum of all the dipoles in this volume, and multiply by $10^{12}\ \text{cm}^{-3}$ to get the dipole moment per unit volume. Although mathematically it would be desirable to proceed to the limit of an infinitesimal volume, this limit makes no sense, because if the volume element is small compared with the atomic size, then there will be no dipoles in it, regardless of whether the volume element is located outside or inside an atom (in the latter case, the volume will possibly contain some charges but no atomic dipole moment). To define **P**, we must take a volume element that is small compared with the size of our macroscopic measuring instruments, but large compared with atomic sizes, and it therefore contains many atoms. Such volume elements are sometimes called physically, rather than mathematically, infinitesimal. The polarization **P** is defined only at the macroscopic level; it has no meaning at a microscopic level.

The electric field $\mathbf{E}(\mathbf{x})$ calculated from **P** according to Eq. (12) is also a macroscopic field. We can, of course, define a microscopic electric field as the field that, say, an electron experiences when placed at a given point; the macroscopic electric field is simply an average over the microscopic field in a small, but not too small, volume element. To emphasize the macroscopic, or average, nature of the field $\mathbf{E}(\mathbf{x})$ in Eq. (12), the notation $\overline{\mathbf{E}}(\mathbf{x})$ is sometimes used. But we will not use this notation because it will be obvious from the context whether the microscopic or the macroscopic electric field is meant. Unless otherwise indicated, whenever we speak of the field in a material, we mean the macroscopic field. For example, it is the latter field that we have in mind when we say that the electric field in a conductor is zero; the microscopic field is certainly not zero in the interior of the atoms making up the conductor (in the vicinity of a nucleus, the field strength is typically 10^{17} statvolts/cm!).

The electric displacement field $\mathbf{D}(\mathbf{x})$ defined by Eq. (14) makes sense only at the macroscopic level. Furthermore, it is worth noting that Eq. (14) is not exact because quadrupole and higher multipole moments have been neglected. If the quadrupole moment of atoms is taken into account, then Eqs. (8) and (14) must be modified to read (using component notation)

$$\rho_{\mathrm{P}} = -\frac{\partial}{\partial x^k} P^k + \frac{\partial}{\partial x^k}\frac{\partial}{\partial x^l} Q^{kl} \tag{17}$$

$$D^k = E^k + 4\pi P^k - 4\pi \frac{\partial}{\partial x^l} Q^{kl} \tag{18}$$

where $Q^{kl}(\mathbf{x})$ is the quadrupole moment per unit volume. The quadrupole terms are usually much smaller than the dipole terms.

Before we can solve the equations for the electric field produced by a given distribution of free charges in the presence of some given dielectrics, we must discover the dependence of the polarization on the electric field. For many materials, the polarization is approximately proportional to the electric field:

$$\mathbf{P} = \chi \mathbf{E} \tag{19}$$

The constant χ is called the *electric susceptibility.* We emphasize that Eq. (19) is an approximation. The right side can be regarded as the first term in a Taylor-series expansion of **P** in terms of **E**. For strong electric fields, higher-order terms become important, and the relation between **P** and **E** becomes nonlinear (this is of crucial importance in nonlinear optics). Furthermore, in most crystals and other anisotropic materials, Eq. (19) fails even for weak fields, because **P** and **E** do not necessarily have the same direction; the electric susceptibility must then be regarded as a matrix multiplying **E**, but we will not deal with that case in this book.

If Eq. (19) does hold, the dielectric is said to be *linear* and *isotropic.* The electric displacement is then directly proportional to the electric field:

$$\mathbf{D} = \mathbf{E} + 4\pi\mathbf{P} = \mathbf{E} + 4\pi\chi\mathbf{E} \tag{20}$$

That is,

$$\mathbf{D} = \varepsilon \mathbf{E} \tag{21}$$

where

$$\varepsilon = 1 + 4\pi\chi \tag{22}$$

Table 4·1 Dielectric constants of some materials*

Material	ε
Helium	1.0000637
Air	1.000550
Carbon dioxide	1.000918
Water, distilled	80.36
Carbon tetrachloride	2.238
Benzene	2.284
Beeswax, yellow	2.9
Epoxy resin	3.6
Glass, Corning 0080	6.75
Nylon	3.5
Plexiglas	3.4
Polyethylene	2.3
Porcelain	6.0-8.0
Rubber, hard	2.8

* At room temperature (20°C) and 1 atm.

The constant ε is called the *dielectric constant* of the material. Both χ and ε are determined by the atomic properties of the dielectric. In a later section, we will examine how these constants depend on the atomic structure. Table 4.1 lists the dielectric constants of some materials.

4·3 The Boundary-Value Problem with Dielectrics

Equation (7) gives us the potential in terms of the distribution of free and bound charges. Unfortunately, this does not take us very far in a calculation of the potential produced by some known distribution of free charges in the presence of some known dielectrics. In most such problems, the polarization—and consequently, the distribution of bound charges—is initially unknown and can be determined only after the potential and the electric fields have been calculated [see Eq. (19)]. We can circumvent this difficulty by expressing **P** in terms of **E**, thereby eliminating **P** from our equations.

If the dielectric is linear, it is quite easy to set up a system of equations in which **P** does not appear explicitly. For this purpose, it is more convenient to start with Eq. (15) rather than with Eq. (7). In a linear dielectric, with $\mathbf{D}=\varepsilon\mathbf{E}$ we can express Eq. (15) in terms of the electric field,

$$\nabla\cdot(\varepsilon\mathbf{E}) = 4\pi\rho_{\mathrm{F}} \tag{23}$$

or in terms of the potential,

$$\nabla\cdot(\varepsilon\nabla\Phi) = -4\pi\rho_{\mathrm{F}} \tag{24}$$

If the dielectric is inhomogeneous, then ε will depend on position. For the sake of simplicity, we will assume that the dielectric consists of several homogeneous pieces, so ε varies from one piece to another but is constant within any given piece (that is, ε is a piecewise constant function with discontinuities at the dielectric interfaces). The equation we must solve is then

$$\nabla^2\Phi = -\frac{4\pi}{\varepsilon}\rho_F \tag{25}$$

We will adopt the following method of solution: first solve Eq. (25) within each homogeneous piece of dielectric; then tie these solutions together by imposing boundary conditions at the interfaces of the different pieces of dielectric.

To derive the boundary conditions that **E** and **D** must satisfy, consider the interface between two dielectrics of dielectric constants ε_1 and ε_2, respectively. The boundary condition on **E** is obtained by evaluating $\oint \mathbf{E}\cdot d\mathbf{l}$ around a small rectangular path of dimensions $l \times l'$, which is half in one dielectric and half in the other (see Fig. 4.6). The contributions from the short sides can be ignored in the limit $l' \to 0$, and therefore

$$\oint \mathbf{E}\cdot d\mathbf{l} = lE_{2\parallel} - lE_{1\parallel} \tag{26}$$

where $E_\parallel$ designates the component of the field parallel to the interface. Since the integral of **E** around any closed path is zero, it follows that

$$E_{1\parallel} = E_{2\parallel} \tag{27}$$

This means that the component of **E** parallel to the interface is continuous. We already became acquainted with a special case of Eq. (27) when we examined the boundary conditions on metallic surfaces; in that case, the continuity of the parallel component implied the vanishing of this component at the surface [see Eq. (3.10)].

Figure 4·6

Path of integration for Eq. (26).

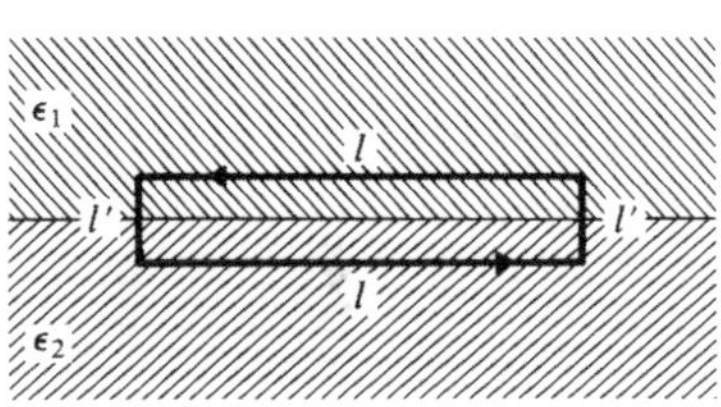

The boundary condition on **D** is obtained by evaluating $\int \mathbf{D} \cdot d\mathbf{S}$ over the surface of a small pillbox of base area A and height h, which is half in one dielectric and half in the other (see Fig. 4.7). The contribution to the surface integral due to the sides of the box can be ignored in the limit $h \to 0$, and hence

$$\int \mathbf{D} \cdot d\mathbf{S} = AD_{1\perp} - AD_{2\perp} \tag{28}$$

where $D_\perp$ designates the component of the field perpendicular to the surface. Assuming that there is no *free* surface charge density* at the interface,

$$\int \mathbf{D} \cdot d\mathbf{S} = \int \nabla \cdot \mathbf{D}\, dV = 0 \tag{29}$$

and hence it follows from Eq. (28) that

$$D_{1\perp} = D_{2\perp} \tag{30}$$

This says that the normal component of **D** is continuous at the interface.

Note that, in general, $D_{\|}$ and $E_\perp$ will *not* be continuous. In fact, Eqs. (27) and (30) imply

$$\frac{1}{\varepsilon_1} D_{1\|} = \frac{1}{\varepsilon_2} D_{2\|} \tag{31}$$

and

$$\varepsilon_1 E_{1\perp} = \varepsilon_2 E_{2\perp} \tag{32}$$

Figure 4·7

Surface of integration for Eq. (28).

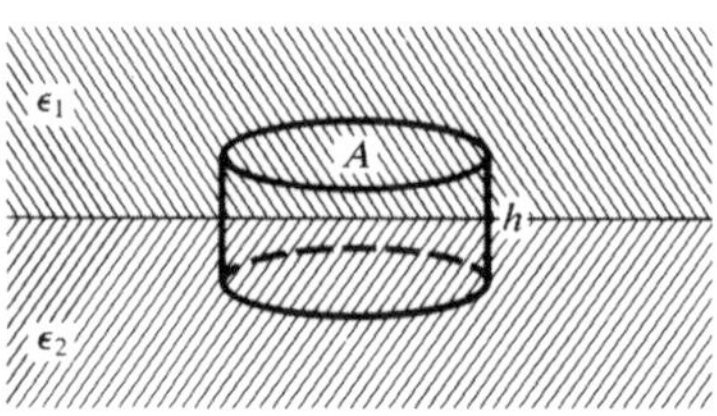

*There is necessarily a polarization surface charge density on *each* dielectric at the interface [see Eq. (9)].

■ ***Exercise 1.*** Show that if there exists a *free* surface charge density σ_F at the interface, then the boundary condition (30) is modified to

$$D_{2\perp} - D_{1\perp} = 4\pi\sigma_F \qquad (33)$$

■

EXAMPLE: *electric field in small cavities in dielectric*

Let us consider a simple application of these boundary conditions. Suppose we have a dielectric of dielectric constant ε with an electric field $\mathbf{E}_0$ in its interior. A small, thin elongated (needle-shaped) cavity oriented along $\mathbf{E}_0$ is cut from the dielectric (see Fig. 4.8). What is the electric field in this empty cavity? The answer may be obtained directly from the boundary condition (27) on $\mathbf{E}$. Since almost all regions of the cavity are near surfaces (the sides of the needle) where $\mathbf{E}$ is continuous, the electric field in the interior is simply $E_\parallel = E_{0\parallel} = E_0$. There is no $E_\perp$ inside the cavity, because $E_\perp = \varepsilon E_{0\perp} = 0$.*

Suppose that, instead of having a needle-shaped cavity, we have a small disk-shaped cavity, oriented perpendicular to $\mathbf{E}_0$ (see Fig. 4.9). What is the electric field in this case? Now, almost all regions of the cav-

Figure 4·8

Needle-shaped cavity in a dielectric. The long axis of the cavity is parallel to the electric field.

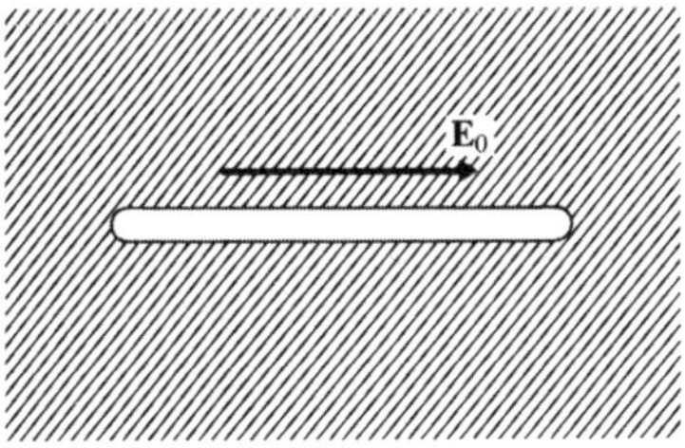

Figure 4·9

Disk-shaped cavity in a dielectric. The flat surfaces are perpendicular to the electric field.

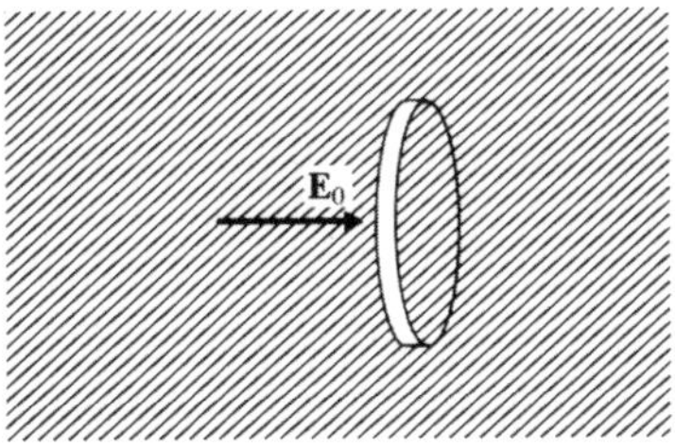

* These arguments fail near the ends of the needle; we will ignore these regions.

ity are near surfaces (the faces of the disk) where $D_\perp$ is continuous. Hence the electric field is $E_\perp = \varepsilon E_{0\perp} = \varepsilon E_0$. There is no $E_\parallel$, because $E_\parallel = E_{0\parallel} = 0$.

To summarize: in the needle-shaped cavity, the electric field is the same as in the dielectric; in the disk-shaped cavity, the electric field is ε times as large.

■ ***Exercise 2.*** What is the (bound) surface charge density on the faces of the disk-shaped cavity oriented perpendicular to $\mathbf{E}_0$? How much does this surface charge density contribute to the field in the cavity? ■

EXAMPLE: *dielectric sphere in uniform electric field*

As an example of a boundary-value problem with dielectrics, we will calculate the potential surrounding a dielectric sphere of radius R, dielectric constant ε_2 when placed in a medium of dielectric constant ε_1 (for example, a fluid) in which there is an initially uniform electric field $\mathbf{E}_0$ (see Fig. 4.10). If the z axis is taken along the field $\mathbf{E}_0$, then the potential associated with this uniform field is

$$-E_0 z = -E_0 r\cos\theta \tag{34}$$

This gives us one boundary condition on Φ:

$$\Phi \to -E_0 r\cos\theta \qquad \text{for } r \to \infty \tag{35}$$

The other boundary conditions are the continuity of $E_\parallel$ and of $D_\perp$ at $r = R$:

Figure 4·10

A dielectric sphere is placed in a medium with an initially uniform electric field.

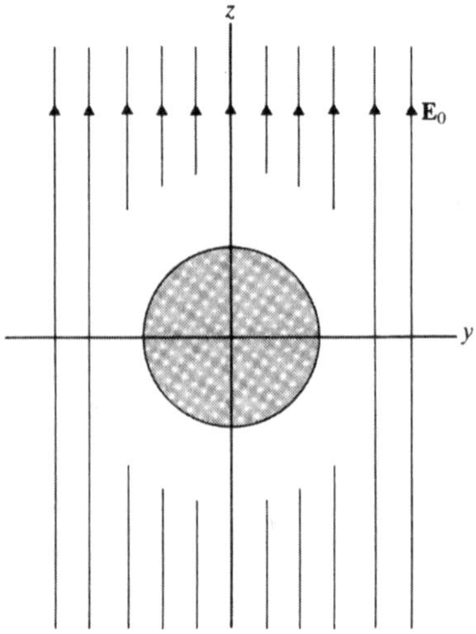

$$\left.\frac{\partial\Phi_1(r,\theta)}{\partial\theta}\right|_{r=R}=\left.\frac{\partial\Phi_2(r,\theta)}{\partial\theta}\right|_{r=R} \tag{36}$$

and

$$\varepsilon_1\left.\frac{\partial\Phi_1(r,\theta)}{\partial r}\right|_{r=R}=\varepsilon_2\left.\frac{\partial\Phi_2(r,\theta)}{\partial r}\right|_{r=R} \tag{37}$$

where Φ_1 and Φ_2 are, respectively, the potential functions for the regions $r > R$ and $r < R$. It is convenient to replace the boundary condition (36) by

$$\Phi_1(r,\theta)\big|_{r=R}=\Phi_2(r,\theta)\big|_{r=R} \tag{38}$$

This new condition implies Eq. (36), as may be seen by differentiating Eq. (38) with respect to θ. Hence Eq. (38) is certainly sufficient. To show that it is necessary, we observe that Eq. (38) expresses the continuity of Φ, which is required by **E**= $\nabla\Phi$, because according to this equation, any discontinuity in Φ would generate a singularity in **E** (the derivative of Φ would be infinite at the discontinuity).

The differential equation that Φ must satisfy in the regions $r < R$ and $r > R$ is Eq. (25) with $\rho_F = 0$ (no free charges):

$$\nabla^2\Phi=0 \tag{39}$$

This is simply Laplace's equation. The solutions of Eq. (39) in spherical coordinates were given in Section 3.4. Within the region $r > R$, the potential function for our problem must accordingly have the form

$$\Phi_1(r,\theta)=-E_0 r\cos\theta+\sum_{l=1}^{\infty}A_l\frac{P_l(\cos\theta)}{r^{l+1}} \tag{40}$$

In this equation, all terms with higher positive powers of r have been excluded because of the boundary condition (35); furthermore, the term A_0/r has been excluded because it would indicate the presence of some net electric charge on the dielectric sphere. Within the region $r < R$, the potential function has the form

$$\Phi_2(r,\theta)=\sum_{l=0}^{\infty}B_l r^l P_l(\cos\theta) \tag{41}$$

Here, negative powers of r have been excluded, because otherwise the potential function would diverge at $r = 0$.

The boundary condition (38) demands

$$E_0 R\cos\theta + \sum_{l=1}^{\infty} A_l \frac{P_l(\cos\theta)}{R^{l+1}} = \sum_{l=0}^{\infty} B_l R^l P_l(\cos\theta) \tag{42}$$

■ ***Exercise 3.*** Show that if this equation is to be valid for all values of θ, then we need

$$\begin{aligned} 0 &= B_0 & -E_0 R + \frac{A_1}{R^2} &= B_1 R \\ \frac{A_2}{R^3} &= B_2 R^2 & \frac{A_3}{R^4} &= B_3 R^3 \end{aligned} \tag{43}$$

and so on. ■

The boundary condition (37) demands

$$\varepsilon_1\left[-E_0\cos\theta - \sum_{l=1}^{\infty}(l+1)A_l \frac{P_l(\cos\theta)}{R^{l+2}}\right] = \varepsilon_2 \sum_{l=1}^{\infty} l B_l R^{l-1} P_l(\cos\theta) \tag{44}$$

■ ***Exercise 4.*** Show that Eq. (44) implies

$$\varepsilon_1\left(-E_0 - \frac{2A_1}{R^3}\right) = \varepsilon_2 B_1 \qquad \varepsilon_1\left(\frac{-3A_2}{R^4}\right) = \varepsilon_2(2B_2 R) \qquad \varepsilon_1\left(\frac{-4A_3}{R^5}\right) = \varepsilon_2(3B_3 R^2) \tag{45}$$

and so on. ■

Equations (43) and (45) are a linear system of equations for the unknown $A_1, A_2, A_3, \ldots, B_0, B_1, B_2, B_3, \ldots$. The solution of this system is

$$\begin{aligned} A_1 &= R^3 E_0 \frac{\varepsilon_2 - \varepsilon_1}{\varepsilon_2 + 2\varepsilon_1} & A_2 &= A_3 = \ldots = 0 \\ B_1 &= -E_0 \frac{3\varepsilon_1}{\varepsilon_2 + 2\varepsilon_1} & B_0 &= B_2 = B_3 = \ldots = 0 \end{aligned} \tag{46}$$

■ ***Exercise 5.*** Check these solutions. ■

The final result for the potential is then

$$\Phi_1(r,\theta) = -E_0 r\cos\theta + R^3 E_0 \frac{\varepsilon_2 - \varepsilon_1}{\varepsilon_2 + 2\varepsilon_1}\frac{\cos\theta}{r^2} \qquad \text{for } r > R \tag{47}$$

$$\Phi_2(r,\theta) = -E_0 \frac{3\varepsilon_1}{\varepsilon_2 + 2\varepsilon_1} r\cos\theta \qquad \text{for } r < R \tag{48}$$

This shows that the electric field in the interior of the sphere ($r < R$) is uniform and of magnitude

$$\frac{3\varepsilon_1}{\varepsilon_2 + 2\varepsilon_1} E_0$$

The electric field in the exterior ($r > R$) is that of a pure dipole of moment $R^3 E_0 (\varepsilon_2 - \varepsilon_1)/(\varepsilon_2 + 2\varepsilon_1)$, plus, of course, a constant field $\mathbf{E}_0$. Figures 4.11 and 4.12 show the field lines of the electric field **E** and the displacement field **D**. Note that the lines of **D** are continuous, but the lines of **E** have sources and sinks at the (bound) polarization charges on the surface of the sphere.

In terms of the fields, the polarization is

$$\mathbf{P} = \frac{\mathbf{D} - \mathbf{E}}{4\pi} = \frac{\varepsilon - 1}{4\pi}\mathbf{E} = -\frac{\varepsilon - 1}{4\pi}\nabla\Phi \tag{49}$$

In the interior region, this gives

$$\mathbf{P}_2 = \frac{\varepsilon_2 - 1}{4\pi}\frac{3\varepsilon_1}{\varepsilon_2 + 2\varepsilon_1} E_0 \hat{\mathbf{z}} \tag{50}$$

Hence the (bound) polarization surface charge density on the sphere is

$$\sigma_{\mathrm{P}_2} = \mathbf{P}_2 \cdot \hat{\mathbf{r}} = P_2 \cos\theta = \frac{\varepsilon_2 - 1}{4\pi}\frac{3\varepsilon_1}{\varepsilon_2 + 2\varepsilon_1} E_0 \cos\theta \tag{51}$$

Figure 4·11

Field lines of **E** in and around a dielectric sphere. Note that some of the field lines end on the polarization charges on the surface of the sphere.

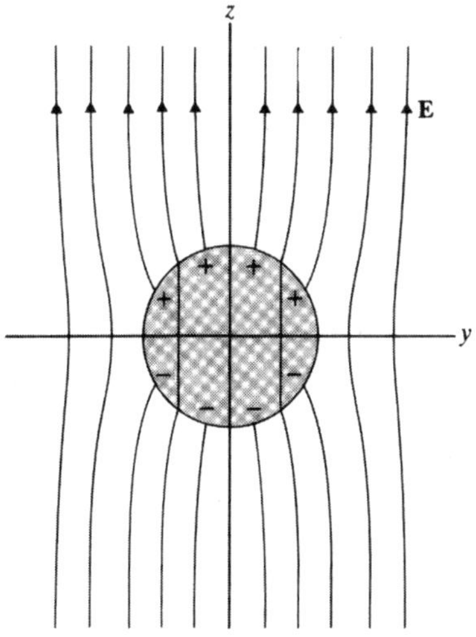

Figure 4·12

Field lines of **D** in and around the dielectric sphere.

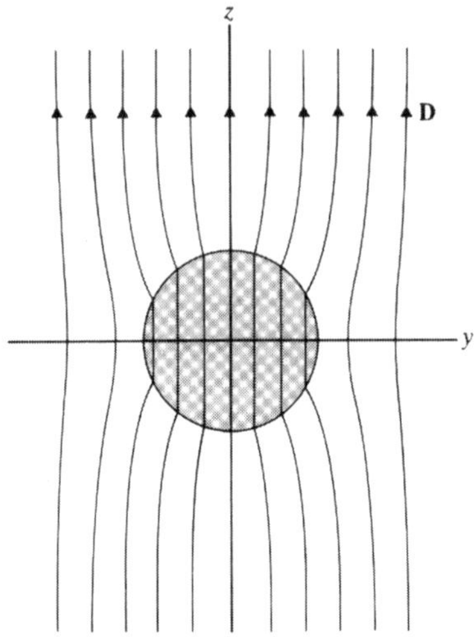

This charge density is positive on the upper hemisphere, negative on the lower.

However, (51) is *not* the only surface charge density. The surface $r = R$ is the interface of two dielectrics, and *each* dielectric has a polarization surface charge density. The surface of the outer dielectric forms a spherical cavity with a normal vector $-\hat{\mathbf{r}}$ and a charge density

$$\sigma_{P_1} = -\mathbf{P}_1 \bullet \hat{\mathbf{r}} \tag{52}$$

To evaluate σ_{P_1}, it is convenient to express $\mathbf{P}_1$ in terms of $\mathbf{D}_1$ and then use the boundary condition on the normal component of **D**, that is, $\mathbf{D}_1 \bullet \hat{\mathbf{r}} = \mathbf{D}_2 \bullet \hat{\mathbf{r}}$. This permits us to express σ_{P_1} in terms of σ_{P_2}:

$$\begin{aligned}\sigma_{P_1} &= -\frac{1-1/\varepsilon_1}{4\pi}\mathbf{D}_1 \bullet \hat{\mathbf{r}} = -\frac{1-1/\varepsilon_1}{4\pi}\mathbf{D}_2 \bullet \hat{\mathbf{r}} \\ &= -\frac{1-1/\varepsilon_1}{4\pi}\frac{4\pi}{1-1/\varepsilon_2}\mathbf{P}_2 \bullet \hat{\mathbf{r}} = -\frac{1-1/\varepsilon_1}{1-1/\varepsilon_2}\sigma_{P_2}\end{aligned} \tag{53}$$

Hence the total surface charge density on the interface is

$$\sigma_P = \sigma_{P_1} + \sigma_{P_2} = \left(1 - \frac{1-1/\varepsilon_1}{1-1/\varepsilon_2}\right)\frac{\varepsilon_2 - 1}{4\pi}\frac{3\varepsilon_1}{\varepsilon_2 + 2\varepsilon_1}E_0 \cos\theta \tag{54}$$

■ ***Exercise 6.*** Show that, in the special case of an empty spherical cavity ($\varepsilon_2 = 1$) in a dielectric, Eq. (54) gives

$$\sigma_P = -\frac{1}{4\pi}\frac{3(\varepsilon_1 - 1)}{1+2\varepsilon_1}E_0 \cos\theta$$ ■

Note that, in the above example, the dielectric has a polarization charge density on its surface but none in its volume. This absence of polarization charge density in the volume of the dielectric can be regarded as a consequence of the following general result, valid for any linear homogeneous dielectric: The polarization charge density in the volume is proportional to the free charge density. This result emerges directly from the proportionality between **P** and **D** [see Eq. (49)]:

$$\mathbf{P} = \left(\frac{1-1/\varepsilon}{4\pi}\right)\mathbf{D}$$

which, when substituted into the expression for the polarization charge density, yields

$$\rho_P = -\nabla\cdot\mathbf{P} = -\left(\frac{1-1/\varepsilon}{4\pi}\right)\nabla\cdot\mathbf{D}$$

$$= -\left(\frac{1-1/\varepsilon}{4\pi}\right)4\pi\rho_F = -\left(1-\frac{1}{\varepsilon}\right)\rho_F \qquad (55)$$

Incidentally: this equation shows that for dielectrics with $\varepsilon > 1$, the polarization charge ρ_P and the free charges ρ_F always have opposite signs.

4·4 A Microscopic View of Dielectrics

Both the electric susceptibility χ and the dielectric constant ε are determined by the atomic structure of the dielectric. Consider, for example, a gas of atoms that have no permanent dipole moments. The polarization in this gas then depends on the induced dipole moments that the electric field generates when acting on the atoms. To calculate the effect of the electric field on an atom, let us adopt a very simple classical model: the atom consists of an electron bound to a spherically symmetric core by an elastic force with spring constant κ. The induced dipole moment is then due to the displacement of the electron from its equilibrium position. If an electric field **E** is applied to the electron, the amount by which the electric force stretches the spring and displaces the electron from equilibrium is

$$\Delta\mathbf{x} = -\frac{e\mathbf{E}}{\kappa} \tag{56}$$

The corresponding dipole moment is

$$\mathbf{p} = -e\Delta\mathbf{x} = \frac{e^2\mathbf{E}}{\kappa} \tag{57}$$

If there are n atoms per unit volume, then the dipole moment per unit volume is

$$\mathbf{P} = n\mathbf{p} = \frac{ne^2\mathbf{E}}{\kappa} \tag{58}$$

The susceptibility is therefore

$$\chi = \frac{ne^2}{\kappa} \tag{59}$$

and, by Eq. (22), the dielectric constant is

$$\varepsilon = 1 + \frac{4\pi ne^2}{\kappa} \tag{60}$$

Expressed in terms of the mass of the electron and the vibration frequency, the spring constant κ is $m_e\omega^2$, so

$$\varepsilon = 1 + \frac{4\pi ne^2}{m_e\omega^2} \tag{61}$$

Despite the oversimplifications involved in our model, the result (61) is essentially correct. A more precise, quantum-mechanical calculation shows that the frequency ω must be taken to be the resonant frequency (absorption frequency) of the atom. Furthermore, each resonant frequency of the atom contributes a term of the form $4\pi ne^2/m_e\omega^2$ to the right side of Eq. (61), with a coefficient that equals the probability for an electron to make a resonant transition. But even without these refinements, Eq. (61) gives reasonable results. For example, for helium, the typical resonant frequency is $\omega = 3.71 \times 10^{16}$/s;* and with a density $n = 2.69 \times 10^{19}/\text{cm}^3$ (for normal STP conditions), Eq. (61) then gives a dielectric constant

EXAMPLE: *dielectric constant of helium*

$$\varepsilon = 1 + \frac{4\pi \times 2.69\times10^{19}/\text{cm}^3 \times (4.8\times10^{-10}\,\text{esu})^2}{9.1\times10^{-28}\,\text{g}\times(3.71\times10^{16}/\text{s})^2} = 1 + 6.2\times10^{-5}$$

*This is the principal-series limit.

This is in fair agreement with the experimental value, $\varepsilon = 1 + 6.84 \times 10^{-5}$.

If the gas consists of molecules with a permanent dipole moment, then the polarization depends (mainly) on the alignment of these molecules. If the dipole moment of one molecule is **p**, then the potential energy of this molecule in an electric field is $U(\theta) = -pE\cos\theta$, where θ is the angle between **p** and **E**. In the absence of thermal motion, all the molecules would settle into the configuration of least energy, with **p** exactly aligned along **E**. The thermal motions destroy this perfect alignment; however, there remains some partial alignment, and on the average, **p** has a positive component along **E**.

The calculation of this average component is a simple problem of statistical mechanics. According to statistical mechanics, the probability that a molecule is oriented at an angle between θ and $\theta + d\theta$ with respect to the electric field is proportional to $e^{-U(\theta)/kT}$, where T is the absolute temperature and k is Boltzmann's constant, $k = 1.38 \times 10^{-16}$ erg/K. The probability is also proportional to the amount of solid angle between θ and $\theta + d\theta$:

$$\begin{pmatrix} \text{Probability for} \\ \text{orientation between} \\ \theta \text{ and } \theta + d\theta \end{pmatrix} \propto e^{pE\cos\theta/kT} \times 2\pi \sin\theta d\theta \tag{62}$$

For notational convenience, let us choose our z axis along **E**. Then the component of **p** along **E** is $p_z = p \cos\theta$, and the average component of **p** along **E** is

$$\langle p_z \rangle = \frac{\int_0^\pi p\cos\theta\, e^{pE\cos\theta/kT}\, 2\pi \sin\theta d\theta}{\int_0^\pi e^{pE\cos\theta/kT}\, 2\pi \sin\theta d\theta} \tag{63}$$

With the change of variable $\mu = \cos\theta$, both these integrals can be readily evaluated, leading to the result

$$\langle p_z \rangle = p\left(\coth\frac{pE}{kT} - \frac{kT}{pE}\right) \tag{64}$$

This is the *Langevin equation.*

■ ***Exercise 7.*** Check this result. ■

■ ***Exercise 8.*** For molecules at room temperature (or above), pE/kT is a small quantity even if the electric field is as large as 10^7 statvolts/cm. Show that, for $pE/kT \ll 1$, the formula (64) can be approximated by

$$\langle p_z \rangle \simeq \frac{p^2 E}{3kT} \tag{65}$$

■

If there are n molecules per unit volume, then the approximation (65) gives a dipole moment per unit volume of

$$\mathbf{P} = \frac{np^2\mathbf{E}}{3kT} \tag{66}$$

Consequently, the susceptibility and the dielectric constant are

$$\chi = \frac{np^2}{3kT} \tag{67}$$

and

$$\varepsilon = 1 + \frac{4\pi np^2}{3kT} \tag{68}$$

For gases consisting of molecules with permanent dipole moments—such as HCl, HBr, H_2S—the dielectric constants calculated from Eq. (68) are somewhat less than the experimental values; this is to be expected because these molecules acquire an extra induced dipole moment when placed in an electric field.

4·5 The Local Field

The theoretical calculation of the dielectric properties of liquids and solids is much more difficult than that of gases. What complicates the picture in liquids and in solids is that neighboring atoms interact—the state of every atom depends on the state of its neighbors. Thus a knowledge of the response of an isolated atom to an electric field does not directly permit us to calculate the response of that atom when it is surrounded by its neighbors.

And there is a further complication: in liquids and in solids, the electric field that acts on an atom or a molecule is not the same as the macroscopic electric field **E.** The reason is that the macroscopic electric field evaluated at the position of some molecule includes a contribution generated by that molecule, a contribution that we must discount if we seek to evaluate the electric field that acts on the molecule. This latter electric field is called the *local* field, to distinguish it from the macroscopic field.

The following calculation, which we owe to Lorentz, gives the value of this local field in terms of the macroscopic field **E** and the polarization **P**. Imagine that we surround the given molecule with a mathematical sphere of radius r; we choose this sphere large enough so that it contains many other molecules but small enough so that the electric field

E is approximately constant over the entire sphere (in brief, the sphere is microscopically large but macroscopically small). The sphere then divides the dielectric into two portions: the portion outside the sphere and the portion inside the sphere. The calculation of Lorentz treats these two portions separately (see Fig. 4.13). The field contributed by the charges (bound and free) outside the sphere is to be evaluated by integration over these charges. The field contributed by the charges inside the sphere is to be evaluated by summation over the individual molecules inside the sphere, excluding the molecule at the very center on which these fields act.

In order to evaluate these several contributions, let us concentrate on the *difference* between the Lorentz calculation and the usual calculation. In the usual calculation, we would integrate over the entire dielectric, and the result would be **E**, the macroscopic electric field. The Lorentz calculation removes the sphere of radius R from the integration and therefore adds three corrections to **E**: (1) remove the volume of the sphere from the volume integration over the dielectric; (2) add the surface area of the sphere to the surface integration over the dielectric; (3) add the discrete summation over the molecules in the sphere. The first item can be neglected because our sphere is so small that we can regard **P** as approximately constant within this sphere; thus the contribution to the electric field from the bound charge ($\rho_P = -\nabla \cdot \mathbf{P}$) is approximately zero. The second item makes a contribution (see Fig. 4.13) of

$$\int -\frac{\sigma_P \hat{\mathbf{r}}}{R^2} dS \tag{69}$$

to the electric field at the center of the sphere. And the third item makes a contribution of

Figure 4·13

The portion of the dielectric outside the spherical surface is regarded as consisting of a continuous charge distribution. The portion inside the spherical surface is regarded as consisting of individual molecules. The molecule at the center (dashed outline) has been omitted.

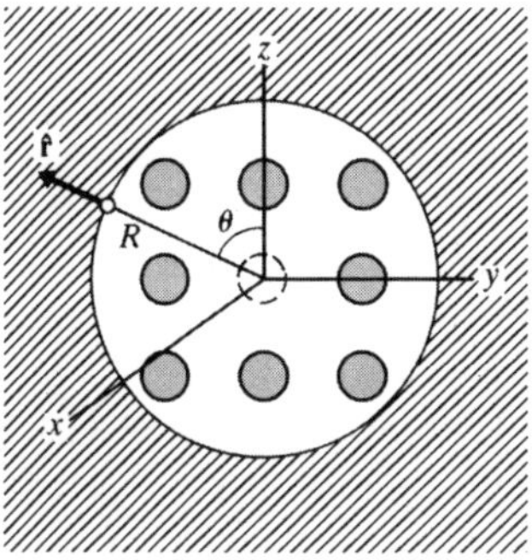

$$\sum_i \mathbf{E}_i \tag{70}$$

where $\mathbf{E}_i$ is the electric field that an individual molecule produces at the center of the sphere. Taking these corrections into account, we find that the local field is

$$\mathbf{E}_{\text{local}} = \mathbf{E} + \int -\frac{\sigma_{\text{P}}\hat{\mathbf{r}}}{R^2}dS + \sum_i \mathbf{E}_i \tag{71}$$

The surface charge density is $\sigma_{\text{P}} = -\hat{\mathbf{r}}\bullet\mathbf{P}$. With the z axis along $\mathbf{P}$, this become $\sigma_{\text{P}} = -P\cos\theta$. In the evaluation of the surface integral in Eq. (71), only the component parallel to $\mathbf{P}$ survives. This component is (see Fig. 4.13)

$$\int -\frac{\sigma_{\text{P}}\cos\theta}{R^2}dS = \int_0^{\pi} \frac{P\cos^2\theta}{R^2} 2\pi R^2 \sin\theta d\theta = \frac{4\pi}{3}P \tag{72}$$

Hence

$$\mathbf{E}_{\text{local}} = \mathbf{E} + \frac{4\pi}{3}\mathbf{P} + \sum_i \mathbf{E}_i \tag{73}$$

The evaluation of the discrete sum in Eq. (73) requires some knowledge of the positions of the individual molecules. If we are dealing with a liquid—with random positions of the molecules—then the contribution from these molecules will average to zero. Consequently,

$$\mathbf{E}_{\text{local}} = \mathbf{E} + \frac{4\pi}{3}\mathbf{P} \tag{74}$$

If we are dealing with a solid with a cubic lattice, then the symmetry of the placement of the molecules leads to cancellations among the individual contributions of the molecules, and the final result will again be zero. For solids with other lattices, the cancellations are not total, and the discrete sum in Eq. (73) will not be zero, although it will usually be small and can often be neglected.

Equation (74) is, of course, also valid for a gas. Under normal conditions, the polarization in a gas is small compared with the electric field; hence $\mathbf{E}_{\text{local}} \simeq \mathbf{E}$. We already made use of this approximation in Section 4.4, where we assumed that the local field that acts on a molecule is simply the macroscopic field [see Eq. (56)]. However, if the gas is under very high pressure, then its density will be comparable with that of a liquid and the difference between $\mathbf{E}_{\text{local}}$ and $\mathbf{E}$ cannot be neglected. Measurements on gases under very high pressure permit a direct experimental test of Eq. (74). The dipole moment that a molecule of the gas acquires when acted upon by the electric field $\mathbf{E}_{\text{local}}$ is proportional to this electric field [see Eqs. (57) and (65)]:

$$\mathbf{p} = \alpha \mathbf{E}_{\text{local}} \tag{75}$$

The constant of proportionality α is called the *molecular polarizability.* The value of this constant depends on the mechanism responsible for the generation of the dipole moment; for instance, $\alpha = e^2/\kappa$ for the induced dipole moment of Eq. (57), and $\alpha = p^2/3kT$ for the permanent dipole moment of Eq. (65). In terms of the constant α, the dipole moment per unit volume is

$$\mathbf{P} = n\mathbf{p} = n\alpha \mathbf{E}_{\text{local}} \tag{76}$$

With the expression (74) for $\mathbf{E}_{\text{local}}$, this becomes

$$\mathbf{P} = n\alpha\left(\mathbf{E} + \frac{4\pi}{3}\mathbf{P}\right) \tag{77}$$

But according to Eqs. (19) and (22),

$$\mathbf{P} = \chi \mathbf{E} = \frac{1}{4\pi}(\varepsilon - 1)\mathbf{E} \tag{78}$$

so

$$\frac{1}{4\pi}(\varepsilon - 1)\mathbf{E} = n\alpha[\mathbf{E} + \frac{1}{3}(\varepsilon - 1)\mathbf{E}] \tag{79}$$

Here, we can cancel **E** and rearrange the equation to read

$$\frac{1}{4\pi}\frac{\varepsilon - 1}{\varepsilon + 2} = \frac{n\alpha}{3} \tag{80}$$

This is called the *Clausius-Mossotti equation.* In essence, it states that $(\varepsilon - 1)/(\varepsilon + 2)$ is directly proportional to the density of the gas. This proportionality has been confirmed in experiments with helium and argon over a wide range of densities, extending from normal density to a thousand times normal density. In practice, Eq. (80) is often used to determine the molecular polarizability from the measured value of the dielectric constant.

Problems

1. Starting with Eqs. (8) and (9), prove that the net polarization charge in the volume and on the surface of a dielectric body is always zero.

2. Derive Eq. (17) from the expression (3.116), which includes the quadrupole contribution to the potential. Derive the following equation for the polarization surface charge density:

$$\sigma_P = \hat{n}^k P^k - \hat{n}^k \frac{\partial}{\partial x^l} Q^{kl}$$

Also show that Eq. (17) leads to Eq. (18).

3. Two long, concentric conducting cylinders of sheet metal have radii R and $3R$, respectively. The space between them is filled with a gas of dielectric constant ε. If the potential difference between the cylinders is V_0, find the **E** and **D** fields in the region between the cylinders.

4. At the interface between two dielectrics (of constants ε_1 and ε_2, respectively), the electric field in the first dielectric has magnitude E_1 and makes an angle α_1 with the normal to the interface (see Fig. 4.14). What is the angle α_2 that the electric field makes in the second dielectric, and what is the magnitude of this electric field?

5. A thin sheet of paper of dielectric constant $\varepsilon = 2.5$ is placed in a uniform electric field of magnitude 20 statvolts/cm. Use the boundary conditions to obtain an approximate value for the electric field inside the paper.
   (a) Assume that the paper is parallel to the electric field.
   (b) Assume that the paper is perpendicular to the electric field.

6. (a) What are the electric field and the electrostatic potential of a point charge q immersed in a medium of dielectric constant ε?
   (b) When a positive arsenic ion is implanted in a piece of silicon, the electric field surrounding the ion can be treated as the electric field of a point charge $q = +e$ immersed in a dielectric medium with the dielectric constant of silicon, $\varepsilon = 11.7$. According to the Bohr model, what is the radius of the smallest orbit of an electron orbiting around the ion? What is the binding energy of the electron? (*Note:* For a more accurate treatment of this problem, it would be necessary to take into account the modified effective mass of the electron; but ignore this correction.)

Figure 4·14

Interface between two dielectrics.

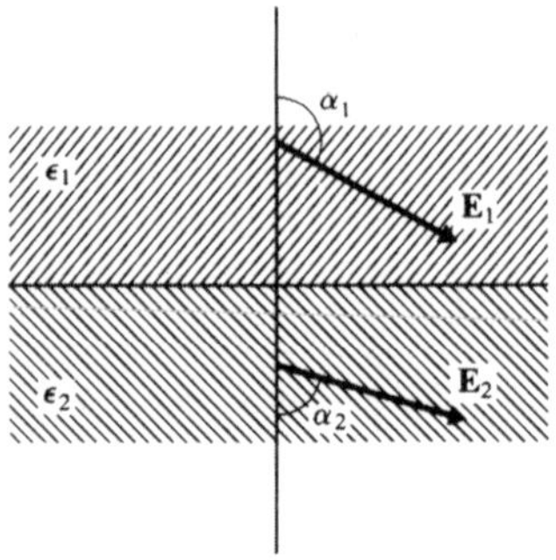

7. A dielectric sphere of radius R with a dielectric constant ε has a free charge Q uniformly distributed over its volume. The sphere is surrounded by empty space. Find **E** and **D** inside and outside the sphere. Find the polarization volume charge density and surface charge density. Evaluate the net polarization volume charge and surface charge.

8. A point charge q is placed at a distance d above the surface of a large lake of dielectric liquid of constant ε. Use the method of images to find the electrostatic potential in the region above the surface and in the region below the surface. [*Hint:* To deal with the region above the surface, try an image charge q' at a distance d below the surface; to deal with the region below the surface, discard the charge q and try an image charge q'' at a distance d above the surface. Then use the boundary conditions (27) and (30) at the dielectric surface to determine the appropriate values of q' and q''.]

9. At the plane interface between two dielectrics with $\varepsilon_1 = 2.0$ and $\varepsilon_2 = 6.0$, the electric field in the first dielectric has a magnitude of 80 statvolts/cm and a direction making an angle of 40ϒ with the normal to the plane (see Fig. 4.14).
   (a) What are the magnitude and the direction of the electric field and the electric displacement in the second dielectric?
   (b) Suppose that the interface carries a free surface charge density of +5.0 esu/cm^2. In this case, what are the magnitude and the direction of the electric field and the electric displacement in the second dielectric?

10. A large flat plate of glass of thickness d and dielectric constant ε is placed in a uniform electric field $\mathbf{E}_0$. The plate is inclined at an angle θ with respect to the electric field.
   (a) Show that if $\theta \ll 1$, then the electric field inside the plate makes an angle $\theta' = \theta / \varepsilon$ with the surface of the plate and has a magnitude of

$$E = E_0\left[1 - \left(1 - \frac{1}{\varepsilon^2}\right)\frac{\theta^2}{2}\right]$$

   (b) Sketch the lines of the electric displacement **D** inside and outside the plate. Sketch the lines of the electric field **E** inside and outside the plate. Assume that $\varepsilon > 1$.

11. A conducting sphere of radius R is half submerged in a lake of oil of dielectric constant ε. The sphere carries a free charge Q.
   (a) Find **E** and **D** in the oil and in the vacuum above the oil.
   (b) Find the free surface charge density on the upper and the lower hemispheres of the conductor, and calculate the net electric force on the sphere. Does this force pull the sphere into the oil or out of it?

12. A parallel-plate capacitor is filled with a dielectric material of a dielectric constant that increases linearly from one plate to the other, that is,

$$\varepsilon = a + bz$$

where z is the distance from one plate to a point in the dielectric. The potential difference between the plates is V_0, and the distance between the plates is d. Find the electrostatic potential, the electric field, and the polarization charge density as a function of z. Find the polarization surface charge density at the surfaces of the dielectric adjacent to the plates. Ignore edge effects.

13. A long copper wire of diameter 4 mm is stretched horizontally in the Earth's atmosphere. The wire is encased in a rubber insulator of thickness 1 mm and dielectric constant $\varepsilon = 3$. The original atmospheric electric field has a strength of 100 volts/m, or 0.33 statvolt/m, and points downward. Find the electric field at the top of the wire and at the bottom of the wire; assume that the wire is not grounded. Qualitatively, how will your answers change if the ends of the wire are grounded?

14. Within a large volume of transformer oil of dielectric constant $\varepsilon = 3$, there is a spherical air bubble. What is the maximum permissible strength of the original electric field in the transformer oil if the air in the bubble is not to suffer electric breakdown? The critical strength for electrical breakdown in air is 3×10^6 volts/m, or 1×10^4 statvolts/m. Qualitatively, how does your answer change if the bubble is elongated in the direction of the electric field? If the bubble is flattened?

15. A long dielectric cylinder of radius R_1 is surrounded by a concentric dielectric shell of inner radius R_1 and outer radius R_2. The dielectric constants of the cylinder and of the shell are ε_1 and ε_2, respectively. The cylinder and shell are placed in an initially uniform electric field $\mathbf{E}_0$ at right angles to the axis of the cylinder. Find the electrostatic potential in the cylinder and the shell. Find the maximum values of the electric field in the cylinder and the shell.

16. A very large piece of dielectric material of dielectric constant ε has an initially uniform electric field $\mathbf{E}_0$ in its interior. A long cylindrical hole of radius R is drilled through this dielectric at right angles to the electric field. Find the electrostatic potential and the electric field in the hole and in the dielectric. Find the polarization surface charge density on the walls of the hole.

17. A very long dielectric cylinder of dielectric constant ε and radius R is placed in an initially uniform electric field $\mathbf{E}_0$ at right angles to the axis of the cylinder. The cylinder has an amount of free charge λ per unit length uniformly distributed over its surface. Find the potential and the electric field inside and outside the cylinder.

18. A very large, horizontal conducting plate has a long ridge of dielectric of constant ε attached to it. The ridge has a semicircular cross section with radius R; the flat side of the ridge is in contact with the plate. The plate is in an (initially) uniform vertical electric field $\mathbf{E}_0$. Find the electrostatic potential in the dielectric and in the empty space above the plate.

19. A dielectric sphere of radius R consists of an upper half of dielectric constant ε_1 and a lower half of dielectric constant ε_2. The sphere is placed in an initially uniform electric field $\mathbf{E}_0$ in the vertical direction. Find the electrostatic potential inside and outside the sphere.

20. (a) Throughout space, the electrostatic potential initially has the value

$$\Phi(r,\theta) = Ar^2\left(\frac{3\cos^2\theta - 1}{2}\right)$$

Find the corresponding electric field.

(b) Now, suppose that a uniform dielectric sphere of radius R and dielectric constant ε is placed in this electric field, at the origin. Find the new potential for both $r > R$ and $r < R$.

21. A conducting sphere of radius R at potential zero is surrounded by a concentric spherical shell of dielectric material of inner radius R, outer radius $2R$, and dielectric constant $\varepsilon = 7/5$. Suppose that the sphere and the dielectric shell are placed in an initially uniform electric field $\mathbf{E}_0$ in the z direction.

(a) Show that the potential is given by

$$\Phi(r,\theta) = -E_0 r\cos\theta + \frac{2E_0R^3}{r^2}\cos\theta \qquad \text{for } r \ge 2R$$

$$\Phi(r,\theta) = -\frac{6}{7}E_0\left(r - \frac{R^3}{r^2}\right)\cos\theta \qquad \text{for } R \le r \le 2R$$

(b) Find the (free) surface charge density induced on the surface of the conducting sphere (as a function of θ).

(c) Find the (bound) charge density induced on the surface of the dielectric at $r = R$.

22. A spherical shell of dielectric material has inner radius R_1, outer radius R_2, and dielectric constant ε. The shell is placed in an initially uniform electric field $\mathbf{E}_0$. Find the electrostatic potential in the three regions $r < R_1$, $R_1 < r < R_2$, and $r > R_2$. What is the electric field in the region $r < R_1$?

23. The electric dipole moment of the water molecule is 1.8×10^{-18} esu $\cdot$ cm. Calculate the dielectric constant of pure steam at atmospheric pressure and a temperature of 140°C. Compare with the measured value of 1.00785.

CHAPTER

Electric Energy 5

We know from Chapter 2 that the potential energy of two point charges q_1 and q_2 separated by a distance r is q_1q_2/r. This is a *mutual* potential energy that belongs to both of these point charges jointly. This potential energy equals the work that we must do on the charges to bring them to a distance r from one another, starting at infinite distance. In the present a chapter, we will develop a general expression for the potential energy of an arbitrary system of many point charges, and we will discover that this potential energy can be expressed as an integral of an *energy density* which is proportional to $\mathbf{E}^2$.

In the interpretation of the electric potential energy, we are faced with a question: Is the potential energy located at the point charges or is it located in the electric fields? This question has no answer in the context of electrostatics, because the outcome of experiments in electrostatics is independent of where the energy is located. In principle, we could decide this question with a gravitational experiment—all forms of energy exert gravitational attractions, and if the electrostatic energy is distributed over the electric field, then the electric field should exert gravitational attractions. Unfortunately, these gravitational effects are so small that a direct measurement is impossible. To decide where the electric energy is located, we must go beyond electrostatics. We must look at electromagnetic waves, such as light or radio waves. These waves carry energy from one place to another; thus there can be no doubt that the electric fields of these waves contain energy. For consistency, we will then suppose that the electric fields of a static distribution of charges also contain energy. This means that fields are a form of matter. If we reckon solids, liquids, gases, and plasmas as the first four states of matter, then fields must be reckoned as the fifth state of matter.

5·1 The Energy in the Electric Field

Suppose that we have a collection of point charges, with charges q_1, q_2, q_3, . . . , q_n at positions $\mathbf{x}_1$, $\mathbf{x}_2$, $\mathbf{x}_3$, . . . , $\mathbf{x}_n$. What is the potential energy of this charge distribution? That is, how much work must you do to bring all these charges to their positions, starting with all of them infinitely separated? To calculate this amount of work, let us imagine that you bring the charges to their positions in succession: first, you bring q_1 to $\mathbf{x}_1$

(from infinity), then you bring q_2 to $\mathbf{x}_2$, and so on. When you bring charge q_j to join the $j-1$ preceding charges, the potential produced by the latter is

$$\Phi(\mathbf{x}) = \sum_{i=1}^{j-1} \frac{q_i}{|\mathbf{x} - \mathbf{x}_i|} \tag{1}$$

The increase in potential energy that occurs when you bring in charge q_j to $\mathbf{x}_j$ is therefore

$$q_j \Phi(\mathbf{x}_j) = \sum_{i=1}^{j-1} \frac{q_j q_i}{|\mathbf{x}_j - \mathbf{x}_i|} \tag{2}$$

The total increase in potential energy that occurs when you bring all the charges to their positions is then

$$U = \sum_{j=1}^{n} \sum_{i=1}^{j-1} \frac{q_j q_i}{|\mathbf{x}_j - \mathbf{x}_i|} \tag{3}$$

This expression is simply the sum of all the mutual potential energies $q_i q_j / |\mathbf{x}_j - \mathbf{x}_i|$ for all possible pairs of charges q_i and q_j.

Equation (3) can be written as

$$U = \frac{1}{2} \sum_{j=1}^{n} \sum_{\substack{i=1 \\ i \pi j}}^{n} \frac{q_j q_i}{|\mathbf{x}_j - \mathbf{x}_i|} \tag{4}$$

where the factor of ½ is needed because each term of the summation in Eq. (3) occurs twice in the summation in Eq. (4). [Note that in Eq. (4), the term with $i = j$ is to be excluded from the summation.] Alternatively, Eq. (4) can be written as

$$U = \frac{1}{2} \sum_{j=1}^{n} q_j \Phi_{j,\text{other}}(\mathbf{x}_j) \tag{5}$$

where $\Phi_{j,\text{other}}(\mathbf{x}_j)$ is the electrostatic potential produced at $\mathbf{x}_j$ by all the charges other than q_j,, that is,

$$\Phi_{j,\text{other}}(\mathbf{x}_j) = \sum_{\substack{i=1 \\ i \pi j}}^{n} \frac{q_i}{|\mathbf{x}_j - \mathbf{x}_i|}$$

For a continuous charge distribution with a charge density $\rho(\mathbf{x})$, Eqs. (4) and (5) become

$$U = \frac{1}{2}\iint \frac{\rho(\mathbf{x})\rho(\mathbf{x}')}{|\mathbf{x}-\mathbf{x}'|}dVdV' \tag{6}$$

and

$$U = \frac{1}{2}\int \rho(\mathbf{x})\Phi(\mathbf{x})dV \tag{7}$$

Note that in Eqs. (4) and (5) we have explicitly indicated the omission of the term with $j = i$, but in Eqs. (6) and (7) we have made no corresponding indication. There is no need to omit the points $\mathbf{x} = \mathbf{x}'$ from these integrations, because the integrand in, say, Eq. (6) is well behaved at such points, provided that the charge density is everywhere finite. To see this, take a given point $\mathbf{x}$ as the origin of coordinates for the integration over $\mathbf{x}'$. Then $\mathbf{x} - \mathbf{x}' = r'$, and the integral over $\mathbf{x}'$ gives

$$\int \frac{\rho(0)\rho(\mathbf{x}')}{r'} r'^2 dr' d\Omega$$

The integrand vanishes in the limit $r' \to 0$, and this point makes no contribution to the integral.*

We can transform the integral in Eq. (7) by means of $\rho(\mathbf{x}) = \nabla \cdot \mathbf{E}/4\pi$:

$$U = \frac{1}{2}\int \frac{1}{4\pi}\nabla\cdot\mathbf{E}\Phi\, dV = \frac{1}{8\pi}\int [\nabla\cdot(\mathbf{E}\Phi) - \mathbf{E}\cdot\nabla\Phi]dV$$

$$= \frac{1}{8\pi}\int_S (\mathbf{E}\Phi)\cdot d\mathbf{S} + \frac{1}{8\pi}\int \mathbf{E}\cdot\mathbf{E}\, dV \tag{8}$$

Here we used Gauss' theorem to convert a volume integral into a surface integral over a surface S surrounding the volume where there are fields. In general, the fields will not be confined to a finite volume, and hence S must be taken as a surface at infinity. Since, at large distances from an arbitrary charge distribution, the electric field falls off as $1/r^2$ (or faster) and the potential as $1/r$, the integrand of the surface integral approaches zero as $1/r^3$ in the limit $r \to \infty$. The area of the surface goes to infinity as r^2 in this limit, and hence the surface integral tends to zero. We are then left with

$$U = \frac{1}{8\pi}\int E^2 dV \tag{9}$$

*For a point charge, $\rho(\mathbf{x})$ is not everywhere finite (the charge density is infinite at the position of the point charge) and the integral in (6) does pick up an extra (infinite) contribution from $\mathbf{x} = \mathbf{x}'$. This extra contribution is the *self-energy* of the point charge, which was omitted in Eq. (4). We will see how to deal with the problem of the self-energy of a charge in an example at the end of this section.

This is an alternative expression for the total electrostatic potential energy of a charge distribution. The field appearing in Eq. (9) is, of course, the microscopic field, and the integral gives the total potential energy stored in all the electric fields of all the charges, whether free or bound.

The integrand appearing in Eq. (9) may be regarded as the *energy density* of the electric field,

$$T^{00} = \frac{1}{8\pi} E^2 \tag{10}$$

We use the notation T^{00} for this energy density because it turns out that the energy density is one component of the energy-momentum tensor (see Chapter 8). Equation (9) suggests that the energy is concentrated in the electric fields, whereas Eq. (6) suggests that it is concentrated in the charges. Because these two equations are mathematically equivalent, electrostatics does not tell us in which of these places the energy is located. As stated in the introduction to this chapter, we can resolve the ambiguity in the location of the energy only by going beyond electrostatics. We will see in Chapter 12 that the fields of electromagnetic waves contain energy, and for consistency, we will therefore adopt the point of view that the fields of a static distribution of charges also contain energy, with an energy density given by Eq. (10).

EXAMPLE: *electrostatic energy of uniformly charged sphere*

To illustrate the concept that the energy is stored in the fields, let us consider the example of a uniformly charged sphere. If the sphere has a radius R and a charge Q uniformly distributed over its volume, the electric field outside and inside the sphere is [see Eq. (2.33)]

$$\mathbf{E} = \frac{Q}{r^2}\hat{\mathbf{r}} \qquad \text{for } r \geq R \tag{11}$$

and

$$\mathbf{E} = \frac{Qr}{R^3}\hat{\mathbf{r}} \qquad \text{for } r \leq R \tag{12}$$

The energy in this electric field is

$$U = \frac{1}{8\pi}\int E^2 dV = \frac{1}{8\pi}\int_0^R\int_0^{2\pi}\int_0^{\pi}\frac{Q^2r^2}{R^6}r^2\sin\theta\, d\theta d\phi dr$$

$$+\frac{1}{8\pi}\int_R^{\infty}\int_0^{2\pi}\int_0^{\pi}\frac{Q^2}{r^4}r^2\sin\theta d\theta d\phi dr$$

$$= \frac{Q^2}{10R} + \frac{Q^2}{2R} = \frac{3}{5}\frac{Q^2}{R} \tag{13}$$

EXAMPLE: *electrostatic energy of uranium nucleus*

This formula has an important practical application: it can be used to calculate the electrostatic energy of a nucleus. For instance, the uranium nucleus has a charge $92e$ and a radius 7.4×10^{-13} cm. If we regard the uranium nucleus as a uniformly charged sphere, Eq. (13) gives

$$U = \frac{3}{5}\times\frac{(92\times4.8\times10^{-10}\ \text{esu})^2}{7.4\times10^{-13}\ \text{cm}} = 1.6\times10^{-3}\ \text{erg} = 990\ \text{MeV}$$

The energy released in the fission of uranium comes from this large amount of electrostatic energy stored in the nucleus. The electrostatic energy of the uranium nucleus not only supplies the kinetic energy of the fission products but also compensates for the increase of nuclear "strong" energy that occurs during the fission process (the "strong" forces oppose the fission, and the "strong" energy of the fission products is larger than that of the uranium nucleus). Thus the energy released in nuclear reactors and nuclear bombs is really electrostatic energy.

EXAMPLE: *electrostatic energy of two point charges*

As another example, let us consider the case of two point charges q and q'. What is the energy calculated from T^{00} as given above? If one charge is at the origin and the other at $\mathbf{x}'$ (see Fig. 5.1), then the net electric field is

$$\mathbf{E}(\mathbf{x}) = \frac{q\mathbf{x}}{|\mathbf{x}|^3} + \frac{q'(\mathbf{x}-\mathbf{x}')}{|\mathbf{x}-\mathbf{x}'|^3} \tag{14}$$

and thc energy density is

$$T^{00} = \frac{1}{8\pi}E^2(\mathbf{x}) = \frac{1}{8\pi}\left[\frac{q^2}{|\mathbf{x}|^4} + \frac{q'^2}{|\mathbf{x}-\mathbf{x}'|^4} + 2qq'\frac{\mathbf{x}\cdot(\mathbf{x}-\mathbf{x}')}{|\mathbf{x}|^3\,|\mathbf{x}-\mathbf{x}'|^3}\right] \tag{15}$$

The integral of this energy density gives

Figure 5·1

A point charge q at the origin, and a point charge q' at $\mathbf{x}'$.

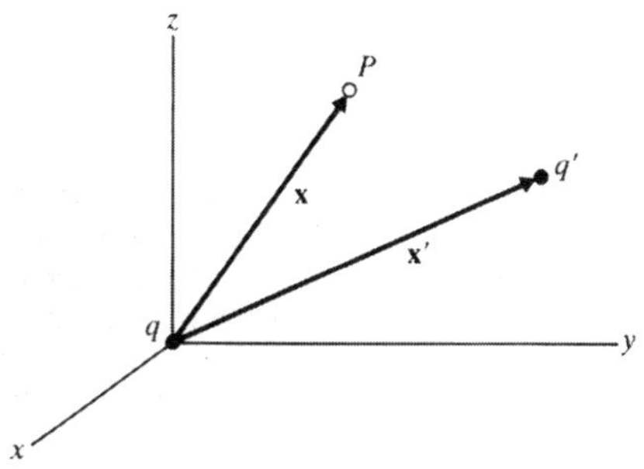

$$U=\frac{1}{8\pi}\int\frac{q^2}{|\mathbf{x}|^4}dV+\frac{1}{8\pi}\int\frac{q'^2}{|\mathbf{x}-\mathbf{x}'|^4}dV+\frac{2qq'}{8\pi}\int\frac{\mathbf{x}\cdot(\mathbf{x}-\mathbf{x}')}{|\mathbf{x}|^3|\mathbf{x}-\mathbf{x}'|^3}dV \qquad (16)$$

Let us examine the first term; this term is the electrostatic energy associated with a single, isolated point charge q, called the *self-energy* of the charge. We can easily calculate the integral in polar coordinates:

$$\frac{1}{8\pi}\int\frac{q^2}{r^4}dV=\frac{1}{8\pi}\iiint\frac{q^2}{r^4}r^2\sin\theta\, d\theta\, d\phi\, dr=\frac{1}{2}\int_0^\infty\frac{q^2}{r^2}dr=-\frac{q^2}{2}\left[\frac{1}{r}\right]_0^\infty \qquad (17)$$

This integral diverges at the lower limit: the self-energy of a single, isolated point charge is infinite.* A more realistic calculation of the self-energy of a point charge should be carried out by means of quantum electrodynamics rather than by classical electrodynamics. But quantum electrodynamics also leads to an infinite self-energy for a point charge. One way to avoid this catastrophe is to pretend that point charges do not exist. This is probably wishful thinking—to the best of our knowledge, electrons are point charges; nobody has ever found any hard (or soft) core in an electron. How an electron keeps its self-energy finite remains a mystery. One suggestion that has been made to solve this mystery is that perhaps the positive, infinite electrostatic energy is canceled by negative, infinite gravitational energy. Some calculations lend support to this idea, but our understanding of the problem of self-energies is inadequate for any firm conclusions.

Fortunately, in practice, we do not have to worry much about this problem. The self-energy (17) is a *constant*, and additive constants in the energy can be ignored. What we really care about are those terms in the energy that change as the position of the charges changes. The first integral in (16) is the self-energy of the charge q, the second integral is the self-energy of the charge q', and only the third integral depends on the position of the charges. We therefore can write

$$U=\text{constant}+\frac{qq'}{4\pi}\int\frac{\mathbf{x}\cdot(\mathbf{x}-\mathbf{x}')}{|\mathbf{x}|^3|\mathbf{x}-\mathbf{x}'|^3}dV \qquad (18)$$

and concentrate our attention on this last integral. The integrand is the product of two gradients:

---

*Note that the numerical factors appearing in Eqs. (17) and (13) are somewhat different. Thus a point charge is not quite the same thing as a uniformly charged ball of vanishing radius. The difference between the numerical factors in Eqs. (17) and (13) is due to the fact that the ball of charge has some energy in its interior, whereas the point charge has energy only in its exterior. If we want to regard the point charge as a limiting case of a finite charge distribution of vanishing radius, we must start with a thin, hollow, spherical shell.

$$\frac{qq'}{4\pi}\int\frac{\mathbf{x}\cdot(\mathbf{x}-\mathbf{x}')}{|\mathbf{x}|^3|\mathbf{x}-\mathbf{x}'|^3}dV=\frac{qq'}{4\pi}\int\nabla\frac{1}{|\mathbf{x}|}\cdot\nabla\frac{1}{|\mathbf{x}-\mathbf{x}'|}dV$$

An integration by parts yields

$$-\frac{qq'}{4\pi}\int\left(\nabla^2\frac{1}{|\mathbf{x}|}\right)\frac{1}{|\mathbf{x}-\mathbf{x}'|}dV$$

Since $\nabla^2(1/|\mathbf{x}|) = 0$ except at $\mathbf{x} = 0$, the integral receives a contribution only from the immediate neighborhood of $\mathbf{x} = 0$.* We can therefore shrink the volume of integration to a small sphere centered on the origin. And we can replace the factor $1/|\mathbf{x} - \mathbf{x}'|$ by $1/|\mathbf{x}'|$ and treat it as a constant. Thus the integral becomes

$$-\frac{qq'}{4\pi}\frac{1}{|\mathbf{x}'|}\int\nabla\cdot\nabla\frac{1}{|\mathbf{x}|}dV \tag{19}$$

The integrand is now a divergence, and by Gauss' theorem, the volume integral can be transformed into a surface integral over a small spherical surface centered on the origin. This surface integral can be evaluated in spherical coordinates, leading to the final result

$$U=\text{constant}+\frac{qq'}{|\mathbf{x}'|} \tag{20}$$

■ ***Exercise 1.*** Evaluate the integral in (19). ■

The result (20) shows that the usual potential energy between two charges is nothing but field energy. The repulsive force between, say, like charges is due to an increase of the field energy as the charges are brought together. Of course, the argument is somewhat circular: our derivation of the expression for the field energy had, as its starting point, the usual expression for the potential energy of point charges. We could avoid this circularity of the argument if we could derive the expression (10) for the field energy without appealing to the potential of point charges. This can be done but requires a general discussion of energy in fields. We cannot give such a discussion without trespassing into the realm of field theory.

---

*Readers familiar with delta functions will be aware that $-(1/4\pi)\nabla^2(1/|\mathbf{x}|) = \delta(\mathbf{x})$, and they can proceed directly to the final result (20).

5·2 The Energy in Dielectrics

The expression (10), calculated with the exact, microscopic field **E**, gives the correct value for the electric field energy in all cases. However, sometimes the field energy is not what is relevant. Suppose you have some dielectric bodies and you place a distribution of free charge on or near these bodies. Then you will usually not care about the field energy, but only about the amount of energy that you must supply in order to put the charges where you want them. This work that you must do to assemble the charge distribution includes not only field energy but also work done in deforming atoms, work done to change the kinetic energy of electrons, work done to heat the dielectric,* and so forth.

To find this total amount of work, we can proceed as follows: Suppose that the charge distribution of free charge that we want is characterized by a density $\rho_F(\mathbf{x})$. To assemble this charge distribution, we begin with zero charge and gradually increase the charge to ρ_F by successively bringing in small amounts of charge $\delta\rho_F$. At some stage of this charging process, the potential will be $\Phi(\mathbf{x})$, and consequently, the work that we must do to bring in a small amount of charge $\delta\rho_F$ will be

$$\delta W = \int \delta\rho_F(\mathbf{x})\Phi(\mathbf{x})dV \tag{21}$$

Here we can use $\delta\rho_F(\mathbf{x}) = \nabla\bullet(\delta\mathbf{D})/4\pi$ to transform the integral by a calculation analogous to that done in Eq. (8). The result is

$$\delta W = \frac{1}{4\pi}\int \mathbf{E}\bullet\delta\mathbf{D}dV \tag{22}$$

The final step is to integrate all these small amounts of work, which means to integrate over **D** from the initial value zero (at zero free charge) to its final value (at a free charge ρ_F). This integration requires a knowledge of the dependence of **E** on **D**. If all the dielectrics are *linear* with given dielectric constants, then **E**(**x**) is proportional to **D**(**x**) and

$$\mathbf{E}\bullet\delta\mathbf{D} = \tfrac{1}{2}\delta(\mathbf{E}\bullet\mathbf{D}) \tag{23}$$

Thus

$$\delta W = \frac{1}{8\pi}\int \delta(\mathbf{E}\bullet\mathbf{D})dV$$

and

$$W = \frac{1}{8\pi}\int \mathbf{E}\bullet\mathbf{D}dV \tag{24}$$

---

*The charging of a thermally isolated dielectric usually leads to a small increase in temperature.

In this final equation, the fields **E** and **D** are, of course, the fields corresponding to the final charge distribution ρ_F.

The work done in assembling the charge distribution can be regarded as the potential energy stored in the fields and in the dielectric material. According to Eq. (24), this potential energy is

$$U = \frac{1}{8\pi}\int \mathbf{E}\cdot\mathbf{D}dV \tag{25}$$

so the energy density is

$$u = \frac{1}{8\pi}\mathbf{E}\cdot\mathbf{D} \tag{26}$$

It is important to keep in mind that Eqs. (25) and (26) are valid only for linear dielectrics. For any other kind of dielectric, we must return to Eq. (22) and evaluate the work by integration over **D**. In such other dielectrics, the relationship between **D** and **E** is nonlinear [Eq. (23) fails], and furthermore, this relationship often depends on the past history of the dielectric (hysteresis). A dependence on the past history indicates that the forces in the dielectric are nonconservative—the dielectric dissipates some energy by friction. It is then not possible to interpret the work done in assembling the charge distribution as potential energy.

The potential energy for a system with linear dielectrics must, of course, be evaluated with the macroscopic electric field. To emphasize this, we will use the notation $\overline{\mathbf{E}}$ for this field, and we will write Eq. (25) as

$$U = \frac{1}{8\pi}\int \overline{\mathbf{E}}\cdot\mathbf{D}dV \tag{27}$$

or, if the dielectric is not only linear but also isotropic,

$$U = \frac{1}{8\pi}\int \varepsilon\overline{\mathbf{E}}^2 dV \tag{28}$$

Let us compare this energy with the microscopic field energy of Eq. (9). To carry out such a comparison, we use the fact that ordinary materials are systems of particles held together by the inverse-square Coulomb force.* For any system of particles interacting only by an inverse-square force, the average kinetic and potential energies are related by the *virial theorem*: The average kinetic energy of the particles equals $-\frac{1}{2}$ times the potential energy,†

---

*This is not quite exact. There are also magnetic forces in materials. But we will ignore these forces.

†This theorem is also valid in quantum mechanics.

$$T_{\text{av}} = -\tfrac{1}{2}V_{\text{av}} \tag{29}$$

■ ***Exercise 2.*** Check that Eq. (29) holds for a charged particle in a circular orbit in a Coulomb field. ■

The total energy of the system of particles is $T_{\text{av}} + V_{\text{av}} = \tfrac{1}{2}V_{\text{av}}$, and hence the change in the total energy is

$$\Delta W = \tfrac{1}{2}\Delta V_{\text{av}} \tag{30}$$

However, we know that the change in the total energy equals

$$\Delta W = \frac{1}{8\pi}\Delta\int \bar{\mathbf{E}}\cdot\mathbf{D}\,dV$$

and that the change in the potential energy equals, according to the discussion following Eq. (20), the change in the field energy:

$$\Delta V_{\text{av}} = \frac{1}{8\pi}\Delta\int \mathbf{E}^2\,dV$$

Substituting these expressions for ΔW and ΔV_{av} into Eq. (30), we find

$$\frac{1}{8\pi}\Delta\int \bar{\mathbf{E}}\cdot\mathbf{D}\,dV = \frac{1}{2}\frac{1}{8\pi}\Delta\int \mathbf{E}^2\,dV \tag{31}$$

This shows that as the dielectric is charged, the change in the microscopic field energy is twice as large as the macroscopic work done during the charging process. The extra energy delivered to the electric fields comes from a decrease in the kinetic energy of the electrons of the atoms of the material (the electrons make transitions to quantum orbits of lower kinetic energy but higher potential energy, that is, transitions to larger orbits). Note that if we had carelessly ignored the distinction between the macroscopic and microscopic fields **E** that must be used in Eqs. (9) and (28), respectively, then we would have reached the (wrong) conclusion that the change in the microscopic field energy is smaller than the work done by the factor ε.

EXAMPLE: *parallel plates with dielectric, at constant charge*

The expression (24) for the work done to assemble a charge distribution can be quite useful in calculations of the macroscopic forces acting on charged bodies. For example, consider two large parallel metallic plates of area A, separated by a small distance l; the region between the plates contains a dielectric material (say, a gas) of dielectric constant ε (see Fig. 5.2). If the charges on the plates are $\pm Q$, then the surface charge densities are approximately uniform, $\sigma = \pm Q/A$. Neglecting the fringing field at the edges, we find that the displacement field is

Figure 5·2

Two large, parallel metallic plates separated by a dielectric material. The plates carry charges $\pm Q$.

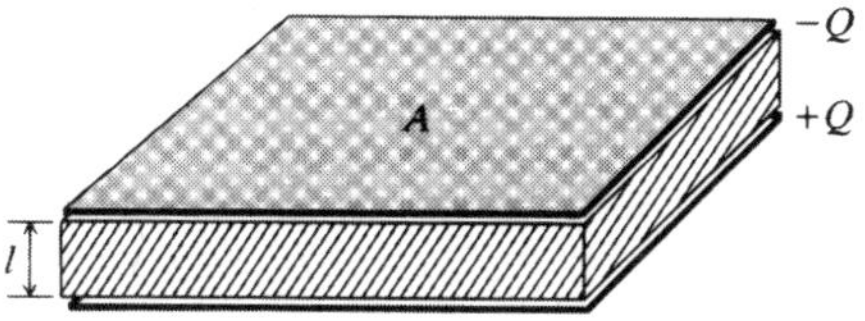

$$D = 4\pi\sigma = \frac{4\pi Q}{A} \tag{32}$$

■ ***Exercise 3.*** Derive Eq. (32). ■

The work required to charge the system is

$$W = \frac{1}{8\pi\varepsilon}\int \mathbf{D}^2 dV = \frac{1}{8\pi\varepsilon}\left(\frac{4\pi Q}{A}\right)^2 Al = \frac{2\pi Q^2 l}{\varepsilon A} \tag{33}$$

This means that for a constant value of Q, the energy increases with l. The parallel plates therefore attract each other with a force

$$F = -\frac{\partial W}{\partial l} = -\frac{2\pi Q^2}{\varepsilon A} \tag{34}$$

EXAMPLE: *parallel plates with dielectric, connected to battery*

On the other hand, if the plates are connected to a battery that maintains the potential difference between them at a constant value $\Delta\Phi$ (see Fig. 5.3), then $E = \Delta\Phi/l$, and the work required to charge the plates is

$$W = \frac{\varepsilon}{8\pi}\int \mathbf{E}^2 dV = \frac{\varepsilon}{8\pi}\left(\frac{\Delta\Phi}{l}\right)^2 Al = \frac{\varepsilon A(\Delta\Phi)^2}{8\pi l} \tag{35}$$

This shows that for a constant value of $\Delta\Phi$, the energy *decreases* with increasing l, which would suggest that the force between the plates is *repulsive*. However, in the calculation of the force, we must also take into account the work that has to be done against the battery as the plates are moved. Since the charge on the plates is

$$Q = \frac{AD}{4\pi} = \frac{A}{4\pi}\frac{\varepsilon\Delta\Phi}{l} \tag{36}$$

Figure 5·3

Two large, parallel metallic plates separated by a dielectric material. The plates are connected to a battery supplying a constant potential difference.

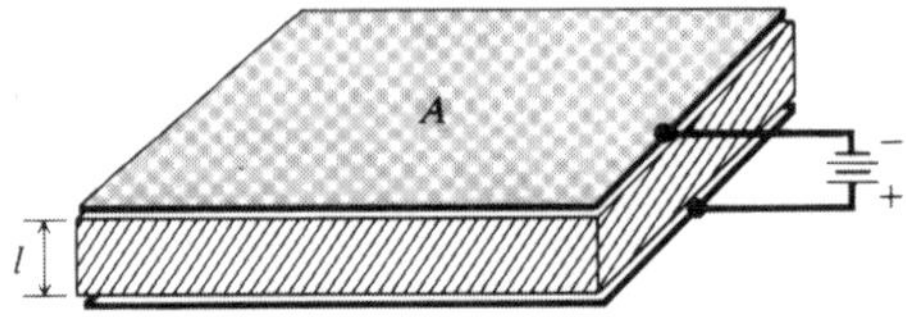

a change in l leads to a change in charge

$$dQ = -\frac{A}{4\pi}\frac{\varepsilon\Delta\Phi}{l^2}dl \tag{37}$$

and the work required to transfer this amount of charge to the battery is

$$-\Delta\Phi dQ = \frac{\varepsilon A}{4\pi}\left(\frac{\Delta\Phi}{l}\right)^2 dl \tag{38}$$

The total work required to increase the separation of the plates by dl is then

$$dW - \Delta\Phi dQ = \left[-\frac{\varepsilon A}{8\pi}\left(\frac{\Delta\Phi}{l}\right)^2 + \frac{\varepsilon A}{4\pi}\left(\frac{\Delta\Phi}{l}\right)^2\right]dl = \frac{\varepsilon A}{8\pi}\left(\frac{\Delta\Phi}{l}\right)^2 dl \tag{39}$$

Consequently, the parallel plates attract each other with a force

$$F = -\frac{(dW - \Delta\Phi dQ)}{dl} = -\frac{\varepsilon A}{8\pi}\left(\frac{\Delta\Phi}{l}\right)^2 \tag{40}$$

Note that this final expression for the force has the same magnitude as, but the opposite sign to, what we would have obtained from Eq. (35) had we naively used the formula $F = -\partial W / \partial l$. This change of sign comes about because in Eq. (39), the term $-\Delta\Phi dQ$ is twice as large as the term dW and has the opposite sign. The expression for the force given in Eq. (40) can be readily shown to be identical to Eq. (34).

■ ***Exercise 4.*** Show that if the potential difference $\Delta\Phi$ is expressed in terms of the corresponding charge Q on the plates, then the right sides of Eqs. (34) and (40) become identical. ■

5·3 Energy of a Charged Body in an External Field

By an external field is meant the field of a given, fixed charge distribution. Suppose that we have such an external field and that we place an extra charged body in it. We will regard this charged body as a test body; that is, we will pretend that the presence of the extra body does not disturb the given external field. Under these conditions, what is the force on the charged body?

If the charged body is a point charge q, then the force on it is obviously

$$\mathbf{F} = q\mathbf{E} \tag{41}$$

where $\mathbf{E}$ is the given external field. But what if the charged body has some finite size and has some complicated charge distribution? In the latter case, we can calculate the force from the energy. Suppose that the charge distribution of the body is $\rho(\mathbf{x})$ and that the potential associated with the external field is $\Phi(\mathbf{x})$. Then the potential energy of the charged body in the external field is

$$U = \int \rho(\mathbf{x})\Phi(\mathbf{x})dV$$

where the integral extends over the volume of the body. Let us designate the position of the center of mass of the body by $\mathbf{x}_0$.* Then a Taylor-series expansion about $\mathbf{x}_0$ gives

$$\begin{aligned}\Phi(\mathbf{x}) &= \Phi(\mathbf{x}_0) + (\mathbf{x} - \mathbf{x}_0)\cdot\nabla\Phi(\mathbf{x}_0) + \dots \\ &= \Phi(\mathbf{x}_0) - (\mathbf{x} - \mathbf{x}_0)\cdot\mathbf{E}(\mathbf{x}_0) + \dots\end{aligned} \tag{42}$$

and

$$\begin{aligned}U &= \int \rho(\mathbf{x})\Phi(\mathbf{x}_0)dV - \int \rho(\mathbf{x})(\mathbf{x} - \mathbf{x}_0)\cdot\mathbf{E}(\mathbf{x}_0)dV + \dots \\ &= Q\Phi(\mathbf{x}_0) - \mathbf{p}\cdot\mathbf{E}(\mathbf{x}_0) + \dots\end{aligned} \tag{43}$$

where $\mathbf{p}$ is the dipole moment of the charged body calculated with respect to the point $\mathbf{x}_0$,

$$\mathbf{p} = \int \rho(\mathbf{x})(\mathbf{x} - \mathbf{x}_0)dV \tag{44}$$

The force exerted on the body is then

---

*Whether $\mathbf{x}_0$ designates the center of mass of some other fixed point on the body is irrelevant for what follows, but the center of mass is usually a convenient choice.

$$F^k = -\frac{\partial U}{\partial x_0^k} = -Q\frac{\partial \Phi}{\partial x_0^k} + \mathbf{p}\cdot\frac{\partial \mathbf{E}}{\partial x_0^k} + \ldots$$

$$= QE^k(\mathbf{x}_0) + \mathbf{p}\cdot\frac{\partial \mathbf{E}}{\partial x_0^k} + \ldots \tag{45}$$

The second term on the right side of Eq. (45) shows that even a neutral body ($Q = 0$) can experience a force, provided it has a charge distribution with a dipole moment and provided the electric field is not uniform. This second term has the same mathematical form as the force on a simple dipole consisting of a pair of opposite charges [see Eq. (2.64)].

We can also calculate the torque that the field exerts on the charged body. If θ is the angle between **p** and **E**, then Eq. (43) yields

$$U = Q\Phi - |\mathbf{p}||\mathbf{E}|\cos\theta \tag{46}$$

The torque about an axis perpendicular to both **p** and **E** is then

$$\tau = -\frac{\partial U}{\partial \theta} = |\mathbf{p}||\mathbf{E}|\sin\theta \tag{47}$$

that is,

$$\tau = \mathbf{p}\times\mathbf{E} \tag{48}$$

Thus the charged body in an electric field experiences a torque that tends to align its dipole moment with the field. This result is a generalization of the result we derived for the torque on a simple dipole [see Eq. (2.65)].

5·4 Capacitors

EXAMPLE: *spherical capacitor*

A capacitor (or condenser) is an arrangement of conductors and dielectrics used for the storage of electric charge and electric energy. The simplest possible capacitor consists of a single piece of conducting material. The *capacitance* of this system is defined as the ratio of charge to potential,

$$C = \frac{Q}{\Phi} \tag{49}$$

where the potential is taken relative to infinity. In the special case of a metallic sphere of radius R,

$$\Phi = \frac{Q}{R} \tag{50}$$

and

$$C = R \tag{51}$$

Note that the capacitance depends only on the radius of the sphere. In general, the capacitance of a system depends on the shape, dimension, and dielectric constant of the capacitor but is independent of the amount of charge stored and the potential.

As is obvious form Eq. (51), the unit of capacitance in the CGS electrostatic system is the centimeter. Thus for a sphere, the capacitance is numerically equal to the radius in centimeters.*

The most common type of capacitor consists of two separate conductors, one carrying charge $+Q$ and the other charge $-Q$. The parallel-plate capacitor shown in Fig. 5.4 is of this type; other possible arrangements are concentric spherical shells, concentric cylindrical shells, and so on. The capacitance of such a system is defined as

$$C = \frac{Q}{\Delta\Phi} \tag{52}$$

EXAMPLE: *parallel-plate capacitor*

where $\Delta\Phi$ is the potential difference between the conductors. For instance, for a parallel-plate capacitor with large plates and no dielectric, the electric field is $E = 4\pi\sigma = 4\pi Q/A$ and the potential difference is $\Delta\Phi = El = 4\pi Ql/A$, so

$$C = \frac{Q}{\Delta\Phi} = \frac{A}{4\pi l} \tag{53}$$

If we fill the capacitor with dielectric, the displacement field will be $D = 4\pi Q/A$ [see Eq. (32)], and the electric field will be smaller, $E =$

Figure 5·4

A parallel-plate capacitor.

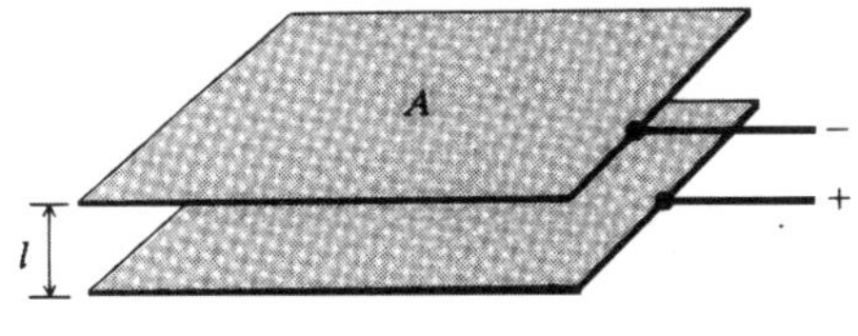

*The SI unit of capacitance is the farad, 1 farad $\Leftrightarrow 9\times10^9$ cm. We write this relationship between the SI and CGS units of capacitance as an equivalence (rather than as an equality) because, in view of the difference between the definitions of the units of charge, these units of capacitance do not have the same dimensions.

$D/\varepsilon = 4\pi Q/\varepsilon A$. This gives a smaller potential difference and a larger capacitance:

$$C = \frac{\varepsilon A}{4\pi l} \tag{54}$$

As a general rule, whenever a capacitor is completely filled with a homogeneous dielectric, its capacitance increases by a factor of ε.

■ ***Exercise 5.*** Consider a capacitor consisting of a long, cylindrical conducting shell of radius R_1 surrounded by a concentric cylindrical shell of radius R_2. The space between the shells is filled with dielectric. Find the capacitance for a segment of length l of this capacitor, and show that the capacitance per unit length is

$$\frac{C}{l} = \frac{\varepsilon}{2\ln(R_2/R_1)} \tag{55}$$

■

The electric energy stored in a two-conductor capacitor is, according to Eq. (7),

$$U = \tfrac{1}{2}Q_1\Phi_1 + \tfrac{1}{2}Q_2\Phi_2 = \tfrac{1}{2}Q\Phi_1 + \tfrac{1}{2}(-Q)\Phi_2 = \tfrac{1}{2}Q\Delta\Phi \tag{56}$$

We can also express this energy as

$$U = \tfrac{1}{2}C(\Delta\Phi)^2 \tag{57}$$

or as

$$U = \frac{1}{2}\frac{Q^2}{C} \tag{58}$$

■ ***Exercise 6.*** Show that, for a parallel-plate capacitor filled with dielectric, the expression (58) for the stored electric energy coincides with Eq. (33). ■

For a general capacitor system consisting of n separate conductors, we define *coefficients of capacitance* as follows: Suppose that the conductors carry charges $Q_1, Q_2, \ldots, Q_n$ and are at potentials $\Phi_1, \Phi_2, \ldots, \Phi_n$. Then the charges and the potentials must be linearly related. The reason is that **D** is linearly proportional to ρ_F [Eq. (4.15)], **E** is linearly proportional **D** [Eq. (4.21)], and $\mathbf{\Phi}$ is linearly proportional to **E**.* We can express this linear relationship as

$$Q_k = \sum_{l=1}^{n} c_{kl}\Phi_l \tag{59}$$

---

*Note that this argument fails if the dielectric constant depends on the field strength (nonlinear material). It also fails if any conductor is of infinite extent (infinite flat plate or infinite long wire), because then both sides of Eq. (59) become infinite and meaningless. For these cases the formalism of coefficients of capacitance is not applicable.

The constants c_{kl} are called *coefficients of capacitance*; these constants are entirely determined by the shape, size, and dielectric constants of the capacitor arrangement. The coefficients satisfy the symmetry relation

$$c_{kl} = c_{lk} \tag{60}$$

Equation (60) is a corollary of the following theorem, known as *Green's reciprocation theorem*:

• ***Theorem:*** Suppose that the potentials on the conductors are $\Phi_1, \Phi_2, \ldots, \Phi_n$ when the charges are $Q_1, Q_2, \ldots, Q_n$. Suppose that the potentials change to $\Phi'_1, \Phi'_2, \ldots, \Phi'_n$ when the charges are changed to $Q'_1, Q'_2, \ldots, Q'_n$. Then

$$\sum_{k=1}^{n} Q_k \Phi'_k = \sum_{k=1}^{n} Q'_k \Phi_k \tag{61}$$

We will give a simple proof of this theorem, with the assumption that there are no dielectrics.* For this purpose, we subdivide each of the charges Q_k on the conductors into N small charge elements of magnitudes $q_k = Q_k/N$. If the positions of these charge elements are $\mathbf{x}_1, \mathbf{x}_2, \ldots, \mathbf{x}_n$, then the potential at the position of a charge element q_k is

$$\Phi_k = \sum_{l=1}^{nN} \frac{q_l}{|\mathbf{x}_l - \mathbf{x}_k|} \tag{62}$$

In this expression, the charge element q_k apparently makes a divergent contribution to the potential at its own position. But in the limiting case $N \to \infty$, with infinitesimal charge elements and a continuous charge distribution, the contribution that an infinitesimal charge element makes to the potential at its own position tends to zero and can be ignored [see the argument given in connection with Eq. (7)]. Likewise, after we change the charges to Q'_k, the potential at the position of a charge element $q'_k = Q'_k/N$ is

$$\Phi'_k = \sum_{l=1}^{nN} \frac{q'_l}{|\mathbf{x}_l - \mathbf{x}_k|} \tag{63}$$

Hence

$$\sum_{k=1}^{nN} q_k \Phi'_k = \sum_{k=1}^{nN} \sum_{l=1}^{nN} \frac{q_k q'_l}{|\mathbf{x}_l - \mathbf{x}_k|} = \sum_{l=1}^{nN} \sum_{k=1}^{nN} \frac{q'_l q_k}{|\mathbf{x}_l - \mathbf{x}_k|} = \sum_{l=1}^{nN} q'_l \Phi_l \tag{64}$$

---

*A general proof, in the presence of dielectrics, is given by J. Jeans, *The Mathematical Theory of Electricity and Magnetism*, Section 185.

This establishes the theorem in terms of the small charge elements q_k and q'_k. As the final step of the proof, we note that for all the charge elements on a given conductor, the value of the potential is the same; hence we can perform the summations over the charge elements on each of the conductors in Eq. (64). This leaves us with an expression containing only the net charges Q_k on each of the conductors, that is, it leaves us with Eq. (61).

To derive Eq. (60), we write Eq. (61) in terms of the coefficients of capacitance:

$$\sum_{k=1}^{n}\sum_{l=1}^{n} c_{kl}\Phi_l\Phi'_k = \sum_{k=1}^{n}\sum_{l=1}^{n} c_{kl}\Phi'_l\Phi_k \qquad \textbf{(65)}$$

Taking the partial derivatives $\partial^2/\partial\Phi_r\,\partial\Phi'_s$ on both sides of this equation gives us the result

$$c_{sr} = c_{rs} \qquad \textbf{(66)}$$

as was to be shown.

EXAMPLE: *symmetry of coefficients of capacitance and potential*

This symmetry relation has the following interesting consequence. Suppose that, say, conductor 1 initially is grounded and carries a charge Q, while conductor 2 is at potential Φ. If now conductor 2 is grounded while conductor 1 is raised to the potential Φ (all other conductors being left at their original potentials), then the charge on conductor 2 will become Q.

Equation (59) can also be solved for the potentials Φ_l in terms of the charges Q_k:

$$\Phi_l = \sum_{k=1}^{n} p_{lk}Q_k \qquad \textbf{(67)}$$

The quantities p_{lk} are called *coefficients of potential.* If we regard Φ_l and Q_l as vectors in an n-dimensional space, then c_{kl} and p_{kl} are $n \times n$ matrices in this space, and these matrices are the inverses of each other. Since c_{kl} is symmetric, p_{kl} must also be symmetric,

$$p_{kl} = p_{lk} \qquad \textbf{(68)}$$

This relation has the following consequence: Suppose that, say, conductor 1 initially has charge zero and potential Φ, while conductor 2 has charge Q. Then if the charge Q is shifted to conductor 1 (the charges on all other conductors being held fixed), the potential on conductor 2 will become Φ. A nice illustration of this reciprocity is the determination of the potential of an uncharged spherical conductor placed in the electric field of a point charge Q, at a center-to-center distance r. For this purpose, we treat the point charge as a small conducting sphere (conductor 1) that initially has charge zero, while the spherical conductor (conductor 2) initially has charge Q. The potential at the point charge will then be Q/r. From this we immediately conclude that if the charge Q is now shifted to the point charge, the potential of the spherical conductor will become Q/r, a result that would have been rather more difficult to establish by solution of the corresponding boundary value problem.

EXAMPLE: *coefficients for concentric spherical shells*

As an example of a capacitor system described by coefficients of capacitance and potential, consider a conducting shell of radius R_1 placed concentrically within a larger conducting shell of radius R_2 in otherwise empty space. Suppose the shells carry charges Q_1 and Q_2, respectively (in contrast to an ordinary spherical capacitor, we do *not* assume that $Q_1 = -Q_2$). The potential of the outer shell is then

$$\Phi_2 = \frac{Q_1}{R_2} + \frac{Q_2}{R_2} \tag{69}$$

and that of the inner shell is

$$\Phi_1 = \Phi_2 + Q_1\left(\frac{1}{R_1} - \frac{1}{R_2}\right) = \frac{Q_1}{R_1} + \frac{Q_2}{R_2} \tag{70}$$

From these equations, we see that the coefficients of potential are

$$\begin{pmatrix} p_{11} & p_{12} \\ p_{21} & p_{22} \end{pmatrix} = \begin{pmatrix} 1/R_1 & 1/R_2 \\ 1/R_2 & 1/R_2 \end{pmatrix} \tag{71}$$

and by inverting this matrix [or by solving Eqs. (69) and (70) for the charges in terms of the potentials], we find that the coefficients of capacitance are

$$\begin{pmatrix} c_{11} & c_{12} \\ c_{21} & c_{22} \end{pmatrix} = \frac{1}{1/R_1R_2 - 1/R_2^2}\begin{pmatrix} 1/R_2 & -1/R_2 \\ -1/R_2 & 1/R_1 \end{pmatrix} \tag{72}$$

As expected, these matrices are symmetric.

The electric energy stored in the capacitor system is

$$U = \frac{1}{2}\sum_{k=1}^{n} Q_k \Phi_k$$

By means of Eqs. (59) and (67), we can rewrite this energy in the two alternative forms

$$U = \frac{1}{2}\sum_{k=1}^{n}\sum_{l=1}^{n} c_{kl}\Phi_k\Phi_l \tag{73}$$

and

$$U = \frac{1}{2}\sum_{k=1}^{n}\sum_{l=1}^{n} p_{kl}Q_k Q_l \tag{74}$$

These two equations are analogous to Eqs. (57) and (58) for the two-capacitor system.

Problems

1. According to the alpha-particle model of the nucleus, some nuclei supposedly consist of a regular geometric arrangement of alpha particles. For example, the nucleus of ^{16}O supposedly consists of four alpha particles arranged at the vertices of a tetrahedron. If the edge of the tetrahedron is 3.0×10^{-13} cm, what is the electrostatic potential energy of the configuration? Treat the alpha particles as small spheres, and ignore their self-energy.

2. Two identical thin rods of plastic of length L carry equal charges Q distributed uniformly along their lengths. The rods are aligned, with a distance d between their nearest ends. Calculate their mutual potential energy; ignore the self-energy of each rod. Calculate the repulsive force from the potential energy.

3. The electrostatic energy of an ionic crystal, such as NaCl, can be calculated by summing the potential energies of all the ions in the lattice. Since the ions are spherical, they can be treated as point charges. Calculate the electrostatic energy of a small "crystal" of NaCl consisting of four ions of Na^+ and four ions of Cl^- arranged alternately on the vertices of a cube of edge 0.281 nm.

4. A charge distribution with spherical symmetry has a charge density $\rho(r) = kr^{-n}$, where k and n are positive constants and $0 \le r \le R$. Calculate the electrostatic potential energy of this charge distribution. For what values of n is the energy finite?

5. The scattering of high-energy electrons by protons has been used to explore the charge distribution within protons. From such experiments, it is known that the charge density of the proton is approximately an exponential function of distance, $\rho = (e/8\pi a^3)e^{-r/a}$, with $a = 0.23\times10^{-13}$ cm. For this charge distribution, find the electric field as a function of radius. At what radius is the electric field maximum? Find the net electrostatic self-energy, and compare it with the rest-mass energy of the proton.

6. The nucleus of ^{238}U is a sphere of radius 7.4×10^{-13} cm and charge $92e$. Suppose that this nucleus fissions into two equal spherical fragments, each of which is a nucleus of ^{119}Pd of radius 5.9×10^{-13} cm.

(a) What is the electrostatic energy of the ^{238}U nucleus before fission? Treat the nucleus as a sphere of uniform charge density.

(b) What is the electrostatic energy immediately after fission? Assume that at this instant, the two fragments are separated by a distance of 1.5 times their diameter, and remember to take into account their self-energy.

(c) What is the electrostatic energy a long time after fission, when the fragments have separated by a large distance?

7. According to a crude model, the neutron consists of an inner core of positive charge surrounded by an outer shell of negative charge. Suppose that the positive charge has a magnitude $+e$ and is uniformly distributed over a sphere of radius 0.5×10^{-13} cm; suppose that the negative charge has a magnitude $-e$ and is uniformly distributed over a concentric shell of inner radius 0.5×10^{-13} cm and outer radius 1.0×10^{-13} cm. Find the electric field and the energy density (as a function of radius). By integration of this energy density, find the electrostatic energy in a neutron; express your answer in MeV.

8. A sphere of radius R with a charge Q uniformly distributed over its volume is placed at the origin. A point charge q is placed on the z axis, at a distance $d > R$ from the center of a sphere.

(a) What is the net electric field produced jointly by the sphere and the point charge?

(b) By integrating the energy density $E^2/(8\pi)$, find the electrostatic potential energy stored in the field. Show that this energy can be expressed in the form U = constant + Qq/d.

9. A conducting sphere of radius R is placed in an initially uniform electric field $\mathbf{E}_0$. What is the energy density in the space surrounding the sphere? What is the difference between the electrostatic energy of the uniform electric field (with the sphere absent) and the electrostatic energy of the field in the presence of the sphere?

10. A point charge q is placed in a uniform electric field $\mathbf{E} = E_0\hat{\mathbf{z}}$. Assuming that the point charge is on the z axis, find the energy density of the net electric field consisting of the sum of thc uniform electric field and the field of the point charge. Show that the electrostatic energy obtained by integration of the energy density is of the form U = constant $- q E_0 z$. [*Hint:* Imitate the method of calculation employed to obtain Eq. (20).]

11. A flat sheet of paper of dielectric constant $\varepsilon = 2.5$ is placed in a uniform electric field of 20 statvolts/cm (compare Problem 4.5). The paper has a surface area of 22 cm $\times$ 28 cm and a thickness of 0.006 cm. Initially, the paper is parallel to the electric field. Calculate (approximately) the work that you must do to turn the paper so it is perpendicular to the electric field.

12. A dielectric sphere of radius R and dielectric constant ε_2 is placed in a medium of dielectric constant ε_1 with an (initially) uniform electric field $\mathbf{E}_0$. Calculate the difference between the electrostatic energy of the uniform field (with the dielectric sphere absent) and the electrostatic energy of the field in the presence of the dielectric sphere.

13. A parallel-plate capacitor consists of two conducting plates of area A separated by a distance d. A slab of dielectric of thickness $d' < d$ and area $A' < A$ has been inserted between the plates, parallel to them. What is the capacitance? Ignore edge effects.

14. A single-conductor capacitor consists of a sphere of metal of radius R that is half-submerged in a large lake of oil of dielectric constant ε. What is the capacitance of this sphere?

15. A cylindrical capacitor consists of two long concentric cylindrical tubes of sheet metal of radii R_1 and R_2, respectively. The space between the tubes is half filled with dielectric of constant ε, so the dielectric occupies a $180°$ sector of the cylinder. Calculate the capacitance per unit length.

16. A spherical capacitor consists of two concentric spherical shells of radii R and $3R$, respectively. The space between the shells is filled with a dielectric of constant ε.
(a) Find the capacitance of this capacitor.
(b) Suppose that the potential difference between the shells is Φ. Find the electric field between the shells, and find the electrostatic energy by integration of the energy density; compare your answer with $U = \frac{1}{2}C(\Delta\Phi)^2$.

17. A spherical capacitor consists of two concentric shells of sheet metal of radii a and b. The space between these shells is filled with two kinds of dielectric: the upper hemisphere is filled with a dielectric of constant ε_1, and the lower hemisphere is filled with a dielectric of constant ε_2. What is the capacitance of this contraption?

18. The plates of a parallel-plate capacitor have an area of 10 cm $\times$ 10 cm each, and they are separated by a distance of 1 mm. A sheet of dielectric of the same area, thickness 1 mm, and $\varepsilon = 3.5$ is free to slide between the plates. Suppose the dielectric sheet has been slid partially out of the capacitor at one side, and suppose that a potential difference of 10 statvolts has been applied to the plates. Show that the plates will tend to pull the sheet into the capacitor, and calculate the magnitude of the pull. Ignore edge effects.

19. A capacitor consists of two long, vertical concentric, conducting cylindrical shells of height h and radii a and b, respectively. The inner conductor carries a net charge Q, and the outer conductor carries a net charge $-Q$. A solid piece of dielectric, with a dielectric constant ε, fits snugly between these cylinders and can be moved up and down. Suppose the dielectric is displaced downward so it fills only a lower length l of the capacitor, leaving the upper length $h - l$ empty.
(a) Find the electric field **E** in the filled region and in the empty region of the capacitor. Find the free surface charge density on the part of the inner cylinder surrounded by dielectric and on the part not surrounded. Neglect edge effects at the ends of the capacitor.
(b) Find the energy needed to charge to capacitor.
(c) Find the force with which the conducting cylinders pull upward on the dielectric.

20. A capacitor system consists of three concentric, spherical shells of radii R, $2R$, and $3R$, respectively. The space around the shells and the space between the shells is empty.
(a) Calculate the coefficients of potential and the coefficients of capacitance for this three-conductor system.
(b) What is the energy stored in this capacitor system if the charges on the shells are (from inner to outer) Q, Q, and $-2Q$?

21. Consider a two-conductor capacitor consisting of two thin concentric spherical shells of radii R_1 and R_2 with $R_2 > R_1$.
(a) Supposes that shell 1 is at potential 0 (relative to infinity) and shell 2 is at potential V_0. What are the charges on the shells?
(b) Suppose that shell 1 is at potential V_0 and shell 2 is at potential 0. What are the charges on the shells in this case? Are your results in agreement with the reciprocation theorem?

22. Repeat the calculation of the coefficients of potential and capacitance for two concentric spherical shells [Eqs. (71), (72)] if the space between the shells is completely filled with dielectric.

23. Repeat the calculation of the coefficients of potential and capacitance for two concentric spherical shells [Eqs. (71), (72)] if the space between the shells is half filled with dielectric, so the dielectric occupies only one hemisphere.

24. Two conducting spheres of radius R are separated by a distance l, with $l \gg R$. The space around the spheres is empty.
    (a) Find approximate values for the coefficients of potential and capacitance for this system.
    (b) What is the electrostatic energy deduced from Eq. (74)?

25. Three conducting spheres of radius R are placed at the vertices of an equilateral triangle of side l, with $l \gg R$. The space around the spheres is empty. Find approximate values for the coefficients of potential and capacitance for this system.

26. A conducting sphere of radius R is placed concentrically within a thin spherical shell of radius $2R$. A second conducting sphere of radius R is placed at distance l from the first sphere, where $l \gg R$. Find approximate values for the coefficients of potential and capacitance for this three-conductor system.

27. Prove the following theorem: If a finite number of positive and negative point charges are in a static equilibrium configuration under the action of their mutual Coulomb forces, then the potential energy of the configuration is necessarily zero. Give an example of such an equilibrium configuration with three or four charges.

CHAPTER

6 The Theory of Special Relativity

In the preceding chapters, we dealt with examples of fields generated by static charges; these fields are functions of space. In later chapters, we will deal with fields generated by moving charges; these fields are functions of both space and time. Thus, spacetime is the arena within which the fields perform their dance, and before we study the dynamical laws governing this dance, we had better examine the geometry of spacetime.

In Newtonian physics, length is absolute and time is absolute. This view of the geometry of spacetime is firmly engrained in our intuition, rooted in our everyday experience. It is a view that stood unchallenged for hundreds of years but then was suddenly demolished in 1905 by Albert Einstein with his theory of special relativity. In relativistic physics, neither length nor time is absolute. Instead, there is only one single absolute quantity that characterizes the spacetime geometry: the spacetime interval. This spacetime interval consists of space intervals and time intervals intimately joined together. This was first recognized by H. Minkowski, who in his incisive analysis of the relativistic geometry of spacetime wrote, "Henceforth space by itself, and time by itself are doomed to fade away into mere shadows, and only a kind of union of the two will preserve an independent reality." [1]

Before we enter into the discussion of Einstein's theory, we will briefly review the properties of spacetime and the meaning of relativity in Newtonian physics.

6·1 Spacetime and Relativity in Newtonian Physics

The spacetime geometry of Newtonian physics relies on several assumptions that derive directly from everyday experience. The most basic of these assumptions is that *spacetime is four-dimensional.* This simply means that each "event" or spacetime point can be uniquely located by four coordinates: one time coordinate and three space coordinates.

To determine the coordinates of a spacetime point, we must begin by setting up a *reference frame*. This is simply an array of rigid measuring rods and synchronized clocks (see Fig. 6.1). All the clocks in

Figure 6·1

If we assume that the black boxes are clocks attached to measuring rods, then this lithograph by M.C. Escher shows a reference frame.

the reference frame are supposed to be of identical construction and adjustment; that is, they are all the same kind of standard clock, synchronized by some procedure consistent with Newton's laws of dynamics.* For instance, we can carefully transport an auxiliary clock, such as a marine chronometer, from one stationary clock in the reference frame to any other and thereby transfer the synchronization. For any spacetime point, the measuring rods permit a direct measurement of the rectangular coordinates x,y,z, and the nearest clock gives a direct measurement of the time coordinate t. There is, of course, no need to use a rectangular array. A spherical array of rigid hoops and protractors would serve equally well and give us spherical coordinates r,θ,ϕ. This freedom in the choice of coordinates brings out a distinction between reference frames and coordinate systems. Rectangular, spherical, or cylindrical coordinate systems (and their associated synchronized clocks) are merely different ways of determining the spacetime points within the same reference frame. Only

*The synchronization of the clocks of an inertial reference frame must be such as to ensure the validity of Newton's First *and* Second Laws (for particles moving at low speeds in the selected reference frame). Failure to pay attention to this essential requirement has led to a great deal of confusion and has given rise to the erroneous speculation that the synchronization might not be unique (the "conventionality of synchronization").[2]

if two coordinate systems are in relative motion will we say that there is a change in reference frame. However, since rectangular coordinates are used almost exclusively in special relativity, we do not need to pay much attention to this distinction between reference frames and coordinate systems.

Different reference frames are in translational or rotational motion relative to one another. We will assume that the measuring rods and clocks in all reference frames have been calibrated and adjusted according to the same, standard procedures. It is then meaningful to ask: How do length and time measurements performed in two different reference frames compare? In Newtonian physics, lengths are absolute. This means the following: Consider two points of space whose coordinates, in a given reference frame, are x, y, z and $x + dx, y + dy, z + dz$ taken at the same time t. Then the length, or distance, between these points is given by

$$dl^2 = dx^2 + dy^2 + dz^2 \tag{1}$$

This distance dl is absolute, or invariant, because it does not depend on the reference frame in which it is evaluated. For example, observers in the reference frame of the ground and observers in the reference frame of a fast-moving airplane agree on the distance that they measure from one airport to another.

Likewise, in Newtonian physics, time intervals are absolute, or invariant. This means that the time interval dt between two spacetime points (at the same or at different space points) does not depend on the reference frame in which it is evaluated. Thus observers on the ground and on the airplane agree on the elapsed time that they measure between, say, takeoff and touchdown.

In consequence of this absolute character of length and of time intervals, the geometry of spacetime splits into two disjoint geometries: a three-dimensional Euclidean geometry for space and a one-dimensional Euclidean geometry for time.

For the purposes of kinematics, all reference frames are equally good. But for the purposes of dynamics, some reference frames are better than others. Newton's laws of motion select from the set of all possible reference frames a preferred subset: the *inertial* reference frames. In the latter, a free body remains at rest or in uniform translational motion, and a body subjected to an external force accelerates according to $m\mathbf{a} = \mathbf{F}$.

Two inertial reference frames can differ only by a uniform translational motion.* In all such inertial reference frames, all of Newton's laws of mechanics are equally valid. Thus there exists no intrinsic mechanical experiment that can distinguish one inertial reference frame from another. For instance, we cannot measure the velocity of an air-

*We will ignore the possibility that two reference frames differ by a constant displacement of their origins or by a constant angle of orientation of their axes. For the present purposes, such differences are discounted; we can regard them as different choices of coordinate systems within the same reference frame.

plane (in unaccelerated motion) by experiments within the airplane; the best we can do is to measure the velocity of the airplane *relative* to the air (air speed) by reaching out into the air with our measuring instruments or measure the velocity *relative* to the ground (ground speed) by observing markers on the ground. The velocity of a reference frame is never an absolute velocity, but only a relative velocity measured in relation to some other reference frame.

The assertion that all the laws of mechanics are the same in all inertial reference frames is called the *Galilean principle of relativity*. This principle of relativity is a symmetry principle; it asserts that certain operations (transformations of reference frame) leave the laws of physics invariant.

In Newtonian physics, the transformation of the space and time coordinates from one inertial reference frame to another is called the *Galilean transformation*. For the sake of simplicity, assume that the velocity of the second reference frame ($x'y'z't'$ reference frame) is along the x axis of the first reference frame ($xyzt$ reference frame) and that the origins O' and O of both these reference frames coincide at time zero. If the velocity of the second reference frame relative to the first is V, then the Galilean transformation equations are

$$t' = t \tag{2}$$

$$x' = x - Vt \tag{3}$$

$$y' = y \tag{4}$$

$$z' = z \tag{5}$$

The first of these equations expresses the absolute character of time: in both reference frames, the clocks run at the same rate and both sets of clocks can therefore be permanently synchronized with each other, so the time coordinates t and t' of any spacetime point are identical. The other three equations rely on the absolute character of length. For instance, according to Eq. (3), the horizontal distance x' from the origin O' to the point P equals the horizontal distance x from O to P minus the distance Vt from O to O' (see Fig. 6.2). The distance x' is measured in the second reference frame, whereas the distances x and Vt are measured in the first reference frame; we have equated x' and $x - Vt$ because both supposedly represent the same (absolute) length expressed in different reference frames.

By differentiating Eqs. (3)–(5) with respect to t or t', we obtain

$$\frac{dx'}{dt'} = \frac{dx}{dt} - V \tag{6}$$

Figure 6·2

x and x' coordinates of a point P in two different reference frames.

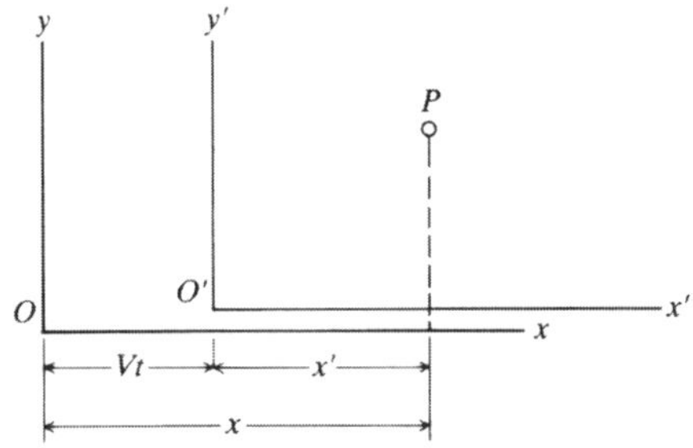

$$\frac{dy'}{dt'} = \frac{dy}{dt} \tag{7}$$

$$\frac{dz'}{dt'} = \frac{dz}{dt} \tag{8}$$

These are the Galilean addition laws for velocity. Intuitively, they seem quite self-evident.

6·2 Einstein's Principle of Relativity; the Michelson-Morley Experiment

If we attempt to apply the Galilean addition law for velocity to a light wave, we are immediately faced with a paradox. Suppose that in some reference frame, the velocity of the light wave is c in empty space, and suppose that an observer moves with velocity V in the same direction as the light wave. According to the Galilean addition law, the velocity of the light wave relative to the observer is then $c - V$. For the exceptional case $V = c$, the velocity of the light wave relative to the observer will then be zero; under these conditions, the observer keeps pace with the wave crests and the wave troughs of the wave and therefore perceives this electromagnetic wave as a *static* electric and magnetic field. But such fields are in conflict with Maxwell's equations—static electric and magnetic fields can exist only in the neighborhood of static charges and steady currents; they cannot exist when space is everywhere empty.

This paradox led Einstein to his theory of special relativity. As he wrote in his autobiographical notes:

> After ten years of reflection such a principle resulted from a paradox upon which I had already hit at the age of sixteen: If I pursue a beam of light with the velocity c (velocity of light in a vacuum), I should observe such a

> beam as an electromagnetic field constant in time, periodic in space. However, there seems to exist no such thing, neither on the basis of experience, nor according to Maxwell's equations. . . .[3]

This paradox left Einstein with two alternatives: either Maxwell's equations are wrong in all reference frames except one (the "ether" frame), or else the spacetime of Newtonian physics is wrong. Einstein chose the second alternative, and he postulated the following Principle of Relativity for all laws of physics:

• ***Principle of Relativity:*** All the laws of physics, and the numerical constants that appear in them, are exactly the same in all inertial reference frames.

Einstein supplemented this principle by a separate postulate for the propagation of light:

• ***Universality of the speed of light:*** The speed of light (in vacuum) is the same in all inertial reference frames.*

The principle of the universality of the speed of light is sometimes regarded as a corollary of the Principle of Relativity, because the validity of Maxwell's equations in all inertial reference frames implies that light waves propagate in the same way in all such reference frames. However, such a deduction hinges on the assumption that Maxwell's equations are valid laws of physics. Although it is true that Einstein was led to relativity by thinking about Maxwell's equations, the theory of relativity is more fundamental than Maxwell's theory. It is therefore undesirable to make relativity logically dependent on Maxwell's equations, and it is preferable to adopt a separate postulate for the universality of the speed of light.

Since the Galilean transformation equations and the addition laws for velocity are incompatible with a universal value of the speed of light, Einstein was compelled to formulate a new set of transformation equations. These new transformation equations are contrived in such a way that the speed of light remains the same in all reference frames, and they have profound implications for our concepts of the geometry of spacetime. The relativistic modification of our concepts of space and time requires a corresponding modification of the laws of mechanics. Newton's laws of mechanics are equally valid in all inertial reference frames if, and only if, spacetime is Newtonian. For the relativistic space-

---

*In 1983, the Conférence Générale des Poids et Mesures adopted a new definition of the meter as the distance traveled by light in vacuum in1/299,792,458 of a second. This means that the speed of light is now a standard of speed, with $c = 299{,}792{,}458$ m/s by definition (in our calculations, we will usually round this number off to $c = 3.00 \times 10^8$ m/s). Note that this adoption of a standard value of the speed of light, without specification of a preferred reference frame, hinges on the validity of the principle of the universality of the speed of light.

time, Einstein and his followers had to develop a new set of equations of mechanics, designed to hold in all inertial reference frames.

Before we proceed to a derivation of the new transformation equations and an examination of the geometry of relativistic spacetime, let us review the experimental evidence for the universality of the speed of light. Long before Einstein, experimenters had sought to detect an effect of the motion of the Earth on the propagation of light measured in their laboratories. Early in the 19th century, physicists had come to regard light as an oscillation in a "luminiferous ether," a tenuous, ghostly medium that supposedly filled all of space and served as a carrier of light waves just as air serves as a carrier of sound waves. It was thought that when in motion relative to the ether, a laboratory on the Earth experiences an "ether wind" that presumably carries the light with it and thereby alters the velocity of light in accordance with the Galilean addition laws.

The earliest attempts at detecting the ether wind were made by M. Hoek in 1868 and by G. B. Airy in 1871, who investigated the propagation of light in water-filled tubes and astronomical telescopes. They found no effect, but this result was inconclusive because according to a version of the ether theory formulated by A. Fresnel, no effect was expected in the Hoek and Airy experiments. Fresnel had proposed that transparent materials, such as water, partially drag the ether along, and he had proposed the Hoek experiment as a means of verifying this ether drag.

The first conclusive experiment capable of detecting the small difference between the parallel and the perpendicular speeds of light was devised by A. A. Michelson in 1881. A few years later, he repeated it in an improved form in cooperation with E. W. Morley, and it became known as the Michelson-Morley experiment. The experiment uses an interferometer with two arms of exactly equal lengths at right angles, one arm oriented parallel to the (presumed) direction of motion of the Earth, one oriented perpendicular to it. The ends of the arms are terminated by mirrors, so beams of coherent light can travel back and forth along the arms several times before they are permitted to interfere. Figure 6.3 shows the interferometer of Michelson and Morley, with mirrors installed on a rigid stone slab, so the diagonals of the slab play the role of "arms" of the interferometer. The light beams originate from a common source *a*, whose light is split into two beams by a half-silvered mirror *b*. The two beams travel back and forth several times along the parallel and perpendicular arms *ee* and *dd*, they are then again reflected into the half-silvered mirror *b*, and they finally come together at *f*, where they interfere constructively or destructively, producing interference fringes.

According to Newtonian physics, the speed of a light beam relative to the laboratory depends on the speed of light relative to the ether and on the speed of the Earth relative to the ether [see Eq. (6)]. Figure 6.4 shows how the velocity vectors **c** for the light and **V** for the Earth are subtracted to obtain the resultant velocity of light relative to the labo-

Figure 6·3

Interferometer of Michelson and Morley. (*a*) The mirrors defining the paths of the light beams (dotted) are mounted on a large stone slab floating on a pool of mercury. (*b*) The diagonals *ee* and *dd* of the stone slab are the "arms" of the interferometer, and the light beams traveling along these arms are at right angles. For the velocity vector **V** shown here, the arm *ee* is parallel to the velocity, and the arm *dd* is perpendicular to it. (From *Philos. Mag.*, December 1887, p. 454.)

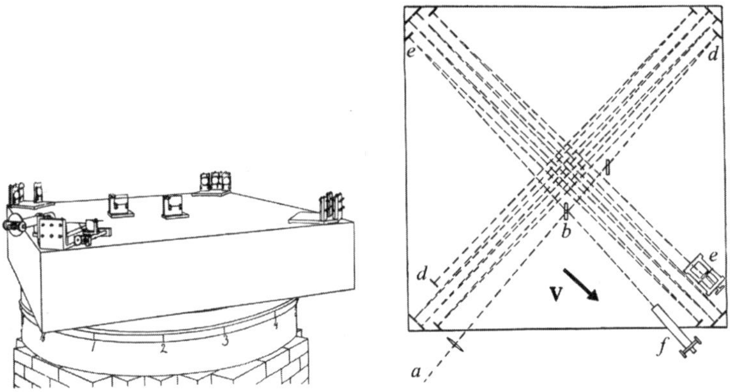

Figure 6·4

The velocity of a light signal relative to the Earth is **c** – **V**. The magnitude of this velocity depends on the direction. (*a*) **c** – **V** is parallel to **V** or (*b*) antiparallel to **V**. (*c*) **c** – **V** is perpendicular to **V**; the magnitude of **c** – **V** is $\sqrt{c^2 - V^2}$.

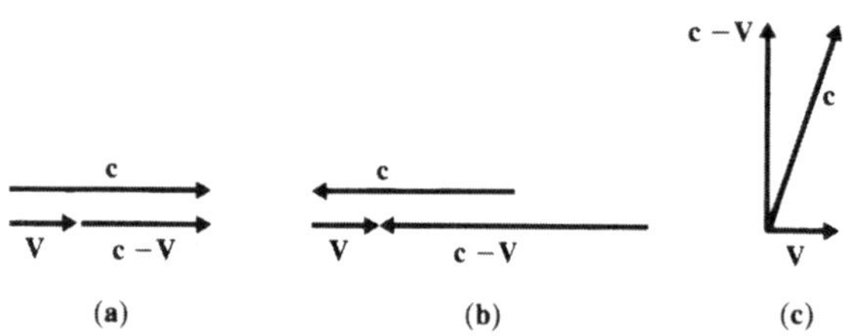

ratory. For the parallel beam, the resultant speed is $c - V$ or $c + V$ (depending on whether the light travels in the same direction as the Earth or in the opposite direction); for the perpendicular beam, the speed is $\sqrt{c^2 - V^2}$. These changes of speed may be thought of as being caused by the "ether wind" that blows through the laboratory and carries the light along. The difference in speeds introduces a phase difference between the parallel and the perpendicular beams reaching the point *f*. The phase difference can be conveniently detected by rotating the interferometer by 90°; this reverses the roles of the parallel and the perpendicular arms and causes interference fringes to march across the point *f*.

■ ***Exercise 1.*** For the sake of simplicity, assume that the light beams return to the half-silvered mirror *b* after just one reflection at *e* or *d* at the upper corners of Fig. 6.3 (*b*). Show that the average speed for the round trip of the parallel beam from *b* to *e* to *b* is $c(1 - V^2/c^2)$. Show that if $V/c \ll 1$, the phase difference between the parallel and the perpendicular beams reaching the point *f* is

$$\Delta\phi = 2\left[\frac{L}{c(1-V^2/c^2)} - \frac{L}{c\sqrt{1-V^2/c^2}}\right]\times\frac{2\pi c}{\lambda};2\pi\frac{L}{\lambda}\frac{V^2}{c^2} \qquad (9)$$

where *L* is the length *be* or *bd*, and λ is the nominal wavelength of the light, $\lambda \equiv 2\pi c/\omega$ (this is the wavelength of the emerging light; the wavelength within the interferometer is different in different arms and in different directions). ■

This prediction of Newtonian physics stands in sharp contrast to the prediction of relativistic physics. According to the principle of universality of the speed of light, the speed is the same along both arms of the interferometer, and there is no phase difference between the two light beams.

Table 6.1 summarizes the results of all the repetitions of the Michelson-Morley experiment, starting with the first such experiment in 1881. To within the experimental errors, no fringe shift was detected by any of these experiments. The last column in the table (V_{eff}) gives the upper limit on the speed of the "ether wind" relative to the Earth consistent with the experimental error in the fringe shift; this speed has been calculated according to Newtonian physics, from Eq. (9). The last and the most precise of the interferometer experiments was that performed by Joos at Jena in 1930,[4] which set a limit of 1.5 km/s on the "ether wind."

The more recent Michelson-Morley experiments listed in Table 6.1 compared not the phases of light beams traveling at right angles but, rather, the frequencies of standing waves in resonating cavities oriented at right angles. The frequency of such a standing wave is directly proportional to the speed of the wave; if the two cavities are of identical length, then any observed difference in the frequencies implies a difference in the speeds of light. The frequency difference, if any, can be accurately detected by superposing waves radiated from the two cavities; the frequency difference will then give rise to beats in the combined wave. The experiment by Jaseja et al.[5] employed lasers of a frequency 3×10^{14} Hz as optical resonant cavities. As in the case of the ordinary Michelson-Morley experiment, the experimenters rotated their apparatus through 90°, reversing the roles of the parallel and perpendicular lasers. They found that this produced no significant effect on the beat frequency, to within the experimental error of ±3000 Hz, and they were able to set a limit of 1 km/s on the "ether wind."

Table 6·1 Michelson-Morley experiments*

Observers	Year	V_{eff}
Michelson	1881	21 km/s
Michelson and Morley	1887	4.7
Morley and Miller	1902–1904	3.3
Miller†	1921	7.7
Miller	1923–1924	7.5
Miller (sunlight)	1924	3.3
Tomashek (starlight)	1924	7.7
Miller	1925–1926	8.3
Kennedy	1926	5.0
Illingworth	1927	2.2
Piccard and Stahel	1927	6.8
Michelson et al.	1929	3.1
Joos[4]	1930	1.5
Essen (frequency of microwave cavities)[5]	1955	2.5
Jaseja et al. (frequency of lasers)[6]	1964	0.95
Brillet and Hall (frequency of etalon)[7]	1979	0.02
Antonini et al. (frequency of optical resonators)[45]	2005	0.009
Stanwix et al. (frequency of microwave resonators)[46]	2005	0.007
Herrmann et al. (frequency of optical resonators)[47]	2005	0.007

*Based on a table by Shankland et al.[8]
†The results for Miller's experiments listed in this table are as calculated by Shankland et al. using Miller's data.

The experiment, by Brillet and Hall,[7] employed a Fabry-Perot etalon as an optical resonant cavity, driven by a laser. This etalon and the laser were mounted on a rotating turntable, and the frequency of the standing wave in the etalon was monitored by comparison with a second laser, fixed in the laboratory, which served as a frequency standard. From the absence of any dependence of the frequency on the orientation of the etalon relative to the fixed stars, the experimenters were able to set a limit of about 15 m/s on the "ether wind."

Even better limits were achieved recently by similar methods using cryogenic resonant cavities of sapphire cooled to about 4 K. The experiments of Antonini et al. and Herrmann et al. operated at optical frequencies, whereas the experiments of Stanwix et al. operated at microwave frequencies. These experiments were able to lower the limits to about 7m/s. This limit is much smaller than the speed of the Earth around the Sun (30 km/s) or the speed of the Sun relative to the cosmic microwave background radiation (370 km/s). Clearly, the expected "ether wind" is not in evidence.

The null result of the Michelson-Morley experiment was viewed with alarm by the 19th-century physicists. They proposed several explanations of this null result: ether drag, ballistic emission of light, and the Lorentz-FitzGerald contraction. According to the ether-drag hypothesis, a large heavy body, such as the Earth, drags along not only the ether in its interior but also the ether in its immediate vicinity. This induced drag eliminates the ether wind in laboratories on the surface of the Earth. This was the explanation favored by Michelson himself, who regarded the outcome of his experiment as confirmation of ether drag. However, the ether-drag hypothesis is in conflict with stellar aberration (that is, the apparent seasonal shift of the observed positions of the stars in the sky). This shift, which can be as large as ±20 arcseconds, is caused by the orbital velocity of the Earth. It is analogous to the tilt of the trajectories of raindrops that you observe when driving through a rain shower. Stellar aberration would be absent if the ether in the vicinity of the Earth were dragged along. Furthermore, if the translational motion of the Earth drags the ether along, then the rotational motion of the Earth should do likewise. But this is not the case, because the rotational motion of the Earth is detectable with optical experiments, as was first demonstrated by Michelson and Gale[9] in 1925 by an experiment with two light beams traveling in opposite directions around a closed path defined by mirrors. Modern laser gyrocompasses, now widely used in navigation systems for aircraft, also exploit such an optical detection of the rotation of the Earth.

Another explanation was proposed by G. F. FitzGerald, who postulated that whenever a body moves through the ether at speed V, it suffers a length contraction along the direction of motion by a factor of $\sqrt{1-V^2/c^2}$. This means that the arm of the interferometer parallel to the direction of motion suffers this length contraction, but the perpendicular arm does not. Accordingly, we must multiply the first length L in Eq. (9) by $\sqrt{1-V^2/c^2}$, which has the consequence that the two terms in brackets become equal, and the phase difference (9) becomes zero. H. A. Lorentz showed that the FitzGerald contraction could be given a dynamical basis by assuming that solid bodies, such as the stone slab of the Michelson-Morley interferometer, consist of charged particles held in equilibrium by electric forces. He was able to show from Maxwell's equations that the electric field of each charged particle contracts in the direction parallel to the motion (we will examine this in Chapter 8; see Fig. 8.2) and that the equilibrium positions of the charged particles therefore also contract. Although this FitzGerald-Lorentz contraction explained the null result obtained by Michelson and Morley, a modified experiment by Kennedy and Thorndike,[10] with an interferometer with arms of unequal lengths, demonstrated that the contraction hypothesis was insufficient. If the arms of the interferometer are unequal—say, L_1 and L_2—then the phase difference is nonzero, but the contraction makes the phase difference independent of the orientation of the arms, and hence rotation of the instrument achieves no observable effect. However, the phase difference contains a term proportional to V^2 (see Problem 3),

$$\Delta\phi \simeq \text{ constant } + \frac{2\pi}{\lambda}(L_1 - L_2)\frac{V^2}{c^2}$$

and this results in diurnal and seasonal variations. Kennedy and Thorndike performed their experiment in 1932, and by then it was known that the Solar System orbits around the center of the Galaxy at a speed of about 250 km/s; it was therefore expected that the ether wind would have this magnitude relative to the Sun. Between midnight and noon, the rotational motion of the Earth changes the forward velocity of a point on the equator by about 1 km/s, and between one time of the year and a time six months later, the orbital motion of the Earth changes the velocity of the Earth by about 60 km/s. Thus the expected diurnal variation in V^2 is about $\Delta V^2 \simeq 2V\Delta V \simeq 2\times 250$ km/s $\times$ 1 km/s = 500 (km/s)^2, and the expected seasonal variation is about $\Delta V^2 \simeq 2\times 250$ km/s $\times 60$ km/s $\simeq$ 30,000 (km/s)^2. These large variations in V^2 would produce easily detectable variations in the phase shift. But Kennedy and Thorndike detected no such variations, and from their experimental errors they concluded that the observed ether wind associated with the motion of the Solar System is not 250 km/s, but at most 15 km/s. This null result demonstrates that the FitzGerald-Lorentz contraction is a failure—it does not achieve its aim of removing the discrepancy between observation and the ether theory of the propagation of light.*

Recent versions of the Kennedy-Thorndike experiment (see Table 6.2) have set better limits on the Solar-System ether wind and confirmed the failure of the FitzGerald-Lorentz contraction. Hils and Hall[11] used a Fabry-Perot etalon as an optical resonant cavity and compared the resonant frequency of this cavity with the frequency of an atomic standard. The period of the standing wave in such a cavity is given by the travel time of light bouncing back and forth between its ends, that is, the period is given by the second term in the brackets in Eq. (9) (or, equivalently, by the first term with an extra length-contraction factor). Thus the period depends on V^2, and it is subject to diurnal and seasonal variations. The experimenters checked for diurnal variations, and from the absence of any such variations they set a limit of 50 m/s on the ether wind. Müller and Soffel[12] exploited the available data from lunar laser ranging, that is, measurements of the travel time of laser pulses sent to the Moon and

---

*Although the FitzGerald-Lorentz contraction by itself cannot explain the Kennedy-Thorndike null result, the contraction in conjunction with a postulated time dilation *can* explain it. However, the combination of length contraction and time dilation leads directly to the Lorentz transformation, which means this merely provides an alternative approach to relativity.

reflected back to Earth by corner reflectors installed on the Moon by the Apollo 11 astronauts. If the effective speed of light for a roundtrip is $\sqrt{c^2 - V^2}$, then the speed has seasonal variations, and the travel time will fluctuate seasonally. From the absence of such seasonal fluctuations, the investigators were able to set a limit of 110 m/s on the ether wind.

Table 6·2 Kennedy-Thorndike experiments

Experiments	Year	V_{eff}
Kennedy and Thorndike[10]	1932	15 km/s
Hils and Hall[11]	1990	50 m/s
Müller and Soffel[12]	1995	110 m/s

Yet another explanation of the null result was attempted with emission theories, or ballistic theories, according to which light has the speed c not with respect to an ether frame but with respect to its source. This makes the emission of light by a source analogous to the firing of a projectile from a gun. Accordingly, the speed of light relative to the observer depends on the motion of the source relative to the observer: if the source is at rest relative to the observer, the speed of light is c; if the source is moving toward or away from the observer with speed V, the speed of light will be $c \pm V$. Since the original Michelson-Morley experiment used a light source at rest relative to the interferometer, no effect is expected. However, some of the later versions of the experiment were performed with sunlight and with starlight, and then an effect *is* expected, but none was observed. Furthermore, astronomers quickly recognized that a dependence of the speed of light on the speed of the source would lead to the observation of weird effects in binary systems of stars, where stars orbit around each other, with a changing velocity. The light emitted by such an orbiting star at two different times during its orbital motion could then arrive at the Earth at the same time, and astronomers on the Earth would then see the star twice—they would see the star and an extra "ghost." But no such weird effects are observed, and this given us strong evidence against emission theories (we will discuss the observational evidence in Section 6.5, in connection with the relativistic formula for transformation of velocities).

6·3 Other Experiments

Limits on the "ether wind" have also been set by experiments that rely on the Doppler shift between an emitter of light and a receiver, both in motion. According to Newtonian physics, the Doppler shift depends on the velocities of the emitter and receiver relative to the ether; thus the Doppler shift can be used to detect the "ether wind," if any. Suppose that the emitter and the receiver have velocities $\mathbf{V}_s$ and $\mathbf{V}_r$ relative to the ether, respectively; then the received frequency can be expressed in terms of the emitted frequency by the formula

$$\nu_r = \nu_s \frac{1-\hat{\mathbf{n}}\cdot\mathbf{V}_r/c}{1-\hat{\mathbf{n}}\cdot\mathbf{V}_s/c} \tag{10}$$

where $\hat{\mathbf{n}}$ is the unit vector perpendicular to the wave fronts of the light.

■ ***Exercise 2.*** Derive Eq. (10). [Hint: Show that the frequency ν in the reference frame of the ether is related as follows to the frequencies in the reference frames of the emitter and the receiver:

$$\nu_s = \nu\left(1-\frac{\hat{\mathbf{n}}\cdot\mathbf{V}_s}{c}\right) \quad \text{and} \quad \nu_r = \nu\left(1-\frac{\hat{\mathbf{n}}\cdot\mathbf{V}_r}{c}\right) \quad]$$ ■

For low velocities, Eq. (10) yields

$$\nu_r \simeq \nu_s\left(1-\frac{\hat{\mathbf{n}}\cdot\mathbf{V}_r}{c}\right)\left(1+\frac{\hat{\mathbf{n}}\cdot\mathbf{V}_s}{c}+\ldots\right)$$

$$\simeq \nu_s\left[1+\frac{\hat{\mathbf{n}}\cdot(\mathbf{V}_s-\mathbf{V}_r)}{c}\right]+\ldots \tag{11}$$

This shows that the Doppler shift depends mainly on the velocity of the emitter relative to the receiver ($\mathbf{V}_s - \mathbf{V}_r$); but it also depends slightly on the velocity relative to the ether, because the direction of propagation $\hat{\mathbf{n}}$ will be altered by the motion of the apparatus (aberration).

The maser experiment of Cedarholm, Bland, Havens, and Townes[13] used a molecular beam with an eastward velocity as emitter. The light, in the form of microwaves, travels northward or southward, perpendicularly away from this beam [the direction of propagation of the waves is determined by the cavity that surrounds the beam; see Fig. 6.5(*a*)]. Looking at the emitted waves from the point of view of the ether reference frame, we find that the direction of propagation of the waves must be tilted slightly forward at an angle of (approximately) V/c, to match the forward speed V of the cavity [see Fig. 6.5(b)]. If the "receiver" is at rest in the laboratory, the velocity $\mathbf{V}_s - \mathbf{V}_r$ is simply the velocity of the emitter molecules relative to the laboratory. The angle between this velocity and the normal to the wave fronts is then $\pi/2 - V/c$, and $\cos(\pi/2 - V/c) = \sin(V/c) \simeq V/c$, so by Eq. (11),

$$\nu_r \simeq \nu_s\left[1+(V_s - V_r)\frac{V}{c^2}\right] \tag{12}$$

Figure 6·5

Emission of microwaves by the molecular beam of a maser. *(a)* If the maser is at rest relative to the ether, the direction of propagation of the microwaves is at right angles to the beam. *(b)* If the maser has a velocity V relative to the ether, the direction of propagation is tilted forward at an angle of V/c.

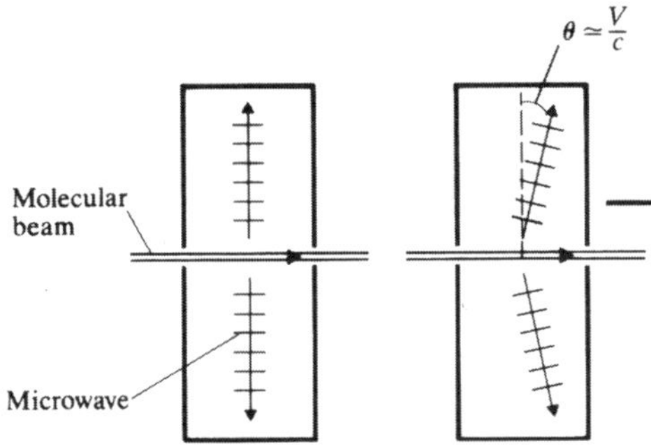

To detect this Doppler shift, the experimenters used a second maser with a molecular beam with westward velocity and compared the frequencies of the microwaves emitted by these two masers. The absence of any observable Doppler shift sets a limit on V. Note that the Doppler shift (12) depends on the first power of V/c, whereas in the Michelson-Morley experiment, the fringe shift depends on the second power of V/c. This circumstance makes the Doppler-shift experiment more sensitive to the "ether wind."

Experiments using the Mössbauer effect have managed to set even tighter limits on the "ether wind" (see Table 6.3). The apparatus used by Champeney, Isaak, and Khan[15] consists of a rotor rapidly spinning about its vertical axis (see Fig 6.6). This rotor carries samples of ^{57}Fe on opposite ends. In one of these samples, the iron nuclei are in an excited state, and they emit gamma rays by transitions without recoil (the absence of recoil of the nucleus when locked in a crystal lattice is, in essence, the Mössbauer effect). These gamma rays can be captured by resonant absorption in the other sample of iron. According to Newtonian physics, we would expect a Doppler shift between emitter and absorber. When the emitter and absorber are instantaneously on the north-south line, one of them moving eastward and one westward, the Doppler shift would be given by Eq. (12); this Doppler shift would reverse sign every one-half revolution of the rotor. The mismatch between the emitter and absorber frequencies would inhibit the resonant absorption of the gamma rays. In the experiment, no such inhibition of the absorption was found, and from the experimental errors an upper bound of 1.6 ± 2.8 m/s was placed on the "ether wind." A later, improved version of this experiment[17] reduced the upper bound to 5 cm/s. This is the tightest bound achieved to date.

Table 6·3 Doppler-shift experiments

Experiments	Year	V_{eff}
Cedarholm et al. (maser)[13]	1958	30 m/s
Cranshaw and Hay (Mössbauer effect)[14]	1963	10
Champeney et al. (Mössbauer effect)[15]	1963	1.6 ± 2.8
Turner and Hill (Mössbauer effect)[16]	1963	2.2 ± 8.4
Isaak (Mössbauer effect)[17]	1970	0.05
Riis et al.[18]	1988	0.9

The development of highly accurate atomic clocks has made feasible yet another type of first-order experiment: the direct measurement of the one-way speed of light over a given distance. These measurements have set limits on the ether wind comparable with those obtained by frequency experiments. The first application of this method was a by-product of Very Long Baseline Interferometry (VLBI) in radioastronomy. In this technique, astronomers use two or more radiotelescopes separated by thousands of kilometers to observe simultaneously a radio source in the sky. They make a digital recording of the radio waves incident on each radiotelescope and tag this recording with precise time signals from a synchronized clock located at each radiotelescope. They can later play back and mix these synchronized recordings and generate interference patterns to determine the location of the radio source in the sky. The time delay and the consequent interference between the radio signals arriving at two separate radiotelescopes depends on the extra path length that the radio wave has to travel, but it also depends on the speed of the wave. If this speed is altered by the component of an ether wind

Figure 6·6

Apparatus of Champeney et al.

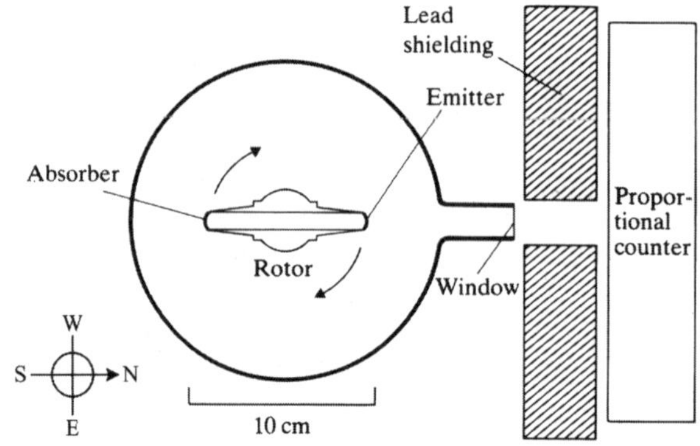

along the direction of travel of the radio wave, then successive determinations of the position of a radio source will display inconsistencies, because at different times of the year the relevant component of the ether wind will be different (for instance, for a radio source in the plane of the Earth's orbit, the wave speed is $c + V$ when the Earth is moving toward the source, it is $\sqrt{c^2 - V^2}$ three months later, and it is $c - V$ six months later). According to Cole,[19] the available data indicate that the determinations of the positions of radio sources at different times of the year are self-consistent to within 0.05 arcsec, and this implies a limit of about 0.07 km/s on the ether wind (see Table 6.4).

A similar limit was obtained by Krisher et al.,[20] who used a pair of synchronized hydrogen-maser clocks to check for a directional variation of the one-way speed of light traveling on an optic fiber between two stations separated by 21 km. Although the two stations were fixed on the Earth, they rotate with the Earth, and this changes the orientation of the path of the light relative to the ether, if any. The experimenters found no effect during observations extending over several days.

A much better limit was recently obtained by Wolf and Petit[21] from an analysis of data on travel times of radio signals transmitted by the satellites of the Global Positioning System (GPS). These satellites carry atomic cesium or rubidium clocks, synchronized to within about a nanosecond. The typical separation of a satellite from one of several ground stations equipped with their own synchronized clocks is about 20,000 km. Hence the one-way speed of a radio signal from the satellite to a ground station can be measured to within a fractional error of about $\delta c / c \simeq \delta t / t \simeq 10^{-9}\,\text{s}/(2\times 10^7\ \text{m} \times 1/c) \simeq 2\times 10^{-8}$. But this error can be reduced by averaging the data from different satellite–ground-station pairs and from repeated passes of the satellites, for a chosen orientation of the path of the radio wave in space (relative to the fixed stars). The error can be reduced still further by selectively extracting data that have the angular dependence expected for an ether wind. The net result is a reduction of the fractional error to about 2×10^{-9}, which corresponds to about 0.6 m/s. To within this observational error, no dependence of the speed of the radio waves on orientation was found.

Table 6·4 One-way Measurements of the Speed of Light

Experiments	Year	V_{eff}
Cole[19]	1976	70 m/s
Krisher et al.[20]	1990	100
Wolf and Petit[21]	1997	0.6

To summarize: All the experiments agree that the propagation of light is unaffected by the motion of the laboratory. The speed of light appears to be the same in all inertial reference frames, exactly as postulated by the principle of universality for this speed.

6·4 The Spacetime Interval

The failure of all explanations of the null result of the Michelson- Morley experiment indicates that we need to question the foundations of Newtonian physics. We must abandon the absolute character of time and of length, and we must abandon the Galilean transformation equations and the addition law for velocities. This involves a far-reaching modification of spacetime, as first recognized by Minkowski soon after Einstein published his work. But in one regard the spacetime of relativistic physics remains the same as in Newtonian physics—the spacetime remains four-dimensional. To determine the coordinates of a spacetime point, we use a reference frame of rigid measuring rods and synchronized clocks, just as in Newtonian physics. The synchronization of clocks can be performed by transport of an auxiliary clock. However, in relativistic physics, a moving clock suffers a time dilation (see Section 6.6), and to achieve a valid synchronization it is necessary to transport the auxiliary clock at low speed with respect to the given reference frame; more precisely, it is necessary to proceed to the limit of zero speed ($v \to 0$) for the clock transport. Such a synchronization ensures that when the reference frame is used to describe dynamics, Newton's laws remain valid for particles moving at low (nonrelativistic) speeds. In practice, the implementation of the limit of zero speed for clock transport is difficult and inconvenient. Because of this complication, an alternative—but equivalent—synchronization procedure is preferable.

The simple alternative synchronization procedure adopted by Einstein relies on the universality of the speed of light. This synchronization procedure involves an exchange of light signals, as follows: At time zero, a light signal is sent from clock A toward clock B; suppose it arrives at clock B when the latter reads T_B. A return signal is immediately sent back to clock A; suppose it arrives at clock A when it reads T_A. If $T_A = 2T_B$, then the clocks are already synchronized; if not, then clock B must be advanced by an amount $\frac{1}{2}T_A - T_B$. Note that this synchronization procedure hinges on the equality of the speeds of the two light signals; that is, the speed must be independent of direction, as postulated by Einstein. Furthermore, this same procedure can be followed in all inertial reference frames. (In Newtonian physics, such a synchronization procedure with light would be valid only for the "ether" frame; for any other reference frame, it would be necessary to take into account the dependence of the speed of light on direction.)

In relativistic physics, it is expedient to adopt a modified time coordinate *ct* that has the dimension of length. Since the speed of light *c* is a universal constant, we can use it as a conversion factor, to convert the time coordinate *t* (measured in time units) to the modified time coordinate *ct* (measured in length units). Points in spacetime are then specified by the coordinates ct, x, y, z, all with dimensions of length.

Suppose now that a flash of light is emitted at the spacetime point ct, x, y, z. This flash is an *event*, or a happening, at a given point of space at a given time. Suppose that the light signal is absorbed at the point $ct + cdt,\ x + dx,\ y + dy,\ z + dz$. This is another event. Figure 6.7 shows the positions of both these events in a *ct-x* diagram, called a spacetime diagram or a Minkowski diagram. For the purpose of this figure, it has been assumed that the light moves in the positive *x* direction. The line connecting the two spacetime points is called the *worldline* of the light signal; this is simply the plot of the position of the light signal as a function of time. Consider the quantity

$$ds^2 \equiv c^2 dt^2 - dx^2 - dy^2 - dz^2 \tag{13}$$

which is called the *spacetime interval.** Since the speed of the light signal is *c*, the distance it travels in a time *dt* is

$$\left(dx^2 + dy^2 + dz^2\right)^{1/2} = cdt \tag{14}$$

Figure 6·7

A spacetime diagram showing the worldline of a light signal between two spacetime points.

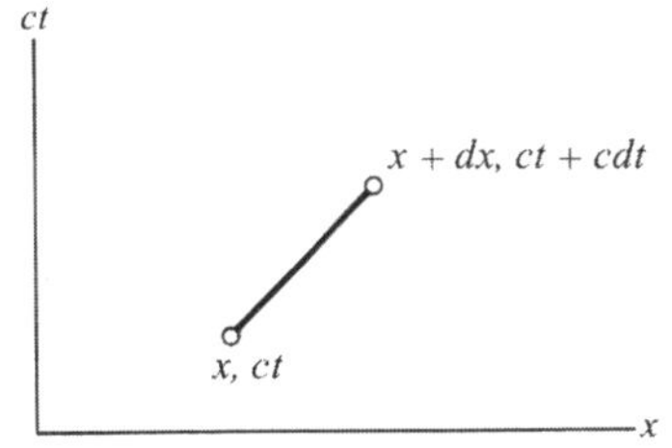

*According to our terminology, ds^2 is called the spacetime interval, and *ds* is the square root of the spacetime interval (note that this square root is imaginary if $ds^2 < 0$). Informally, both ds^2 and *ds* are often called spacetime intervals.

Accordingly, the right side of Eq. (13) is zero,

$$ds^2 = 0 \tag{15}$$

Thus the spacetime interval between any two spacetime points connected by a light signal is zero.

If we look at the events from a different inertial reference frame, then the coordinates of the two events will be ct', x', y', z' and $ct' + cdt'$, $x' + dx'$, $y' + dy'$, $z' + dz'$, respectively. However, the spacetime interval

$$ds'^2 \equiv c^2 dt'^2 - dx'^2 - dy'^2 - dz'^2 \tag{16}$$

measured in this different reference frame will still be zero,

$$ds'^2 = 0 \tag{17}$$

This is obvious because the speed of light is exactly the same as before. Thus we have shown that whenever the spacetime interval is zero in one inertial reference frame, then it is also zero in any other inertial reference frame.

Let us next consider two arbitrary events *not* connected by a light signal. Then the spacetime interval between them is

$$ds^2 \equiv c^2 dt^2 - dx^2 - dy^2 - dz^2 \neq 0 \tag{18}$$

We will show that the spacetime interval as measured in a different inertial reference frame has exactly the same numerical value:

$$ds'^2 = ds^2 \tag{19}$$

or, written out in full,

$$c^2 dt'^2 - dx'^2 - dy'^2 - dz'^2 = c^2 dt^2 - dx^2 - dy^2 - dz^2 \tag{20}$$

To simplify the writing, we will ignore the y and z dimensions and deal only with "one-dimensional relativity" where all displacements and velocities are along the x axis. The inertial character of the reference frames implies that the coordinates t', x' and t, x are related by linear equations with constant coefficients, because a free particle, with a straight worldline in one reference frame, must also have a straight worldline in the other reference frame.* Accordingly, the differentials

*Mathematically, there also exists a class of *nonlinear* rational transformations of the form $x' = (A_1 x + A_2 t)/(C_1 x + C_2 t + C_3)$, $t' = (B_1 x + B_2 t)/(C_1 x + C_2 t + C_3)$ that map straight lines in one reference frame into straight lines in the other reference frame. But these transformations are not physically acceptable, because they have singularities (they map points with $C_1 x + C_2 t + C_3 = 0$ into $x' = t' = \infty$).

dt', dx' and dt, dx are also related by linear equations with constant coefficients, and this implies that the squares of the differentials are likewise related by equations with constant coefficients. If we express $c^2dt'^2 - dx'^2$ in terms of dt and dx, we therefore obtain some sum of dx^2, $dxdt$, and dt^2:

$$c^2dt'^2 - dx'^2 = Ac^2dt^2 + Bdx^2 + Dcdxdt \qquad (21)$$

where the coefficients A, B, D can depend on the relative velocity V of the reference frames, but they cannot depend on x, t or on dx, dt. Equation (21) simply says that since the left side is of second order in the differentials, the right side must also be of second order.

We already know that if $dx = \pm cdt$, then $dx' = \pm cdt'$. Hence, in this special case, Eq. (21) reduces to

$$0 = Adx^2 + Bdx^2 \pm Ddx^2 \qquad (22)$$

Since this equation must hold irrespective of the sign in front of D, we need $D = 0$. Equation (22) then implies $A = -B$, and hence Eq. (21) becomes

$$c^2dt'^2 - dx'^2 = A(c^2dt^2 - dx^2) \qquad (23)$$

It only remains to evaluate the coefficient A. We already know that A can depend only on the relative velocity between the two reference frames. Furthermore, A cannot depend on the direction of the velocity. We can see this as follows: a change in the velocity from V to $-V$ is equivalent (by isotropy of space) to a change in the orientation of the x axis; but such a change of x to $-x$ does not produce any change in dt^2, dx^2, dt'^2, or dx'^2, and hence it cannot change A [see Eq. (23)]. Consequently, A must be some function of the magnitude of the relative velocity between the two reference frames, and Eq. (23) can be written as

$$ds'^2 = A(|V|)ds^2 \qquad (24)$$

This expresses ds' in terms of ds. We can also express ds in terms of ds'. Since the two reference frames are physically equivalent, the coefficient A that accomplishes the conversion from ds to ds' must be the same function of V as before. The only difference is that if the velocity of the $x't'$ frame with respect to the xt frame is V, then the velocity of xt with respect to $x't'$ is $-V$. Thus

$$ds^2 = A(|-V|)ds'^2 = A(|V|)ds'^2 \qquad (25)$$

Comparison of Eqs. (24) and (25) shows that $A(|V|) = 1$, and therefore,

$$ds'^2 = ds^2 \qquad (26)$$

In the derivation of Eq. (26), we have left out the y and z coordinates. But the equation remains true even if we include these coordinates. It is easy to show that

$$dy'^2 + dz'^2 = dy^2 + dz^2 \tag{27}$$

where, as always, we have assumed that the relative velocity between the two inertial reference frames is along the x axis. Equation (27) asserts that lengths transverse to the velocity remain unchanged by the transformation of reference frames. A simple argument for this is the following: Imagine that a piece of pipe is at rest in the $xyzt$ frame and another identical piece of pipe is at rest in the $x'y'z't'$ frame (see Fig. 6.8). If the motion of the second pipe were to cause a contraction of the transverse diameter of the second pipe, then—by the Principle of Relativity—the motion of the first pipe relative to the second would have to cause a contraction of the diameter of the first pipe. These contraction effects are mutually contradictory because, in one case, the second pipe would fit inside the first pipe and, in the other case, it would fit outside. The only logical alternative is that neither pipe contracts; that is, the transverse lengths are unchanged, as asserted by Eq. (27). Taking into account this equation, we see that the y and z differentials do not affect the equality between the two sides of Eq. (20). This establishes the general validity of Eq. (19) for the spacetime interval.

What we have shown is that the spacetime interval between two arbitrary events is *invariant* with respect to changes in reference frame. The crucial distinction between the geometry of relativistic spacetime and the geometry of Newtonian spacetime is that in the latter both the length interval dl (at a given time) and the time interval dt are invariant; whereas, in the former, only the spacetime interval ds^2 is invariant (see Table 6.5).

Figure 6·8

Two pieces of pipe, one at rest in the $xyzt$ frame, the other at rest in the $x'y'z't'$ frame.

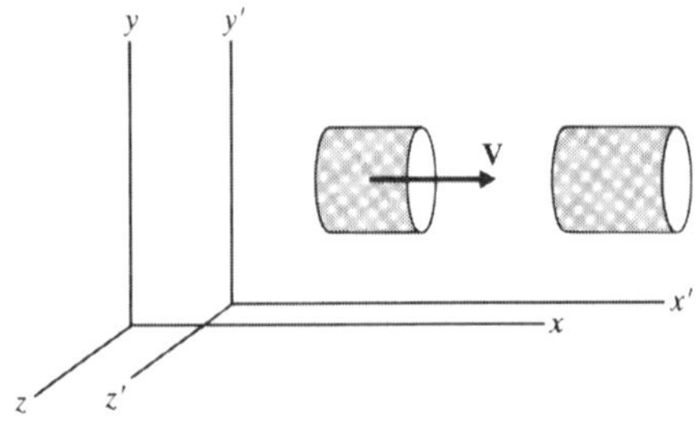

Table 6·5 Comparison of spacetimes

	Newtonian Spacetime	Relativistic Spacetime
Coordinates of an event	t, x, y, z	t, x, y, z
Invariants	$\begin{cases} dl^2 = dx^2 + dy^2 + dz^2 \text{ (at } dt = 0) \\ dt \end{cases}$	$ds^2 = c^2dt^2 - dx^2 - dy^2 - dz^2$

The geometry of Newtonian spacetime consists of two disjoint Euclidean geometries for space and for time. The geometry of relativistic spacetime consists of a single *pseudo-Euclidean* geometry [the terminology pseudo-Euclidean refers to the presence of negative signs in Eq. (13), which means that the spacetime interval is not positive definite].

6·5 The Lorentz Transformation

As we know from Section 6.1, in Newtonian spacetime the transformation from one inertial reference frame xt to a second inertial reference frame $x't'$, moving at velocity V along the x axis of the first, is the Galilean transformation

$$t' = t \qquad x' = x - Vt$$

$$y' = y \qquad z' = z$$

where we have assumed, as usual, that the origins of both sets of coordinates coincide at $t = 0$. This transformation leaves distance (at a given time) and time intervals invariant. In spacetime, the transformation can be described geometrically as a transformation from rectangular to slanted axes, shown in Fig. 6.9. The x' axis coincides with the x axis, and the t' axis coincides with the straight line $x = Vt$. This means that the t' axis is the worldline of the origin ($x' = 0$) of the moving reference frame, as it should be.

To find the transformation appropriate to relativistic spacetime, let us begin by plotting the worldine of a light signal emitted at $ct = 0$, $x = 0$. Since $x = ct$ for a signal emitted in the positive x direction, the wordline of this signal lies halfway between the $+ct$ and $+x$ axes; and for a signal emitted in the $-x$ direction, the worldline lies halfway between the $+ct$ and $-x$ axes (see the worldlines toward the right and the left in Fig. 6.10). How do we have to draw the ct' and x' axes in this spacetime diagram? The ct' axis must, again, coincide with the straight line $x = Vt$, which is the worldline of the origin of the moving reference frame. The

Figure 6·9

Graphical representation of the Galilean transformation in spacetime. By inspection of the diagram, we see that $x' = x - Vt$.

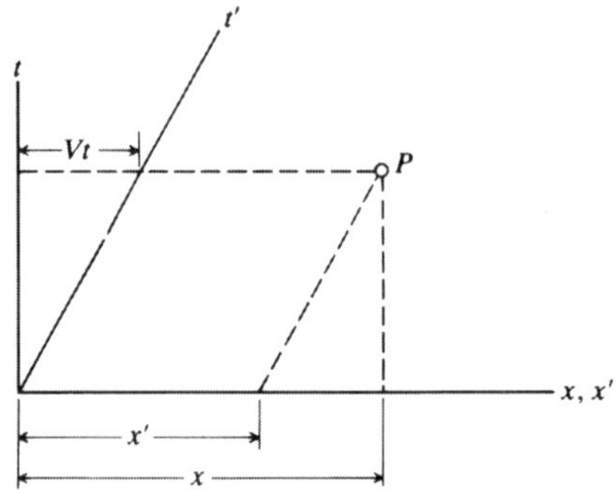

x' axis must also be a straight line, as is evident from the argument given in connection with Eq. (21), which established that the transformation equations between x', ct' and x, ct are linear. Furthermore, this straight line must be arranged so the worldline of the light signal makes equal angles with the ct' and x' axes. The easiest way to understand this requirement of equal angles is to examine what would happen if the angles were not equal, say if the x' axis were to make a larger angle with the worldline of the rightward light signal. Then the $-x'$ axis would make a *smaller* angle with the worldline of the leftward light signal (see Fig. 6.11). This would mean that the speed $|dx'/dt'|$ of the leftward signal would be *larger* than the speed of the rightward signal, in contradiction to the requirement of equal speeds imposed by the principle of universality for the speed of light. Thus, the symmetric arrangement of the ct' and x' axes with respect to the worldline of the light signal is a direct consequence of this principle.

The transformation equations between the rectangular spacetime coordinates x,ct and the symmetrically slanted coordinates x',ct' are then

$$ct' \propto ct - \frac{V}{c}x \qquad (28)$$

$$x' \propto x - \frac{V}{c}ct \qquad (29)$$

The symmetry in these formulas expresses the symmetry in the arrangement of the axes. The formulas are written as proportionalities, rather than equalities, because Fig. 6.10, by itself, requires only that the t' and x' axes be placed symmetrically with respect to the worldline of the light signal, but it does not fix the scale along these axes. Including a constant

Figure 6·10

Worldlines of light signals emitted toward the right and the left from the origin. The ct' and x' axes must be arranged symmetrically in relation to these worldlines.

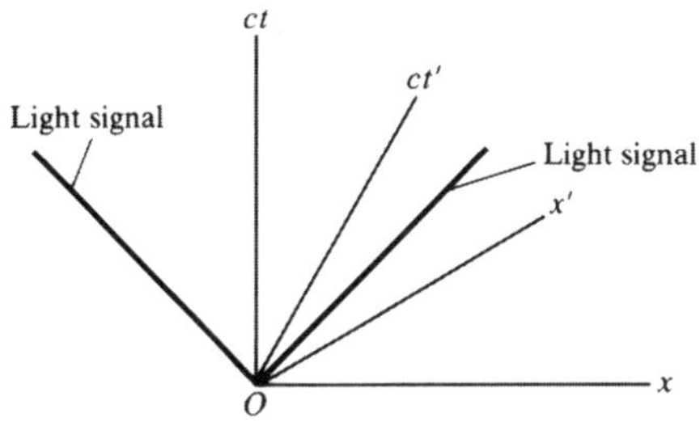

Figure 6·11

If the x' axis makes a larger angle with the worldline of the rightward light signal, then it makes a smaller angle with the worldline of the leftward light signal.

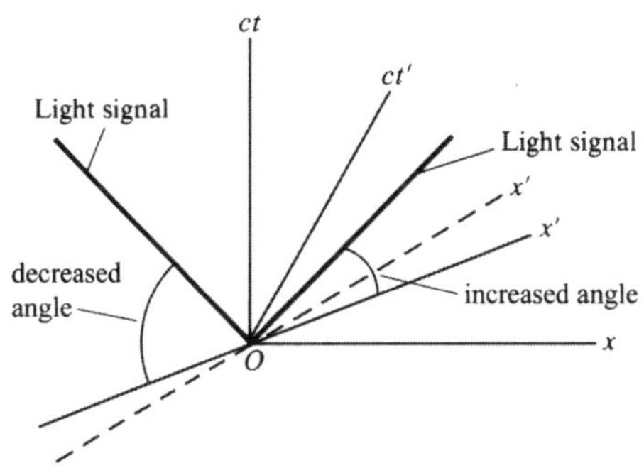

of proportionality in these formulas and canceling a factor of c, we can write

$$t' = \gamma(t - Vx/c^2) \quad \textbf{(30)}$$

$$x' = \gamma(x - Vt) \quad \textbf{(31)}$$

where the constant γ remains to be determined.

The value of γ can be deduced from the invariance of the spacetime interval, as indicated in Eq. (26). If we use Eqs. (30) and (31) to express $c^2 dt'^2 - dx'^2$ in terms of dt and dx, we obtain

$$\gamma^2 c^2 (dt - Vdx/c^2)^2 - \gamma^2 (dx - Vdt)^2 = c^2 dt^2 - dx^2$$

or

$$\gamma^2 c^2 (1 - V^2/c^2) dt^2 - \gamma^2 (1 - V^2/c^2) dx^2 = c^2 dt^2 - dx^2 \quad \textbf{(32)}$$

This equation implies

$$\gamma = \frac{1}{\sqrt{1 - V^2 / c^2}} \tag{33}$$

Hence the required transformations are

$$t' = \frac{1}{\sqrt{1 - V^2 / c^2}} (t - Vx / c^2) \tag{34}$$

$$x' = \frac{1}{\sqrt{1 - V^2 / c^2}} (x - Vt) \tag{35}$$

In three space dimensions, we must also deal with y and z. From Eq. (27), we know that

$$dy'^2 + dz'^2 = dy^2 + dz^2$$

Without any significant loss of generality, we can therefore assume

$$y' = y \tag{36}$$

$$z' = z \tag{37}$$

The only possible alternative to Eqs. (36) and (37) involves a rotation of the y-z plane about the x axis; this alternative is of no interest in the present context.

The transformation (34)–(37) is called the *Lorentz transformation.* It was originally formulated by Lorentz during his investigations of the Maxwell equations, but its implications for relativity were not fully understood until Einstein's work.

Graphically, the presence of the factor γ in Eqs. (30) and (31) means that a slanted line from the unit point $ct = 1$ on the ct axis intercepts the ct' axis at $ct' = \gamma$; hence the unit point on the ct' axis must be closer to the origin than this intercept by a factor $1/\gamma$ [see Fig. 6.12 (*a*)]. This sets the scale of the diagram.

A more elegant way of obtaining the scale of the diagram is to use the invariance of the interval. Consider the hyperbola

$$c^2t^2 - x^2 = 1 \tag{38}$$

This hyperbola passes through the unit point on the ct axis. But since the (finite) spacetime interval $c^2t^2 - x^2$ is invariant, the equation of this hyperbola may also be written as

$$c^2t'^2 - x'^2 = 1 \tag{39}$$

Figure 6·12

Calibration of the ct' axis. (*a*) The ct' coordinate of the point $ct = 1$, $x = 0$ is $ct' = \gamma$. (*b*) The hyperbola $c^2t^2 - x^2 = 1$ (or $c^2t'^2 - x'^2 = 1$) passes through the points $ct = 1$, $x = 0$ and $ct' = 1$, $x' = 0$.

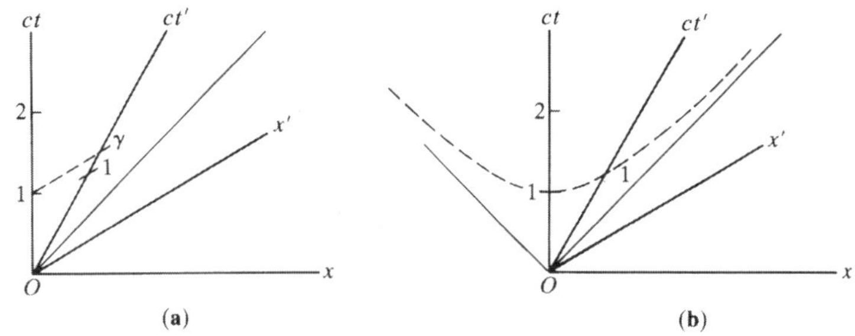

which shows that it passes through the unit point on the ct' axis. Hence, the hyperbola joins the unit points on the time axes [see Fig. 6.12(*b*)].

In the limiting case of low velocity, the Lorentz transformation reduces to

$$x' = x - Vt \tag{40}$$

$$t' = t - Vx/c^2 \tag{41}$$

Furthermore, the term Vx/c^2 in Eq. (41) can often be neglected. For example, if $V = 10^{-7}c$ (corresponding to ~100 km/h) and $x = 10^7$ m, then $Vx/c^2 \simeq 3\times10^{-9}$ s, and such a small time can usually be neglected. In this case, Eqs. (40) and (41) reduce to the Galilean equations. We should, however, keep in mind that if x is large, then Vx/c^2 can be important even if V is small—taking the limit of low velocity is not always enough to eliminate all relativistic effects.

EXAMPLE: *reversal of time order*

One of the most astonishing consequences of the Lorentz transformation is that the time order between two events is not necessarily absolute. For example, consider the events P (at $ct = 0$, $x = 0$) and Q of Fig. 6.13. In the xt coordinate system, Q is clearly later than P. But in the $x't'$ system, Q has a negative ct' coordinate and is *earlier* than P. A simple change of reference frame has reversed the time order between two events.

It is clear that such a reversal can be arranged only if Q is below the 45° line (light worldline) drawn from P; this line represents the limiting position for an x' axis. In relation to the point P, all of spacetime may be divided into three regions: the *absolute future*, the *absolute past*, and the *neutral region* (see Fig. 6.14). Anything that happens in the absolute future or past will be, respectively, later or earlier than P as seen from every reference frame. Anything that happens in the *neutral region* can be later than P, simultaneous with P, or earlier than P, depending on the

Figure 6·13

The ct coordinate of Q is positive; the ct' coordinate of Q is negative.

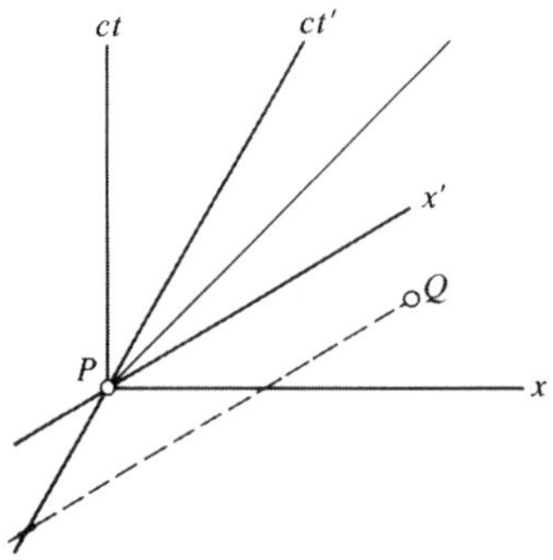

Figure 6·14

The absolute future, the absolute past, and the neutral region in relation to the point P.

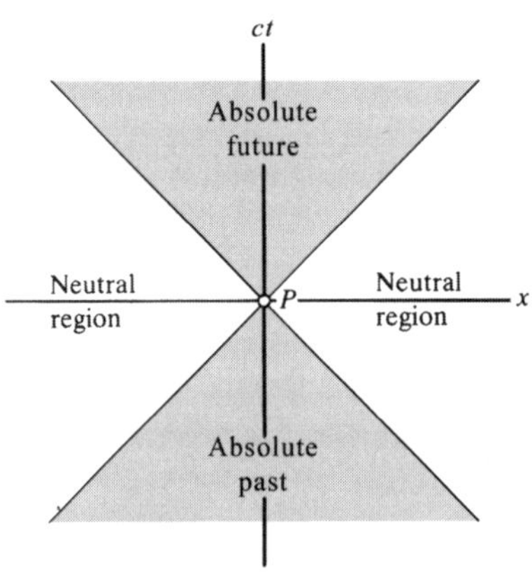

choice of reference frame. The boundary between these regions consists of the worldlines of all possible light signals leaving P or arriving at P. This boundary is called the *light cone.* If we draw a spacetime diagram with two space and one time axes (see Fig. 6.15), then this boundary indeed has the shape of a cone; for three space dimensions, it is a cone in four dimensions, and it cannot be visualized.

There can be no causal connection between what happens at P and what happens somewhere in the neutral zone. Causes must precede their effect (by definition), but whether a point Q in the neutral zone is later or earlier than P depends entirely on the choice of reference frame. In contrast, if Q is in the absolute future of P, then P could be the cause of Q, and if Q is in the absolute past of P, then Q could be the cause of P.

Figure 6·15

The light cone.

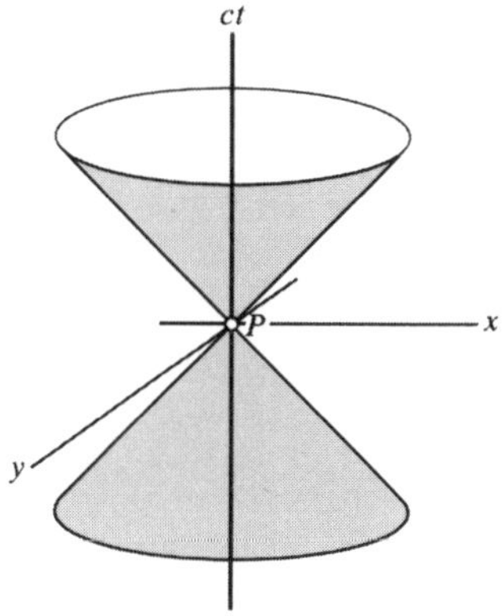

Since the existence of a signal between two events allows them to be causally related, it follows that no signals can be sent from P into the neutral region; therefore, no signals can be sent whose speed exceeds the speed of light. Note that if such a signal could be sent from P to Q in Fig. 6.13, then in the $x't'$ reference frame, the signal would be received at Q before it was sent out from P.

EXAMPLE: *implications of speed greater than c*

Even more is true: if there exist signals that propagate at a speed greater than that of light, then you can send signals into your own past. For simplicity, suppose we have available signals of infinite speed (a signal of speed exceeding that of light will be a signal of infinite speed in some other reference frame). Figure 6.16 shows your worldline (a vertical line) and also the worldline of such a signal emitted by you at P. The signal travels along the x axis and is received by some device at Q in a spaceship moving away from P at high speed. If, in the reference frame of this spaceship, a similar infinite-speed signal is sent back along the $-x'$ axis, then the signal intercepts the worldline on which P lies at Q'. The

Figure 6·16

Tachyon signals from P to Q and from Q to Q'.

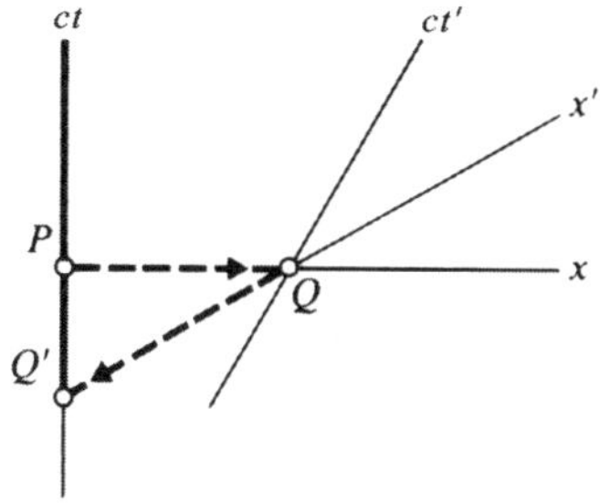

point Q' is in the past of P—you receive a signal from yourself before you send it. This leads to logical contradictions, because you could program yourself to send the signal if and only if you do not receive the signal.

There have been some speculations about *tachyons*, that is particles that move faster than light. According to the preceding arguments, the existence of such particles would lead to logical contradictions. Some experimental searches for tachyons have been performed, but with no success.

The nonexistence of signals of speed larger than that of light implies the nonexistence of rigid bodies. A sudden push on a rod made of absolutely rigid material could be used to send a signal from one end of the rod to the other with infinite speed. Since, in fact, such a signal would consist of a sound wave (compressional shock wave) traveling down the rod, we can conclude that the maximum permissible rigidity of a material is the rigidity that gives sound waves a speed equal to that of light.

The relativistic formula for the transformation of velocity is obtained by differentiating Eqs. (34) and (35):

$$dx' = \gamma(dx - Vdt) \qquad dt' = \gamma(dt - Vdx/c^2) \tag{42}$$

Hence

$$\frac{dx'}{dt'} = \frac{dx - Vdt}{dt - Vdx/c^2} = \frac{(dx/dt) - V}{1 - V(dx/dt)/c^2} \tag{43}$$

If we write $u_x \equiv dx/dt$ and $u'_x \equiv dx'/dt'$, then

$$u'_x = \frac{u_x - V}{1 - u_x V/c^2} \tag{44}$$

We can solve this equation for u_x and obtain the inverse transformation:

$$u_x = \frac{u'_x + V}{1 + u'_x V/c^2} \tag{45}$$

These equations give the transformation formulas for the x component of the velocity; it is easy to derive corresponding formulas for the y and the z components of the velocity.

■ ***Exercise 3.*** Show that

$$u'_y = u_y \frac{\sqrt{1 - V^2/c^2}}{1 - u_x V/c^2} \tag{46}$$

$$u'_z = u_z \frac{\sqrt{1-V^2/c^2}}{1-u_x V/c^2} \tag{47}$$

■

According to Eq. (45), a light signal with $u'_x = c$ sent out from a light source approaching the laboratory at velocity V has a velocity

$$u_x = \frac{c+V}{1+cV/c^2} = c \tag{48}$$

in the laboratory. This result confirms that the relativistic formulas for the combination of velocities satisfy the basic requirement that the speed of light be invariant.

The velocity-combination formula has been tested experimentally with moving sources of light. According to Eq. (48), the light emitted by a moving source has the same speed as the light from a stationary source. This independence of the speed of light from the speed of the source is, of course, also true in the ether theory. Hence an experimental test of this independence cannot discriminate between relativity and ether theory, but it can discriminate between relativity and emission theories, in which a source with a speed V produces light with a speed $c \pm V$ (for an approaching or a receding source, respectively). As we noted earlier, such a dependence of the speed of light on the speed of the source would lead to weird effects in the observations of binary star systems. The most stringent observational constraints emerge from an analysis of data on X-ray pulses emitted by binary, pulsating X-ray sources, such as Hercules X-1.[22] This source consists of a neutron star, which is orbiting around a normal star. The component of the velocity of the neutron star along our line of sight changes by about 300 km/s as this star moves around its orbit. If this change of velocity were to affect the speed of the X rays emitted by the neutron star, then pulses from different portions of the orbit would travel toward the Earth at different speeds, and some pulses emitted at different times would arrive at the Earth at the same time. Thus an observer on the Earth would see several sets of overlapping pulses, or "ghosts." From the absence of such overlapping pulses in the observational data, it is possible to infer an upper limit on the variation of the speeds of the X rays caused by the orbital motion: the speeds of the X rays emitted at different times vary by no more than 6×10^{-7} km/s, compared with the variation of 300 km/s expected according to emission theory.

Direct laboratory tests of the velocity-combination formula have been performed with gamma rays emitted by high-speed π^0 mesons. These mesons spontaneously decay into two gamma rays,

$$\pi^0 \rightarrow \gamma + \gamma$$

each of which has a speed c in the reference frame of the π^0. In an experiment by Alväger et al.,[23] the speed of the π^0 mesons was in excess of $0.99975c$. Nevertheless, direct measurement of the speeds of gamma-ray photons emitted in the (nearly) forward direction gave a result of $(0.99992 \pm 0.00013)c$ in the reference frame of the laboratory—equal to the speed of light to within the experimental errors. For comparison, according to emission theory, the Galilean addition law for velocities would have predicted that photons emitted in the forward direction have a speed of about $2c$ with respect to the laboratory.

6·6 Time Dilation

Suppose that the reference frame $x'y'z't'$ is moving at high speed through the laboratory. How does the rate of the clocks belonging to this reference frame compare with the rate of laboratory clocks?

Figure 6.17 shows the axes of the $x'y'z't'$ frame and of the laboratory frame, or $xyzt$ frame. A clock at rest in the laboratory has a worldline that coincides with the ct axis. The clock ticks at uniform intervals of one second. Each tick is an event. What is the time interval between these events as measured in the $x'y'z't'$ frame? If we draw the slant lines toward the ct' axis, they intercept the t' axis above the calibration marks. Hence, as measured by the $x'y'z't'$ clocks, the laboratory clock goes slow. From the scale of our diagram, we know that the calibration marks on the ct' axis differ from the slant-line intercepts by a factor γ. It follows that when the laboratory clock shows an elapsed time Δt, the $x'y'z't'$ clocks measure this time as

$$\Delta t' = \gamma \Delta t = \frac{\Delta t}{\sqrt{1 - V^2/c^2}} \qquad \text{(clock at rest in } xyzt \text{ frame)} \qquad (49)$$

Figure 6·17

Ticking of a clock at rest in the laboratory.

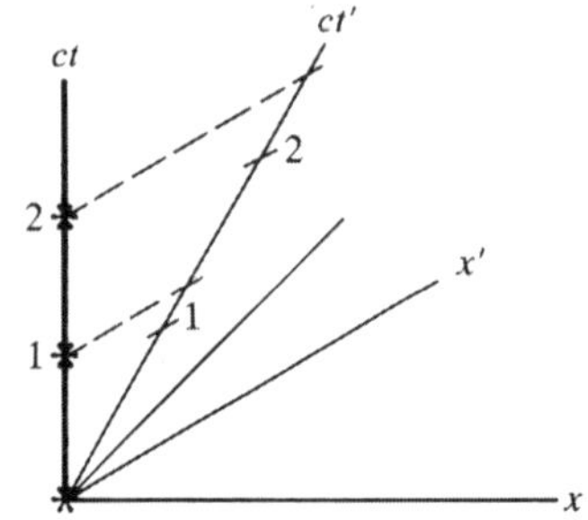

Another, quicker way to obtain this result is by appeal to the invariance of the spacetime interval between one tick of the laboratory clock and the next,

$$c^2\Delta t^2 - \Delta x^2 = c^2\Delta t'^2 - \Delta x'^2 \tag{50}$$

In the laboratory frame, this clock is at rest, $\Delta x = 0$; in the $x'y'z't'$ frame, this clock moves backward, $\Delta x' = -V\,\Delta t'$. Hence

$$c^2\Delta t^2 = c^2\Delta t'^2 - V^2\Delta t'^2$$

which is equivalent to Eq. (49).

We might also ask: If a clock in the $x'y'z't'$ frame shows an elapsed time $\Delta t'$, how long is this time according to the laboratory clocks? Now our diagram looks as shown in Fig. 6.18, and the $x'y'z't'$ clock is slow. We can easily check that the factor between the time intervals is again γ,

$$\Delta t = \frac{\Delta t'}{\sqrt{1 - V^2/c^2}} \qquad \text{(clock at rest in } x'y'z't' \text{ frame)} \tag{51}$$

This shows that the time-dilation effect is symmetric: as measured in a given, inertial reference frame, the moving clock always suffers time dilation.

We can summarize both Eqs. (49) and (51) by writing

$$dt = \frac{d\tau}{\sqrt{1 - V^2/c^2}} \tag{52}$$

where $d\tau$ is the time interval measured by a clock in its own reference frame (time shown by the clock) and dt is the time measured in an inertial reference frame in which that clock has velocity V. The quantity

Figure 6·18

Ticking of a clock at rest in the $x'y'z't'$ frame.

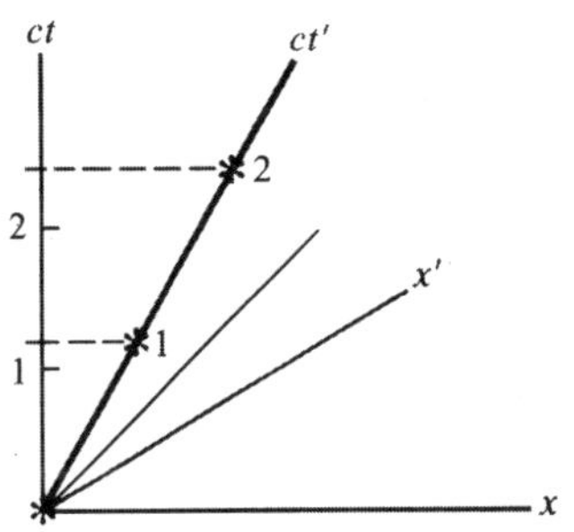

$d\tau$ is called the *proper time* interval. Since $d\tau = dt\sqrt{1-V^2/c^2} = \sqrt{dt^2 - dx^2/c^2} = ds/c$, the proper time is proportional to the square root of the spacetime interval evaluated along the worldline of the clock.

There is some advantage in writing Eq. (52) in differential form because it then may also be applied to a clock in accelerated motion, provided V is taken to be the instantaneous velocity, $V = V(t)$. Equation (52) remains valid in this case because an accelerated motion (with a curved worldline) can be approximated by a sequence of uniform-velocity segments. The total time registered by the moving clock may be obtained by integrating Eq. (52):

$$\Delta\tau = \tau_2 - \tau_1 = \int_{t_1}^{t_2} \sqrt{1-V^2(t)/c^2}\,dt \qquad (53)$$

In using the time-dilation formula for accelerated motion, we must keep in mind that the symmetry of the effect is now lost—we cannot measure an inertial clock with an accelerated clock and find a dilation. We will discuss this asymmetry some more in the context of the twin "paradox" at the end of this section.

Note that the relativistic time dilation spoils the Newtonian procedure for synchronization of clocks by means of clock transport. The transported clock is in motion relative to the clocks of the inertial reference frame; it therefore runs slow and cannot be used to transfer the synchronization from one location to another. However, synchronization by transport remains valid if we proceed to the limit of very low speed. In this case the deviation between the transported clock and the synchronized clocks of the reference frame remains small even if the transport distance is large. To establish this, assume that the transported clock has a speed *V(t);* the deviation between the elapsed proper time $\Delta\tau$ of the transported clock and the elapsed time Δt of the clocks of the reference frame will then be

$$\Delta\tau - \Delta t = \int \sqrt{1-V^2/c^2}\,dt - \Delta t \simeq -\int \frac{1}{2}\frac{V^2}{c^2}\,dt = -\frac{1}{2c^2}\int V dl$$

If we designate the path-average of the speed by $\bar{V}$ and the length of the path by L, then we can expressed this as

$$\Delta\tau - \Delta t \simeq -\bar{V}L/2c^2$$

This approaches zero for $\bar{V} \to 0$ for any path of a finite length. From this we see that instead of synchronizing the clocks of the reference frame by light signals according to the procedure described in Section 6.4, we can equally well synchronize them by transport, provided that the transported clocks move at low speeds.

EXAMPLE: time dilation for high-speed particle

The slowing down of the rate at which time elapses applies to all physical processes. It leads to a drastic dilation of the lifetimes of elementary particles moving at very high speed. For example, an unstable particle such as a muon lives only 2.2×10^{-6} s when at rest in the laboratory. But if such a muon is moving through the laboratory at a speed 0.995c, then its lifetime will be dilated by a factor

$$\gamma = \frac{1}{\sqrt{1-V^2/c^2}} = \frac{1}{\sqrt{1-(0.995)^2}} \simeq 10 \tag{54}$$

An increase of lifetime of this order of magnitude has been observed for muons produced by the impact of primary cosmic rays on the top of the atmosphere. These muons travel 10 or 20 km from the top of the atmosphere to the ground; if they did not benefit from time dilation, they could travel at most $2.2 \times 10^{-6}\ s \times c \simeq$ 0.6 km before their decay, and they could never reach the ground. Quantitative measurements of the lifetime of these muons as a function of speed were first carried out by Rossi and Hall in 1941.[25] A later version of this experiment has confirmed the relativistic formula to within the experimental uncertainty of 4%.[30]

The most precise measurements of the increase of lifetime have been performed with muons and with π mesons produced by accelerators;[33,35] the experimental uncertainties in these measurements were of the order of 0.1%.

The time-dilation effect gives rise to a transverse Doppler shift, which is a frequency shift of the light emitted by a moving source when the line of sight is at right angles to the direction of motion of the source. According to Newtonian physics, there should be no Doppler shift in this case [see Eq. (10)]. But according to relativistic physics, the time dilation reduces the frequency by a factor of $\sqrt{1-V^2/c^2}$; thus, if the frequency of the light is ν_0 in the reference frame of the source, it will be

$$\nu = \nu_0\sqrt{1-V^2/c^2} \tag{55}$$

in the reference frame of the laboratory. If $V/c \ll 1$, we can make the approximation $\sqrt{1-V^2/c^2} \simeq 1 - V^2/2c^2$, and Eq. (55) then leads to a fractional frequency shift

$$\frac{\nu-\nu_0}{\nu_0} \simeq -\frac{V^2}{2c^2} \tag{56}$$

Note that this transverse Doppler shift is of second order in *V*, in contrast to the ordinary Doppler shift, which is of first order in *V*.

This second-order frequency shift was first measured in the light emitted by a beam of high-speed hydrogen atoms with $V/c = 3 \times 10^{-3}$. For this speed,

$$\frac{\nu - \nu_0}{\nu_0} \simeq -4\times10^{-6}$$

Because the effect is so small, the direct detection of the second-order Doppler shift in the light received at right angles to the beam is difficult; even a very slight deviation from exact perpendicularity would introduce a much larger additional first-order Doppler shift, which would completely overwhelm the second-order shift we seek to detect. This difficulty was bypassed with the ingenious method adopted by Ives and Stilwell,[24] who measured the average of the Doppler shifts for the light emitted in the forward and the backward directions along the beam of atoms. It is easy to show that this average Doppler shift coincides with the transverse Doppler shift. An improved version of this experiment[29] confirmed the relativistic time dilation to within about 5%.

More precise tests of the relativistic time dilation have been performed by laser spectroscopy, in which a sample of gas is exposed to the beam from a tunable laser, and the frequency of the latter is adjusted to achieve resonant absorption of the laser light by the atoms. This resonance condition indicates that the frequencies are exactly matched and permits an extremely precise determination of the frequency of the atomic transition. Snyder and Hall[34] used laser spectroscopy to compare the resonant frequency of neon atoms in a discharge cell with the resonant frequency of such atoms in a fast atomic beam crossing the laser beam at right angles. With atoms of a speed of about $10^{-3}c$, they were able to verify the time dilation to within about 0.5%. Juncar et al.[37] obtained similar results using a laser beam parallel to a fast atomic beam. MacArthur et al.[39] verified the time dilation to within 0.03% by means of a laser beam intersecting a very fast atomic hydrogen beam of a speed of $0.84c$ at various angles.

Kaivola et al.[38] used a two-laser technique, in which the atoms in the fast atomic beam resonantly absorb two photons in succession while completing a single transition. One of these photons is supplied by a laser beam aimed in a direction parallel to the atomic beam, and the other photon is supplied by a laser beam aimed in the opposite direction, antiparallel to the atomic beam. Thus the average Doppler shift of these two photons is simply the transverse Doppler shift, as in the Ives-Stilwell experiment. The experiment of Kaivola et al. verified the relativistic time dilation to within 0.004%, and another version of this experiment, by McGowan et al.,[41] verified the time dilation to within 0.0002%. Klein et al.[40] and Grieser et al.[42,43] used an improved version of this two-laser technique with a beam of Li+ ions of higher speed ($0.064c$), and they were able to verify the time dilation to within 7×10^{-7}%. This is the most precise result obtained to date.

Several tests of the relativistic time dilation have been performed by means of the Mössbauer effect. Although the precision of the Mössbauer experiments has by now been surpassed by the laser experiments, the former experiments remain of some interest because they provide empirical evidence that acceleration does not directly influence the time dilation. In the experiments by Hay et al.[26] and by Kündig et al.,[31] the samples of ^{57}Fe serving as emitter and absorber of gamma rays were mounted, respectively, at the center and at the rim of an ultracentrifuge rotor. For a 15-cm rotor spinning at 35,000 rpm, the speed of the absorber is 5.5×10^4 cm/s relative to the emitter. The time dilation then introduces a frequency shift between the resonant frequencies of vibration of the absorber and the emitter. This frequency shift is simply the transverse Doppler shift given by Eqs. (55) and (56). Since $V/c = (5.5 \times 10^4 \text{ cm/s})/(3 \times 10^{10} \text{ cm/s}) = 1.8 \times 10^{-6}$, the predicted, fractional frequency shift is

$$\frac{\nu - \nu_0}{\nu_0} \simeq -2 \times 10^{-12}$$

Although this frequency shift is quite small, it produces a detectable reduction in the resonant absorption of the gamma rays. This experiment has checked the time dilation to within 1.1%.

Another test of the time dilation is provided by the thermal Mössbauer shift. Suppose that the samples of ^{57}Fe that act as emitter and as absorber of the gamma rays are both at rest in the laboratory. The individual atoms and nuclei will then still have a thermal motion. Let us calculate the effect of this thermal motion by means of Eq. (53). This equation requires that the calculations be done from the point of view of an inertial reference frame, say, the laboratory frame. Then, for a given time interval Δt, the elapsed proper time of an atom with speed $V(t)$ is

$$\Delta\tau = \int_0^{\Delta t} \sqrt{1 - V^2(t)/c^2}\, dt \simeq \int_0^{\Delta t} \left[1 - \frac{V^2(t)}{2c^2}\right] dt \qquad (57)$$

The average thermal speed is related to the temperature by

$$\overline{V^2}(t) = \frac{3kT}{m} \qquad (58)$$

where m is the mass of the atom. Hence the average proper time is

$$\Delta\tau = \left(1 - \frac{3}{2}\frac{kT}{mc^2}\right)\Delta t \qquad (59)$$

Thus $(1 - 3kT/2mc^2)$ can be regarded as the time-dilation factor for an atom in thermal motion. This time dilation will reduce the frequency of vibration of the nucleus, as measured in the laboratory. If the temperatures of the emitting and the absorbing samples are T_s and T_a, respectively, then the frequency difference between the nuclei is

$$\nu_s - \nu_a = \left(1 - \frac{3}{2}\frac{kT_s}{mc^2}\right)\nu_0 - \left(1 - \frac{3}{2}\frac{kT_a}{mc^2}\right)\nu_0 = \frac{3}{2}\frac{k}{mc^2}(T_a - T_s)\nu_0 \qquad (60)$$

and the fractional frequency difference is

$$\frac{\nu_s - \nu_a}{\nu_0} = \frac{3}{2}\frac{k}{mc^2}(T_a - T_s) \qquad (61)$$

For samples of ^{57}Fe, this frequency difference amounts to $\simeq 10^{-15}$ per kelvin. The Mössbauer effect does permit the detection of such a small frequency shift. Measurements of the thermal Mössbauer shift by Pound and Rebka[27] confirmed the time dilation to within about 10%. And these measurements also confirmed that the time dilation does not depend (directly) on the acceleration. The nuclei in the samples of iron suffer violent accelerations of up to $10^{16}\,g$ in their thermal motions, but these accelerations have no detectable effect on their rate of vibration.

The time-dilation effect has also been tested with atomic clocks. Häfele and Keating[28] took atomic clocks on trips around the world on commercial aircraft and then compared them with a clock left on the ground. As seen from an inertial reference frame (not rotating with the Earth), clocks carried eastward have a greater speed than clocks on the ground, and clocks carried westward have a lesser speed (see Fig. 6.19). Hence clocks carried eastward should suffer a time dilation, and clocks carried westward a time contraction, as compared with a clock on the ground. In agreement with this prediction, the experiment showed that the east-going clocks lost time compared with the clock on the ground,

Figure 6·19

Velocities of clocks traveling in aircraft and of a clock on the ground (not to scale).

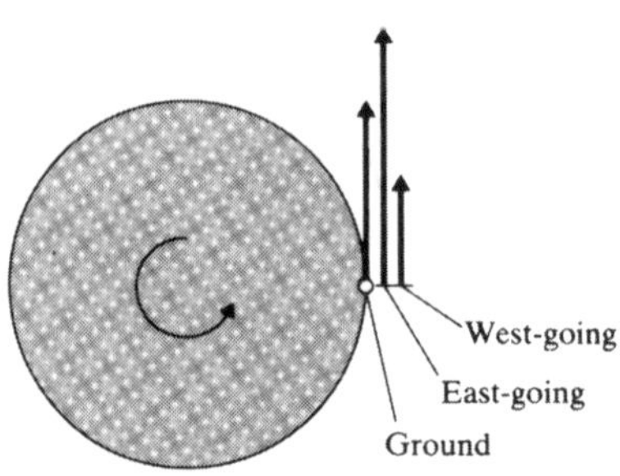

and the west-going clocks gained time. The magnitude of the loss or gain was of the order of 10^{-7}s for one complete trip around the Earth. These experimental measurements cannot be directly compared with the prediction of special relativity because the clocks carried in the aircraft also suffer from an extra gravitational effect: a clock at high altitude,* where the gravitational potential is high, will run somewhat faster than a clock on the ground. The experiment with traveling clocks was, in fact, intended primarily as a test of this gravitational effect predicted by general relativity. But the experiment also served as a test of the time-dilation effect of special relativity, because east-going and west-going clocks suffer the same gravitational effect (if at the same altitude); hence the difference between these clocks must be attributed to their difference in speeds.

EXAMPLE: *time dilation for atomic clock*

The calculation of the predicted time dilation was made from the flight data of the aircraft. To get some feeling for the quantities involved, let us assume that the aircraft flies along the equator, in the eastward direction. The proper time registered by the aircraft clock is then

$$\Delta\tau_1 = \int_0^t \sqrt{1 - V_1^2/c^2}\,dt \qquad (62)$$

(where V_1 is the velocity as measured from an inertial frame), and the proper time for a clock on the ground at the equator is

$$\Delta\tau_2 = \int_0^t \sqrt{1 - V_2^2/c^2}\,dt \qquad (63)$$

Hence

$$\Delta\tau_2 - \Delta\tau_1 \simeq \int_0^t \left(\frac{-V_2^2}{2c^2} + \frac{V_1^2}{2c^2} \right) dt \simeq \int_0^t \frac{1}{2c^2}(V_1 - V_2)(V_1 + V_2)\,dt$$
$$\simeq \frac{1}{2c^2} t (V_1 - V_2)(V_1 + V_2) \qquad (64)$$

where it is assumed that V_1 has approximately a constant magnitude. Since $V_1 - V_2$ is the ground speed of the aircraft, $(V_1 - V_2)t$ is the distance traveled by the aircraft relative to the Earth; thus, for a trip around the equator of the Earth,

$$\Delta\tau_2 - \Delta\tau_1 \simeq \frac{(2\pi R)(V_1 + V_2)}{2c^2} \qquad (65)$$

where R is the radius of the Earth.† With $V_2 = 4.6 \times 10^4$ cm/s for the rotational speed of the Earth and $V_1 = V_2 +$ (ground speed of aircraft) =

---

*Cruising altitude for commercial aircraft is about 30,000 ft.

†Note that, for a clock transported at low velocity relative to the ground ($V_1 - V_2 \ll V_2$), the right side of Eq. (65) becomes approximately $2\pi R V_2/c^2$, independent of V_1.

4.6×10^4 cm/s $+ 2.5\times10^4$ cm/s, this equation predicts a time loss of about 2×10^{-7}s for the aircraft clock. In the experiments, the time loss (or gain) was measured to within about 3%, and it agreed very well with the predicted time loss.

A more precise test of the time dilation with atomic clocks was performed by Vessot et al.[36] In this test, a Scout rocket was used to launch several hydrogen-maser clocks into a nearly vertical trajectory, with a maximum speed of about 8 km/s and a maximum height of about 10,000 km. The clocks were destroyed in the terminal crash of the rocket, so they could not be compared with the ground clocks at the end of the flight. Instead they were compared during flight, by means of radio signals transmitted both ways between the ground station and the rocket. This experiment, like the flying-clock experiment, was intended primarily to test the general-relativistic time dilation, but it also tested the special-relativistic time dilation (the special-relativistic effect dominates during the initial and final portions of the trajectory, at high speed, and the general-relativistic effect dominates when the rocket is near apogee, at high altitude). The rocket experiment confirmed the time dilation to within the experimental uncertainties of about 0.007%.

Table 6.6 summarizes the experimental evidence for time dilation.

Table 6·6 Time-dilation experiments

Experiments	Year	Experimental Uncertainty, %
Ives and Stilwell (Doppler shift)[24]	1938	3
Rossi and Hall (muon lifetime)[25]	1941	~50
Hay et al. (Mössbauer, rotor)[26]	1960	2
Pound and Rebka (Mösbauer, thermal)[27]	1960	10
Häfele and Keating (atomic clocks)[28]	1962	3
Mandelberg and Witten (Doppler shift)[29]	1962	5
Frisch and Smith (muon lifetime)[30]	1963	4
Kündig (Mössbauer, rotor)[31]	1963	1
Champeney et al.[32]	1963	3
Ayres et al. (pion lifetime)[33]	1971	0.4
Snyder and Hall (Doppler shift, laser)[34]	1975	0.5
Bailey et al. (muon lifetime)[35]	1977	0.1
Vessot et al. (atomic clocks)[36]	1980	0.007
Juncar et al. (Doppler shift, laser)[37]	1985	0.5
Kaivola et al. (Doppler shift, two lasers)[38]	1985	0.004
MacArthur et al. (Doppler shift, two lasers)[39]	1986	0.03
Klein et al. (Doppler shift, two lasers)[40]	1992	0.001
McGowan et al. (Doppler shift, two lasers)[41]	1993	0.0002
Grieser et al. (Doppler shift, two lasers)[42,43]	1994	7×10^{-7}

EXAMPLE: *twin paradox*

Finally, let us discuss the celebrated twin paradox. Suppose that Stella and Terra are identical twins, 30 years old. Stella leaves the Earth (regarded as an inertial frame) traveling in a rocket ship at $V/c = ½$; she then turns around (almost instantaneously) and returns to the Earth when Terra is 40 years old. How old is Stella? The proper time (in years) elapsed along Stella's worldline is

$$\Delta\tau = \int_{30}^{40} \sqrt{1-V^2/c^2}\,dt = \int_{30}^{40} \sqrt{1-(½)^2}\,dt \simeq 8.7 \qquad (66)$$

Hence Stella is only 38.7 years old. However, as seen from the reference frame of Stella, Terra was moving at $V/c = ½$, and it would seem that Terra should have suffered time dilation. This is the twin paradox.

The resolution of the paradox hinges on the fact that Stella's reference frame is not inertial—to reverse her velocity and return to Earth, Stella must accelerate. Although this acceleration has no effect on the aging of Stella as measured by Terra, it does have an effect on the aging of Terra as measured by Stella. Figure 6.20 shows the worldlines of Stella and Terra. The two reference frames $x't'$ and $x''t''$ are, respectively, the rest frames for Stella on the outward and return portions of her trip. The reference frame $x't'$ has velocity V along the x axis of the xt frame; the reference frame $x''t''$ has velocity $-V$. The origins of these reference frames have been placed such that the turnaround point P corresponds to $t' = 0$, $t'' = 0$.

As far as Stella is concerned, the trip begins at A, at time $t'_A < 0$. Stella reaches the turnaround point P at time $t' = 0$. Meanwhile, Terra moves along her worldline and reaches the point B at $t' = 0$. Between A and B, Terra's clock registers an increase $t_B - t_A$ in proper time. According to Stella, this time interval is dilated:

Figure 6·20

Worldlines of Stella (*APD*) and of Terra (*ABCD*).

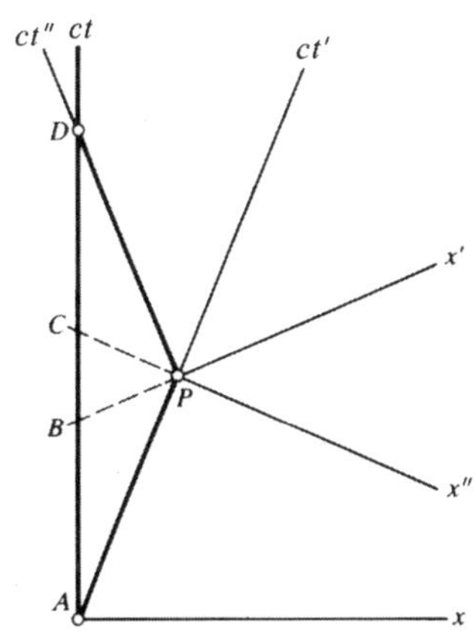

$$t_B - t_A = \sqrt{1 - V^2/c^2}\,|t'_A| \tag{67}$$

After passing the turnaround point P, Stella is at rest in the $x''t''$ frame, and she returns to Earth at time t''_D. In the time interval from $t'' = 0$ to $t'' = t''_D$, Stella observes Terra to move from C to D. The time shown by Terra's clock increases by $t_D - t_C$, and Stella judges this interval to be dilated:

$$t_D - t_C = \sqrt{1 - V^2/c^2}\,t''_D \tag{68}$$

However, the total time registered by Terra's clock includes not only the intervals AB and CD but also the interval BC. This latter interval does not correspond to negative values of t' nor to positive values of t''. The finite time interval $t_C - t_B$ elapses on Terra's clock in the *one instant* that Stella needs to transfer from one inertial reference frame to another at $t' = t'' = 0$. According to Stella, in the instant between $t' = 0^-$ and $t'' = 0^+$, the hands on Terra's clock jump ahead by $t_C - t_B$. Of course, such a discontinuity is unphysical, but the blame for this must be placed on the unphysical worldline of Stella—we have assumed an instantaneous change in velocity at the point P, with infinite acceleration. If we make a more realistic assumption, with a gradual change in Stella's velocity, then we find that Terra's clock does not jump, but simply speeds up during the time that Stella accelerates. It is this speeding up of Terra's clock which more than compensates for the time dilation along the other parts of the worldline [see Eqs. (67) and (68)] and makes Terra older, no matter what reference frame is used to calculate the aging.

6·7 Length Contraction

Suppose that a meter stick is at rest in the reference frame $x'y'z't'$ moving through the laboratory at high speed. Figure 6.21 shows the worldlines of the front end and of the rear end of this meter stick. The meter stick occupies the region of spacetime that is shown hatched in Fig. 6.21. How long is this meter stick as seen in the laboratory frame? At a given instant ($t = 0$), the rear end of the meter stick is at A, and the front end is at B. Since the point B is nearer the origin than the 1-meter calibration mark, its length is less than 1 meter as measured in the laboratory frame. By the construction of the calibration on the axes, the length contraction factor is $1/\gamma$:

$$\Delta x = \frac{\Delta x'}{\gamma} = \sqrt{1 - V^2/c^2}\,\Delta x' \quad \text{(rod at rest in } x'y'z't' \text{ frame)} \tag{69}$$

Figure 6·21

Worldlines of the front end and of the rear end of a meter stick at rest in the $x'y'z't'$ frame.

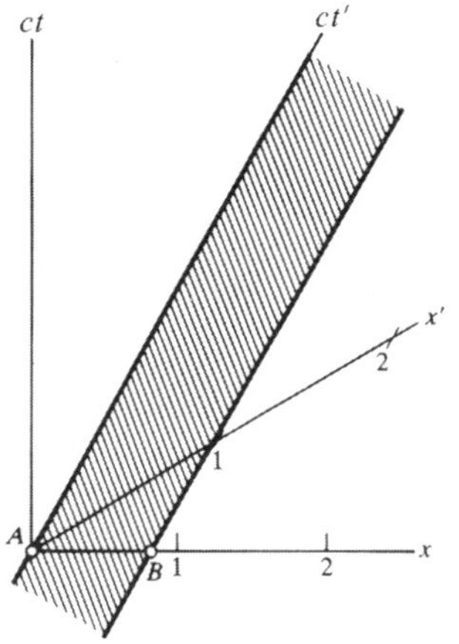

For a meter stick at rest in the laboratory, the situation is as shown in Fig. 6.22. The measured length in the $x'y'z't'$ frame is AB, and this length is again shorter than 1 meter:

$$\Delta x' = \sqrt{1 - V^2/c^2}\,\Delta x \qquad \text{(rod at rest in } xyzt \text{ frame)} \tag{70}$$

The general conclusion is, then, that moving bodies are always shortened in the direction of their motion. Note that the dimensions in the trans-

Figure 6·22

Worldlines of the front end and of the rear end of a meter stick at rest in the laboratory.

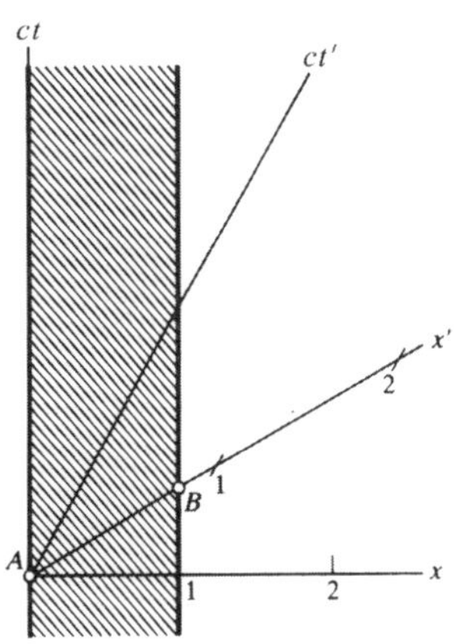

Figure 6·23

Simulated photograph of a set of four parallelepipeds moving by at high speed.

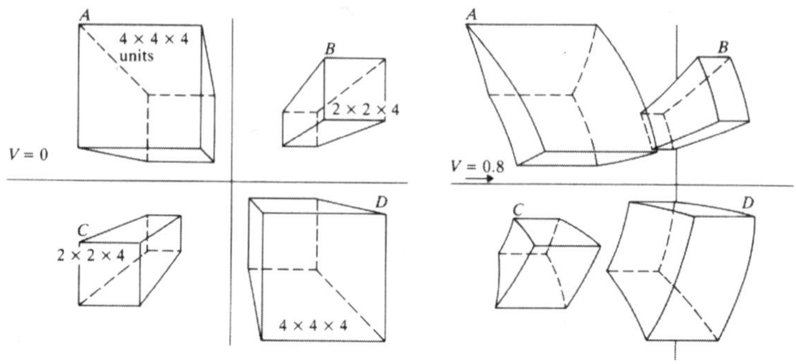

verse directions remain unchanged. One consequence of this contraction is that if a body has a volume $\Delta x \Delta y \Delta z$ as measured in its own reference frame, then it will have a volume

$$\Delta x' \Delta y' \Delta z' = \sqrt{1 - V^2 / c^2}\, \Delta x \Delta y \Delta z \tag{71}$$

as measured in an inertial reference frame in which the body has velocity V.

Although the length contraction is real, it would be difficult to observe, even if we had available a macroscopic body passing by us at relativistic speed. If we were to photograph such a body, the contraction effect would be obscured by time-delay effects. A photograph of a body does not show where the body is at a given time; rather, it shows the light that reaches the camera at a given time. This means that the light from those parts of the body that are farther away from the camera must be emitted earlier than the light from those parts that are nearer. Hence, quite apart from any length contraction, the photograph of a fast-moving body will always show some distortion.

Figure 6.23 shows what the photograph of a set of four parallelepipeds would look like;[44] both the length contraction and the distortion arising from the time delays associated with light propagation have been taken into account. The two parallelepipeds on the front show an obvious contraction, but the other two appear dilated.

References

1. H. A. Lorentz, A. Einstein, H. Minkowski, and H. Weyl, *The Principle of Relativity* (Dover, New York), p. 75.
2. H. C. Ohanian, *Am. J. Phys.* **72**, 141 (2004).
3. A. Einstein, in *Albert Einstein: Philosopher-Scientist*, ed. P. Schilpp (Tudor, New York, 1951), p. 53.
4. G. Joos, *Ann. Phys. (Leipzig)* **7**, 385 (1930); *Naturwissenschaften* **38**, 784 (1931).
5. L. Essen, *Nature* **175**, 793 (1955).
6. T. S. Jaseja, A. Javan, J. Murray, and C. H. Townes, *Phys. Rev.* A **133**, 1221 (1964).
7. A. Brillet and J. L. Hall, *Phys. Rev. Lett.* **42**, 549 (1979).
8. R. S. Shankland, S. W. McCuskey, F. C. Leone, and G. Kuerti, *Rev. Mod. Phys.* **27**, 167 (1955).
9. A. A. Michelson and H. G. Gale, *Astrophys. J.* **61**, 240 (1925).
10. R. J. Kennedy and E. M. Thorndike, *Phys. Rev.* **42**, 400 (1932).
11. D. Hils and J. L. Hall, *Phys. Rev. Lett.* **64**, 1697 (1990).
12. J. Müller and M. H. Soffel, *Phys. Lett.* A **198**, 71 (1995).
13. J. P. Cedarholm, G. F. Bland, B. L. Havens, and C. H. Townes, *Phys. Rev. Lett.* **1**, 342 (1958).
14. T. Cranshaw and H. Hay; quoted by Champeney et al., Ref. 15.
15. D. C. Champeney, G. R. Isaak, and A. M. Khan, *Phys. Lett.* **7**, 241 (1963).
16. K. C. Turner and H. A. Hill, *Bull. Am. Phys. Soc.* **8**, 28 (1963).
17. G. R. Isaak, *Phys. Bull.* **21**, 255 (1970).
18. E. Riis, L. A. Andersen, N. Bjerre, O. Poulsen, S. A. Lee, and J. L. Hall, *Phys. Rev. Lett.* **60**, 81 (1988).
19. T. W. Cole, *Mon. Not. R. Astr. Soc.* **175**, 93P (1976).
20. T. P. Krisher, L. Maleki, G. F. Lutes, L. E. Primas, R. T. Logan, and J. D. Anderson, *Phys. Rev.* D, **42**, 731 (1990).
21. P. Wolf and G. Petit, *Phys. Rev.* A, **56**, 4405 (1997).
22. K. Brecher, *Phys. Rev. Lett.* **39**, 1051 (1977).
23. T. Alväger, F. J. M. Farley, J. Kjellman, and I. Wallin, *Phys. Lett.* **12**, 260 (1964).
24. H. E. Ives and G. R. Stilwell, *J. Opt. Soc. Am.* **28**, 215 (1938).
25. B. Rossi and D. B. Hall, *Phys. Rev.* **59**, 223 (1941).
26. H. J. Hay, J. P. Schiffer, T. E. Cranshaw, and P. A. Egelstaff, *Phys. Rev. Lett.* **4**, 165 (1960).
27. R. V. Pound and G. A. Rebka, *Phys. Rev. Lett.* **4**, 274 (1960).

28. J. C. Häfele and R. E. Keating, *Science* **177**, 168 (1962).

29. H. I. Mandelberg and L. Witten, *J. Opt. Soc. Am.* **52**, 529 (1962).

30. D. H. Frisch and J. H. Smith, *Am. J. Phys.* *31*, 342 (1963).

31. W. Kündig, *Phys. Rev.* **129**, 2371 (1963).

32. D. C. Champeney, G. R. Isaak, and A. M. Khan, *Nature* **198**, 1186 (1963); *Proc. Phys. Soc. London* **85**, 583 (1965).

33. D. S. Ayres, A. M. Cormack, A. J. Greenberg, R. W. Kenney, D. O. Cladwell, V. B. Elings, W. P. Hesse, and R. J. Morrison, *Phys. Rev. D.* **3**, 1051 (1971).

34. J. J. Snyder and J. L. Hall, in *Laser Spectrosopy*, ed. S. Haroche et al. (Springer-Verlag, New York, 1975).

35. J. Bailey, K. Borer, K. Combley, H. Drumm, F. Krienen, F. Lange, E. Picasso, W. von Ruden, F. J. Farley, J. H. Field, and P. M. Hattersley, *Nature* **268**, 301 (1977).

36. R. F. C. Vessot, M. W. Levine, E. M. Mattison, E. L. Blomberg, T. E. Hoffman, G. U. Nystrom, B. F. Farrel, R. Decher, P. B. Eby, C. R. Baugher, J. W. Watts, D. L. Teuber, and F. D. Wills, *Phys. Rev. Lett.* **45**, 2081 (1980).

37. P. Juncar, C. R. Bingham, J. A. Bounds, D. J. Pegg, H. K. Carter, R. L. Mlekodaj, and J. D. Cole, *Phys. Rev. Lett.* **54**, 11 (1985).

38. M. Kaivola, O. Poulsen, E. Riis, and S. A. Lee, *Phys. Rev. Lett.* **54**, 255 (1985).

39. D. W. MacArthur, K. B. Butterfield, D. A. Clark, J. B. Donahue, P. A. M. Gram, H. C. Bryant, C. J. Harvey, W. W. Smith, and G. Comtet, *Phys. Rev. Lett.* **56**, 282 (1986).

40. R. Klein, R. Grieser, I. Hoog, G. Huber, I. Klaft, P. Merz, T. Kühl, S. Schröder, M. Grieser, D. Habs, W. Petrich, and D. Schwalm, *Z. Phys.* A **342**, 455 (1992).

41. R. W. McGowan, D. M. Giltner, S. J. Sternberg, and S. A. Lee, *Phys. Rev. Lett.* **70,** 251 (1993).

42. R. Grieser, R. Klein, G. Huber, S. Dickopf, I. Klaft, P. Knobloch, P. Merz, F. Albrecht, M. Grieser, D. Habs, D. Schwalm, and T. Kühl, *Appl. Phys.* B **59**, 127 (1994).

43. R. Grieser, T. Kühl, and G. Huber, *Am. J. Phys.* **63**, 665 (1995).

44. G. D. Scott and M. R. Viner, *Am. J. Phys.* **33**, 534 (1965).

45. P. Antonini, M. Okhapkin, E. Göklü, and S. Schiller, *Phys. Rev.* A**71**, 050101 (2005).

46. P. L. Stanwix, M. E. Tobar, P. Wolf, M. Susli, C. R. Locke, E. N. Ivanov, J. Winterflood, and F. van Kann, *Phys. Rev. Lett.* **95**, 040404 (2005).

47. S. Herrmann, A. Sanger, E. Kovalchuk, H. Müller, and A. Peters, *Phys. Rev. Lett.* **95**, 150401 (2005).

Problems

1. Since the reference frame of the rotating Earth is not an inertial reference frame, a correct relativistic analysis of the Michelson-Morley experiment requires the use of an inertial frame moving with the Earth, but not rotating with it. In such a reference frame, what is the time taken by a light beam to travel a distance L eastward along one arm of the interferometer? Westward? What is the time taken for a light beam to travel a distance L northward along the other arm? Southward? Is there a difference in the round trip travel time for these light beams? [*Hint*: The lengths L are defined in the reference frame of the rotating Earth; take their length contraction into account.]
2. Consider a closed rectangular (or, more precisely, trapezoidal) path on the surface of the Earth, consisting of four short segments: an eastward segment, followed by a northward segment, followed by a westward segment, and followed by a southward segment. Mirrors placed at an angle at the corners permit light to travel around this path. Show that the difference between the travel times for counterclockwise and clockwise light signals around this path is approximately $4A\omega\sin\Theta/c^2$, where A is the area enclosed in the path, ω the angular velocity of rotation of the Earth, and Θ the latitude at the position of the path. What is the magnitude of the time difference for a path that encloses a horizontal area of 1 m^2 at latitude $45°$? This is called the *Sagnac effect*. [*Hint*: Use an inertial reference frame, not rotating with the Earth.]

3. (a) Calculate the phase difference between the two beams of light in a Michelson-Morley interferometer with arms of unequal lengths, L_1 and L_2, the first of which is perpendicular to the ether wind, and the second parallel. Verify that if a FitzGerald-Lorentz contraction is introduced for the parallel arm, the phase difference becomes

$$\Delta\phi = \frac{2\pi}{\lambda}\frac{2(L_2-L_1)}{\sqrt{1-V^2/c^2}} \simeq \frac{2\pi}{\lambda}[2(L_2-L_1)+(L_2-L_1)V^2/c^2]$$

   (b) Recalculate the phase difference if the arm L_1 is parallel to the ether wind and the arm L_2 perpendicular. Verify that the phase difference is the same as in part (a).
   (c) The second term in the bracket in the equation in part (a) gives the dependence of the phase shift on V^2. The interferometer of Kennedy and Thorndike had $L_2 - L_1 = 16$ cm. With V changing from 220 km/s to 280 km/s in six months, what is the expected change in the phase difference?

4. A straight rigid bar of length $2L$ lies in the x-y plane and rotates about the z axis with uniform angular velocity ω. In the $xyzt$ frame, the bar is perfectly straight at each instant of time, and the points on the bar can be described parametrically by

$$x = R\cos\omega t \qquad y = R\sin\omega t$$

   where R is a parameter that ranges from $-L$ to L. Consider this bar from the point of view of the $x'y'z't'$ frame moving along the x axis with velocity V. Where are the points on the bar at one instant of t' time? Is the bar still straight? Roughly sketch the shape of the bar at one instant of t' time, say, $t'=0$; assume that $V = 0.8c$ and $\omega L = 0.8c$.

5. Equation (9) gives the phase difference between the two emerging beams of light if one arm of the interferometer is parallel and the other arm is perpendicular to the direction of motion of the Earth through the ether. Derive an equation for the phase difference if one arm is at an angle θ and the other arm is at an angle $90° + \theta$ with the direction of motion.

6. The Mössbauer effect hinges on the recoilless emission of a gamma ray by a nucleus embedded in a crystal lattice that can absorb the recoil momentum without a concomitant reduction of the energy of the photon. By contrast, the emission of a gamma ray by a free nucleus involves a reduction of the energy of the photon, making the resonant absorption of this photon by another nucleus impossible.

   (a) The energy of the photon emitted by an excited nucleus of ^{57}Fe (formed in the radioactive decay of ^{60}Co) is 14.4 keV. The momentum of this photon is 14.4 keV/c. What is the recoil energy of the ^{57}Fe nucleus? The mass of this nucleus is 56.94 u.
   (b) What is the fractional reduction of the energy of the photon, as compared with the energy of a photon emitted under recoilless conditions?
   (c) The fractional width $\Delta E/E$ of the resonant-absorption peak for ^{57}Fe is 3×10^{-13}. Can the photon emitted by a free nucleus be reabsorbed by another nucleus?

7. The *geometrodynamic clock* of Marzke and Wheeler is a light clock (compare Problem 22) consisting of two mirrors between which a light pulse bounces back and forth. However, instead of the mirrors being held at a fixed distance by a mechanical restraint, they are allowed to drift freely, and their separation is checked by a trial-and-error procedure involving auxiliary light signals and mir-

Figure 6·24

Construction of parallel worldlines for the geometrodynamic clock.

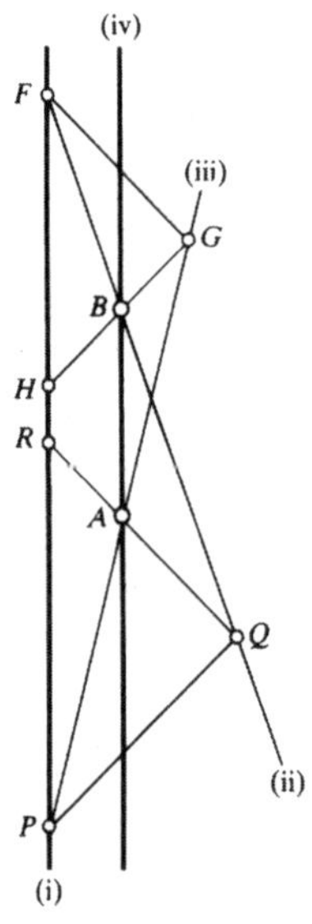

rors. In essence, the procedure hinges on the construction of parallel worldlines, because parallel worldlines in a spacetime diagram imply constant separation in space. Figure 6.24 shows the straight worldline *PF* of one free mirror (i) in uniform motion. To construct a worldline parallel to *PF,* introduce an auxiliary free mirror (ii), with a worldline *QF* that intersects *PF* at some point *F* in the future. A light signal *PQR* is emitted at mirror (i), reflected by (ii), and returned to (i). Introduce another auxiliary mirror (iii), with worldline *PG* that intersects *PF* at *P*. A light signal *HGF* is emitted at mirror (i), reflected by (iii), and returned to (i). Show that if a free mirror has a worldline that passes through the intersections *A* and *B*, then the worldline *AB* of this last mirror is parallel to *PF*.

8. Some quasars, such as 3C 345, have been observed to eject luminous blobs of gas that appear to move away from the quasar with a transverse speed in excess of the speed of light. The explanation of this illusion seems to be that the blob is actually moving at a speed slightly less than the speed of light in a direction making a small angle with the line of sight (see Fig. 6.25). The light emitted when the blob is born at *Q* and the light emitted when the blob arrives at *P* then reach the Earth almost simultaneously, creating the illusion that the blob moves almost instantaneously from *Q* to *P*.

(a) Suppose that the time interval between the birth of the blob at *Q* and its arrival at *P* is Δt_0. Show that if $\theta \ll 1$, the time interval between reception at the Earth of the light emitted at *Q* and the light emitted at *P* is $\Delta t = (1 - V/c)\,\Delta t_0$, where *V* is the speed of the blob.

(b) Show that the apparent transverse speed is

$$\frac{PP'}{\Delta t} = \frac{OP}{QP}\frac{\theta' V/c}{1-V/c}$$

and give a numerical example demonstrating that this speed can exceed the speed of light.

9. The speed of sound in a fluid depends on the derivative of the pressure with respect to the density,

$$v = \sqrt{\frac{dP}{d\rho}}$$

Since the speed of sound must be less than the speed of light, we can deduce the upper bound

$$\frac{dP}{d\rho} \le c^2$$

for the derivative of the pressure with respect to the density. We can use this inequality to set some limits on the maximum pressure permitted in neutron stars, where the equation of state of the material is not well known.

(a) By integration, show that

$$P \le P_0 + (\rho - \rho_0)c^2$$

where P_0 is the pressure at some given density ρ_0.

Figure 6·25

A luminous blob moves from Q to P in a time Δt.

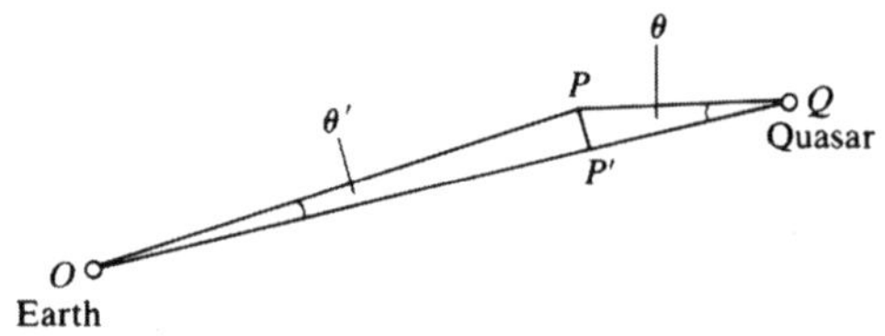

(b) According to calculations by Harrison and Wheeler, the pressure in neutron-star material is 1.22×10^{34} g/cm·s^2 at a density of 4.6×10^{14}g/cm^3. What can you deduce about the pressure at a density of 10^{20} g/cm^3? At a density of 10^{30} g/cm^3?

10. (a) Show that the Lorentz transformation can be written as

$$x' = x\cos\theta - ct\sinh\theta$$

$$ct' = -x\sinh\theta + ct\cosh\theta$$

where θ is the velocity parameter defined by $V/c = \tanh\theta$.

(b) Show that the combination of successive Lorentz transformations with velocity parameters θ_1 and θ_2 is equivalent to a single Lorentz transformation with velocity parameter $\theta = \theta_1 + \theta_2$.

11. (a) Show that the combination of two successive Lorentz transformations with velocities V_1 and V_2 along the same direction (x axis) is equivalent to a single Lorentz transformation with a velocity

$$V = \frac{V_1 + V_2}{1 + V_1V_2/c^2}$$

(b) Show that the set of all Lorentz transformations with this composition rule forms a group.

12. Astronomers on the surface of the Earth (regarded as an inertial reference frame) see two novas flare up simultaneously. One of the novas is at a distance of 1.0×10^4 light-years in the constellation Draco; the other nova is at the same distance in the constellation Tucana in a direction (as seen from the Earth) exactly opposite to that of the first nova. According to astronomers aboard an aircraft traveling at 750 km/h along the line from Draco to Tucana, these novas are not simultaneous. According to these astronomers, which nova happened first? By how many hours?

13. At $14^h0^m0^s$ GMT, lightning strikes in New York City; and at $14^h0^m0.003^s$, lightning strikes in Chicago, at a distance of 1140 km from the first stroke. Find the velocity, relative to the Earth, of a reference frame in which the first stroke occurs after the second.

14. Derive the following Lorentz-transformation formulas for acceleration:

$$a'_x = (1 - V^2/c^2)^{3/2} \frac{a_x}{(1 - Vu_x/c^2)^3}$$

$$a'_y = (1 - V^2/c^2) \frac{a_y}{(1 - Vu_x/c^2)^2} + (1 - V^2/c^2) \frac{Vu_y a_x/c^2}{(1 - Vu_x/c^2)^3}$$

$$a'_z = (1 - V^2/c^2) \frac{a_z}{(1 - Vu_x/c^2)^2} + (1 - V^2/c^2) \frac{Vu_z a_x/c^2}{(1 - Vu_x/c^2)^3}$$

15. Suppose that a spaceship operates its rocket engines in such a way that, as measured in its own reference frame, the acceleration is constant, a'_x = constant.

(a) Show that the worldline of this spaceship has the equation

$$x = \frac{c^2}{a'_x}\left(\sqrt{1 + (a'_x t/c)^2} - 1\right)$$

[Hint: According to the transformation equations for acceleration (see Problem 14),

$$a_x = (1 - V^2/c^2)^{3/2} \frac{a'_x}{(1 + Vu'_x/c^2)^3}$$

Use this with $u'_x = 0$ and $a_x = dV/dt$.]

(b) Show that for low speeds, the above equation yields the familiar result $x = \frac{1}{2}a'_x t^2$.

16. In the reference frame $x'y'z't'$, a particle moves with speed u' along a straight line in the x'- y' plane; the direction of motion of the particle makes an angle θ' with the x' axis.

(a) Show that in the reference frame *xyzt,* the direction of motion of the particle makes an angle θ with the x axis such that

$$\tan\theta = \frac{u'\sqrt{1 - V^2/c^2}\,\sin\theta'}{u'\cos\theta' + V}$$

(b) Suppose that a source at the origin of the $x'y'z't'$ reference frame emits particles in all directions. Show that for $V \rightarrow 1$, all these particles, except those emitted exactly in the backward direction (θ' = 180°), will form a narrow beam in the forward direction ($\theta = 0$) in the *xyzt* reference frame.

17. In a transparent medium of index of refraction n, the speed of light relative to the medium is c/n. Use the velocity-combination formula to show that if the medium has a speed V relative to the laboratory, then the speed of light relative to the laboratory is approximately given by the formula

$$c' = \frac{c}{n} \pm \left(1 - \frac{1}{n^2}\right)V$$

where the plus sign applies for **c** parallel to **V** and the minus sign for **c** antiparallel to **V**. [This formula was first obtained by Fresnel, who thought that the moving medium partially drags the ether along, giving it a drag speed $(1 - 1/n^2)V$.]

18. In the rest frame of a (flat) mirror, the law of reflection states that the angle of reflection of a light ray equals the angle of incidence.

(a) Suppose that the mirror is moving relative to the laboratory with a velocity V in a direction perpendicular to its face. If the angle of incidence of a light ray is θ (measured in the laboratory), what is the angle of reflection?

(b) Suppose that the mirror is moving relative to the laboratory with a velocity V in a direction parallel to its face. If the angle of incidence of a light ray is θ (measured in the laboratory), what is the angle of reflection?

19. Suppose that we place a sequence of n clocks along the equator of the Earth, with small distances Δx between one clock and the next. Since the Earth rotates, these clocks are not in an inertial reference frame, and they cannot be synchronized by the standard procedure with light signals. However, two adjacent clocks have approximately the same rotational velocity $V = \omega R$ (in magnitude and direction), and they can be regarded as having an instantaneous uniform eastward translational motion relative to the inertial reference frame centered on the Earth (and not rotating with it). This suggests that to synchronize the clocks, we ought to adjust them in such a way that there is a time shift $-V\Delta x/\sqrt{1-V^2/c^2}/c^2 \simeq -V\Delta x/c^2 = -\omega R\Delta x/c^2$ between adjacent clocks as observed from the inertial reference frame.

(a) If we proceed eastward around the Earth and successively adjust adjacent clocks according to this prescription, what is the net time shift between the nth clock and the first?

(b) What is the time shift between the first clock and the nth if, instead of proceeding by n steps eastward, we proceed by one step westward? (The conflict between the results obtained in (a) and (b) implies that it is not possible to attain consistent synchronization in the rotating reference frame by pretending adjacent clocks have an instantaneous uniform translational motion. In practice, physicists avoid this synchronization conflict in the rotating reference frame of the Earth by giving the clocks on the rotating Earth the same synchronization as the clocks of the nonrotating inertial reference frame centered on the Earth; that is, they simply ignore the time shift associated with the rotational motion.)

20. At the Fermilab accelerator, protons of energy 1000 GeV, or 1 TeV, travel around a circular ring of radius 1 km. What is the period of this motion, expressed in proper time?

21. The third crew of the *Skylab* space station remained aboard for 84 days. In this time interval, how many seconds did the special-relativistic time-dilation effect cause their clocks to fall behind clocks on the Earth? Assume that the speed of *Skylab* was 7.7 km/s and that the speed of a clock on the ground is the equatorial speed of rotation of the Earth.

22. A simple "light clock" consists of two parallel mirrors separated by a fixed distance L. A light pulse bounces back and forth between these mirrors, each bounce counting as a tick of the clock. In the rest frame of the clock, the light pulse moves at right angles to the mirrors, and the time between ticks is then L. By direct calculation, show that, in a reference frame with respect to which this clock is in motion with velocity V in a direction parallel to the mirrors (see Fig. 6.26), the time between ticks is $L/\sqrt{1-V^2/c^2}$.

23. A freshly laid chicken egg is placed in a spaceship that is launched from the Earth with a speed of $V = 0.80c$. When the chick hatches, the spaceship sends a radio signal back to Earth. How many days (as measured by the clocks on Earth) after launching does this signal reach the Earth? The normal incubation period of a chicken egg is 21 days.

24. For a precise synchronization of the atomic clocks at the National Institute of Standards and Technology at Boulder, Colorado, and the Bureau International de l'Heure at Paris, France, scientists carried portable atomic clocks from one location to the other. Roughly estimate how many seconds such a traveling clock falls behind the clocks fixed on the surface of the Earth during the eastward trip from Boulder to Paris.

25. Consider the spaceship with the worldline described in Problem 15. Show that the elapsed proper time (starting at $t = 0$) registered by a clock on board this spaceship is

$$\Delta\tau = \frac{c}{a'_x}\sinh^{-1}(a'_x t/c)$$

26. Figure 6.27 shows the worldlines of two identical clocks that move from A to B. The first clock has a straight worldline $x = V_0 t$, where V_0 is a constant. The second clock has the worldline $x = \frac{1}{2}a_0 t^2$, where a_0 is a constant. For each clock, find the elapsed proper time between A and B. Which clock shows the longer proper time?

27. Two spacetime points, A and B, have a timelike separation. Consider the straight worldline connecting A and B, and consider curved worldlines. Show that the proper time between A and B,

$$\Delta\tau = \int_A^B d\tau = \int_A^B \sqrt{1 - V^2/c^2}\,dt$$

is largest when evaluated along the straight worldline. Hence the worldline of a free particle can be characterized as the spacetime curve of *maximum* proper time. [*Hint*: Use a reference frame in which the particle moving along the straight worldline is at rest.]

28. The star Capella is at a distance of 46 light-years from the Earth. Suppose that an astronaut wants to travel from the Earth to Capella in a time of no more than 20 years as reckoned by clocks aboard her spaceship. At what speed would she have to travel? How long would the trip take as reckoned by clocks on the Earth?

Figure 6·26

Two parallel mirrors moving toward the right. The dashed line shows the trajectory of a light signal.

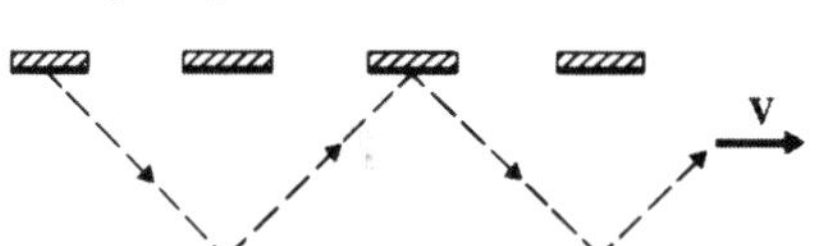

29. Assume that a spaceship operating with matter-antimatter fuel leaves the Earth on an excursion to Barnard's star. The position of the spaceship as a function of time is (see Problem 15)

$$x = \frac{c^2}{g}\left(\sqrt{1+\frac{g^2t^2}{c^2}}-1\right)$$

where t is the time measured in the reference frame of the Earth, with $t = 0$ at the start of the trip, and g is the acceleration of the spaceship in its own reference frame. The value of the acceleration has been chosen as $g = 981$ cm/s^2 to ensure the comfort of the crew.

(a) Show that

$$v = \frac{gt}{\sqrt{1+g^2t^2/c^2}}$$

(b) When 2 years have elapsed in the reference frame of the Earth, what time has elapsed in the reference frame of the spaceship?

(c) If the spaceship sends out a radio signal toward the Earth at this time, when will this signal reach the Earth?

(d) If the signal is emitted with a frequency of 1×10^9 Hz (in the reference frame of the spaceship), what will be the frequency with which this signal is received (in the reference frame of the Earth)?

30. Show that if an emitter of radio waves is receding from the receiver with a speed V, then the frequency ν_0 of the radio wave in the reference frame of the emitter and the frequency ν in the reference frame of the receiver are related by

$$\nu = \nu_0\sqrt{\frac{1-V/c}{1+V/c}}$$

[*Hint*: In Newtonian physics, the frequencies would be related by the Doppler formula for a moving emitter and a stationary receiver, $\nu = \nu_0/(1 + V/c)$. Modify this formula to take into account the relativistic time-dilation effect.]

Figure 6·27

Worldlines of two clocks.

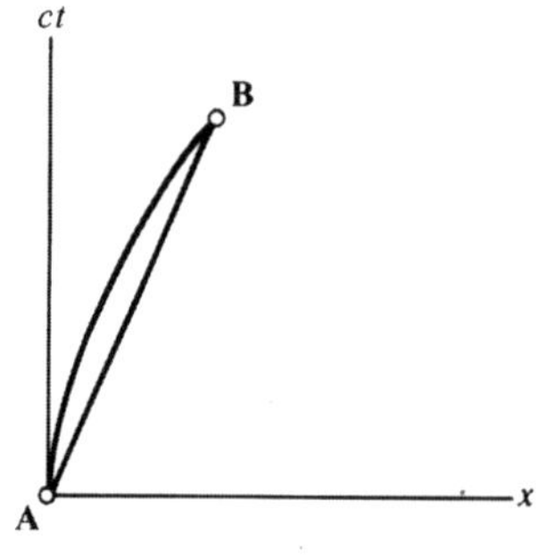

31. The Earth moves around the Sun at a speed of 30 km/s. Assume that in its own reference frame, the Earth is a perfect sphere of radius 6.4×10^6 m. In the reference frame of the Sun, the Earth suffers a relativistic length contraction along the east-west direction. By how many meters is the east-west diameter of the Earth contracted? By how many cubic meters is the volume of the Earth contracted?

32. The *pole-and-barn* paradox involves the following: A very fast runner carrying a long horizontal pole runs into a barn with an open door at one end and a rigid wall on the other end. The length of the pole (in its own rest frame) is, say, 6 m; and the length of the barn (in *its* own rest frame) is, say, 5 m. In the reference frame of the barn, the pole is length contracted, and all of the pole can fit into the barn at one instant of time. However, in the reference frame of the pole, the barn is length contracted, and it would seem that the pole can never fit into the barn. Resolve this apparent paradox by drawing a spacetime diagram displaying the worldlines of the front and rear ends of the moving pole and of the stationary barn. On the basis of this diagram, show that in the reference frame of the barn, the compressional sound wave that travels backward along the pole when the front end strikes the wall reaches the rear end of the pole only *after* this end is within the barn. Thus, in the reference frame of the barn, the pole fits because it necessarily crumples.

33. (a) Consider two reference frames with cylindrical coordinates ρ, ϕ, z, t and ρ', ϕ', z', t'. The z axes are parallel, and the second reference frame is moving at speed V in the direction of the z axis. Write the Lorentz transformation in terms of these cylindrical coordinates.

(b) Consider two reference frames with spherical coordinates r, θ, ϕ, t and r', θ', ϕ', t'. If the orientation of the coordinates and the motion are as described in part (a), write the Lorentz transformation in terms of these spherical coordinates.

34. In the laboratory frame $xyzt$, a straight horizontal rod of length L is in free fall. The rod has no horizontal velocity, and its vertical position is $y = -\frac{1}{2}gt^2$. Consider the configuration of this rod at one instant of time in the $x'y'z't'$ frame which has a velocity V along the x axis of the laboratory frame. What is the equation $y' = y'(x', t')$ that describes the instantaneous position of points on the rod in the $x'y'z't'$ frame? Plot y' as a function of x' for $t' = 0$. Is the rod straight? Does the shape of the rod remain fixed as it continues to fall? Does the length of the rod remain fixed?

CHAPTER

7

Vector Calculus in Spacetime

In this chapter, we will become acquainted with vector calculus in four-dimensional spacetime. Many of the equations in this chapter are obvious four-dimensional analogs of the three-dimensional equations of Chapter 1. The formalism developed in this chapter will prove very advantageous in expressing relativistic equations of motion for particles or for fields and in expressing relativistic conservation laws of energy and momentum for the collisions and reactions of high-energy particles.

7·1 Four-Vectors and Tensors

In a given inertial reference frame, the coordinates of an event, or spacetime point, are ct, x, y, z. The four-components object (ct, x, y, z) is the position vector of the spacetime point; this is a four-dimensional vector, or a *four-vector*. It is convenient to introduce the notation (x^0, x^1, x^2, x^3) for the position four-vector, with

$$x^0 = ct \qquad x^1 = x \qquad x^2 = y \qquad x^3 = z \tag{1}$$

In compact notation, we will often write the position four-vector as

$$x^\mu \tag{2}$$

where it is understood that = 0, 1, 2, 3. Note the distinction between the three-dimensional position vector x^k of Chapter 1 and the four-dimensional position vector x^μ introduced here: the former is written with a Latin superscript, the latter with a Greek superscript.

The Lorentz transformation for velocity V along the x axis [see Eqs. (6.34)–(6.37)] can now be written as

$$x'^\mu = \sum_{\nu=0}^{3} a^\mu{}_\nu \, x^\nu \tag{3}$$

where $a^\mu{}_\nu$ is the matrix

$$a^{\mu}{}_{\nu} = \begin{pmatrix} \frac{1}{\sqrt{1-V^2/c^2}} & \frac{-V/c}{\sqrt{1-V^2/c^2}} & 0 & 0 \\ \frac{-V/c}{\sqrt{1-V^2/c^2}} & \frac{1}{\sqrt{1-V^2/c^2}} & 0 & 0 \\ 0 & 0 & 1 & 0 \\ 0 & 0 & 0 & 1 \end{pmatrix} \tag{4}$$

Here the first index (μ) gives the row; the second (ν) gives the column.

The Einstein-summation convention for Greek indices says that whenever a Greek vector index appears twice in a product, then summation over this index is taken for granted. Thus Eq. (3) is the same as

$$x'^{\mu} = a^{\mu}{}_{\nu} x^{\nu} \tag{5}$$

Note the difference between the Einstein-summation convention for Latin and for Greek indices: Latin indices are summed from 1 to 3, Greek indices from 0 to 3.

We know from Section 6.4 that the spacetime interval between two arbitrary events is invariant. In our new notation, we can write this spacetime interval as

$$d\tau^2 = (dx^0)^2 - (dx^1)^2 - (dx^2)^2 - (dx^3)^2 \tag{6}$$

or as

$$d\tau^2 = \eta_{\mu\nu} dx^{\mu} dx^{\nu} \tag{7}$$

where $\eta_{\mu\nu}$ is the tensor, or matrix,

$$\eta_{\mu\nu} = \begin{pmatrix} 1 & 0 & 0 & 0 \\ 0 & -1 & 0 & 0 \\ 0 & 0 & -1 & 0 \\ 0 & 0 & 0 & -1 \end{pmatrix} \tag{8}$$

The tensor $\eta_{\mu\nu}$ is called the *metric tensor*, or the *Minkowski tensor*, of spacetime; it determines the invariant "distance" in spacetime.

We saw in Chapter 1 that any matrix describing rotations in three dimensions must satisfy the condition (1.38). There exists a corresponding condition for any matrix $a^{\mu}{}_{\nu}$ describing a Lorentz transformation that reads

$$a^{\alpha}{}_{\mu} \eta_{\alpha\beta} a^{\beta}{}_{\nu} = \eta_{\mu\nu} \tag{9}$$

■ ***Exercise 1.*** Check explicitly that the matrix for the special Lorentz transformation given by Eq. (4) satisfies Eq. (9). ■

To derive the general result, we suppose that $a^\mu{}_\nu$ represents some arbitrary Lorentz transformation (not necessarily with velocity along the x axis). The transformation must leave any spacetime interval invariant; for the finite interval from the origin to x^μ, this means

$$\eta_{\alpha\beta} x'^\alpha x'^\beta = \eta_{\mu\nu} x^\mu x^\nu \tag{10}$$

Substituting $x'^\alpha = a^\alpha{}_\mu x^\mu$ in this equation, we obtain

$$\eta_{\alpha\beta} a^\alpha{}_\mu x^\mu a^\beta{}_\nu x^\nu = \eta_{\mu\nu} x^\mu x^\nu \tag{11}$$

or

$$x^\mu x^\nu (a^\alpha{}_\mu \eta_{\alpha\beta} a^\beta{}_\nu - \eta_{\mu\nu}) = 0 \tag{12}$$

From this, Eq. (9) follows by adopting various choices for the arbitrary vector x^μ or, alternatively, by taking the partial derivative $\partial^2 / \partial x^\kappa \partial x^\lambda$ [see the argument given in connection with Eq. (1.32)].

Note that Eq. (1.38) is a special case of Eq. (9) if the matrix $a^\mu{}_\nu$ has $a^0{}_0 = 1$ and $a^\mu{}_0 = 0$.

■ ***Exercise 2.*** Check this. ■

Transformations of this special type, which leave the time coordinate unchanged ($x'^0 = x^0$), are therefore ordinary spatial rotations. We will call any transformation satisfying Eq. (9) a general Lorentz transformation. Spatial rotations are then also regarded as Lorentz transformations, and to distinguish spatial rotations from the "true" Lorentz transformations with nonzero velocity, we will call the latter *pure Lorentz transformations*, or *boosts*.

It is convenient to define the *covariant* components of the position vector by the following equation:

$$x_\mu = \eta_{\mu\nu} x^\nu \tag{13}$$

Written out in full, these components are

$$x_0 = ct, \qquad x_1 = -x, \qquad x_2 = -y, \qquad x_3 = -z \tag{14}$$

For contrast, the components x^μ, given by Eq. (1), are called the *contravariant* components of the position vector. Our notation distinguishes between these two ways of representing the components of a vector by the placement of the index: contravariant components have an upper index; covariant, a lower. Care must be taken with these indices; for example, $x^1 = x$, $x_1 = -x$, and so on.

We can invert Eq. (13) as follows:

$$x^{\nu} = \eta^{\nu\mu} x_{\mu} \tag{15}$$

where

$$\eta^{\nu\mu} = \begin{pmatrix} 1 & 0 & 0 & 0 \\ 0 & -1 & 0 & 0 \\ 0 & 0 & -1 & 0 \\ 0 & 0 & 0 & -1 \end{pmatrix} \tag{16}$$

■ ***Exercise 3.*** Show that Eq. (15) agrees with Eq. (13). ■

The spacetime interval can now be expressed as

$$d\tau^2 = dx^{\mu} dx_{\mu} \tag{17}$$

For a finite displacement, from the origin to the point x^{μ}, the spacetime interval is

$$\Delta\tau^2 = x^{\mu} x_{\mu} \tag{18}$$

The vector x^{μ} is called *timelike, spacelike,* or *lightlike* according to whether $x^{\mu}x_{\mu}$ is positive, negative, or zero. Examples of vectors of the three types are shown in Fig. 7.1.* Timelike vectors point into the absolute future (or past), spacelike into the neutral region, and lightlike along the light cone.

The transformation law for x_{μ} can be obtained from Eq. (5):

Figure 7·1

Examples of (*a*) timelike, (*b*) spacelike, and (*c*) lightlike position vectors. This diagram shows the time axis and one of the space axes.

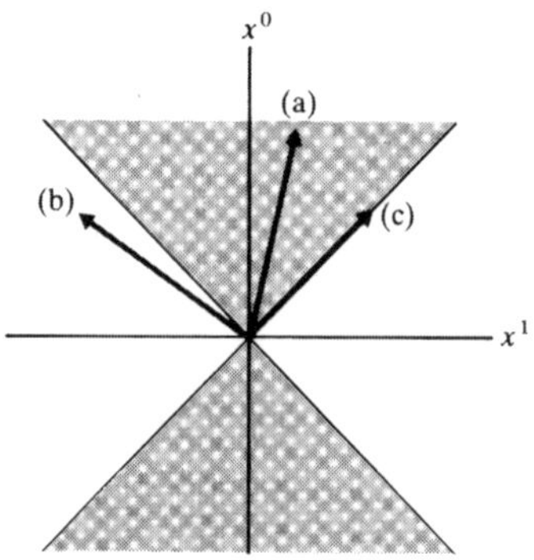

*The *contravariant* components x^{μ} are always shown in such diagrams; the covariant components x_{μ} never are.

$$x'_\mu = \eta_{\mu\nu} x'^\nu = \eta_{\mu\nu} a^\nu{}_\alpha x^\alpha = \eta_{\mu\nu} a^\nu{}_\alpha \eta^{\alpha\beta} x_\beta \tag{19}$$

If we define

$$a_\mu{}^\beta \equiv \eta_{\mu\nu} a^\nu{}_\alpha \eta^{\alpha\beta} \tag{20}$$

then Eq. (19) becomes

$$x'_\mu = a_\mu{}^\beta x_\beta \tag{21}$$

■ ***Exercise 4.*** If the matrix $a^\nu{}_\alpha$ is given by Eq. (4), show that

$$a_\mu{}^\beta = \begin{pmatrix} \dfrac{1}{\sqrt{1-V^2/c^2}} & \dfrac{V/c}{\sqrt{1-V^2/c^2}} & 0 & 0 \\ \dfrac{V/c}{\sqrt{1-V^2/c^2}} & \dfrac{1}{\sqrt{1-V^2/c^2}} & 0 & 0 \\ 0 & 0 & 1 & 0 \\ 0 & 0 & 0 & 1 \end{pmatrix} \tag{22}$$

■

■ ***Exercise 5.*** Show that the matrices appearing in Eqs. (4) and (22) are the inverses of each other. ■

It is easy to show that for any general Lorentz transformation, the matrices $a_\alpha{}^\mu$ and $a^\alpha{}_\nu$ are always inverses, that is,

$$a_\alpha{}^\mu a^\alpha{}_\nu = \delta^\mu_\nu \tag{23}$$

where δ^μ_ν is the Kronecker delta in four dimensions,

$$\delta^\mu_\nu = \begin{cases} 1 & \text{if } \mu = \nu \\ 0 & \text{if } \mu \neq \nu \end{cases} \tag{24}$$

Note that in Eq. (23), the index to be summed is the first index in $a_\alpha{}^\mu$ and the first index in $a^\alpha{}_\nu$.

■ ***Exercise 6.*** Derive Eq. (23). [*Hint:* Multiply both sides of Eq. (9) by $\eta^{\nu\sigma}$, and use the following identity, obtained by multiplication of the matrices (8) and (16):

$$\eta_{\mu\nu}\eta^{\nu\sigma} = \delta^\sigma_\mu \quad]$$

■

Equation (23) is equivalent to

$$(a^T)_\mu{}^\alpha a^\alpha{}_\nu = \delta^\mu_\nu \tag{25}$$

which resembles Eq. (1.32).

If we multiply Eq. (21) by $a^{\mu}{}_{\nu}$, sum over μ, and use Eq. (23), we obtain

$$a^{\mu}{}_{\nu}x'_{\mu} = a^{\mu}{}_{\nu}a_{\mu}{}^{\beta}x_{\beta} = \delta_{\nu}^{\beta}x_{\beta} = x_{\nu} \tag{26}$$

This is the inverse Lorentz transformation for the covariant position vector; it expresses x_{ν} in terms of x'_{μ}.

We can now lay down the general definitions of vectors and tensors in four-dimensional spacetime. As in the three-dimensional case, we regard the position vector as the prototype for all other vectors so that the Lorentz-transformation properties of the position vector serve as model for the transformation properties of all other vectors.

• ***Definition:*** A four-vector is an object with four components that transform as follows under Lorentz transformations:

$$B'^{\mu} = a^{\mu}{}_{\nu}B^{\nu} \tag{27}$$

We have seen how to construct the covariant components of the position vector from the contravariant ones [see Eq. (13)]. In general, if B^{μ} are the contravariant components of a vector, then the covariant components are

$$B_{\mu} = \eta_{\mu\nu}B^{\nu} \tag{28}$$

Conversely,

$$B^{\nu} = \eta^{\nu\mu}B_{\mu} \tag{29}$$

The operations involved in Eqs. (28) and (29) are called *lowering* and *raising* an index, respectively.

The covariant components of a vector have the transformation law

$$B'_{\mu} = a_{\mu}{}^{\nu}B_{\nu} \tag{30}$$

This can be readily established by an argument similar to that leading from Eq. (19) to Eq. (21).

The definition of a tensor is a direct generalization of Eq. (27).

• ***Definition:*** A tensor of the second rank is an object with 16 components that transform as follows under Lorentz transformation:

$$T'^{\mu\nu} = a^{\mu}{}_{\alpha}a^{\nu}{}_{\beta}T^{\alpha\beta} \tag{31}$$

A tensor can be written in terms of its contravariant components, convariant components, or mixed components. These types of components are related as follows:

$$T_{\mu\nu} = \eta_{\mu\alpha}\eta_{\nu\beta}T^{\alpha\beta} \tag{32}$$

$$T_{\mu}^{\ \nu} = \eta_{\mu\alpha}T^{\alpha\nu} \tag{33}$$

and they have the transformation laws

$$T'_{\mu\nu} = a_{\mu}^{\ \alpha}a_{\nu}^{\ \beta}T_{\alpha\beta} \tag{34}$$

$$T'^{\ \nu}_{\mu} = a_{\mu}^{\ \alpha}a^{\nu}_{\ \beta}T_{\alpha}^{\ \beta} \tag{35}$$

EXAMPLE: *some tensors*

For example, $x^{\mu}x^{\nu}$ is a tensor, and we can verify this by simply substituting the known transformation law (5) for x^{μ}:

$$x'^{\mu}x'^{\nu} = (a^{\mu}_{\ \alpha}x^{\alpha})(a^{\nu}_{\ \beta}x^{\beta}) = a^{\mu}_{\ \alpha}a^{\nu}_{\ \beta}x^{\alpha}x^{\beta}$$

The following exercises deal with some more specific examples of tensors, with the transformation laws given by Eqs. (34) and (35), respectively.

■ ***Exercise 7.*** Show that $x_{\mu}x_{\nu}$ is a tensor. ■

■ ***Exercise 8.*** Show that δ^{μ}_{ν} is a tensor, with δ'^{μ}_{ν} defined by the same equation (24) as δ^{ν}_{μ}. ■

■ ***Exercise 9.*** Show that $\eta_{\mu\nu}$ is a tensor, with $\eta'_{\mu\nu}$ defined by the same equation (8) as $\eta_{\mu\nu}$. ■

A second-rank tensor is said to be *symmetric* if $T^{\mu\nu} = T^{\nu\mu}$, and it is said to be *antisymmetric* if $T^{\mu\nu} = -T^{\nu\mu}$.

We can also define tensors of higher rank; for instance, a third-rank tensor is an object with 64 components that transforms as follows under Lorentz transformations:

$$T'^{\mu\nu\sigma} = a^{\mu}_{\ \alpha}a^{\nu}_{\ \beta}a^{\sigma}_{\ \delta}T^{\alpha\beta\delta} \tag{36}$$

We will not have any need for tensors of higher rank, but they play an important role in the relativistic theory of gravitation (General Relativity).

A scalar is, of course, a one-component object that remains unchanged under Lorentz transformation. The spacetime interval $dx^{\mu}dx_{\mu}$ is an obvious example of a scalar. More generally, the "scalar" product $A^{\mu}B_{\mu}$ of two arbitrary vectors is a scalar. Written out in full, this scalar product is

$$A^{\mu}B_{\mu} = A^0B^0 - A^1B^1 - A^2B^2 - A^3B^3 \tag{37}$$

The summation over two tensor indices, one covariant (upper) and one contravariant (lower), is called a *contraction*; and the scalar product (37) is also called a contracted product.

■ **Exercise 10.** Use the transformation laws for A^{μ} and for B_{μ} to show that $A^{\mu}B_{\mu} = A'^{\mu}B'_{\mu}$. ■

■ **Exercise 11.** Show that if B^{μ} is a vector and $A_{\alpha\mu}$ a tensor, then the contracted product $A_{\alpha\mu}B^{\mu}$ is a vector. ■

■ **Exercise 12.** Show that if $T^{\mu\nu}$ is a tensor, then $T^{\mu}{}_{\mu}$ is a scalar. ■

The scalar $T^{\mu}{}_{\mu}$ is called the *trace* of the tensor $T^{\mu}{}_{\nu}$. This is the sum of the diagonal components of $T^{\mu}{}_{\nu}$. If the trace is zero, the tensor is said to be traceless.

7·2 The Energy-Momentum Vector

In Newtonian physics, the momentum of a particle is the product of mass and velocity,

$$\mathbf{p} = m\mathbf{v} \tag{38}$$

The momentum of a system of particles obeys an important conservation law: In the absence of external forces, the net momentum of the system of particles is constant. This conservation law is a direct consequence of Newton's Second and Third Laws. In the simple case of a system with only two interacting particles (A and B), the proof is as follows:

$$\frac{d}{dt}(\mathbf{p}_A + \mathbf{p}_B) = \mathbf{F}_A + \mathbf{F}_B = \mathbf{F}_A + (-\mathbf{F}_A) = 0 \tag{39}$$

where the first equality depends on the Second Law and the next equality depends on the Third Law, $\mathbf{F}_A$ and $\mathbf{F}_B$ being an action-reaction pair.

In relativistic physics, we cannot derive the conservation of momentum in this way. The trouble is that Newton's Third Law fails—the relativity of simultaneity makes the instantaneous balance of action and reaction meaningless. For example, consider two charged particles exerting Coulomb forces on one another while in motion. Action and reaction are then defined at two different points of space (the positions of the particles), and they both change with time. The permanent balance of action and reaction would require that these changes be simultaneous in all reference frames, but this is impossible because the relativity of simultaneity tells us that events at different points of space cannot be simultaneous in all reference frames. It turns out that the imbalance of action and reaction, and the consequent imbalance of conservation of momentum of the system of particles, is compensated by a transfer of momentum to the fields of force surrounding the particles—these fields act as storehouses of momentum in such a way that the net momentum of particles and fields is conserved. However, in order to prove the general conservation law, we need a complete theory of momentum and energy stored in the fields. This would take us beyond the scope of this chapter.

Here, we will be satisfied with a discussion of momentum conservation for the special, simple case of contact forces. In this case, action and reaction are defined at the same point of space (the point of collision), and therefore, the relativity of simultaneity does not interfere with Newton's Third Law. In any collision involving such contact forces, we then expect that the net momentum will be conserved. However, as we will see later, we can only satisfy this expectation if we define momentum in relativistic physics by a formula somewhat more complicated than the simple formula [Eq. (38)] of Newtonian physics.

As the starting point of our derivation of the relativistic formula for momentum, we take the assumption that the direction of the momentum is parallel to the direction of the velocity, that is,

$$\mathbf{p} = f(v^2)m\mathbf{v} \tag{40}$$

This assumption if self-evident because the direction of the velocity is the only available preferred direction associated with a free particle. The scalar function $f(v^2)$ depends only on the magnitude of the velocity and is subject to the restriction that $f(v^2) \to 1$ for $v \to 0$; this restriction ensures that for low (nonrelativistic) velocities, the formula (40) coincides with the Newtonian formula.

We now must choose $f(v^2)$ in such a way that the net momentum of a system of colliding particles is conserved. For this purpose, it is simplest to consider a collision between two smooth, identical elastic balls of mass m. The balls have momenta with y components of equal magnitudes and opposite directions, but ball A differs from B in that its momentum also has a large x component. Figure 7.2 shows the balls in the $xyzt$ reference frame before the collision and after the collision. Figure 7.3 shows the balls in the $x'y'z't'$ reference frame, which is moving to the right with a velocity equal to the x velocity of A. In this reference frame, B has a large momentum in the negative x direction. Note the symmetry between the two reference frames: except for a contrary direction, whatever ball A does in the $xyzt$ frame, ball B does in the $x'y'z't'$ frame, and vice versa. This symmetry is an expression of the Principle of Relativity, and it ensures that if momentum is conserved in one of these reference frames, it is also conserved in the other.

During the collision, the y component of the momentum of each ball is reversed. If the momentum changes reckoned in the $xyzt$ frame are Δp_{Ay} and Δp_{By}, then momentum conservation requires

Figure 7·2

Momentum vectors of the balls as seen in the *xyzt* frame. (*a*) Before the collision. (*b*) After the collision.

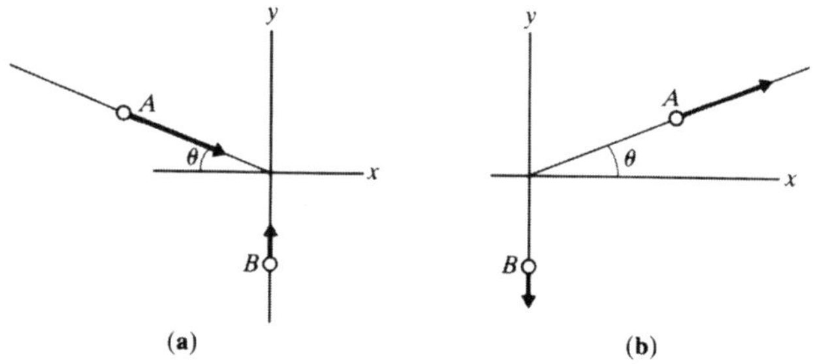

Figure 7·3

Momentum vectors of the balls as seen in the *x'y'z't'* frame. (*a*) Before the collision. (*b*) After the collision.

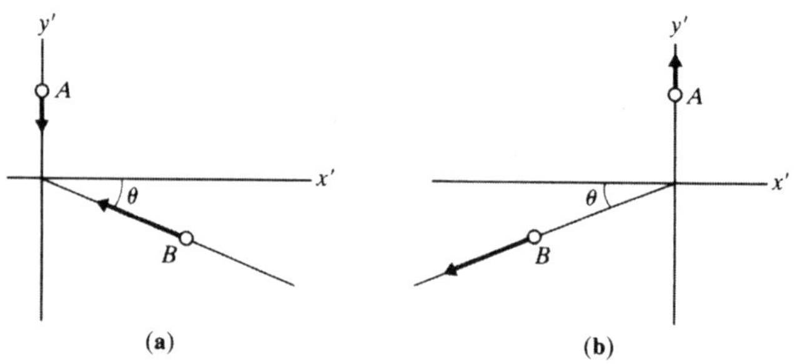

$$\Delta p_{A,y} = -\Delta p_{B,y} \tag{41}$$

Because of the symmetry between the balls and the reference frames, the momentum change of ball B in the *xyzt* frame matches the momentum change of ball A in the *x'y'z't'* frame, except for sign (see Figs. 7.2 and 7.3):

$$\Delta p_{B,y} = -\Delta p'_{A,y} \tag{42}$$

Combining this with the preceding equation, we find

$$\Delta p_{A,y} = \Delta p'_{A,y} \tag{43}$$

By means of Eq. (40), we can express this relationship between the transverse components of the momentum in terms of the velocity. Each momentum change in Eq. (43) equals twice the y component of the momentum, so

$$2f(v^2)mv_y = 2f(v'^2)mv'_y$$

Since $v' = v'_y$ and $v'_y = v_y / \sqrt{1 - v_x^2 / c^2}$ [see Eq. (6.46) with $V = u_x = v_x$], this becomes

$$2f(v^2)mv_y = 2f\left(\frac{v_y^2}{1 - v_x^2 / c^2}\right)m\frac{v_y}{\sqrt{1 - v_x^2 / c^2}} \tag{44}$$

If we now proceed to the limit $v_y \to 0$, we obtain

$$2f(v^2)mv_y = 2f(0)m\frac{v_y}{\sqrt{1 - v^2 / c^2}}$$

from which we deduce $f(v^2) = 1/\sqrt{1 - v^2/c^2}$. With this result for the function f, Eq. (40) becomes

$$\mathbf{p} = \frac{m\mathbf{v}}{\sqrt{1 - v^2 / c^2}} \tag{45}$$

This is the relativistic formula for the momentum of a particle. The mass that appears in Eq. (45) is the *rest mass*, that is, the mass that the particle has when measured in a reference frame in which it is at rest. In this book, the symbol m for the mass of a particle will always mean the rest mass.

EXAMPLE: *relativistic vs. Newtonian momentum*

For a given velocity, the momentum (45) is larger than the Newtonian value $m\mathbf{v}$ by the factor $1/\sqrt{1 - v^2/c^2}$, and the momentum diverges to infinity as $v \to c$. For low velocities, we can approximate

$$\mathbf{p} = \frac{m\mathbf{v}}{\sqrt{1 - v^2 / c^2}} \simeq \frac{m\mathbf{v}}{1 - ½v^2 / c^2} \simeq m\mathbf{v}(1 + ½v^2 / c^2)$$

For instance, for the electrons in a 20-keV X-ray tube, $v \simeq 8 \times 10^9$ cm/s and $½v^2 / c^2 \simeq 4 \times 10^{-2}$, so the relativistic momentum is larger than the Newtonian momentum by 4%. This is a significant correction, and it needs to be included when designing the tube.

Note that our arguments prove only that the sum of the momenta calculated from Eq. (45) is conserved in the case of the symmetric elastic collision we dealt with above. What happens in an arbitrary collision involving many particles interacting inelastically? If the forces between pairs of particles are contact forces and obey Newton's Third Law, then the total momentum is conserved—the particles merely exchange momentum, but the total amount of momentum is constant. However, if the particles interact over some distance, by means of electric, magnetic, or other fields, then these fields can absorb and supply momentum, and the sum of the particle momenta, calculated from Eq. (45), is not necessarily conserved. We will examine the momentum stored in electric and magnetic fields in the next chapter.

The three momentum components given by Eq. (45) may be regarded as the last three components of a four-component object p^μ:

$$p^\mu \equiv \left(\frac{mc}{\sqrt{1-v^2/c^2}}, p_x, p_y, p_z \right)$$

$$= \left(\frac{mc}{\sqrt{1-v^2/c^2}}, \frac{m}{\sqrt{1-v^2/c^2}}\frac{dx}{dt}, \frac{m}{\sqrt{1-v^2/c^2}}\frac{dy}{dt}, \frac{m}{\sqrt{1-v^2/c^2}}\frac{dz}{dt} \right) \qquad (46)$$

This can also be written as

$$p_\mu = mu^\mu \qquad (47)$$

where

$$u^\mu = \left(\frac{c}{\sqrt{1-v^2/c^2}}, \frac{1}{\sqrt{1-v^2/c^2}}\frac{dx}{dt}, \frac{1}{\sqrt{1-v^2/c^2}}\frac{dy}{dt}, \frac{1}{\sqrt{1-v^2/c^2}}\frac{dz}{dt} \right)$$

$$= \left(c\frac{dt}{d\tau}, \frac{dx}{d\tau}, \frac{dy}{d\tau}, \frac{dz}{d\tau} \right) = \frac{dx^\mu}{d\tau} \qquad (48)$$

Since dx^μ is a four-vector and $d\tau$ is a scalar, it follows that u^μ is a four-vector, called the *four-velocity*. Hence p^μ is also a four-vector.

■ ***Exercise 13.*** Show that

$$u_\mu u^\mu = c^2 \qquad (49)$$

■

What is the physical interpretation of the component p^0? It turns out that the quantity cp^0 is the *energy* of the particle:

$$cp^0 = E = \frac{mc^2}{\sqrt{1-v^2/c^2}} \qquad (50)$$

To justify the interpretation of cp^0 as energy, we note that in the nonrelativistic limit ($v \ll c$),

$$cp^0 \simeq \frac{mc^2}{1-\tfrac{1}{2}v^2/c^2} \simeq mc^2 + \tfrac{1}{2}mv^2 \qquad (51)$$

The quantity $mv^2/2$ is the familiar nonrelativistic kinetic energy. The quantity mc^2 in Eq. (51) plays the role of an additive constant in the energy; this additive constant is the *rest-mass energy* of the particle (we will discuss the physical significance of this rest-mass energy in more detail later in this section). The relativistic kinetic energy is the difference between the relativistic energy and the rest-mass energy:

$$T = E - mc^2 = \frac{mc^2}{\sqrt{1 - v^2/c^2}} - mc^2 \qquad (52)$$

Incidentally: when particle physicists speak of, say, protons of energy 2GeV, what they mean is *kinetic* energy.*

The relativistic formulas (45) and (52) for the momentum and the energy can be tested directly by measurements on fast particles. However, it is difficult to achieve high precision in such experiments. The best available direct measurements of momentum and speed permit a confirmation of Eq. (45) to within about 0.1%.[1,2] These measurements were performed on protons of a kinetic energy of a few hundred MeV; the momentum was determined from the radius of the circular orbit of the protons in a known magnetic field, and the speed was determined from the period of the orbit.

A few of the consequences of the relativistic formulas are amenable to tests of very high precision. For instance, Eq. (52) implies that the energy diverges as the speed approaches the speed of light; stated otherwise, the speed becomes independent of the energy as the energy becomes large. This prediction has been tested in an elegant experiment at the Stanford Linear Accelerator.[3] Electrons of energies between 15 and 20 GeV were made to race against photons produced by the penetration of the electrons through a thin target (Bremsstrahlung). The electrons and photons emerging from the target were allowed to race a distance of 1000 m, but at the finish, the photons were not perceptibly ahead of the electrons. This experiment established that electrons of 15 to 20 GeV have a speed equal to that of light to within 2 parts in 10^7.

The most precise available test of the relativistic formulas for momentum and energy has emerged indirectly from experiments measuring the intrinsic magnetic moment of the electron (*g* factor of the electron).[4] These high-precision experiments determined the magnetic moment through a comparison of its frequency of precession in a magnetic field with the frequency of orbital motion of the electron in the same magnetic field. One such experiment was performed with electrons of very low energy (5×10^{-4} eV); another was performed with electrons of

*Particle physicists commonly measure energy in GeV (where 1GeV = 10^9 eV = 1.6×10^{-10} J = 1.6×10^{-3} erg), momentum in GeV/c, and mass in GeV/c^2.

relativistic energy (0.11 MeV). A comparison between the magnetic moments found in these two experiments establishes that the quantity $\sqrt{1-v^2/c^2}(p/m)(dp/dE)-1$, which enters into the comparison, is zero to within 5 parts in 10^9. According to the relativistic formulas for energy and momentum, this quantity is *exactly* zero. Thus these experiments provide us with an extremely precise confirmation of relativity. However, besides the formulas for energy and momentum, several other aspects of relativity are inextricably interwoven in the analysis of the experiments. Thus, although these experiments do provide a test of the formulas for energy and momentum, they do not provide a clear-cut test.

After this digression on experiments, let us return to Eq. (46). In terms of the relativistic energy *E*, we can write Eq. (46) as

$$p^\mu = \left(\frac{E}{c}, p_x, p_y, p_z\right) \tag{53}$$

The expression (53) is called the *energy-momentum four-vector*. For later reference, we write the Lorentz transformation for p^μ; if the transformation has velocity *V* along the *x* axis, then

$$E' = \frac{E - Vp_x}{\sqrt{1-V^2/c^2}} \qquad p'_x = \frac{p_x - VE/c^2}{\sqrt{1-V^2/c^2}}$$

$$p'_y = p_y \qquad p'_z = p_z \tag{54}$$

Note that these Lorentz-transformation equations mix energy and momentum. This has an important consequence for energy conservation: in any collision, the total relativistic energy of a system of particles is conserved if the momentum is conserved. To prove this conservation law, suppose that *E* is the total energy and **P** the total momentum of the system of particles,

$$E = \sum_A E_A \qquad \mathbf{P} = \sum_A \mathbf{p}_A$$

where the summation runs over all particles in the system. Since P^μ = $(E/c, P_x, P_y, P_z)$ is a four-vector, the change ΔP^μ in P^μ produced by the collision is also a four-vector and obeys the Lorentz-transformation law

$$\Delta E' = \frac{\Delta E - V\Delta P_x}{\sqrt{1-V^2/c^2}} \tag{55}$$

$$\Delta P'_x = \frac{\Delta P_x - V\Delta E/c^2}{\sqrt{1-V^2/c^2}} \tag{56}$$

Under the assumption that momentum is conserved (in both reference frames), we have $\Delta P_x = \Delta P'_x = 0$, and it follows from Eqs. (55) and (56) that $\Delta E = \Delta E' = 0$. This shows that energy conservation is a consequence of momentum conservation. The conservation laws are not independent, because energy and momentum are nothing but components of the same four-vector P^μ.

This calculation raises a puzzling question: How can an inelastic collision, involving a loss of *kinetic* energy, lead to conservation of *total* energy? The answer is that any loss of kinetic energy by "frictional" processes will be balanced by a gain in rest-mass energy, so the total energy E remains constant. For instance, in a totally inelastic collision between two balls of putty, the kinetic energy is transformed into thermal energy of the balls; this thermal energy contributes to the rest-mass energy of the balls—after the collision, the balls will have a slightly larger mass.

The covariant components of the energy-momentum vector are

(57)

and the scalar product of (53) and (57) is

$$p_\mu p^\mu = \frac{E^2}{c^2} - p_x^{\ 2} - p_y^{\ 2} - p_z^{\ 2} \tag{58}$$

From the relativistic formulas for E and $\mathbf{p}$, we can easily check that

$$\frac{E^2}{c^2} - p_x^{\ 2} - p_y^{\ 2} - p_z^{\ 2} = m^2c^2 \tag{59}$$

so

$$p_\mu p^\mu = m^2c^2 \tag{60}$$

■ ***Exercise 14.*** Check Eq. (59). ■

Physicists, and especially particle physicists, make much use of *momentum space;* this is a four-dimensional space with axes p^0, p_x, p_y, p_z. In this space, the equation

$$(p^0)^2 - p_x^{\ 2} - p_y^{\ 2} - p_z^{\ 2} = m^2c^2 \tag{61}$$

defines a hyperboloid, called the *mass shell* of the particle. The energy-momentum vector p^μ must connect the origin to some point on the mass shell (see Fig. 7.4). Since $p^0 \geq 0$, the energy-momentum vector must always be directed upward, into the region we would designate as the absolute future in a spacetime diagram (compare Fig. 6.13). In the special case of a particle of rest mass zero (photon, graviton, neutrino), $m = 0$ and the hyperboloid degenerates into a cone; the energy-momentum vector of such a particle always makes an angle of 45° with the p^0 axis.

EXAMPLE: *photon emission described in momentum space*

Diagrams in momentum space sometimes allow us to see at a glance that certain reactions are forbidden by energy-momentum conservation. For example, consider the emission of a photon by a moving particle. Figure 7.5 shows the initial (i) and final (f) energy-momentum vectors of the particle. The difference $p^\mu_{(i)} - p^\mu_{(f)}$ also shown in this figure represents the momentum of the (hypothetical) photon. No matter what we choose for $p^\mu_{(i)}$ and $p^\mu_{(f)}$, it is obviously impossible for the difference to make an angle of 45° with the p^0 axis; hence this difference *cannot* represent the momentum of a photon, and the reaction is impossible. It is crucial for this argument that $p^\mu_{(i)}$ and $p^\mu_{(f)}$ lie on the same mass shell (no change in the mass of the particle). Emission of a photon is possible only if there is a mass change in the emitter.

As the speed of a particle approaches the speed of light ($v \to c$), the energy approaches infinity,

$$E = \frac{mc^2}{\sqrt{1 - v^2/c^2}} \to \infty \tag{62}$$

Hence particles of finite rest mass can never be accelerated to a speed equal to that of light. Particles of zero rest mass have $v = c$, and Eq. (62) becomes indeterminate. This simply means that the energy cannot be

Figure 7·4

The mass shell in momentum space. This diagram shows only one of the space axes; it therefore shows a section through the hyperboloid. Only the upper branch of the hyperboloid, with $p^0 > 0$, is of physical significance.

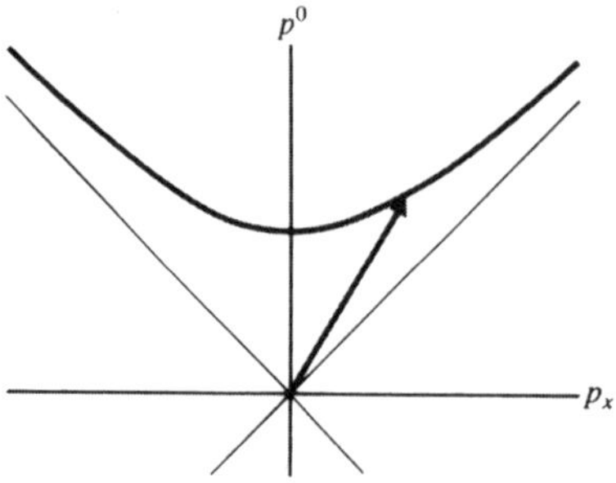

Figure 7·5

The initial and the final energy-momentum vectors for a particle trying to emit a photon. The difference between these vectors represents the energy-momentum vector of the photon.

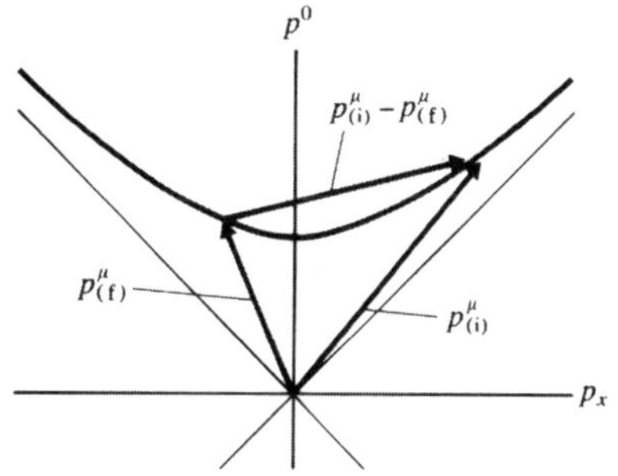

calculated from the velocity; all particles have the same speed c. However, Eq. (59) remains valid and gives

$$E = c|p| \tag{63}$$

Note that Eq. (63), which is exact for zero-rest-mass particles, is a good approximation for particles of nonzero rest mass provided $E \gg mc^2$, that is, $v \simeq c$.

EXAMPLE: *the hypothetical tachyon*

Although Eq. (62) indicates that it is impossible to make an ordinary particle exceed the speed of light, it does not forbid the existence of some particle that has, and always has had, $v > c$. The point is that, although the energy diverges at $v = c$, it is finite for $v > c$. Of course, if the rest mass is real, then $v > c$ gives an imaginary (and therefore meaningless) energy. But for a particle that always moves at $v > c$, we are entitled to assume that the mass is a purely imaginary number,

$$m = i|m| \tag{64}$$

and

$$E = \frac{mc^2}{\sqrt{1 - v^2/c^2}} = \frac{i|m|c^2}{i\sqrt{v^2/c^2 - 1}} = \frac{|m|c^2}{\sqrt{v^2/c^2 - 1}} \qquad \text{for } v > c \tag{65}$$

The energy-momentum vector is now

$$p^\mu = \left(\frac{|m|c}{\sqrt{v^2/c^2 - 1}}, \frac{|m|}{\sqrt{v^2/c^2 - 1}}\frac{dx}{dt}, \frac{|m|}{\sqrt{v^2/c^2 - 1}}\frac{dy}{dt}, \frac{|m|}{\sqrt{v^2/c^2 - 1}}\frac{dz}{dt} \right) \qquad \text{for } v > c \tag{66}$$

and

$$p^\mu p_\mu = (p^0)^2 - p_x^2 - p_y^2 - p_z^2 = -|m|^2 c^2 \qquad \text{for } v > c \tag{67}$$

Particles with $v > c$ have been called *tachyons*. Figure 7.6 shows the mass shell for such a particle. As already pointed out in Section 6.5, transmission of signals by means of tachyons would lead to serious trouble for our notions of causality. No tachyons have been discovered so far.

Let us now discuss the rest-mass energy in some detail. According to Eq. (50), the energy of a particle at rest is

$$E_0 = mc^2 \tag{68}$$

As long as we are dealing with a given particle of some given, fixed mass, the rest-mass energy is nothing but an additive constant in the energy, and it has no physical significance. It acquires physical significance only if the particle undergoes some reaction that changes its rest mass, with a corresponding conversion of rest-mass energy into other kinds of energy. Reactions that change the rest mass are possible if the particle has some internal structure— for instance, if the "particle" is a nucleus, an atom, or a system consisting of many atoms. Such a "particle" can be regarded as a system of pointlike particles. We will show that the relation (68) is always satisfied for such a system; that is, the total rest energy of the system equals its mass.

But first, we must give a precise definition of rest frame and rest energy. By the former we mean the reference frame in which the system has zero momentum; this reference frame is also often called the center-of-momentum, or C.M., frame.* The rest energy E_0 of the system is de-

Figure 7·6

The mass shell for a tachyon.

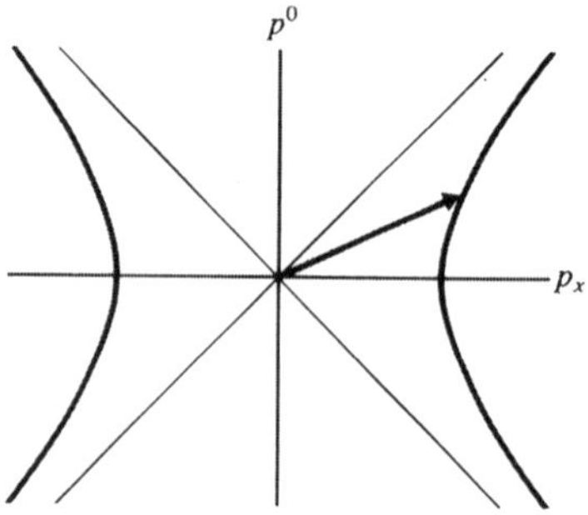

*In nonrelativistic physics, the center-of-momentum frame is the same as the center-of-mass frame. In relativistic physics this is not true, and we will ignore the center of mass entirely.

fined as the energy measured in the rest frame. If the C.M. moves with velocity **v** through the laboratory, then the Lorentz transformations for energy and momentum yield

$$E = \frac{E_0}{\sqrt{1 - v^2/c^2}} \tag{69}$$

$$\mathbf{P} = \frac{E_0 \mathbf{v}/c^2}{\sqrt{1 - v^2/c^2}} \tag{70}$$

for the energy and momentum of the system in the laboratory frame.

■ ***Exercise 15.*** Derive Eqs. (69) and (70). ■

However, if we have a system of mass M moving at velocity **v**, its energy and momentum must be

$$E = \frac{Mc^2}{\sqrt{1 - v^2/c^2}} \tag{71}$$

$$\mathbf{P} = \frac{M\mathbf{v}}{\sqrt{1 - v^2/c^2}} \tag{72}$$

Comparison of Eqs. (69) and (70) with Eqs. (71) and (72) shows that

$$E_0 = Mc^2 \tag{73}$$

In words, the C.M. energy of any system equals its mass.

As an example, let us apply Eq. (73) to a bound system such as an atom or a nucleus consisting of particles of masses m_1, m_2, m_3,... with velocities $\mathbf{v}_1$, $\mathbf{v}_2$, $\mathbf{v}_3$, ... The total C.M. energy of the system is the sum of the particle energies and potential energies, and therefore the mass of the system is

$$M = \sum_i \frac{m_i}{\sqrt{1 - v_i^2/c^2}} + U/c^2 \tag{74}$$

where the summation is over all particles in the system. It is convenient to separate the particle energies into kinetic and rest-mass energies:

$$M = \sum_i m_i + \left(\sum_i T_i/c^2 + U/c^2 \right) \tag{75}$$

where

$$T_i = \frac{m_i c^2}{\sqrt{1 - v_i^2/c^2}} - m_i c^2 \tag{76}$$

The term in parentheses in Eq. (75) is negative for a bound system; it is the binding energy. Hence the total mass is less than the sum of the particle rest masses. For nuclei, the binding energy is very large, and the difference between the nuclear mass and the sum of particle masses amounts to several percent.*

For a system consisting of *free* particles, the mass is given by Eq. (75), with $U = 0$. The total mass of such a system is larger than or equal to the sum of the particle rest masses. To evaluate Eq. (75), we must go to the C.M. frame. This is sometimes not convenient, and a quicker method for calculating the mass is the following: If we designate the total energy momentum in the C.M. frame by P'^{μ}, then

$$P'^{\mu} = (Mc, 0, 0, 0) \tag{77}$$

If we designate the energy-momentum vector in the laboratory by P^{μ}, then

$$P'^{\mu} P'_{\mu} = P^{\mu} P_{\mu} \tag{78}$$

The left side of this equation is M^2c^2, and hence

$$M^2 c^2 = P^{\mu} P_{\mu} \tag{79}$$

This equation expresses the mass of the system in terms of laboratory quantities. Because of Eq. (79), $(P^{\mu} P_{\mu} / c^2)^{1/2}$ is often called the *invariant mass* of the system.

For example, if the particle system consists of a proton with an energy-momentum vector

$$p^{\mu}_{(\mathrm{p})} = \left(\frac{E}{c}, p_x, p_y, p_z \right) \tag{80}$$

and a neutron at rest with

$$p^{\mu}_{(\mathrm{n})} = (m_{\mathrm{n}} c, 0, 0, 0) \tag{81}$$

then the mass of the system is given by

$$\begin{aligned} M &= \frac{(P^{\mu} P_{\mu})^{1/2}}{c} = \frac{[(p^{\mu}_{(\mathrm{p})} + p^{\mu}_{(\mathrm{n})})(p_{\mu(\mathrm{p})} + p_{\mu(\mathrm{n})})]^{1/2}}{c} \\ &= \frac{[p^{\mu}_{(\mathrm{p})} p_{\mu(\mathrm{p})} + p^{\mu}_{(\mathrm{n})} p_{\mu(\mathrm{n})} + 2 p^{\mu}_{(\mathrm{p})} p_{\mu(\mathrm{n})}]^{1/2}}{c} = \left(m_{\mathrm{p}}^2 + m_{\mathrm{n}}^2 + \frac{2Em_{\mathrm{n}}}{c^2} \right)^{1/2} \end{aligned} \tag{82}$$

*For a quantum-mechanical system, such as an atom or a nucleus, the kinetic energies should be expressed in terms of momenta rather than velocities, because momenta rather than velocities appear in the quantum-mechanical equations. But in any case, Eq. (75) remains valid.

EXAMPLE: *pion production*

If we wanted to produce a pion in the collision of this proton and neutron, according to the reaction

$$\mathrm{p} + \mathrm{n} \rightarrow \mathrm{p} + \mathrm{p} + \pi^- \tag{83}$$

then the minimum proton energy we would need is such that the reaction products p, p, π^- are at rest in the C.M. frame. In this case, the C.M. energy, after the reaction, is

$$(m_\mathrm{p} + m_\mathrm{p} + m_\pi)c^2 \tag{84}$$

By energy conservation, the C.M. energy before the reaction must have the same value—that is, the expression (84) must equal the value of Mc^2 given by Eq. (82), so

$$\left(m_\mathrm{p}^2 + m_\mathrm{n}^2 + \frac{2Em_\mathrm{n}}{c^2} \right)^{1/2} = m_\mathrm{p} + m_\mathrm{p} + m_\pi \tag{85}$$

or

$$E = \frac{(m_\mathrm{p} + m_\mathrm{p} + m_\pi)^2 - m_\mathrm{p}^2 - m_\mathrm{n}^2}{2m_\mathrm{n}} c^2 \tag{86}$$

This equation gives the minimum energy (threshold energy) for the reaction.

EXAMPLE: *C.M. energy for particle production*

The C.M. energy Mc^2 given by Eq. (82) tells us how much energy is available for particle production. Note that in the extreme relativistic limit ($E \gg m_\mathrm{p}c^2$), we find

$$Mc^2 \simeq \sqrt{2Em_\mathrm{n}c^2} \tag{87}$$

Since $m_\mathrm{n} \simeq 1$ GeV, protons of an energy $E \simeq 1000$ GeV incident on neutrons give a C.M. energy of only about 45 GeV. This shows that it is very advantageous to use *colliding beams* of particles of opposite momenta; then the C.M. frame is the laboratory frame, and the C.M. energy is the sum of the particle energies.

7·3 Differential Operators in Spacetime

A general electric or magnetic field is a function of both space and time—it is a field in spacetime. To formulate the relativistic equations for such a spacetime field, we will need to take derivatives with respect to space and time. In this, the four-dimensional gradient operator will play a crucial role. The four-dimensional gradient operator has four components:

$$\frac{\partial}{\partial x^{\mu}} = \left(\frac{1}{c}\frac{\partial}{\partial t}, \frac{\partial}{\partial x}, \frac{\partial}{\partial y}, \frac{\partial}{\partial z}\right) \tag{88}$$

We will often use the abbreviation ∂_μ for $\partial/\partial x_\mu$.

The transformation law for ∂_μ is

$$\partial'_\mu = a_\mu{}^\nu \partial_\nu \tag{89}$$

To derive this, we note that multiplication of

$$x'^\alpha = a^\alpha{}_\nu x^\nu \tag{90}$$

by $a_\alpha{}^\mu$ and summation over α yields, according to Eq. (23),

$$a_\alpha{}^\mu x'^\alpha = a_\alpha{}^\mu a^\alpha{}_\nu x^\nu = \delta^\mu_\nu x^\nu = x^\mu \tag{91}$$

that is,

$$x = a_\alpha{}^\mu \; x'^\alpha \tag{92}$$

This result simply expresses the fact that $a_\alpha{}^\mu$ gives the transformation inverse to $a^\alpha{}_\nu$. By the chain rule for derivatives, it follows that

$$\frac{\partial}{\partial x'^\alpha} = \frac{\partial x^\mu}{\partial x'^\alpha}\frac{\partial}{\partial x^\mu} = a_\alpha{}^\mu \frac{\partial}{\partial x^\mu}$$

which is equivalent to Eq. (89).

According to Eq. (89), the components of the operator ∂_μ transform covariantly, as suggested by their lower index. We can also introduce the operator

$$\partial^\mu = \eta^{\mu\nu}\partial_\nu = \left(\frac{1}{c}\frac{\partial}{\partial t}, -\frac{\partial}{\partial x}, -\frac{\partial}{\partial y}, -\frac{\partial}{\partial z}\right) \tag{93}$$

whose components transform contravariantly,

$$\partial'^\mu = a^\mu{}_\nu \partial^\nu$$

The scalar product

$$\partial_\mu \partial^\mu = \frac{1}{c^2}\frac{\partial^2}{\partial t^2} - \frac{\partial^2}{\partial x^2} - \frac{\partial^2}{\partial y^2} - \frac{\partial^2}{\partial z^2} \tag{94}$$

is then invariant:

$$\partial'_\mu \partial'^\mu = \partial_\mu \partial^\mu$$

The second-order differential operator $\partial_\mu \partial^\mu$ is called the *D'Alembertian* or the *wave operator.* This operator is the four-dimensional analog of the Laplacian operator [see Eq. (1.74)]. Correspondingly, the four-dimensional analog of Laplace's equation is

$$\partial_\mu \partial^\mu \Phi(t, \mathbf{x}) = 0 \qquad (95)$$

This is a wave equation. For instance, if the function Φ depends only on t and x, then Eq. (95) reduces to

$$\left(\frac{1}{c^2} \frac{\partial^2}{\partial t^2} - \frac{\partial^2}{\partial x^2} \right) \Phi(t, x) = 0 \qquad (96)$$

which has the well-known solution

$$\Phi(t,x) = f(ct \pm x)$$

where f is an arbitrary function of one variable. The function $f(ct - x)$ represents a wave traveling in the positive x direction, and the function $f(ct + x)$ represents a wave traveling in the negative x direction.

■ ***Exercise 16.*** Show that for any differentiable function f, $f(ct \pm x)$ is a solution of Eq. (96). ■

A field in spacetime will be either a scalar, a vector, or a tensor field. At any given point of spacetime, the field must satisfy the appropriate transformation law, as specified by the definitions of the preceding section. For instance, at any given point P of spacetime, a scalar field Φ must remain invariant under Lorentz transformations:

$$\Phi'(P) = \Phi(P) \qquad (97)$$

If we describe the spacetime point P by its coordinates x^μ, then Eq. (97) becomes

$$\Phi'(x'^\mu) = \Phi(x^\mu) \qquad (98)$$

or

$$\Phi'(x'^\mu) = \Phi(a_\alpha{}^\mu x'^\alpha) \qquad (99)$$

EXAMPLE: *Lorentz transformation of a scalar field*

If the function $\Phi(x^\mu)$ is known, then Eq. (99) completely determines the function $\Phi'(x'^\mu)$. Note that Φ' is not the same function of x'^μ as Φ is of x^μ. For instance, if $\Phi(x^\mu) = ax^0 + bx^1$, then the Lorentz transformation (4) gives the more complicated expression

$$\Phi'(x'^{\mu}) = a\frac{x'^{0} + Vx'^{1}/c}{\sqrt{1-V^2/c^2}} + b\frac{x'^{1} + Vx'^{0}/c}{\sqrt{1-V^2/c^2}} \tag{100}$$

However, at any given point—with equivalent coordinates x^{μ} and x'^{μ} —the functions Φ and Φ' have the same numerical value.

Whenever the differential operator ∂_{μ} acts on a spacetime field, it generates another spacetime field. If ∂_{μ} acts on a scalar field Φ, it generates a vector field

$$\partial_{\mu}\Phi \tag{101}$$

If ∂_{μ} acts on a vector field B_{ν}, it generates a tensor field

$$\partial_{\mu} B_{\nu} \tag{102}$$

And if ∂_{μ} acts on a tensor field $F_{\alpha\beta}$, it generates a third-rank tensor field

$$\partial_{\mu} F_{\alpha\beta} \tag{103}$$

In contrast, whenever the differential operator $\partial_{\mu}\partial^{\mu}$—which behaves as a scalar—acts on a scalar, vector, or tensor, it merely generates a new scalar, vector, or tensor (for instance, $\partial_{\mu}\partial^{\mu}\,\Phi$, $\partial_{\mu}\partial^{\mu}B_{\nu}$, or $\partial_{\mu}\partial^{\mu}F_{\alpha\beta}$).

7·4 The Current Four-Vector

In Section 1.6 we introduced the concepts of charge density [$\rho(t, \mathbf{x})$] and current density [$j(t, \mathbf{x})$] to describe a continuous distribution of charge. These densities obey the continuity equation

$$\nabla\cdot\mathbf{j} + \frac{\partial\rho}{\partial t} = 0 \tag{104}$$

which expresses the conservation of electric charge. In this section, we will see how to rewrite Eq. (104) in four-dimensional notation.

First, we will prove that the four-component object

$$j^{\mu} = (c\rho, j_x, j_y, j_z) \tag{105}$$

is a four-vector, or, more precisely, a four-vector field. For this purpose, we begin by writing ρ and $\mathbf{j}$ in terms of the density and velocity of the charged particles that make up the distribution of charge and current [compare Eqs. (1.118) and (1.121)]:

$$\rho = qn \tag{106}$$

$$\mathbf{j} = qn\mathbf{v} \tag{107}$$

where n is the density of particles and $\mathbf{v}$ their velocity; in general, both of these quantities depend on position. Note that we have assumed that only one kind of particle, with charge q, contributes to the charge density and to the current density; if several kinds contribute, then their charge densities must be added and their current densities must be added. If at some spacetime point the velocity of the particles is $\mathbf{v}$, then at that point the density n measured in the laboratory is related to the proper density n_0 measured in the rest frame of the particles by

$$n = \frac{n_0}{\sqrt{1-v^2/c^2}} \tag{108}$$

This relation is an immediate consequence of the volume-contraction formula (6.71)—the volume spanned by some fixed number of particles is smaller when measured in the laboratory than when measured in their rest frame, and the density in the laboratory is therefore higher than in the rest frame. If we express the density and the current density in terms of n_0, we find

$$\rho = \frac{qn_0}{\sqrt{1-v^2/c^2}} \tag{109}$$

$$\mathbf{j} = \frac{qn_0\mathbf{v}}{\sqrt{1-v^2/c^2}} \tag{110}$$

Equation (105) then becomes

$$j^\mu = qn_0\left(\frac{c}{\sqrt{1-v^2/c^2}}, \frac{v_x}{\sqrt{1-v^2/c^2}}, \frac{v_y}{\sqrt{1-v^2/c^2}}, \frac{v_z}{\sqrt{1-v^2/c^2}}\right) \tag{111}$$

The four-component object on the right side of this equation is the four-velocity u^μ; hence

$$j^\mu = qn_0u^\mu \tag{112}$$

Here q and n_0 are scalars (they are both invariant under Lorentz transformations*), and u^μ is a four-vector. The product qn_0u^μ, or j^μ, is then also a four-vector. It is called the *current four-vector.*

---

*If q were not invariant, then the charge of, say, an electron would depend on its state of motion. This can be ruled out on theoretical grounds, because it would result in a violation of charge conservation [violation of Eq. (113)], which would lead to inconsistencies requiring unacceptable modifications of the basic equations of electrodynamics. It can also be ruled out on direct experimental grounds, because a dependence of the electron's charge on its state of motion would cause deviations from neutrality in atoms, in conflict with the experimental measurements mentioned in Section 2.1, which established that atoms are neutral to within better than 1 part in 10^{21}.

In terms of j^μ, the continuity equation (104) takes the concise form

$$\partial_\mu j^\mu = 0 \tag{113}$$

By writing the continuity equation in this way, we immediately recognize that the law of conservation of charge satisfies the Principle of Relativity. A Lorentz transformation changes Eq. (113) into

$$\partial'_\mu j'^\mu = 0 \tag{114}$$

which has the same mathematical form as the original equation (113)—the equations look the same in both reference frames. This is exactly what is required by the Principle of Relativity.

■ ***Exercise 17.*** From the Lorentz-transformation equations for the current four-vector,

$$\rho' = \frac{\rho - Vj_x/c^2}{\sqrt{1-V^2/c^2}} \tag{115}$$

$$j'_x = \frac{j_x - V\rho}{\sqrt{1-V^2/c^2}} \tag{116}$$

show that if a wire is electrically neutral but carries a current along the x axis in the $xyzt$ frame, then this wire will not be neutral in the $x'y'z't'$ frame. ■

References

1. D. J. Grove and J. G. Fox, *Phys. Rev.* ***90***, 378 (1953).
2. V. P. Zrelov, A. A. Tyapkin, and P. S. Farrago, *Sov. Phys. JETP* **7**, 384 (1958).
3. Z. G. T. Guiragossian, G. B. Rothbart, M. R. Yearian, R. A. Gearhart, and J. J. Murray, *Phys. Rev. Lett.* **34**, 335 (1975).
4. D. Newman, G. W. Ford, A. Rich, and E. Sweetman, *Phys. Rev. Lett.* **40**, 1355 (1978).

Problems

1. The transformation matrix for a rotation about the z axis is [see Eq. (1.20)]

$$a_\mu{}^\nu = \begin{pmatrix} 1 & 0 & 0 & 0 \\ 0 & \cos\theta & \sin\theta & 0 \\ 0 & -\sin\theta & \cos\theta & 0 \\ 0 & 0 & 0 & 1 \end{pmatrix}$$

Check explicitly that this matrix satisfies the conditions for a (general) Lorentz transformation.

2. (a) Suppose that an object $A(\mu, \nu)$ with two indices has the property that $A(\mu, \nu)B^\mu$ is a vector whenever B^μ is some arbitrary vector. Show that $A(\mu, \nu)$ is a tensor.
(b) Suppose that an object with two indices has the property that $A(\mu, \nu)C^{\mu\nu}$ is a scalar for any arbitrary tensor $C^{\mu\nu}$. Can you show that $A(\mu, \nu)$ is a tensor?

3. Two vectors have the property that $A^\mu B_\mu = 0$.
(a) Suppose $A^\mu A_\mu > 0$. Show that $B^\mu B_\mu \le 0$.
(b) Suppose $A^\mu A_\mu = 0$. Show that B^μ is either proportional to A^μ (that is, $B^\mu = kA^\mu$) or else $B^\mu B_\mu < 0$.

4. Let X be the matrix

$$X = \begin{pmatrix} x^0 + x^3 & x^1 - ix^2 \\ x^1 + ix^2 & x^0 - x^3 \end{pmatrix}$$

(a) Show that the determinant of this matrix is $\eta_{\mu\nu}x^\mu x^\nu$.
(b) Consider a new matrix $X' = UXU^\dagger$, where U is a complex 2×2 matrix, $U^\dagger$ is its Hermitian conjugate, and the determinant of U equals 1. Define new coordinates x'^μ by

$$X' = \begin{pmatrix} x'^0 + x'^3 & x'^1 - ix'^2 \\ x'^1 + ix'^2 & x'^0 - x'^3 \end{pmatrix}$$

Show that x'^μ and x^μ are then related by a linear transformation,

$$x'^\mu = B^\mu{}_\nu x^\nu$$

where the matrix $B^\mu{}_\nu$ is real and satisfies all the conditions necessary for it to be the matrix of a Lorentz transformation.
(c) To obtain the Lorentz transformation (4), what choice must we make for the components of the matrix U?

5. Suppose that a tensor of rank n (n indices) is symmetric in all pairs of indices, and the trace taken over any pair of indices is zero; that is,

$$A^{\mu\nu\alpha\beta\ldots} = A^{\nu\mu\alpha\beta\ldots} = A^{\mu\alpha\nu\beta\ldots} = A^{\beta\nu\alpha\mu\ldots} = \ldots$$

and

$$A_\mu{}^{\mu\alpha\beta\ldots} = A_\mu{}^{\nu\mu\beta\ldots} = A_\mu{}^{\nu\alpha\mu\ldots} = \ldots = 0$$

Show that this tensor has $(n + 1)^2$ independent components. [*Hint:* Show that a tensor symmetric in all pairs of indices and subject to no other restriction has $(n + 3)!/(3!n!)$ components.]

6. The fourth-rank alternating tensor $\varepsilon^{\mu\nu\alpha\beta}$ is defined as follows:

$\varepsilon^{\mu\nu\alpha\beta} = 0$ unless all indices are different

$\varepsilon^{0123} = 1$

$\varepsilon^{\mu\nu\alpha\beta} = 1$ if $\mu\nu\alpha\beta$ is a permutation of 0123 with an even number of transpositions of adjacent pairs (for example, $\varepsilon^{1032} = 1$)

$\varepsilon^{\mu\nu\alpha\beta} = -1$ if $\mu\nu\alpha\beta$ is a permutation of 0123 with an odd number of transpositions of adjacent pairs (for example, $\varepsilon^{1023} = -1$)

Show that this tensor transforms into itself under Lorentz transformations.

7. (a) A particle moves along a straight line so the acceleration is always parallel to the velocity. In this case, we can define an "effective mass" m_L as the constant of proportionality between $d\mathbf{p}/dt$ and $d\mathbf{v}/dt$,

$$\frac{d\mathbf{p}}{dt} = m_L \frac{d\mathbf{v}}{dt}$$

Show that $m_L = m/(1 - v^2/c^2)^{3/2}$, which is called the *longitudinal mass.*

(b) A particle moves along a circle, with uniform circular motion, so the acceleration is always perpendicular to the velocity. Show that, in this case, the effective mass is $m_T = m/(1 - v^2/c^2)^{1/2}$, which is called the *transverse mass.*

(c) Consider a particle that has an acceleration that is neither parallel nor perpendicular to the velocity. Show that, in this case, the acceleration can be expressed as follows in terms of the parallel and perpendicular components of **dp**/dt:

$$\frac{d\mathbf{v}}{dt} = \frac{1}{m_L}\frac{d\mathbf{p}}{dt}\bigg|_{\parallel} + \frac{1}{m_T}\frac{d\mathbf{p}}{dt}\bigg|_{\perp}$$

Show that if the speed is near the speed of light, then the direction of the acceleration $d\mathbf{v}/dt$ is nearly perpendicular to the velocity, regardless of the direction of $d\mathbf{p}/dt$. What is the maximum angle between the directions of $d\mathbf{v}/dt$ and $d\mathbf{p}/dt$?

8. Suppose that a tachyon of infinite velocity ($v = \infty$) strikes a proton at rest. Show that the proton cannot absorb this tachyon. [*Hint:* Draw the mass hyperbola for the proton in momentum space. What are the initial and final momentum vectors for the proton?]

9. (a) Show that for an extremely relativistic particle, with $v \simeq c$, the difference between the speed of the particle and the speed of light is

$$c - v = \frac{c}{2}\left(\frac{mc^2}{E}\right)^2$$

(b) At the Fermilab accelerator, protons are given an energy of 1000 GeV, or 1 TeV. What is the difference $c - v$ for such a proton?

(c) At the Stanford Linear Accelerator, electrons are given an energy of 40 GeV. What is the difference $c - v$ for such an electron?

10. The velocity of the center of mass of a system of particles is

$$\mathbf{v}_{\text{C.mass}} = \frac{m_1\mathbf{v}_1 + m_2\mathbf{v}_2 + \ldots + m_n\mathbf{v}_n}{m_1 + m_2 + \ldots + m_n}$$

The velocity of the center of momentum is $v_{\text{C.M.}} = c^2\mathbf{P}/E$ [see Eqs. (69) and (70)]. By means of a specific counterexample, show that for relativistic particles, these two velocities are different; therefore, the center-of-mass frame does not coincide with the center-of-momentum frame.

11. At the CERN Super Proton Synchrotron, protons and antiprotons of 270 GeV each were made to collide head on. The available C.M. energy was then 540 GeV. Suppose we wanted to attain the same C.M. energy in a collision of a moving proton with a stationary proton. What would have to be the energy of the moving proton?

12. At the Fermilab Main Ring, protons of kinetic energy 1000 GeV, or 1 TeV, were made to collide with stationary protons. What is the velocity of the center-of-momentum frame?

13. A beam of π^- mesons is incident on a stationary hydrogen target. Find the minimum kinetic energy that the π^- must have in order to initiate the reaction

$$\pi^- + \mathrm{p} \rightarrow \mathrm{K}^0 + \Sigma^0$$

Express your result in terms of the rest masses m_1, m_2, m_3, m_4 of the particles π^-, p, K^0, Σ^0 respectively.

14. A high-energy beam of K^- mesons is used to bombard a stationary hydrogen target. Find the minimum K^--meson kinetic energy needed for the reaction

$$\mathrm{K}^- + \mathrm{p} \rightarrow \Lambda + \pi^0 + \pi^+ + \pi^-$$

The masses of the particles p, K^-, Λ, and $\pi^{0,+,-}$ are, respectively, 0.94 GeV/c^2, 0.49 GeV/c^2, 1.11 GeV/c^2, and 0.14 GeV/c^2.

15. A physicist wants to produce tau particles in proton collisions by the reaction

$$\mathrm{p} + \mathrm{p} \rightarrow \mathrm{p} + \mathrm{p} + \tau + \bar{\tau}$$

where τ stands for the tau particle and $\bar{\tau}$ for the antitau particle. The proton mass is 0.94 GeV/c^2, and the tau (and antitau) mass is 1.8 GeV/c^2.

(a) If a proton target is bombarded with a proton beam, what is the minimum kinetic energy of the protons in the beam that will make the reaction possible?

(b) If two proton beams are used in a head-on collision ("colliding beams"), what is the minimum kinetic energy of the protons?

16. Find the threshold kinetic energies of the incident particles in each of the following reactions:

(a) $\pi^+ + \mathrm{p} \rightarrow \pi^+ + \pi^0 + \mathrm{p}$

(b) $\mathrm{p} + \mathrm{n} \rightarrow \mathrm{p} + \mathrm{n} + \pi^0 + \pi^0$

The incident particle is the first particle listed in the reaction; the target particle is at rest. The masses of the π^+ and π^0 pions are 139.567 MeV/c^2 and 134.963 MeV/c^2, respectively.

17. (a) A particle of mass m at rest decays into n particles of masses $m_1 + m_2 + \cdots + m_n$. Show that the maximum kinetic energy that particle i can ever have is

$$\Delta mc^2\left(1 - \frac{m_i}{m} - \frac{\Delta m}{2m}\right)$$

where

$$\Delta m = m - \sum_{i=1}^{n} m_i$$

[*Hint:* Particle i acquires maximum energy when all the *other* particles remain together and recoil in a direction opposite to that of i.]

(b) Find the maximum possible kinetic energy of *each* of the decay products in the reaction

$$\pi^+ \rightarrow \pi^0 + e^+ + \nu$$

The masses of the π^+ and π^0 pions are 139.567 MeV/c^2 and 134.963 MeV/c^2, respectively.

18. A π^0 meson of total energy 250 MeV decays into two gamma rays, one of which is emitted in the direction of motion of the meson and the other in the opposite direction. What are the energies of the two gamma rays?

19. Two gamma rays, each of energy $E > m_e c^2$, collide. If the angle between their directions of propagation is $\phi = 180°$ (a head-on collision), electron-positron creation is possible. If the angle is $\phi = 0$, creation is not possible. What is the critical angle ϕ at which creation is *barely* possible?

20. (a) Show that, according to relativistic physics, the final velocity v of a rocket accelerated by its rocket motor in empty space is given by

$$\frac{M_i}{M} = \left(\frac{c+v}{c-v}\right)^{c/2v_{ex}}$$

where M_i is the initial mass of the rocket at launch (including the fuel), M is the final mass at burnout, and v_{ex} is the velocity, relative to the rocket, with which the exhaust gases emerge from the nozzle of the rocket motor. [*Hint:* If the rocket ejects a small mass dm of exhaust, conservation of energy and momentum in the instantaneous rest frame $x'y'z't'$ of the rocket imply

$$c^2 dM = \frac{-c^2 dm}{\sqrt{1 - v_{ex}^2/c^2}} \quad \text{and} \quad M dv' = \frac{v_{ex} dm}{\sqrt{1 - v_{ex}^2/c^2}}$$

Furthermore, the velocity-combination formula implies

$$v + dv = \frac{v + dv'}{1 + v dv'/c^2}$$]

(b) Typical chemical fuels yield $v_{ex} = 3 \times 10^5$ cm/s. What mass of fuel is required to attain a final velocity of $0.1c$ for a final mass of 1 ton?

(c) Matter-antimatter fuel yields $v_{ex} = c$ (the exhaust consists of photons). What mass of fuel is required in this case?

21. Given that $A^\mu(t, x)$ is a vector field with components

$$A^\mu(t, x) = (0, \cos(5x - 2ct), \cos(5x - 2ct), 0)$$

find the vector field $A'^\mu(t', x')$, where t, x and t', x' are related by the Lorentz transformation (4). Express your answer in the form of a function of t', x'.

22. Given that $\Phi(t, \mathbf{x})$ is a scalar field, show that the integral

$$\int \Phi(t, \mathbf{x}) dV dt$$

evaluated over a given four-dimensional volume of spacetime is a scalar. [*Hint:* To transform $dVdt$, use volume contraction and time dilation; alternatively, use the Jacobian determinant.]

23. Given that the current density j^μ differs from zero only in a volume of finite extent in the x,y,z directions, show that the total charge

$$Q = \int j^0 dV$$

is a scalar under Lorentz transformations. [*Hint:* Integrate $\partial_\mu j^\mu = 0$ over the volume of a four-dimensional world tube that encloses the region of interest and extends from x^0 = constant to x'^0 = constant. Convert the four-dimensional volume integral into a surface integral, by Gauss' theorem.]

24. Given that a vector field $J^\mu(t, \mathbf{x})$ satisfies the differential equation

$$\partial_\mu J^\mu = 0$$

and that J^μ is different from zero only in a volume of finite extent in the x,y,z directions, show that

$$\frac{d}{dt}\int J^0 dV = 0$$

[*Hint:* Integrate $\partial_\mu J^\mu = 0$ over the volume of a four-dimensional world tube that encloses the region of interest and extends from x^0 = constant to x'^0 = constant. Convert the four-dimensional volume integral into a surface integral, by Gauss' theorem.]

25. A thin, straight wire carries a current of 3.0×10^{11} esu/s (100 ampere); the wire is electrically neutral in the reference frame *xyzt* of the laboratory. What is the charge per unit length on this wire in the reference frame *x'y'z't'* moving at $V = 0.8c$ in a direction parallel to the wire?

26. In the rest frame of a circular loop of wire, a constant current I flows around the loop. Suppose that the loop moves past you with velocity $\mathbf{V}$; this velocity is in the plane of the loop (see Fig. 7.7). What charge density (charge per unit length) will you observe on the loop? Express your answer as a function of the central angle θ defined in the rest frame of the loop.

27. Two large, horizontal, parallel conducting plates carry currents in opposite directions, the positive x direction and the negative x direction. The current is uniformly distributed over each plate, with k esu of charge per second flowing across each centimeter measured along the plate perpendicularly to the current. Show that in a reference frame moving at speed V in the positive x direction, the plates carry both a current and a charge density, and find the electric field generated by this charge density.

Figure 7·7

Circular loop of wire with current I moving at velocity $\mathbf{V}$.

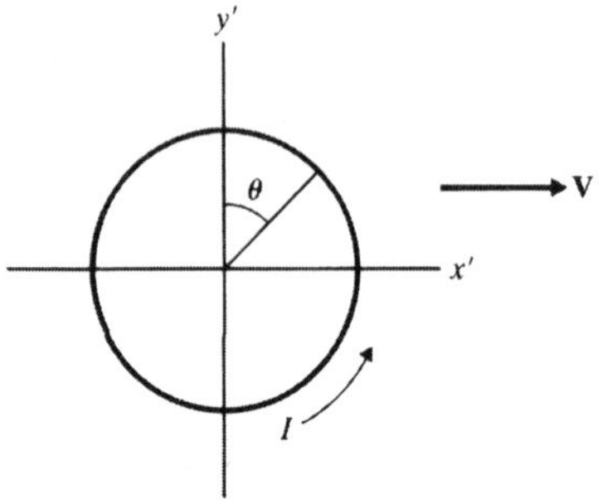

CHAPTER

8

Relativistic Electrodynamics

We will now construct a relativistic theory of electric interactions. This means that we will take the nonrelativistic electrostatic equations of Chapter 2 and ask how these equations must be generalized to satisfy the Principle of Relativity. It turns out that this question has a unique answer: there is one and only one set of relativistic equations that reduces to the electrostatic equations in the static limit. As a by-product of our investigations, we will discover that the electric field is insufficient for describing the interactions among charges. An extra field—the magnetic field—is needed whenever we deal with charges in motion.

8·1 The Four-Vector Potential

In the electrostatic case, the charge density ρ is time independent, and the current density $\mathbf{j}$ is zero. Under these conditions, the equations in Chapter 2 give us a complete description of the electric interactions in terms of a single function of space: the potential Φ. If $\mathbf{j}$ is not zero, then this potential function is insufficient to describe the electric interactions. The following simple example will make this clear.

Consider a long straight wire carrying a uniform current density but no charge density. For a concrete picture, we may think of the wire as carrying a flow of conduction electrons whose negative contribution to the charge density is compensated by a positive contribution from the positive ions of the crystal lattice of the wire. Consider, next, a point charge q that is moving with velocity $\mathbf{v}$ parallel to the wire [see Fig. 8.1(a)]. Since the net charge on the wire is zero, there is no electric field in the laboratory frame, or $xyzt$ frame. Nevertheless, the wire will exert a force on the moving charge. To see this, we go to the rest frame of the charge q, or the $x'y'z't'$ frame [see Fig. 8.1 (b)]. The transformation equations for the charge and current densities [Eqs. (7.115) and (7.116)] tell us that in this reference frame,

Figure 8·1

(a) In the laboratory reference frame, the wire is at rest and the point charge q moves toward the right. The wire carries a current, but the wire is electrically neutral. (b) In the $x'y'z't'$ frame, the point charge q is at rest and the wire moves toward the left. The wire now has a negative charge density.

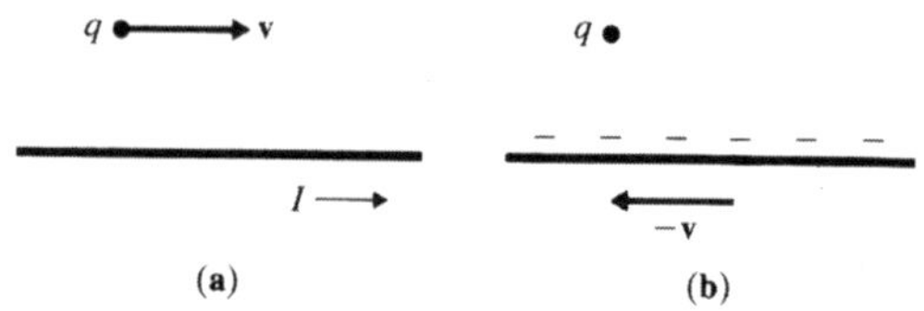

EXAMPLE: *charge density resulting from Lorentz transformation of current*

$$\rho' = \frac{\rho - vj_x/c^2}{\sqrt{1-v^2/c^2}} = -\frac{vj_x/c^2}{\sqrt{1-v^2/c^2}} \tag{1}$$

$$j'_x = \frac{j_x - v\rho}{\sqrt{1-v^2/c^2}} = \frac{j_x}{\sqrt{1-v^2/c^2}} \tag{2}$$

According to Eq. (1), the charge density ρ' is nonzero, and hence an electric field will exist in this reference frame. This electric field will exert a force on the charge q.

■ ***Exercise 1.*** Show that in the rest frame of the charge q, the wire is surrounded by a radial electric field of magnitude

$$E' = \frac{2\pi R^2 \rho'}{r} = -\frac{2\pi R^2 v j_x/c^2}{r\sqrt{1-v^2/c^2}} \tag{3}$$

where R is the radius of the wire. ■

If the charge q is positive and its velocity is in the same direction as **j**, then the force on this charge is attractive [see Fig. 8.1(b)]. The existence of such a radial force in the $x'y'z't'$ frame implies the existence of a corresponding force in the original laboratory frame. Since there is no charge density in the latter frame, the force must be generated in some way by the current. Thus the conclusion we can draw from this example is that currents exert forces on moving charges. The potential Φ does not take these new kinds of forces into account—in the preceding example, the force cannot be calculated from the potential in the laboratory frame. These new forces are, of course, magnetic forces. It will be our objective to derive the equations describing these forces from a relativistic theory of electric interactions.

What we need is a new set of equations that involve not only the charge density but also the current density. To find these equations, we begin with the electrostatic equation

$$-\nabla^2\Phi = 4\pi\rho \qquad (4)$$

and try to construct a generalization that meets three requirements: it satisfies the principle of relativity, it permits a time dependence in ρ and Φ, and it reduces to Eq. (4) if ρ and Φ are time independent. As it stands, Eq. (4) does not satisfy the principle of relativity—if the equation holds in one reference frame, it cannot have the same form in a second frame because the Lorentz transformation will convert the purely spatial derivatives appearing in Eq. (4) into space and *time* derivatives. The first thing we must do to obtain a set of equations that keep their form under Lorentz transformations is to replace the differential operator $-\nabla^2$ by a Lorentz-invariant differential operator. The correct Lorentz-invariant operator is obviously $\partial_\mu\partial^\mu$, because this is the operator that reduces to $-\nabla^2$ in the time-independent case. With this modification, Eq. (4) becomes

$$\partial^\mu\partial_\mu \Phi = 4\pi\rho \qquad (5)$$

Since $c\rho$ is the j^0 component of the four-vector current, we can also write this equation as

$$\partial^\mu\partial_\mu\Phi = \frac{4\pi}{c} j^0 \qquad (6)$$

If this equation is to be valid in all reference frames, both sides must transform in the same way, that is, both sides must be tensor components of the same kind. The right side is the time component of a four-vector; consequently, the left side must also be the time component of a four-vector. Since the differential operator $\partial^\mu\partial_\mu$ is a scalar, Φ must necessarily be the time component of some four-vector. We will designate this four-vector by A^μ. By definition, $A^0 \equiv \Phi$, so

$$A^\mu = (A^0, A^1, A^2, A^3) = (\Phi, A_x, A_y, A_z) \qquad (7)$$

This four-vector is called the *four-vector potential*. Its space components form a three-dimensional vector $\mathbf{A} = A_x\hat{\mathbf{x}} + A_y\hat{\mathbf{y}} + A_z\hat{\mathbf{z}}$, called the *vector potential*. At this stage, we do not yet have any physical interpretation for the space components. However, we expect that the forces exerted by currents must somehow be related to these components.

Expressed in terms of A^0, our equation for the potential becomes

$$\partial^\mu\partial_\mu A^0 = \frac{4\pi}{c} j^0 \qquad (8)$$

Since the Lorentz transformations mix the components of a four-vector, this equation for the time component cannot be valid in all reference frames unless all other components satisfy similar equations:

$$\partial^\mu \partial_\mu A^1 = \frac{4\pi}{c} j^1 \qquad \partial^\mu \partial_\mu A^2 = \frac{4\pi}{c} j^2 \qquad \partial^\mu \partial_\mu A^3 = \frac{4\pi}{c} j^3 \tag{9}$$

■ ***Exercise 2.*** Show that if one of the equations in (9) is *not* satisfied in a given reference frame, then there exists some other reference frame in which the equation for the time component is not satisfied, that is, $\partial'^\mu \partial'_\mu A'^0 \neq 4\pi j'^0/c$. ■

Combining Eqs. (8) and (9), we obtain the following differential equation for the four-vector potential:

$$\partial^\mu \partial_\mu A^\nu = \frac{4\pi}{c} j^\nu \tag{10}$$

This equation keeps its form under Lorentz transformations; that is, the only change the equation suffers under a Lorentz transformation is that all quantities acquire primes:

$$\partial'^\mu \partial'_\mu A'^\nu = \frac{4\pi}{c} j'^\nu \tag{11}$$

Thus the equation satisfies the Principle of Relativity.

Equation (10) is the fundamental field equation for relativistic electrodynamics. It is a wave equation, with the current four-vector acting as the source of waves—the charge density acts as the source of potential [Eq. (8)], and the current density acts as the source of vector potential [Eq. (9)]. Intuitively, such equations make good sense; whenever the charge or current density changes in some way, we expect that the electrodynamic fields in the surrounding space will suffer a disturbance, and we expect that such disturbances take the form of waves propagating outward from their source at the speed of light.

In addition to Eq. (10), we will assume that the four-vector potential also satisfies the equation

$$\partial_\nu A^\nu = 0 \tag{12}$$

This is known as the *Lorentz condition*. To see where it comes from, let us apply the differential operator ∂_ν to both sides of Eq. (10):

$$\partial^\mu \partial_\mu (\partial_\nu A^\nu) = \frac{4\pi}{c} \partial_\nu j^\nu \tag{13}$$

The conservation law for electric charge tells us that $\partial_\nu j^\nu = 0$ [see Eq. (7.113)], and hence

$$\partial^\mu \partial_\mu (\partial_\nu A^\nu) = 0 \tag{14}$$

This is a third-order differential equation, and the simplest way to satisfy this equation is to impose the Lorentz condition (12).

To gain further understanding of Eq. (12), let us check what happens in the case of electrostatics. If $A^k = 0$, Eq. (12) reduces to $\partial A^0/\partial x^0 = 0$, that is,

$$\frac{\partial \Phi}{\partial t} = 0 \tag{15}$$

This simply says that Φ is time independent. Equation (15) is *not* trivial, because even in electrostatics, the potential could be time dependent. The reason is that, although electrostatics requires the electric field to be time independent, this does not demand a time-independent potential. For example, an electric field of magnitude E_0 along the x axis can be represented by the potential

$$\Phi(x) = -E_0 x \tag{16}$$

and also by the potential

$$\Phi(x) = -E_0 x + f(t) \tag{17}$$

where $f(t)$ is an arbitrary function of time.

■ ***Exercise 3.*** Check that both (16) and (17) give the same electric field. The transformation $\Phi(x) \to \Phi(x) + f(t)$ is an instance of a *gauge transformation*; such a transformation changes the potential but does not change the electric field. ■

In our earlier discussions of electrostatics, we always assumed implicitly that there is no additive function of time in the potential. Equation (15) makes this assumption explicit, and Eq. (12) can be regarded as the relativistic generalization of Eq. (15).

The Lorentz condition (12) makes our relativistic equation for the potential consistent with the conservation law for electric charge. There is, however, another way in which we can achieve this consistency. Suppose we modify Eq. (10) by the addition of an extra term $-\partial^\nu \partial_\mu A^\mu$ on the left side; our new equation then reads

$$\partial^\mu \partial_\mu A^\nu - \partial^\nu \partial_\mu A^\mu = \frac{4\pi}{c} j^\nu \tag{18}$$

If we now apply the differential operator ∂_ν to both sides, we find

$$\partial^\mu \partial_\mu \partial_\nu A^\nu - \partial_\nu \partial^\nu \partial_\mu A^\mu = \frac{4\pi}{c} \partial_\nu j^\nu \tag{19}$$

Since both μ and ν are dummy indices (both are repeated, and summation on both is understood), the left side *vanishes identically*. Hence we achieve consistency with current conservation without having to require $\partial_\nu A^\nu = 0$.

The modified equation (18) obviously satisfies the Principle of Relativity; furthermore, its time component ($\nu = 0$) agrees with the electrostatic equation [Eq. (4)] if A^0 is time independent. Thus Eq. (18) is an acceptable alternative equation for the four-vector potential. This means that the question of what is the relativistic generalization of the equation for the potential has two answers. But this ambiguity need not trouble us—it turns out that both equations for the potential lead to the same equations for the electric and magnetic fields. As we will see, these fields are unique and free of any ambiguities.

8·2 The Field Tensor and the Equation of Motion

We must now explore the physical significance of the components of the four-vector potential A^μ. At this stage, we know only the significance of A^0: it coincides with the electric potential. In the static case, the negative gradient of the electric potential equals the electric field,*

$$E^1 = \partial^1 A^0 \quad E^2 = \partial^2 A^0 \quad E^3 = \partial^3 A^0 \tag{20}$$

These equations tell us something about the Lorentz-transformation properties of the electric field—they tell us that E^1, E^2, and E^3 are, respectively, the 10, 20, and 30 components of the tensor $\partial^\mu A^\nu$. This may come as somewhat of a surprise. In *three* dimensions, we are used to thinking of **E** as a vector quantity, and we naively might have expected that in *four* dimensions, **E** somehow generalizes into a four-vector quantity. But the components $\partial^k A^0$ appearing in Eq. (20) are part of a tensor, and hence so are the components E^k.

We must keep in mind that Eq. (20) holds only in the case of a static charge distribution. In the general case of a time-dependent charge distribution, we do not, as yet, know how to calculate the electric field from the four-vector potential. In general, all we can safely conclude from our discussion is that E^k is part of some tensor $F^{\mu\nu}$,

$$E^k = F^{k0} \tag{21}$$

The tensor $F^{\mu\nu}$ is called the *field tensor*. What can we say about this tensor?

Consider a particle of charge q moving in the neighborhood of some arbitrary distribution of charge and current. The charges and currents will then exert electric forces and also extra forces on the particle; these extra forces are the magnetic forces. To make a clean separation

*Note that $\partial^1 = -\partial/\partial x$, etc. Hence the signs in Eq. (20) are correct.

between electric and magnetic forces, let us go to the instantaneous rest frame of the particle q. We will regard the force that the particle experiences in its own rest frame as purely electric, and we will define the electric field in this reference frame as the force per unit charge. This definition coincides with the definition we used in Chapter 2, but it is now restricted to the rest frame of the particle (if we want to know the electric field in some other reference frame, then we must use a particle at rest in that other reference frame.) The force on the particle, or the rate of change of its momentum, in its instantaneous rest frame is then

$$\frac{dp'^k}{dt'} = qE'^k \tag{22}$$

Since $dt' = d\tau$ in this reference frame, we can also write Eq. (22) as

$$\frac{dp'^k}{d\tau} = qE'^k \tag{23}$$

or as

$$\frac{dp'^k}{d\tau} = qF'^{k0} \tag{24}$$

This may be regarded as an operational definition of the components F'^{k0} of the field tensor $F'^{\mu\nu}$. By analogy with Eq. (24), we will define the component F'^{00} by the equation

$$\frac{dp'^0}{d\tau} = qF'^{00} \tag{25}$$

Note that in the instantaneous rest frame, $v = 0$, and therefore

$$\frac{dp'^0}{d\tau} = \frac{d}{d\tau}\left[\frac{mc^2}{(1-v^2/c^2)^{1/2}}\right] = \frac{mv}{(1-v^2/c^2)^{3/2}}\frac{dv}{d\tau} = 0$$

Hence Eq. (25) implies $F'^{00} = 0$, but this does not invalidate the definition.

Taken together, Eqs. (24) and (25) amount to the following four-component equation:

$$\frac{dp'^\mu}{d\tau} = qF'^{\mu 0} \tag{26}$$

Since $p'_\nu = m(c, 0, 0, 0)$ in the rest frame, this equation is the same as

$$\frac{dp'^\mu}{d\tau} = \frac{q}{mc} p'_\nu F'^{\mu\nu} \tag{27}$$

The advantage of Eq. (27) is that its Lorentz-transformation properties are obvious: the left side is the product of a four-vector dp'^{μ} and a scalar $1/d\tau$, and the right side is the contracted product of a four-vector p'_{ν} and a tensor $F'^{\mu\nu}$; hence each side is a four-vector. Transforming from the rest frame to the laboratory frame, we obtain

$$\frac{dp^{\mu}}{d\tau} = \frac{q}{mc} p_{\nu} F^{\mu\nu} \tag{28}$$

This is the relativistic equation of motion for a particle in an electromagnetic field. We will take a closer look at this equation later on.

If we multiply Eq. (28) by p_{μ} and sum over μ, we find

$$p_{\mu} \frac{dp^{\mu}}{d\tau} = \frac{q}{mc} p_{\mu} p_{\nu} F^{\mu\nu} \tag{29}$$

The left side of this equation is identically zero.

■ ***Exercise 4.*** Show that $p_{\mu}\, dp^{\mu}/d\tau \equiv 0$. [*Hint:* $p_{\mu} p^{\mu} = m^2c^2 =$ constant.] ■

Hence the right side of Eq. (29) must also be identically zero. That is,

$$p_{\mu} p_{\nu} F^{\mu\nu} = 0 \tag{30}$$

for arbitrary p_{μ}, which is possible only if

$$F^{\mu\nu} + F^{\nu\mu} = 0 \tag{31}$$

This equation says that $F^{\mu\nu}$ is an *antisymmetric* tensor.

■ ***Exercise 5.*** Derive Eq. (31). [*Hint:* Differentiate Eq. (30) with respect to p_{α} and p_{β}. Note that the constraint $p^{\alpha} p_{\alpha} = m^2c^2$ can now be ignored because Eq. (30) must be true for all possible values of the mass.] ■

We have, by now, made considerable progress in our construction of relativistic electrodynamics: we know how the current generates the four-vector potential [Eq. (10)], and we know how the field tensor exerts a force on a charged particle [Eq. (28)]. The one missing link is the exact relationship between $F^{\mu\nu}$ and A^{ν}; concerning this, we know only that $F^{k0} = \partial^k A^0$ in the static case and that $F^{\mu\nu}$ is antisymmetric in general. However, it turns out that this is just about enough information to determine the relationship completely. The only extra assumption we need is that the expression for the field tensor should be *linear* in the potential; this is a very natural assumption, which implies that the electrodynamic force [Eq. (28)] is linearly related to the current that produces the field [Eq. (10)]. We will accept this assumption on the grounds of its simplicity.* We can then argue as follows: To construct the tensor $F^{\mu\nu}$, we have available only the vector A^{μ} and the "vector" ∂^{ν}. The only antisymmetric tensor that we can build out of these two quantities is

$$\partial^{\mu} A^{\nu} - \partial^{\nu} A^{\mu} \tag{32}$$

or this multiplied by some scalar. The reason is that if the indices μ and ν appear on the same object (for example, $A^{\mu}A^{\nu}$ or $\partial^{\mu}\partial^{\nu}$), then antisymmetrization gives zero. The only scalars by which we might multiply the expression (32) are $A^{\alpha}A_{\alpha}$, $\partial^{\alpha}A_{\alpha}$, $A_{\alpha}\partial^{\alpha}$, $\partial^{\alpha}\partial_{\alpha}$, or some function of these, for example, $1 + (A^{\alpha} A_{\alpha})^{3/2}$. The first three of these are unacceptable because they would destroy the linearity of the expression (32); the last is unacceptable because, with an extra operator $\partial^{\alpha}\partial_{\alpha}$, the expression (32) would fail to give the required static limit $\partial^{k}A^{0}$. We therefore conclude that

$$F^{\mu\nu} = \partial^{\mu} A^{\nu} - \partial^{\nu} A^{\mu} \tag{33}$$

This is the required connection between $F^{\mu\nu}$ and A^{ν}.

Written out in full, the components of $F^{\mu\nu}$ are

$$F^{\mu\nu} = \begin{pmatrix} 0 & \partial^0 A^1 - \partial^1 A^0 & \partial^0 A^2 - \partial^2 A^0 & \partial^0 A^3 - \partial^3 A^0 \\ \partial^1 A^0 - \partial^0 A^1 & 0 & \partial^1 A^2 - \partial^2 A^1 & \partial^1 A^3 - \partial^3 A^1 \\ \partial^2 A^0 - \partial^0 A^2 & \partial^2 A^1 - \partial^1 A^2 & 0 & \partial^2 A^3 - \partial^3 A^2 \\ \partial^3 A^0 - \partial^0 A^3 & \partial^3 A^1 - \partial^1 A^3 & \partial^3 A^2 - \partial^2 A^3 & 0 \end{pmatrix} \tag{34}$$

or

$$F^{\mu\nu} = \begin{pmatrix} 0 & \frac{1}{c}\frac{\partial A_x}{\partial t} + \frac{\partial \Phi}{\partial x} & \frac{1}{c}\frac{\partial A_y}{\partial t} + \frac{\partial \Phi}{\partial y} & \frac{1}{c}\frac{\partial A_z}{\partial t} + \frac{\partial \Phi}{\partial z} \\ -\frac{\partial \Phi}{\partial x} - \frac{1}{c}\frac{\partial A_x}{\partial t} & 0 & -(\nabla \times \mathbf{A})_z & (\nabla \times \mathbf{A})_y \\ -\frac{\partial \Phi}{\partial y} - \frac{1}{c}\frac{\partial A_y}{\partial t} & (\nabla \times \mathbf{A})_z & 0 & -(\nabla \times \mathbf{A})_x \\ -\frac{\partial \Phi}{\partial z} - \frac{1}{c}\frac{\partial A_z}{\partial t} & -(\nabla \times \mathbf{A})_y & (\nabla \times \mathbf{A})_x & 0 \end{pmatrix} \tag{35}$$

From Eq. (35), we see that the electric field $E^{k} = F^{k0}$ is given by

---

*Instead of relying on simplicity, we could justify the linear relationship between field and current by regarding it as an approximation, valid for the limiting case of weak currents. In this limiting case, the linear relationship represents the lowest-order term in a Taylor-series expansion of the field in terms of the current.

$$\mathbf{E} = -\nabla\Phi - \frac{1}{c}\frac{\partial \mathbf{A}}{\partial t} \tag{36}$$

Furthermore, the presence of the $\nabla \times \mathbf{A}$ terms in Eq. (35) makes it expedient to define a vector field **B** as follows:

$$\mathbf{B} \equiv \nabla \times \mathbf{A} \tag{37}$$

This is the *magnetic field.* In the next section, we will justify this definition by showing that $\nabla \times \mathbf{A}$ has all the properties expected of the magnetic field: it is generated by electric currents, and it exerts forces on moving charges. In terms of **E** and **B**, we can then express the field tensor as

$$F^{\mu\nu} = \begin{pmatrix} 0 & -E_x & -E_y & -E_z \\ E_x & 0 & -B_z & B_y \\ E_y & B_z & 0 & -B_x \\ E_z & -B_y & B_x & 0 \end{pmatrix} \tag{38}$$

We can summarize this field tensor concisely as

$$F^{k0} = E^k \qquad F^{mn} = -\varepsilon^{mnl}B^l \tag{39}$$

■ ***Exercise 6.*** Check Eq. (39). Remember that the first index in $F^{\mu\nu}$ indicates the column and the second indicates the row; also remember that $E^1 = E_x$, etc. ■

Note that, although the components of **B**, like those of **E**, constitute a vector in three dimensions, they are components of a tensor in four dimensions.

What we see in Eq. (38) is that the electric and magnetic fields are closely related; they are not independent, three-dimensional entities but merely different components of a single four-dimensional entity, the field tensor. Correspondingly, the differential equations for the electric and magnetic fields are merely different components of an equation for the field tensor. The relationship between the electric and magnetic fields is as close as that between the time and space components of the position vector. Special relativity unifies space and time into spacetime, and it unifies electric and magnetic fields into a tensor field $F^{\mu\nu}$. As we will see later, this unification shows up very clearly in the Lorentz transformation of the fields—just as the Lorentz transformations mix time and space components of the position vector x^μ, they mix electric and magnetic components of the electromagnetic tensor $F^{\mu\nu}$.

The CGS electrostatic unit of magnetic field is called the *gauss.* This unit is dimensionally the same as statvolt/cm. Thus, in the CGS electrostatic system, both electric and magnetic fields are measured in essentially the same units. The use of statvolt/cm for electric fields and of gauss for magnetic fields merely reflects historical tradition. We could use statvolts/cm to measure both these fields; alternatively, we could use gauss to measure both fields (in fact, in astrophysics, the gauss is occasionally used for both fields). Since **E** and **B** are merely different components of the single tensor $F^{\mu\nu}$, it makes very good sense to measure them in the same units.*

Let us now see what happens to the **E** and **B** fields when we perform a Lorentz transformation. The transformation law of $F^{\mu\nu}$ is that of a tensor:

$$F'^{\mu\nu} = a^{\mu}{}_{\alpha}\, a^{\nu}{}_{\beta}\, F^{\alpha\beta} \tag{40}$$

For a pure Lorentz transformation along the x axis, the matrix $a^{\mu}{}_{\alpha}$ is (with the abbreviation $\gamma = 1/\sqrt{1-V^2/c^2}$)

$$a^{\mu}{}_{\alpha} = \begin{pmatrix} \gamma & -\gamma V/c & 0 & 0 \\ -\gamma V/c & \gamma & 0 & 0 \\ 0 & 0 & 1 & 0 \\ 0 & 0 & 0 & 1 \end{pmatrix} \tag{41}$$

We can treat the right side of Eq. (40) as a product of three matrices, with summations over *adjacent* indices, which can be evaluated by the familiar row-by-column multiplication rule:

$$F'^{\mu\nu} = a^{\mu}{}_{\alpha} F^{\alpha\beta} (a^{T})^{\beta}{}_{\nu}$$

$$= \begin{pmatrix} \gamma & -\gamma V/c & 0 & 0 \\ -\gamma V/c & \gamma & 0 & 0 \\ 0 & 0 & 1 & 0 \\ 0 & 0 & 0 & 1 \end{pmatrix} \begin{pmatrix} 0 & -E_x & -E_y & -E_z \\ E_x & 0 & -B_z & B_y \\ E_y & B_z & 0 & -B_x \\ E_z & -B_y & B_x & 0 \end{pmatrix}$$

$$\times \begin{pmatrix} \gamma & -\gamma V/c & 0 & 0 \\ -\gamma V/c & \gamma & 0 & 0 \\ 0 & 0 & 1 & 0 \\ 0 & 0 & 0 & 1 \end{pmatrix}$$

*In contrast, in the SI system, the units of **E** and **B** are not the same; and this requires the insertion of some extra factors of c in expression (38) for the field tensor. The SI unit of magnetic field is the tesla; 1 T $\Leftrightarrow$ 10^4 gauss. For a comparison of these units, see Appendix 2.

$$= \begin{pmatrix} 0 & -E_x & -\gamma\left(E_y - \frac{VB_z}{c}\right) & -\gamma\left(E_z + \frac{VB_y}{c}\right) \\ E_x & 0 & -\gamma\left(B_z - \frac{VE_y}{c}\right) & \gamma\left(B_y + \frac{VE_z}{c}\right) \\ \gamma\left(E_y - \frac{VB_z}{c}\right) & \gamma\left(B_z - \frac{VE_y}{c}\right) & 0 & -B_x \\ \gamma\left(E_z + \frac{VB_y}{c}\right) & -\gamma\left(B_y + \frac{VE_z}{c}\right) & B_x & 0 \end{pmatrix} \tag{42}$$

■ ***Exercise 7.*** Carry out the matrix multiplications needed to obtain Eq. (42). ■

Equation (42) states that $E'_x = E_x$, $E'_y = \gamma(E_y - VB_z/c)$, and so on. The relationship between the electric and magnetic fields in the two reference frames can be summarized as follows:

$$\mathbf{E}'_\parallel = \mathbf{E}_\parallel \qquad \mathbf{E}'_\perp = \gamma\left(\mathbf{E}_\perp + \frac{\mathbf{V}}{c}\times\mathbf{B}\right) \tag{43}$$

$$\mathbf{B}'_\parallel = \mathbf{B}_\parallel \qquad \mathbf{B}'_\perp = \gamma\left(\mathbf{B}_\perp - \frac{\mathbf{V}}{c}\times\mathbf{E}\right) \tag{44}$$

where $\mathbf{E}_\parallel$, $\mathbf{B}_\parallel$ and $\mathbf{E}_\perp$, $\mathbf{B}_\perp$ are, respectively, the components of the fields parallel and perpendicular to the velocity; that is, the parallel component is the x component, and the perpendicular components are the y and z components ($\mathbf{E}_\parallel = \hat{\mathbf{x}}E_x$, $\mathbf{E}_\perp = \hat{\mathbf{y}}E_y + \hat{\mathbf{z}}E_z$, and so on). These equations show that under Lorentz transformations, electric and magnetic fields transmute into one another. The three-dimensional vector fields **E** and **B** are merely different aspects of the four-dimensional field tensor $F^{\mu\nu}$. What observers see as electric and what as magnetic depends on their reference frame.

■ ***Exercise 8.*** Derive Eqs. (43) and (44) from (42). ■

EXAMPLE: *fields of a moving point charge*

As an application of Eqs. (43) and (44), we will calculate the electric and magnetic fields of a point charge moving at uniform velocity **v** through the laboratory. In the rest frame $x'y'z't'$ of the point charge, the field is purely electric; if the charge is at the origin,

$$\mathbf{E}' = q\frac{\mathbf{x}'}{|\mathbf{x}'|^3} \tag{45}$$

To find the **E** and **B** fields in the laboratory frame, we start with the equations inverse to (43) and (44), that is, the equations with primes interchanged and **V** replaced by –**v**. We then obtain

$$\mathbf{E}_{\parallel} = \mathbf{E}'_{\parallel} \qquad \mathbf{E}_{\perp} = \gamma\left(\mathbf{E}'_{\perp} - \frac{\mathbf{v}}{c}\times\mathbf{B}'\right) = \gamma\mathbf{E}'_{\perp} \tag{46}$$

$$\mathbf{B}_{\parallel} = \mathbf{B}'_{\parallel} = 0 \qquad \mathbf{B}_{\perp} = \gamma\left(\mathbf{B}'_{\perp} + \frac{\mathbf{v}}{c}\times\mathbf{E}'\right) = \gamma\frac{\mathbf{v}}{c}\times\mathbf{E}' \tag{47}$$

These equations show that at some point at a given proper distance from the charge, say $|\mathbf{x}'| = 1$, the parallel electric field is unchanged by the Lorentz transformation, whereas the perpendicular electric field is increased by a factor γ. This means that the pattern of electric field lines of a moving charge differs from that of a charge at rest in being compressed longitudinally by a factor $1/\gamma$ (see Fig. 8.2). Such a "compression" clearly increases the density of the transverse field lines. Figure 8.2 also suggests that the density of longitudinal field lines is *decreased.* But this is not so because we must compare the densities of lines in Figs. 8.2(*a*) and 8.2(*b*) at a given *proper distance* from the charge q, for example, at some given point P at a distance $|\mathbf{x}'| = 1$; there, the densities of horizontal lines are the same.

To obtain an explicit expression for the electric field of the moving charge, we separate Eq. (45) into parallel and perpendicular components; in the x-y plane, these are simply the x and y components:

Figure 8·2

Electric field of a point charge moving along the x axis. (*a*) In the rest frame, the pattern of electric fields lines is spherically symmetric. (*b*) In the laboratory frame, the pattern of electric field lines is compressed along the direction of motion.

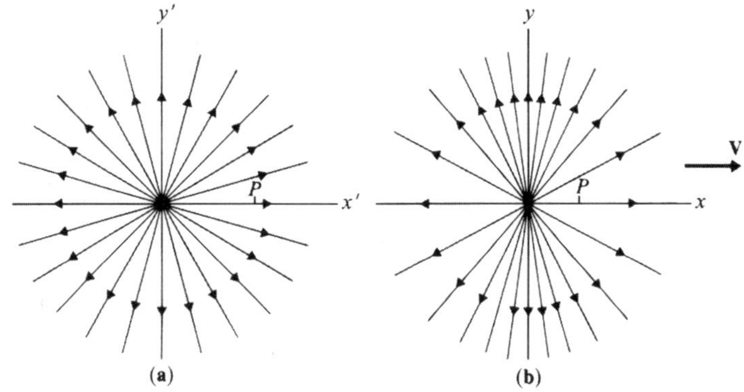

$$E'_x = q\frac{x'}{(x'^2 + y'^2)^{3/2}} \qquad E'_y = q\frac{y'}{(x'^2 + y'^2)^{3/2}}$$

Then we use Eq. (46) and substitute the Lorentz-transformation equations $x' = \gamma(x - vt) = \gamma(x - x_q)$ and $y' = y$, where $x_q = vt$ is the instantaneous position of the point charge. This gives us the electric field

$$E_x = q\frac{\gamma(x - x_q)}{[\gamma^2(x - x_q)^2 + y^2]^{3/2}} \qquad E_y = q\frac{\gamma y}{[\gamma^2(x - x_q)^2 + y^2]^{3/2}} \tag{48}$$

Note that the factor of γ in the numerator of E_y arises from the transformation of the perpendicular component of the electric field according to Eq. (46), whereas the same factor in E_x arises from the Lorentz transformation of the x coordinate. From Eq. (48) we see that directly ahead of (or behind) the point charge, the electric field is the same as that for a charge at rest; that is, $E_x = q/(x - x_q)^2$ for $y = 0$.

The moving point charge is endowed with a magnetic field, which, according to Eq. (47), is everywhere perpendicular to the velocity **v**. The field lines of the magnetic field are closed circles centered on the line of motion. We will leave the detailed discussion of magnetic fields for the next chapter.

8·3 Maxwell's Equations

Equation (18) or (10) is the field equation for A^μ; it determines how a given four-current j^μ generates a four-vector potential. We now want to look for the field equations satisfied by $F^{\mu\nu}$. Since

$$F^{\mu\nu} = \partial^\mu A^\nu - \partial^\nu A^\mu \tag{49}$$

we have

$$\partial_\mu F^{\mu\nu} = \partial_\mu \partial^\mu A^\nu - \partial^\nu \partial_\mu A^\mu$$

This is exactly the left side of Eq. (18). Hence

$$\partial_\mu F^{\mu\nu} = \frac{4\pi}{c} j^\nu \tag{50}$$

This is one of the field equations satisfied by $F^{\mu\nu}$. Another equation satisfied by $F^{\mu\nu}$ is the following:

$$\partial^\sigma F^{\mu\nu} + \partial^\mu F^{\nu\sigma} + \partial^\nu F^{\sigma\mu} = 0 \tag{51}$$

This equation can be checked directly by substituting for $F^{\mu\nu}$ according to Eq. (49).

■ ***Exercise 9.*** Make this substitution in Eq. (51), and show that all the terms cancel identically. ■

■ ***Exercise 10.*** Show that if we use Eqs. (10) and (12) instead of Eq. (18), the result is again Eq. (50). ■

The above field equations are expressed in terms of the tensor field $F^{\mu\nu}$. We can also express these equations in terms of the fields **E** and **B**. For instance,

$$\partial_\mu F^{\mu 0} = \frac{4\pi}{c} j^0 \tag{52}$$

is the same as

$$\frac{\partial}{\partial x^k} E^k = 4\pi\rho \tag{53}$$

or

$$\nabla \cdot \mathbf{E} = 4\pi\rho \tag{54}$$

This equation is, of course, familiar from Chapter 2.

Next, consider

$$\partial_\mu F^{\mu 3} = \frac{4\pi}{c} j^3 \tag{55}$$

Expressed in terms of **E** and **B**, this becomes

$$-\frac{\partial E_z}{\partial x^0} + \frac{\partial B_y}{\partial x^1} - \frac{\partial B_x}{\partial x^2} = \frac{4\pi}{c} j^3 \tag{56}$$

that is,

$$(\nabla \times \mathbf{B})_z = \frac{4\pi}{c} j_z + \frac{1}{c}\frac{\partial E_z}{\partial t} \tag{57}$$

As a final example, consider

$$\partial^3 F^{12} + \partial^1 F^{23} + \partial^2 F^{31} = 0 \tag{58}$$

This is equivalent to

$$\left(-\frac{\partial}{\partial x^3}\right)(-B_z) + \left(-\frac{\partial}{\partial x^1}\right)(-B_x) + \left(-\frac{\partial}{\partial x^2}\right)(-B_y) = 0 \tag{59}$$

or

$$\nabla \cdot \mathbf{B} = 0 \tag{60}$$

Upon taking into account all components of Eqs. (50) and (51), we obtain a total of eight differential equations involving **E** and **B**, which can be written as follows:

$$\nabla \cdot \mathbf{E} = 4\pi\rho \tag{61}$$

$$\nabla \times \mathbf{B} = \frac{4\pi}{c}\mathbf{j} + \frac{1}{c}\frac{\partial \mathbf{E}}{\partial t} \tag{62}$$

$$\nabla \cdot \mathbf{B} = 0 \tag{63}$$

$$\nabla \times \mathbf{E} = -\frac{1}{c}\frac{\partial \mathbf{B}}{\partial t} \tag{64}$$

■ ***Exercise 11.*** Derive Eq. (64) from Eq. (51). ■

Equations (61)–(64) are *Maxwell's equations* for the electric and magnetic fields. The first of these equations is familiar from Chapter 2; it shows how a charge distribution generates an electric field. The last of these equations [Eq. (64)] shows that an electric field can also be generated by a time-dependent magnetic field (electromagnetic induction). The second equation [Eq. (62)] tells us how magnetic fields are generated by current distributions and by time-dependent electric fields. Finally, the third equation [Eq. (63)] expresses the absence of magnetic "charges." All of these equations were originally discovered independently, on the basis of experimental investigations stretching over two centuries (see the chart on the back end papers of this book). The equations were put in their final form (61)–(64) by Maxwell in 1861, but it took another 40 years before Einstein discovered that all these equations are related by Lorentz transformations. Our derivation makes it clear that the three-dimensional Maxwell equations (61)–(64) are merely different components of the four-dimensional tensor equations (50) and (51).

The remaining chapters of this book can all be regarded as corollaries of Maxwell's equations. In the next chapters, we will use these equations to calculate the magnetic fields generated by currents and to calculate the electric fields generated by electromagnetic induction. In later chapters, we will see how these equations imply the existence of electromagnetic waves. In the calculation of the fields generated by a given distribution of charges and currents, the Maxwell equations (61)–(64) are usually more convenient than the tensor equations (50) and (51). However, Eqs. (61)–(64) suffer from the disadvantage that their behavior under Lorentz transformations is not at all obvious. For transforming Maxwell's equations into a different reference frame, the best procedure is first to replace them by the equivalent equations (50) and (51) for the tensor $F^{\mu\nu}$. Then perform the Lorentz transformation on the latter equations according to the obvious Lorentz-transformation law indicated by the vector and tensor indices and obtain the corresponding equations for $F'^{\mu\nu}$. Finally, the equations for $F'^{\mu\nu}$ can be converted into equations for $\mathbf{E}'$, $\mathbf{B}'$. The result is a set of equations such as (61)–(64) but with primes on everything.

We will next express the equation of motion [Eq. (28)] in terms of **E** and **B**. Consider the $\mu = 1$ component of this equation,

$$\begin{aligned}\frac{dp^1}{d\tau} &= \frac{q}{mc} p_\nu F^{1\nu} \\ &= \frac{q}{mc}(p_0 F^{10} + p_1 F^{11} + p_2 F^{12} + p_3 F^{13})\end{aligned} \tag{65}$$

This amounts to

$$\begin{aligned}\frac{dp^1}{d\tau} &= \frac{q}{mc}[p_0 E^1 + (-p^2)(-B_z) + (-p^3)B_y] \\ &= \frac{q}{mc}\left(\frac{mc}{\sqrt{1-v^2/c^2}}E_x + \frac{mv_y}{\sqrt{1-v^2/c^2}}B_z - \frac{mv_z}{\sqrt{1-v^2/c^2}}B_y\right)\end{aligned} \tag{66}$$

or, with $d\tau = \sqrt{1-v^2/c^2}\,dt$,

$$\frac{dp_x}{dt} = qE_x + q\left(\frac{\mathbf{v}}{c}\times\mathbf{B}\right)_x \tag{67}$$

■ ***Exercise 12.*** Show that taking into account the other components results in the equation of motion

$$\frac{d\mathbf{p}}{dt} = q\mathbf{E} + q\frac{\mathbf{v}}{c}\times\mathbf{B} \tag{68}$$

■

The momentum **p** appearing in this equation is, of course, the relativistic momentum, $\mathbf{p} = m\mathbf{v}/\sqrt{1-v^2/c^2}$. In the nonrelativistic approximation, $\mathbf{p} \simeq m\mathbf{v}$, and the equation reduces to

$$m\frac{d\mathbf{v}}{dt} = q\mathbf{E} + q\frac{\mathbf{v}}{c}\times\mathbf{B} \tag{69}$$

Note that the form of the expression for the force exerted on the particle [the right side of Eqs. (68) and (69)] is the same in the relativistic and nonrelativistic cases; what changes is the expression for the momentum. The right side of Eqs. (68) and (69) is called the *Lorentz force*:

$$\mathbf{F} = q\mathbf{E} + q\frac{\mathbf{v}}{c}\times\mathbf{B} \tag{70}$$

The force extended by the magnetic field is always at right angles to **B** and **v**. Hence this force cannot change the magnitude of the momentum; it changes only the direction of motion.

8·4 The Energy-Momentum Tensor

Consider a distribution of particles of charge q with a particle density n. The force that the electromagnetic field exerts on each particle in this distribution is given by Eq. (70). If we sum all such forces exerted on the particles in a small volume and divide by the volume, we obtain the *force density*

$$\mathbf{f} = n\mathbf{F} \tag{71}$$

This force density is simply the total force exerted by the fields on the particles per unit volume. According to Eq. (70),

$$\mathbf{f} = \rho\mathbf{E} + \frac{\mathbf{j}}{c}\times\mathbf{B} \tag{72}$$

Let us write the x component of this equation in four-tensor notation:

$$\begin{aligned} f^1 = f_x = \rho E^1 + \frac{j^2B^3 - j^3B^2}{c} &= \rho F^{10} + \frac{-j^2F^{12} - j^3F^{13}}{c} \\ &= \frac{1}{c}\left(F^{10}j_0 + F^{12}j_2 + F^{13}j_3\right) \end{aligned} \tag{73}$$

If we add to the right side a term $F^{11}j_1/c$, which is identically zero, then

$$f^1 = \frac{1}{c}F^{1\nu}j_\nu \tag{74}$$

and, in general,

$$f^k = \frac{1}{c}F^{k\nu}j_\nu \tag{75}$$

We will now *define*

$$f^0 \equiv \frac{1}{c}F^{0\nu}j_\nu \tag{76}$$

so that

$$f^\mu = \frac{1}{c}F^{\mu\nu}j_\nu \tag{77}$$

The four-component object f^μ is a vector; this is obvious because it consists of the contracted product of the tensor $F^{\mu\nu}$ with the vector j_ν.

What is the meaning of f^0? If we evaluate f^0 according to Eq. (77), we find

$$f^0 = \frac{1}{c} F^{0\nu} j_\nu = \frac{1}{c} F^{0k} j_k = \frac{1}{c}(-E^1 j_1 - E^2 j_2 - E^3 j_3) = \frac{1}{c} \mathbf{j} \cdot \mathbf{E} \qquad (78)$$

Let us compare this with the dot product of **v** and **f**:

$$\mathbf{v} \cdot \mathbf{f} = \mathbf{v} \cdot \left(\rho \mathbf{E} + \rho \frac{\mathbf{v}}{c} \times \mathbf{B} \right) = \rho \mathbf{v} \cdot \mathbf{E} + 0 = \mathbf{j} \cdot \mathbf{E} \qquad (79)$$

From this we see that cf^0 is the same as $\mathbf{v} \cdot \mathbf{f}$. But the latter quantity is the work that the fields do on the particles per unit time and unit volume. We can therefore summarize the meaning of f^μ as follows:

$$f^0 = \frac{1}{c} \times (\text{rate of change of particle energy per unit volume}) \qquad (80)$$

$$f^k = (\text{rate of change of particle momentum per unit volume}) \qquad (81)$$

Next, we want to carry out some transformations on the right side of Eq. (77). The objective of these (somewhat tedious) transformations is to express this right side as the four-dimensional divergence of a product of field tensors. We begin by substituting the field equation

$$j_\nu = \frac{c}{4\pi} \partial_\alpha F^\alpha{}_\nu \qquad (82)$$

and get

$$f^\mu = \frac{1}{4\pi} F^{\mu\nu} \partial_\alpha F^\alpha{}_\nu \qquad (83)$$

Furthermore, on the right side we can use the identity

$$F^{\mu\nu} \partial_\alpha F^\alpha{}_\nu = \partial_\alpha (F^{\mu\nu} F^\alpha{}_\nu) - \partial_\alpha F^{\mu\nu} F^\alpha{}_\nu \qquad (84)$$

where the last term can be expressed as

$$\begin{aligned} \partial_\alpha F^{\mu\nu} F^\alpha{}_\nu &= \partial^\alpha F^{\mu\nu} F_{\alpha\nu} = \partial^\nu F^{\mu\alpha} F_{\nu\alpha} = \partial^\nu F^{\alpha\mu} F_{\alpha\nu} \\ &= \tfrac{1}{2}(\partial^\nu F^{\alpha\mu} + \partial^\alpha F^{\mu\nu}) F_{\alpha\nu} \end{aligned} \qquad (85)$$

According to Eq. (51), the term in parentheses equals $-\partial^\mu F^{\nu\alpha}$. Hence

$$\begin{aligned} \partial_\alpha F^{\mu\nu} F^\alpha{}_\nu &= -\tfrac{1}{2} \partial^\mu F^{\nu\alpha} F_{\alpha\nu} = \tfrac{1}{2} \partial^\mu F^{\nu\alpha} F_{\nu\alpha} \\ &= \tfrac{1}{4} \partial^\mu (F^{\nu\alpha} F_{\nu\alpha}) \end{aligned} \qquad (86)$$

Putting Eqs. (84) and (86) together, we find that Eq. (83) is equivalent to

$$f^{\mu} = -\frac{1}{4\pi}\left[\partial_{\alpha}(F^{\mu\nu}F^{\alpha}{}_{\nu})-\frac{1}{4}\partial^{\mu}(F^{\nu\alpha}F_{\nu\alpha})\right] \qquad \textit{(87)}$$

We can also write this as

$$f^{\mu} = -\frac{\partial}{\partial x^{\alpha}}T^{\mu\alpha} \qquad \textit{(88)}$$

where $T^{\mu\alpha}$ is defined by

$$T^{\mu\alpha} = -\frac{1}{4\pi}\left(F^{\mu\nu}F^{\alpha}{}_{\nu}-\frac{1}{4}\eta^{\alpha\mu}F^{\nu\beta}F_{\nu\beta}\right) \qquad \textit{(89)}$$

The tensor $T^{\mu\alpha}$ is called the *energy-momentum tensor* of the electromagnetic field. Before we deal with the physical interpretation of this tensor, let us express it in terms of **E** and **B**, for later reference.

■ ***Exercise 13.*** Show that $T^{\mu\alpha}$ is a symmetric tensor. ■

■ ***Exercise 14.*** Obtain the following from Eq. (89):

$$T^{00} = \frac{1}{8\pi}(E^2+B^2) \qquad \textit{(90)}$$

$$T^{01} = \frac{1}{4\pi}(\mathbf{E}\times\mathbf{B})_x \qquad \textit{(91)}$$

$$T^{11} = \frac{1}{8\pi}(E_y^2+E_z^2-E_x^2+B_y^2+B_z^2-B_x^2) \qquad \textit{(92)}$$

$$T^{12} = -\frac{1}{4\pi}(E_xE_y+B_xB_y) \qquad \textit{(93)}$$

■

The general expression for T^{mn} (for $m, n = 1, 2, 3$) is

$$T^{mn} = \frac{1}{4\pi}\left[-E^mE^n-B^mB^n+\frac{1}{2}\delta^{mn}(E^2+B^2)\right] \qquad \textit{(94)}$$

The quantity T^{mn} is known as the *Maxwell stress tensor.*

The complete expression for $T^{\mu\nu}$ is then

$$T^{\mu\nu} = \begin{pmatrix} \frac{1}{8\pi}(E^2+B^2) & \frac{1}{4\pi}(\mathbf{E}\times\mathbf{B})_x & \frac{1}{4\pi}(\mathbf{E}\times\mathbf{B})_y & \frac{1}{4\pi}(\mathbf{E}\times\mathbf{B})_z \\ \frac{1}{4\pi}(\mathbf{E}\times\mathbf{B})_x & & & \\ \frac{1}{4\pi}(\mathbf{E}\times\mathbf{B})_y & & T^{mn} & \\ \frac{1}{4\pi}(\mathbf{E}\times\mathbf{B})_z & & & \end{pmatrix} \tag{95}$$

The T^{0k} (or T^{k0}) components of this tensor constitute a vector in three dimensions; this vector, with an extra factor of c, is called the *Poynting vector:*

$$\mathbf{S} = \frac{c}{4\pi}\mathbf{E}\times\mathbf{B} \tag{96}$$

To see what the tensor $T^{\mu\nu}$ has to do with energy and momentum, we will use our expression for the force per unit volume [Eq. (88)] to calculate the exchange of energy and momentum between particles and fields. The space components of Eq. (88) are

$$f^m = -\frac{\partial T^{m\nu}}{\partial x^\nu} = -\frac{\partial T^{mn}}{\partial x^n} - \frac{1}{c}\frac{\partial T^{m0}}{\partial t}$$

$$= -\frac{\partial T^{mn}}{\partial x^n} - \frac{1}{c^2}\frac{\partial S^m}{\partial t} \tag{97}$$

If we integrate this equation over a volume that contains some particles and fields, we obtain

$$\int f^m \, dV = -\int \frac{\partial}{\partial x^n} T^{mn} \, dV - \frac{1}{c^2}\frac{d}{dt}\int S^m \, dV \tag{98}$$

But since f^m is the force density, its integral is the total force, which must equal the rate of change of particle momentum,

$$\int f^m \, dV = F^m = \frac{dP^m}{dt} \tag{99}$$

Let us define

$$G^m \equiv \frac{1}{c^2}\int S^m \, dV \tag{100}$$

Then Eq. (98) reduces to

$$\frac{d}{dt}(P^m + G^m) = -\int \frac{\partial}{\partial x^n} T^{mn} \, dV \tag{101}$$

By Gauss' theorem, we can transform the right side into a surface integral over the surface σ that forms the boundary of the volume of integration:

$$\frac{d}{dt}(P^m + G^m) = -\int_\sigma T^{mn}\, d\sigma^n \qquad (102)$$

Suppose now that our volume of integration is so large that it includes all the fields in our system. In this case, **E** and **B** are zero on the boundary, and hence

$$\frac{d}{dt}(P^m + G^m) = 0 \qquad (103)$$

This says that the rate of change of the particle momentum is compensated by a change in G^m; that is, any momentum lost by the particles appears in G^m. The latter quantity must therefore be the momentum in the electromagnetic fields. That fields can contain momentum should not surprise us; we already know (from Chapter 5) that fields contain energy, and relativistically, anything with energy will have momentum when in motion. Our conclusion is then

$$\text{(momentum in electromagnetic fields)} = G^m = \frac{1}{c^2}\int S^m\, dV \qquad (104)$$

In three-dimensional vector notation, we can write G^m as

$$\mathbf{G} = \frac{1}{c^2}\int \mathbf{S}\, dV = \int \frac{1}{4\pi c}\mathbf{E}\times\mathbf{B}\, dV \qquad (105)$$

The integrand

$$\frac{1}{c^2}\mathbf{S} = \frac{1}{4\pi c}\mathbf{E}\times\mathbf{B} \qquad (106)$$

can be identified as the *density of momentum* in the fields.

Now let us return to Eq. (102). If the volume of integration does not include all the fields, then $d/dt(P^m + G^m) \neq 0$, and the total momentum in the volume under consideration changes in time. The right side of Eq. (102) represents the momentum gained by the volume, and therefore $T^{mn} d\sigma^n$ must be the amount of m momentum that flows out through the area $d\boldsymbol{\sigma}$ in unit time. The flux of momentum, or flow per unit area and unit time, must be T^{mn}:

$$T^{mn} = \text{(flux of } m \text{ component of momentum in the } n \text{ direction)} \qquad (107)$$

For instance, T^{12} is the amount of x momentum that flows in the y direction across unit area in unit time, and so on.

We can carry out a similar calculation with the time component of Eq. (88):

$$cf^0 = -c\frac{\partial T^{0\nu}}{\partial x^\nu} = -\frac{\partial T^{00}}{\partial t} - \frac{\partial S^n}{\partial x^n} \tag{108}$$

Integration gives

$$\int cf^0\,dV = -\frac{d}{dt}\int T^{00}\,dV - \int \frac{\partial S^n}{\partial x^n}dV \tag{109}$$

The quantity on the left side is the rate of change of the particle energy. We can therefore rewrite Eq. (109) as

$$\frac{d}{dt}(P^0 + U) = -\int_\sigma S^n\,d\sigma^n \tag{110}$$

where U is defined as

$$U \equiv \int T^{00}\,dV \tag{111}$$

Again, we first suppose that the volume of integration includes all the fields. The right side of Eq. (110) is then zero, and we can identify the quantity U as the energy in the fields,

$$\text{(energy in electromagnetic fields)} = U = \int T^{00}dV \tag{112}$$

Hence the *energy density* in the fields is

$$T^{00} = \frac{1}{8\pi}(E^2 + B^2) \tag{113}$$

The contribution made by the electric fields to the energy density of course agrees with what we found in Chapter 5 [see Eq. (5.10)].

Returning to Eq. (110), we recognize that $S^n d\sigma^n = \mathbf{S}\boldsymbol{\cdot} d\boldsymbol{\sigma}$ must be the energy that flows out through the area $d\boldsymbol{\sigma}$ in unit time; that is, the Poynting vector

$$\mathbf{S} = \frac{c}{4\pi}\mathbf{E}\times\mathbf{B} \tag{114}$$

is the energy flux in the fields.

To conclude, we can summarize what we have learned about the physical interpretation of the components of $T^{\mu\nu}$:

$$T^{\mu\nu} = \begin{pmatrix} \text{energy density} & \begin{matrix}\text{(energy flux)}/c\text{, or} \\ c\times\text{ (momentum density)}\end{matrix} \\ \\ \begin{matrix}\text{(energy flux)}/c\text{, or} \\ c\times\text{ (momentum density)}\end{matrix} & \text{momentum flux} \end{pmatrix} \qquad (115)$$

EXAMPLE: *energy-momentum tensor for parallel-plate capacitor*

As a simple example of the application of the energy-momentum tensor, let us calculate the force that the electric field between the plates of a parallel-plate capacitor exerts on these plates. Suppose the plates are horizontal and parallel to the x-y plane, and the electric field is vertical, $\mathbf{E} = E_0\hat{\mathbf{z}}$. Then the energy-momentum tensor (95) is

$$T^{\mu\nu} = \begin{pmatrix} \frac{1}{8\pi}E_0^2 & 0 & 0 & 0 \\ 0 & \frac{1}{8\pi}E_0^2 & 0 & 0 \\ 0 & 0 & \frac{1}{8\pi}E_0^2 & 0 \\ 0 & 0 & 0 & -\frac{1}{8\pi}E_0^2 \end{pmatrix} \qquad (116)$$

The force per unit area acting on one of the plates—say, the upper plate—is simply the flux of momentum into the plate; that is, it is the T^{33} component of the energy-momentum tensor:

$$\frac{F}{A} = T^{33} = -\frac{1}{8\pi}E_0^2 \qquad (117)$$

In terms of the surface charge density, the electric field is $E_0 = 4\pi\sigma$ [see Eq. (3.13)]; therefore, we can express the force per unit area as $F/A = -2\pi\sigma^2$. This expression for the force agrees with the result obtained in Section 3.1. Note that in this calculation of the force from the energy-momentum tensor, we did not have to separate the electric field near the plate into a "self" contribution and an "other" contribution, as we did in the calculation in Section 3.1.

Problems

1. The transformation $A^\mu \rightarrow A^\mu + \partial^\mu\chi$, where χ is some arbitrary scalar function, is called a *gauge transformation.*

(a) Show that Eq. (18) is invariant under gauge transformations; that is, show that whenever A^μ is a solution of Eq. (18), then $A^\mu + \partial^\mu\chi$ is also a solution.

(b) Show that the electric and magnetic fields are invariant under gauge transformations; that is, show that the electric and magnetic fields calculated from A^μ and from $A^\mu + \partial^\mu\chi$ according to Eq. (33) are exactly the same.

2. A particle of charge q and mass m, initially at rest, is acted upon by a uniform electric field **E** directed along the x axis. Show that the relativistic equation of motion has the solution

$$x = \frac{c}{qE}\sqrt{m^2c^2 + (qEt)^2} - \frac{mc^2}{qE}$$

Check that $dx/dt \rightarrow c$ as $t \rightarrow \infty$. Check that in the nonrelativistic limit, this solution has the expected form.

3. Equation (38) gives $F^{\mu\nu}$ as a matrix. Write out corresponding matrices for $F_\mu{}^\nu$, $F^\mu{}_\nu$, and $F_{\mu\nu}$ in terms of the electric and magnetic fields.

4. A cosmic-ray proton with an energy of 10^{13} GeV passes near an electron at rest near the top of the atmosphere. The proton travels on a straight line, and the distance of closest approach to the electron is 10 cm. Find the magnitude of the electric and magnetic fields acting on the electron (in the electron rest frame) at the instant of closest approach. Draw a diagram showing the direction of the fields.

5. In a long, cylindrical region of radius R, there flows a gas of charged particles. In the laboratory, the charge density of the gas is ρ and the current density is **j**; both ρ and **j** are constant in the interior of the cylinder and zero in the exterior.

(a) Find the velocity of a reference frame $x'y'z't'$ such that in this frame, the current density **j**$'$ is exactly zero. Find the charge density ρ' in this frame.

(b) Find the electric field **E**$'$ that the charge density ρ' produces in the $x'y'z't'$ frame. Consider the interior and the exterior regions of the cylinder.

(c) Using the Lorentz-transformation laws for the electric and magnetic fields, find the values of **E** and **B** in the laboratory frame from the known value of **E**$'$ in the $x'y'z't'$ frame.

6. In its rest frame, a long cylindrical solenoid of radius R has a uniform magnetic field **B** in its interior and no magnetic field in its exterior. If this solenoid is moving at velocity **V** perpendicular to its axis, what will be the electric field of the solenoid? Show the field lines of this field in a diagram. Is the electric field conservative?

7. A pure electric dipole of moment **p** is moving at uniform velocity **V** in a direction perpendicular to **p**. What is the instantaneous magnetic field of this dipole? Assume that the motion is nonrelativistic.

8. Show that, for the electric field **E** and magnetic field **B** of the moving point charge [see Eq. (46) and (47)], the magnitude of **E** at every point is larger than the magnitude of **B**.

9. (a) Find the potential Φ, or A^0, of a point charge q moving at uniform velocity through the laboratory. Roughly describe the shape of the equipotential surfaces at one instant of time.
(b) Find the vector potential **A** for this point charge.

10. Assume that an electron is in the smallest Bohr orbit (of radius 0.0529 nm) around a proton. Treating the motion classically, calculate the magnetic field at the position of the electron generated by the motion of the proton in the instantaneous rest frame of the electron.

11. (a) In the laboratory, a very long cylindrical solenoid with a uniform magnetic field is moving in a direction perpendicular to its axis. Show that, according to the Lorentz-transformation equations for lengths and for magnetic fields, the magnetic flux intercepted by the cross section of the solenoid is the same in the laboratory frame as in the rest frame.
(b) In the laboratory, a very large parallel-plate capacitor with a uniform electric field is moving in a direction parallel to its plates. Show that the electric flux intercepted by a plane between the plates is the same in the laboratory frame as in the rest frame.

12. A very long cylinder of dielectric material of radius R has charge λ per unit length uniformly distributed over its volume. The cylinder moves at velocity **V** perpendicular to its axis. Assume that the motion is nonrelativistic. What is the instantaneous magnetic field for $r < R$ and for $r > R$?

13. Suppose that in the laboratory frame, there is an electric field of magnitude 500 statvolts/cm that points in the z direction and a magnetic field of 300 statvolts/ cm that points in the y direction.
(a) Find the velocity (magnitude and direction) of a reference frame $x'y'z't'$ in which the magnetic field is zero. What is the magnitude of the electric field in this reference frame?
(b) Can you find a different reference frame such that the electric field is zero? If your answer is affirmative, find the magnitude of the magnetic field in that frame. If your answer is negative, explain.

14. When at rest, a parallel-plate capacitor has a uniform electric field $\mathbf{E}_0$ between its plates. Suppose that this capacitor is made to move at velocity **V** in a direction parallel to its plates. Use the Lorentz-transformation equations for the field, the potential, and the charge and current densities in parts (a), (b), and (c).
(a) Find the electric and magnetic fields between the plates of the moving capacitor.
(b) Assume that in the rest frame of the capacitor, the electrostatic potential is zero on one of the plates. Find the potential and the vector potential between the plates of the moving capacitor, and verify explicitly that differentiation of these potentials leads to electric and magnetic fields in agreement with those found in part (a).
(c) Find the charge density and the current density on the plates of the moving capacitor. Is the electric field found in part (a) related to the charge density in the expected way?

15. (a) Prove that $E^2 - B^2$ and $\mathbf{E}\cdot\mathbf{B}$ are invariant under the special Lorentz transformation (41).

(b) Prove that both $E^2 - B^2$ and $\mathbf{E}\cdot\mathbf{B}$ are invariant under a general Lorentz transformation with a transformation matrix $a_\mu{}^\nu$. [*Hint:* Evaluate $F_{\mu\nu}F^{\mu\nu}$ and evaluate $\varepsilon_{\alpha\beta\mu\nu}F^{\alpha\beta}F^{\mu\nu}$, where the tensor $\varepsilon_{\alpha\beta\mu\nu}$ is as defined in Problem 7.6.]

16. Consider the Maxwell equations (61)–(64). By manipulating these equations, derive the law of conservation of electric charge,

$$\frac{\partial\rho}{\partial t}+\nabla\cdot\mathbf{j}=0$$

17. Show that the trace of the energy-momentum tensor $T^{\mu\nu}$ defined by Eq. (89) is zero, that is, $T^\mu{}_\mu = 0$.

18. Given that the electric and magnetic fields in some region of space are uniform and at right angles, with $\mathbf{E} = E_0\hat{\mathbf{x}}$ and $\mathbf{B} = B_0\hat{\mathbf{y}}$, find all components of the energy- momentum tensor.

19. A point charge q is placed at the origin of coordinates. Calculate the components of the energy-momentum tensor at any arbitrary point **x** with coordinates x, y, z.

20. Suppose that in the $xyzt$ reference frame, there is an electrostatic field with an energy density $E^2/8\pi$. If we observe this field from a reference frame $x'y'z't'$ moving at a low (nonrelativistic) velocity **V**, we naively might expect that the energy flux we observe is approximately the product of the velocity and the energy density, $-\mathbf{V}E^2/8\pi$. Is this true or false? If true, prove it; if false, give a counterexample.

21. Consider a uniform electric field $\mathbf{E} = E_0\hat{\mathbf{x}}$.

(a) Find all components of the energy-momentum tensor of this field.

(b) Find the fields $\mathbf{E}'$ and $\mathbf{B}'$ observed in a reference frame $x'y'z't'$ moving at velocity V in the x direction. Find the components of the energy-momentum tensor in this reference frame.

(c) Repeat part (b) for a reference frame $x'y'z't'$ moving at velocity V in the y direction.

22. Find the Poynting vector at a point in the x-y plane for the electromagnetic field of a moving point charge [see Eqs. (46) and (47)] when the point charge is instantaneously at the origin.

23. Consider the electric field generated jointly by two point charges $\pm Q$ separated by a distance d. Using the Maxwell stress tensor, calculate the rate at which the electric field at the midplane between these two charges delivers momentum across the midplane. Verify that this rate of delivery of momentum agrees with the value of the Coulomb force acting between the two charges.

24. A cube of dimension $L \times L \times L$ finds itself in an electric field that is everywhere in the z direction. The field has a constant magnitude E_1 at the bottom face and at the sides and a constant magnitude E_2 at the top face (see Fig. 8.3). Find the force that the field exerts on the cube. *Note:* Nothing is known about the charge distribution or dielectric constant of the cube.

Figure 8·3

A cube of dimension $L \times L \times L$ in a nonuniform electric field.

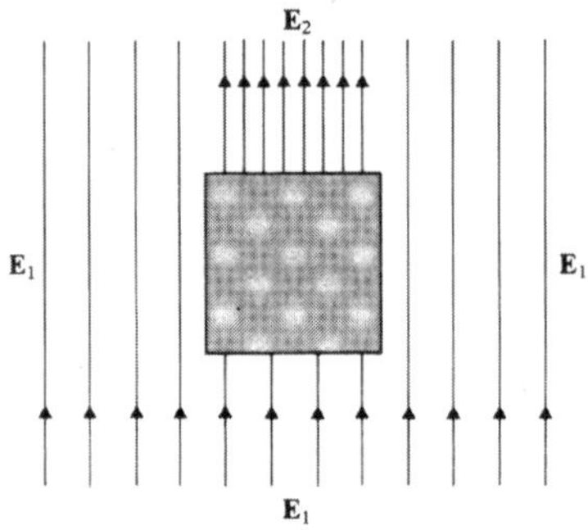

25. The Andromeda nebula is at a distance of 2.2×10^6 light-years from the Earth. Consider an observer passing by the Earth with a velocity of $0.99c$ in the direction of Andromeda.

(a) If the observer measures the apparent angular size of Andromeda and divides this angular size into the (known) transverse dimension of Andromeda, what distance will she infer? [*Hint:* Derive the following formula for the transformation of apparent angular size:

$$\theta' = \theta\sqrt{\frac{1-V/c}{1+V/c}}$$

This is valid for light arriving from directions near the direction of motion, that is, $\theta \ll 1$. This formula can be obtained from the general aberration formula for light, Eq. (12.51); alternatively, it can be obtained from the result stated in Problem 6.16.]

(b) If the observer measures the apparent brightness ($\text{erg/cm}^2\cdot\text{s}$) of Andromeda and then compares it with the (known) intrinsic luminosity (erg/s) of Andromeda, what distance will she infer? [*Hint*: Use the Lorentz-transformation properties of the Poynting vector.]

26. A physical explanation of the length contraction can be contrived by adopting a dynamical model for a rigid rod. The simplest such model assumes that a rod or any other rigid body consists of positive and negative charged particles at rest in equilibrium positions under the action of their attractive and repulsive electric forces (this model is unrealistic, because such an equilibrium would not be stable, but any improvement on this model requires quantum mechanics). Consider a simple arrangement of three positive charges $+q$ at the vertices of an equilateral triangle and a negative charge $-q/\sqrt{3}$ at the center. Show that when at rest, these charges are in equilibrium. Then consider how the equilibrium of the central charge is affected when the arrangement is in motion in a direction perpendicular to one of the sides of the triangle, and the electric fields contract. Without length contraction of the dimensions of the triangle, will the central charge remain in equilibrium? Will it do so with length contraction?

CHAPTER

9

Magnetostatics

In the preceding chapter, we discovered that the magnetic field is a relativistic kin of the electric field. Both of these fields are components of the field tensor $F^{\mu\nu}$, and under Lorentz transformations, an electric field can partially metamorphose into a magnetic field, and vice versa. We also discovered the field equations that determine the magnetic fields generated by given currents. In this chapter, we will examine in detail the magnetic fields generated by charges moving with constant velocity and the magnetic fields generated by steady (time-independent) currents. If the current is time independent, then the magnetic field is also time independent. We will see how to calculate such magnetic fields directly from Biot-Savart's law or from Ampère's law and indirectly via the vector potential. And we will see how to calculate the motion of charged particles in a given magnetic field.

9·1 The Magnetic Field of a Point Charge

Nothing illustrates the kinship between the electric and magnetic fields more clearly than the calculation we did in Section 8.2 of the magnetic field of a moving point charge. In its own rest frame, or $x'y'z't'$ frame, the point charge has only an electric field, the Coulomb field. If the point charge is at the origin,

$$\mathbf{E}' = q' \frac{\mathbf{x}'}{|\mathbf{x}'|^3} \tag{1}$$

From this, the Lorentz-transformation equations for the field tensor permit us to find the electric and magnetic fields in the laboratory frame.

According to Eq. (8.47), the resulting magnetic field is

$$\mathbf{B}_\perp = \gamma \frac{\mathbf{v}}{c} \times \mathbf{E}' \tag{2}$$

where $\mathbf{v}$ is the velocity of the point charge in the laboratory. Since $\mathbf{B}_\perp$ is the only nonzero component of the magnetic field, we can drop the subscript $\perp$ and write

$$\mathbf{B} = \gamma \frac{\mathbf{v}}{c} \times \mathbf{E}' \tag{3}$$

For the sake of simplicity, let us assume that the velocity is nonrelativistic ($v \ll c$). Then, $\gamma \simeq 1$ and $\mathbf{x}' \simeq \mathbf{x} - \mathbf{v}t$, or $\mathbf{x}' \simeq \mathbf{x} - \mathbf{x}_{q'}$, where $\mathbf{x}_{q'} = \mathbf{v}t$ is the instantaneous position of the point charge. With $\mathbf{E}' = q\mathbf{x}'/|\mathbf{x}'|^3 \simeq q(\mathbf{x} - \mathbf{x}_{q'})/|\mathbf{x} - \mathbf{x}_{q'}|^3$, this yields

$$\mathbf{B} = q' \frac{\mathbf{v}}{c} \times \frac{(\mathbf{x} - \mathbf{x}_{q'})}{|\mathbf{x} - \mathbf{x}_{q'}|^3} \tag{4}$$

Along any radial line making a fixed angle with the velocity vector, the strength of this magnetic field decreases as the inverse square of the distance.

Magnetic fields can be represented graphically by field lines. Figure 9.1 shows the field lines for our uniformly moving charge. The field lines are circles concentric with the line of motion. The direction of the field lines is given by a right-hand rule: if the thumb lies along the direction of motion of a positive charge, then the fingers curl in the direction of the field lines. Note that the magnetic field lines for any arbitrary magnetic field always form closed loops; this follows from $\nabla \cdot \mathbf{B} = 0$ [see Eq. (8.63)], which asserts that there are no magnetic "charges" on which field lines could begin or end (see, however, Section 9.7, where we will discuss some recent speculations concerning magnetic "charges," or magnetic monopoles.)

Figure 9·1

Magnetic field lines for a moving, positive point charge. The velocity is perpendicular to the plane of the page, out of this plane. (*a*) Here the charge is behind the plane of the page (it has not yet reached the plane of the page). We see the field lines in front of the charge; the field is zero at the center. (*b*) Now the point charge, marked by the dot, is in the plane of the page; the field is infinite at the center.

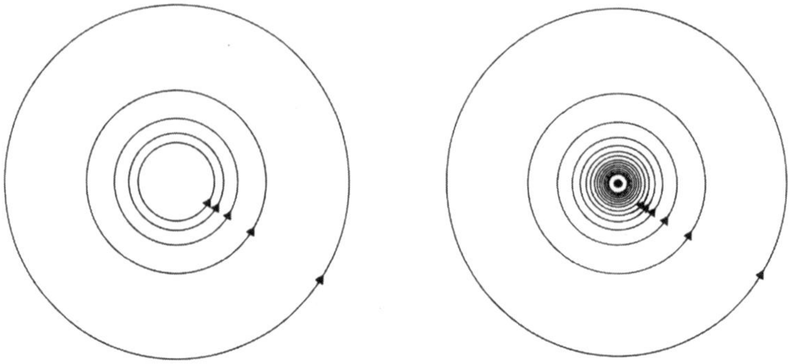

(a) (b)

From Eq. (4), we can calculate the magnetic force that a second point charge q experiences when it moves through the magnetic field generated by the first point charge. If the velocity of the second charge is $\mathbf{u}$, then

$$\mathbf{F}_{\text{on } q} = q\frac{\mathbf{u}}{c}\times\mathbf{B} \tag{5}$$

or

$$\mathbf{F}_{\text{on } q} = \frac{qq'}{c^2}\mathbf{u}\times\left[\mathbf{v}\times\frac{(\mathbf{x}-\mathbf{x}_{q'})}{\left|\mathbf{x}-\mathbf{x}_{q'}\right|^3}\right] \tag{6}$$

This law for the magnetic force exerted by q' on q is the magnetic "Coulomb" law. We could take this force law as the starting point of our calculations in magnetostatics, just as we did with the Coulomb law in electrostatics. However, since we already know the field equations for the magnetic field, we will not adopt this approach.

■ ***Exercise 1.*** Show that the magnetic force exerted by q on q' is

$$\mathbf{F}_{\text{on } q'} = \frac{q'q}{c^2}\mathbf{v}\times\left[\mathbf{u}\times\frac{(\mathbf{x}_{q'}-\mathbf{x})}{\left|\mathbf{x}_{q'}-\mathbf{x}\right|^3}\right] \tag{7}$$

Show that, in general, $\mathbf{F}_{\text{on } q} \neq -\mathbf{F}_{\text{on } q'}$. This means that Newton's Third Law *fails* for magnetic forces between point charges. (The consequent imbalance in momentum conservation is compensated by momentum transferred to the electromagnetic fields.) ■

Of course, in any calculation of the force between point charges, we must also take the electric force into account; this is usually larger than the magnetic force. Furthermore, in a comparison of the magnetic and electric forces, the factor γ in Eq. (8.43) must be treated with circumspection. Since Eq. (7) is of order uv/c^2, consistency demands that we retain terms of order v^2/c^2 in Eq. (8.43); that is, we must approximate $\gamma \simeq 1 + \frac{1}{2}v^2/c^2$.

9·2 Steady Currents; Ohm's Law

The electric current in a metal, a plasma, or an electrolyte consists of a flow of electrons or ions. Collisions between these moving charged particles and the stationary atoms or molecules of the conducting material remove energy and momentum from the particles. If the current is to remain steady, this loss must be made good by the driving force supplied by an electric field within the conductor. Ohm's law states that the current density is directly proportional to this electric field. This proportionality can be written as

$$\mathbf{j} = \sigma \mathbf{E} \tag{8}$$

or as

$$\mathbf{j} = \frac{\mathbf{E}}{\rho} \tag{9}$$

where the constant σ is the *conductivity* and $\rho = 1/\sigma$ is the *resistivity*. These constants depend on the characteristics of the conducting material. The CGS electrostatic unity of resistivity is the statohm·cm,* which coincides with the unit of time (1 statohm·cm = 1 s).

To devise a simple derivation of Eq. (8), we begin with the observation that the charged particles in a conductor have large random speeds; for example, the conduction electrons in a metal form a free-electron gas with random speeds of the order of 10^8 cm/s (this is the typical Fermi velocity for metals). The electric field gives this gas an average drift velocity, but the magnitude of this drift velocity always remains much smaller than the magnitude of the random speed. Hence the rate at which an electron engages in collisions with the atoms of the conducting material is nearly independent of the drift velocity and of the electric field. Let us designate the rate of collision by $1/\tau$; this is the number of collisions per electron per second. If the average drift velocity of an electron is $\mathbf{v}_\mathrm{d}$, then its average momentum is $m_\mathrm{e}\mathbf{v}_\mathrm{d}$. We expect that in each collision, the electron loses all of this momentum and is left with only the random motion. The rate of loss of momentum is then $m_\mathrm{e}\mathbf{v}_\mathrm{d}/\tau$. On the other hand, the rate of gain of momentum due to the electric driving force is $-e\mathbf{E}$. Under steady-state conditions, the rate of loss must match the rate of gain:

$$\frac{m_\mathrm{e}\mathbf{v}_\mathrm{d}}{\tau} = -e\mathbf{E}$$

or

$$\mathbf{v}_\mathrm{d} = \frac{-e\mathbf{E}\tau}{m_\mathrm{e}} \tag{10}$$

If the density of free electrons is n, the current density will be

$$\mathbf{j} = -en\mathbf{v}_\mathrm{d} = \frac{e^2 n\tau}{m_\mathrm{e}}\mathbf{E} \tag{11}$$

This says that $\mathbf{j}$ is proportional to $\mathbf{E}$ and completes our derivation of Ohm's law.

Note that Ohm's law is of restricted validity—it applies only if the current is generated by the electric field via the collision process described above. If the current is generated by some other process, then there is no reason for a proportionality of current and electric field. For

*The SI unit of resistivity is the ohm·m $\Leftrightarrow 1.113 \times 10^{-10}$ statohm·cm.

instance, if the current consists of an electron beam emitted by an electron gun, then it can be of any magnitude and direction we choose, regardless of the magnitude and direction of the electric field. But for the remainder of this section, we will assume that such extra, non-ohmic currents are absent.

In a conductor of uniform cross section—such as a long, uniform wire—the current density is the same at all points within the conductor [see Fig. 9.2(a)], and the electric field is also the same at all points within the conductor. In consequence of this uniformity, we can express the current density as $j = I/A$, where I is the current carried by the wire and A is its cross-sectional area; and we can express the electric field as $E = \Delta\Phi / \Delta l$, where $\Delta\Phi$ is the potential difference between the ends of the wire and Δl is its length. Upon substituting these expressions for j and E into Eq. (9), we find that Ohm's law takes the simple and familiar form

$$I = \frac{\Delta\Phi}{R} \qquad (12)$$

where the constant

$$R = \rho \frac{\Delta l}{A}$$

is the *resistance* of the wire. The CGS electrostatic unity of resistance is the statohm.*

The electric field **E** performs work on the electrons, and this work is dissipated as thermal energy in the electron collisions, causing Joule heating in the conductor. To calculate the power absorbed by Joule heating, we begin with the power $-e\mathbf{v}_d \bullet \mathbf{E}$ that one electron absorbs from

Figure 9·2

Flow lines of the current inside a conductor [by Eq. (8), they coincide with the electric field lines]. (*a*) Conductor of uniform cross section. The current density is constant. (*b*) Conductor of nonuniform cross section (the vertical width of the conductor increases stepwise; the horizontal width, perpendicular to the plane of the page, remains constant). The current density is a function of position.

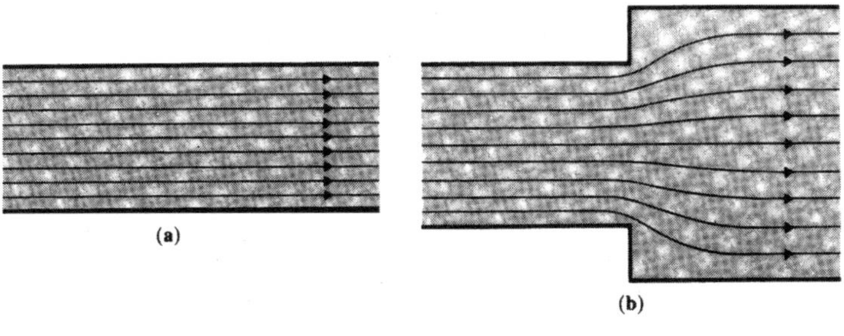

*The SI unit of resistance is the ohm, with 1 ohm $\Leftrightarrow 1.113\times10^{-12}$ statohm.

the electric field, and we multiply by n to obtain the power $-en\mathbf{v}_d\cdot\mathbf{E}$ absorbed per unit volume of conductor. With $\mathbf{j}=-en\mathbf{v}_d$, the total power absorbed in the total volume of conductor is then

$$P=\int \mathbf{j}\cdot\mathbf{E}\,dV=\int jE\,dV \tag{13}$$

In the case of a conductor of uniform cross section A and length Δl, the current density is $j=I/A$, the electric field is $E=\Delta\Phi/\Delta l$, and the total volume is $A\Delta l$, so Eq. (13) reduces to

$$P=I\Delta\Phi$$

which is the familiar expression for the power absorbed by Joule heating.

In a conductor of arbitrary shape, with a nonuniform cross section, the current density will be a function of position [for example, see Fig. 9.2(*b*)]. The exact distribution of the flow of current over the volume of the conductor can be calculated by means of the continuity equation

$$\nabla\cdot\mathbf{j}+\frac{\partial\rho}{\partial t}=0$$

For steady-state conditions, $\partial\rho/\partial t=0$, and

$$\nabla\cdot\mathbf{j}=0$$

With $\mathbf{j}=\sigma\mathbf{E}$, this becomes

$$\nabla\cdot(\sigma\mathbf{E})=0 \tag{14}$$

If we assume that the conductor is homogeneous, then the factor σ in this equation is constant and can be omitted. With $\mathbf{E}=\nabla\Phi$, our equation reduces to

$$\nabla^2\Phi=0 \tag{15}$$

This is Laplace's equation, and it shows that a homogeneous conductor carrying a steady current has no net charge in its interior. In this regard, the current-carrying conductor in steady conditions resembles a conductor in electrostatic conditions. However, the current-carrying conductor has an electric field in its interior, whereas the electrostatic conductor does not; and the current-carrying conductor is not an equipotential, whereas the electrostatic conductor is.

At the lateral surfaces of the conductor, the current density must have a vanishing normal component, because the current cannot flow out of the conductor; thus $\hat{\mathbf{n}}\cdot\mathbf{j}=0$, or $\hat{\mathbf{n}}\cdot\mathbf{E}=0$. We will suppose that at the end faces of the conductor, where the current enters or leaves, the current density is prescribed, so $\hat{\mathbf{n}}\cdot\mathbf{E}=0$ is prescribed, or else the potential is prescribed. This means that the solution of Eq. (15) is subject to Neumann boundary conditions or else mixed Neumann-Dirichlet boundary conditions, and it is therefore unique, up to an additive constant (see Section 3.1). Once the potential Φ has been found, the distribution of current throughout the volume of the conductor follows from $\mathbf{j} = -\sigma\nabla\Phi$. Readers familiar with hydrodynamics will recognize that the calculation of the current distribution in a conductor is mathematically equivalent to the calculation of the velocity field for the flow of an incompressible fluid ("potential" flow).

Furthermore, once the potential has been determined at the surface of the conductor, the electric field *outside* the conductor can be calculated by proceeding to the solution of $\nabla^2\Phi = 0$ in the region outside the conductor. This is a Dirichlet problem: at the surface of this region, Φ is prescribed, but $\hat{\mathbf{n}}\bullet\mathbf{E}$ is not prescribed. The normal component of the electric field usually has a discontinuity across the lateral surface of the conductor; this is due to the accumulation of a static layer of charge on this surface. This layer of charge plays an important role in shaping the field lines in the interior of the conductor, ensuring that they stay within the conductor (see Fig. 9.2). This layer of charge also shields the volume of the conductor from any external electric fields, ensuring that the current remains independent of any external electric disturbances (except when these act on the end faces of the conductor).

The solution for the electric field in the region outside the conductor is often a rather complicated mathematical problem, which requires the complete specification of the boundary conditions on any conducting surfaces surrounding the conductor. Since the current in the conductor is independent of the outside electric field, this problem is of little practical interest, and few exact solutions have been attempted.

EXAMPLE: *electric field outside of a conductor carrying a current*

Figure 9.3 shows one case for which an exact solution is available: a long, cylindrical conductor, or wire, surrounded by a concentric shell. The current density and the electric field are uniform within the cylindrical conductor; their direction is parallel to the axis of the cylinder, or the z axis. The potential on the surrounding shell is assumed to be zero. This poses a well-defined boundary-value problem for the region between the cylinder and the shell: on the shell $\Phi = 0$, and on the cylinder $\Phi = -E_0 z$, where E_0 is the magnitude of the uniform electric field. The solution of the Laplace equation turns out to be fairly simple (see Problem 5). Figure 9.3 shows the field lines of the electric field. Note that the lines of the external field terminate on the surface of the cylinder—the lower half of the cylinder carries a positive surface charge, and the upper half carries a negative surface charge.

EXAMPLE: *approach to equilibrium in a poor conductor*

Finally, we want to deal with an application of Ohm's law in the problem of the approach to electrostatic equilibrium in a conductor. Experience shows that when we place some amount of charge on one end of a conductor, this charge redistributes itself, and the conductor very

Figure 9·3

Electric field lines in and near a cylindrical conductor carrying an electric current. The conductor is surrounded by a concentric shell at zero potential.

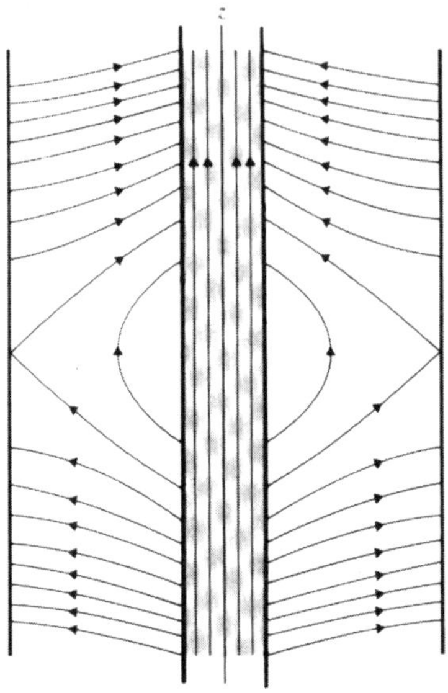

quickly achieves static equilibrium (zero current). Likewise, when we connect the ends of a conductor to a battery that supplies a constant emf, the conductor very quickly achieves steady conditions (constant current). How can we calculate the time that the conductor takes to reach equilibrium, or steady conditions? For a *poor* conductor, the answer is as follows: Begin with the continuity equation

$$\nabla \cdot \mathbf{j} + \frac{\partial \rho}{\partial t} = 0 \qquad (16)$$

and substitute $\mathbf{j} = \sigma \mathbf{E}$ according to Ohm's law, so

$$\sigma \nabla \cdot \mathbf{E} + \frac{\partial \rho}{\partial t} = 0$$

But $\nabla \cdot \mathbf{E} = 4\pi\rho / \varepsilon$; hence

$$\frac{4\pi\sigma}{\varepsilon}\rho + \frac{\partial \rho}{\partial t} = 0 \qquad (17)$$

The solution of this differential equation is

$$\rho(t) = \rho(0)\,\mathrm{e}^{-4\pi\sigma t/\varepsilon} \qquad (18)$$

This solution indicates that the charge everywhere inside the conductor decreases exponentially; that is, the charge flows toward the surface of the conductor and accumulates there. Equilibrium corresponds to $\rho = 0$. The e-folding time of the exponential function is

$$t_0 = \frac{\varepsilon}{4\pi\sigma} \tag{19}$$

This time tells us how fast the conductor approaches electrostatic equilibrium—it is the *relaxation time* of the conductor. For example, in distilled water, $\sigma \simeq 2 \times 10^6 \text{ s}^{-1}$ and $\varepsilon \simeq 80$, so $t_0 \simeq 3\times10^{-6}\text{ s}$.

We must strongly emphasize that, contrary to the assertions made in some electricity textbooks, the preceding calculation *is not valid for good conductors,* such as metals. If we were to commit the naive mistake of substituting the conductivity of a metal into Eq. (19)—say, $\sigma = 5 \times 10^{17} \text{ s}^{-1}$ for copper—we would obtain $t_0 \simeq 1/(4\pi\sigma) \simeq 2\times10^{-19}\text{ s}$, a ridiculously small number. This small number is inconsistent with the assumptions underlying Ohm's law—on a time scale of 10^{-19} s, *Ohm's law is not valid.* The reason is that the momentum balance [Eq. (10)] crucial for Ohm's law cannot be achieved in a time shorter than the collision time τ. For copper, $\tau \simeq 2\times10^{-14}\text{ s}$, and we cannot rely on Ohm's law on a time scale shorter than this.

Once we enter the regime of validity of Ohm's law, we expect that the free charge will flow to the surface of the conductor very quickly. This suggests that the relaxation time for the expulsion of free charges from the volume of the conductor should be about equal to the collision time, a result that is confirmed by detailed calculations of the damping of the plasma oscillations of the free charges. But the expulsion of the free charges from the volume is only the first step toward electrostatic equilibrium. The moving charges constitute a current, which produces a time-dependent magnetic field, which (by induction) generates an electric field, which drives an extra current, and so on. To attain electro- and magnetostatic equilibrium, the conductor must expel these dynamic self-supporting fields and currents. Thus, in the analysis of the approach to equilibrium in a good conductor, we must give due consideration to *all* of Maxwell's equations—and we will do this in Section 12.9. Intuitively, we expect that, in a good conductor, the relaxation time should be proportional to the conductivity; in such a conductor, the relaxation time should be long because there is little "friction" to dissipate the dynamical fields and currents. We will see, in Section 12.8, that the calculations confirm these expectations.

9·3 The Static Magnetic Field; Ampère's Law and Biot-Savart's Law

For a time-independent magnetic field generated by a steady current, the Maxwell equations (8.62) and (8.63) reduce to

$$\nabla \times \mathbf{B}(\mathbf{x}) = \frac{4\pi}{c}\mathbf{j}(\mathbf{x}) \tag{20}$$

$$\nabla \cdot \mathbf{B}(\mathbf{x}) = 0 \tag{21}$$

The first of these equations is called *Ampère's law.* If we integrate both sides of this equation over an arbitrary open surface, we obtain

$$\int_S \nabla \times \mathbf{B} \cdot d\mathbf{S} = \frac{4\pi}{c}\int_S \mathbf{j} \cdot d\mathbf{S} \tag{22}$$

By means of Stokes' theorem, the left side can be converted into an integral along the boundary of the surface. This yields

$$\oint \mathbf{B} \cdot d\mathbf{l} = \frac{4\pi}{c}\int_S \mathbf{j} \cdot d\mathbf{S} \tag{23}$$

This is the integrated form of Ampère's law. It states that the integral of **B** along any closed path is proportional to the current intercepted by the area bounded by the path.

If the distribution of current has a high degree of symmetry, then Eq. (23) leads to a very simple and direct calculation of the magnetic field. The classic example of this is the calculation of the magnetic field generated by a current I on a very long, straight wire. According to Eq. (20), the magnetic field lines must be closed loops. The only patterns of field lines consistent with the cylindrical symmetry of the current are concentric circles, radial lines, and parallel lines in the direction of the current; of these, only the pattern of concentric circles (see Fig. 9.4) is consistent with the requirement that the field lines be closed loops. The integration $\int \mathbf{B} \cdot d\mathbf{l}$ along a circle of radius r then gives $2\pi rB$; furthermore the integration $\int \mathbf{j} \cdot d\mathbf{S}$ gives I, the current on the wire. Hence

EXAMPLE: *magnetic field of a current on a straight wire*

$$2\pi rB = \frac{4\pi}{c}I \tag{24}$$

or

$$B = \frac{2I}{c}\frac{1}{r} \tag{25}$$

Figure 9·4

Magnetic field lines around a very long, straight wire. The wire is perpendicular to the plane of the page.

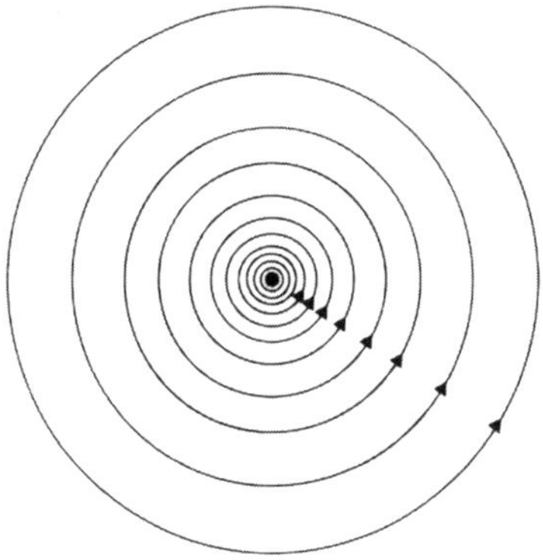

For more complicated current distributions that do not lend themselves to such a simple treatment with the integrated form of Ampère's law, we must find the general solution of Eqs. (20) and (21). The solution is most easily accomplished with the help of the vector potential **A**. In the time-independent case, each component of the vector potential satisfies a Poisson equation [see Eq. (8.9)]:*

$$\nabla^2 A_x(\mathbf{x}) = -\frac{4\pi}{c} j_x(\mathbf{x}) \tag{26}$$

$$\nabla^2 A_y(\mathbf{x}) = -\frac{4\pi}{c} j_y(\mathbf{x}) \tag{27}$$

$$\nabla^2 A_z(\mathbf{x}) = -\frac{4\pi}{c} j_z(\mathbf{x}) \tag{28}$$

We know from Chapter 2 that the Poisson equation for the electrostatic potential

*This neat separation of the equations for the vector potential—with the x, y, and z component of **A** related exclusively to the corresponding x, y, and z component of **j**—is valid only for rectangular coordinates. In spherical and in cylindrical coordinates, the equations mix the components, and each component of **A** is related to all components of **j**. A straightforward, but tedious, way to obtain the equations in, say, spherical coordinates is to express the partial derivatives $\partial/\partial x, \partial/\partial y, \partial/\partial z$ in Eqs. (26)—(28) in terms of $\partial/\partial r, \partial/\partial\theta, \partial/\partial\phi$ by the chain rule and express the x, y, z components of **A** and of **j** in terms of the r, θ, ϕ components. This results in three very messy coupled equations for the spherical components of **A**.

$$\nabla^2\Phi(\mathbf{x}) = -4\pi\rho(\mathbf{x}) \tag{29}$$

has the solution

$$\Phi(\mathbf{x}) = \int \frac{\rho(\mathbf{x}')}{|\mathbf{x}-\mathbf{x}'|} dV' \tag{30}$$

By analogy, we see that the solutions of Eqs. (26)–(28) must be

$$A_x(\mathbf{x}) = \frac{1}{c}\int \frac{j_x(\mathbf{x}')}{|\mathbf{x}-\mathbf{x}'|} dV' \tag{31}$$

and corresponding expressions for $A_y(\mathbf{x})$ and $A_z(\mathbf{x})$. We can combine these solutions in a vector equation:

$$\mathbf{A}(\mathbf{x}) = \frac{1}{c}\int \frac{\mathbf{j}(\mathbf{x}')}{|\mathbf{x}-\mathbf{x}'|} dV' \tag{32}$$

Given this vector potential, the magnetic field is

$$\mathbf{B}(\mathbf{x}) = \nabla\times\mathbf{A}(\mathbf{x}) = \frac{1}{c}\int \nabla\times\frac{\mathbf{j}(\mathbf{x}')}{|\mathbf{x}-\mathbf{x}'|} dV' \tag{33}$$

In this expression, the differential operator ∇ acts only on $\mathbf{x}$; we can therefore rewrite the integrand as follows:

$$\nabla\times\frac{\mathbf{j}(\mathbf{x}')}{|\mathbf{x}-\mathbf{x}'|} = -\mathbf{j}(\mathbf{x}')\times\nabla\frac{1}{|\mathbf{x}-\mathbf{x}'|} = \mathbf{j}(\mathbf{x}')\times\frac{\mathbf{x}-\mathbf{x}'}{|\mathbf{x}-\mathbf{x}'|^3} \tag{34}$$

With this, our expression for $\mathbf{B}(\mathbf{x})$ becomes

$$\mathbf{B}(\mathbf{x}) = \frac{1}{c}\int \mathbf{j}(\mathbf{x}')\times\frac{\mathbf{x}-\mathbf{x}'}{|\mathbf{x}-\mathbf{x}'|^3} dV' \tag{35}$$

In many practical applications, the current flows along one (or several) wires. The integration in Eq. (35) is then restricted to the interior volume of the wire. For a thin wire, we can regard $(\mathbf{x}-\mathbf{x}')/|\mathbf{x}-\mathbf{x}'|^3$ as constant when integrating over the cross section of the wire. The integration of $\mathbf{j}(\mathbf{x}')$ over the cross section simply gives the current I carried by the wire [for an explicit example of this integration, see below, Eqs. (37)–(38)]. The three-dimensional integral in Eq. (35) then reduces to a one-dimensional integral along the wire:

$$\mathbf{B}(\mathbf{x}) = \frac{I}{c}\int d\mathbf{x}' \times \frac{\mathbf{x}-\mathbf{x}'}{|\mathbf{x}-\mathbf{x}'|^3} \tag{36}$$

where $d\mathbf{x}'$ is a displacement along the wire (see Fig. 9.5). This is *Biot-Savart's Law.*

■ ***Exercise 2.*** Show that Eq. (36) follows from Eq. (35) upon integration over the cross section of the wire. ■

EXAMPLE: *vector potential of a current on a straight wire*

The remainder of this section deals with several examples of calculations of the magnetic field. For a first example, let us repeat the calculation of the magnetic field of the very long, straight wire, using the vector potential. Figure 9.6 shows the wire lying along the z axis. Since the wire is thin, $1/|\mathbf{x}-\mathbf{x}'|$ is constant over the cross section of the wire. As in the case of Eq. (35), the three-dimensional integral then reduces to a one-dimensional integral:

Figure 9·5

A small segment $d\mathbf{x}'$ of wire and its contribution to the magnetic field.

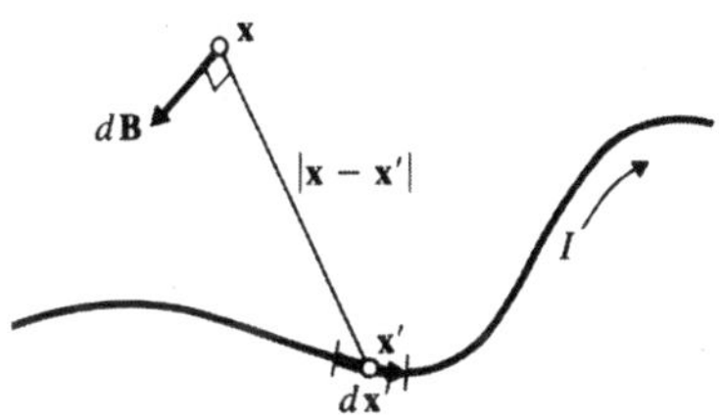

Figure 9·6

Very long, straight wire lying along the z axis.

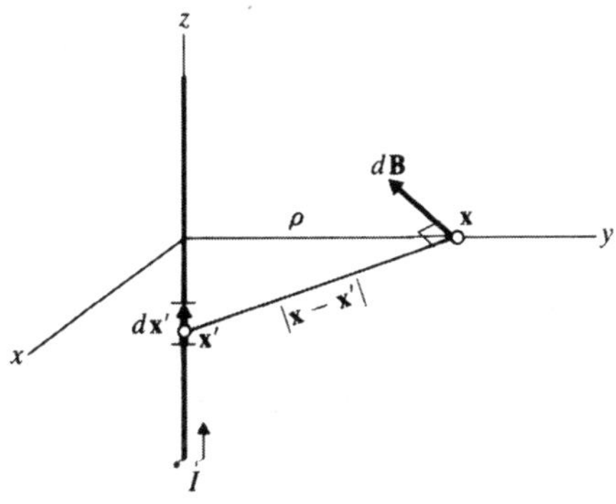

$$\mathbf{A}(\mathbf{x}) = \frac{1}{c}\int \frac{\mathbf{j}(\mathbf{x}')}{|\mathbf{x}-\mathbf{x}'|}dV' \tag{37}$$

$$= \frac{1}{c}\int \frac{1}{|\mathbf{x}-\mathbf{x}'|}\left(\iint \mathbf{j}\,dx'dy'\right)dz'$$

$$= \frac{1}{c}\int \frac{1}{|\mathbf{x}-\mathbf{x}'|} I\hat{\mathbf{z}}\,dz' \tag{38}$$

With the coordinates shown in Fig. 9.6, the integral becomes

$$\mathbf{A}(\mathbf{x}) = \frac{I}{c}\hat{\mathbf{z}}\int_{-\infty}^{\infty} \frac{dz'}{\sqrt{\rho^2 + z'^2}} \tag{39}$$

This integral diverges logarithmically. The trouble is that the solution (32) for the vector potential implicitly assumes that $\mathbf{A} \to 0$ at infinity (in all directions); this is not true for a wire of infinite length. We must therefore readjust the additive constant in the potential. This is most easily accomplished by first evaluating (39) for a wire of finite length:

$$\mathbf{A}(\mathbf{x}) = \frac{I}{c}\hat{\mathbf{z}}\int_{-L}^{L} \frac{dz'}{\sqrt{\rho^2 + z'^2}} = \frac{2I}{c}\hat{\mathbf{z}}\int_{0}^{L} \frac{dz'}{\sqrt{\rho^2 + z'^2}} \tag{40}$$

$$= \frac{2I}{c}\hat{\mathbf{z}}\left[-\ln\rho + \ln(L + \sqrt{\rho^2 + L^2})\right] \tag{41}$$

If $L \gg \rho$, then the second logarithm function in the brackets is nearly constant. We therefore obtain

$$\mathbf{A} = -\frac{2I}{c}\hat{\mathbf{z}}\ln\rho + \text{constant} \tag{42}$$

By means of the expression for the gradient operator in cylindrical coordinates, we then readily find

$$\mathbf{B} = \nabla \times \mathbf{A} = \frac{2I}{c}\frac{1}{\rho}\hat{\boldsymbol{\phi}} \tag{43}$$

in agreement with Eq. (25).

■ ***Exercise 3.*** Evaluate $\nabla \times \mathrm{A}$ and show that the result is Eq. (43). ■

EXAMPLE: *magnetic field of a ring of current*

Another example, of considerable practical importance, is the calculation of the magnetic field of a circular ring of current (see Fig. 9.7). For a start, we will deal only with the magnetic field on the axis of the ring, leaving a more general calculation for the next section. For points on the axis, it is easiest to use the Biot-Savart law directly (to use the vector potential, we would first have to calculate this vector potential at points off the axis so that we can take derivatives; this is very messy).

Figure 9.7 shows the ring, with radius R and current I. According to Eq. (36), a small segment of the ring contributes a magnetic field

$$dB = \frac{I}{c}\frac{dl}{R^2 + z^2} \tag{44}$$

where $\sqrt{R^2+z^2} = |\mathbf{x}-\mathbf{x}'|$ is the distance between $\mathbf{x}$ and $\mathbf{x}'$. This contribution to the magnetic field is at right angles to the line connecting $\mathbf{x}$ and $\mathbf{x}'$. In view of the rotational symmetry, the horizontal components of the magnetic field cancel upon integration, and only a vertical component survives:

$$B_z = \frac{I}{c}\int \frac{\cos\theta\, dl}{R^2+z^2} = \frac{I}{c}\frac{2\pi R\cos\theta}{R^2+z^2} \tag{45}$$

$$= \frac{I}{c}\frac{2\pi R^2}{(R^2+z^2)^{3/2}} \tag{46}$$

EXAMPLE: *magnetic field of a solenoid*

From this expression for the magnetic field of a ring, we can find the magnetic field on the axis of a solenoid. Figure 9.8 shows the solenoid of finite length L. The solenoid can be regarded as a large number of rings stacked close together. Suppose that the number of rings is n per

Figure 9·7

Circular ring of current in the x-y plane, centered on the origin.

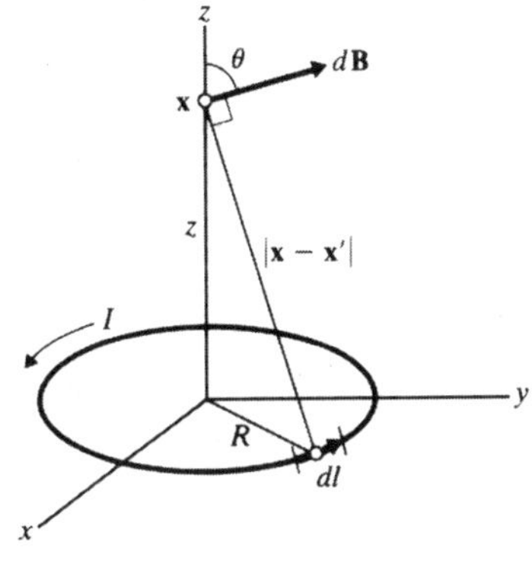

Figure 9·8

A solenoid of finite length centered on the z axis, seen in longitudinal section.

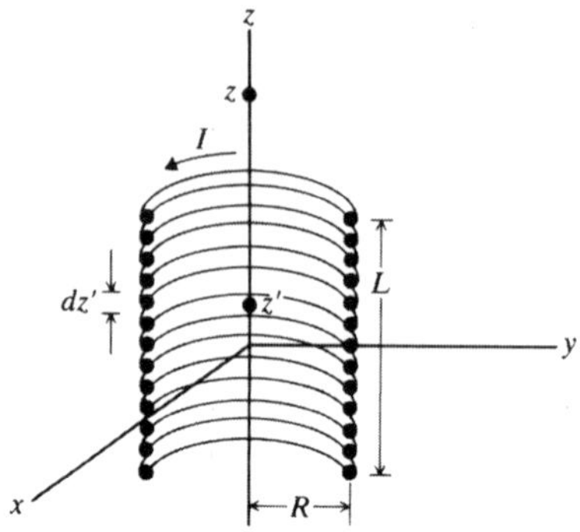

unit length, each with a current I. Then the number of rings in a segment dz' of the solenoid is ndz', and the magnetic field contributed by this segment is

$$dB_z = \frac{I}{c}\frac{2\pi R^2}{[R^2+(z-z')^2]^{3/2}}n\,dz' \tag{47}$$

so

$$B_z = \frac{I}{c}2\pi R^2 n\int_{-L/2}^{L/2}\frac{dz'}{[R^2+(z-z')^2]^{3/2}}$$

$$= \frac{I}{c}2\pi n\left[\frac{L/2-z}{\sqrt{R^2+(z-L/2)^2}}+\frac{L/2+z}{\sqrt{R^2+(z+L/2)^2}}\right] \tag{48}$$

For an infinitely long solenoid, we must take the limit $L \to \infty$, which yields

$$B_z = 4\pi n\frac{I}{c} \tag{49}$$

We can also obtain this result very quickly from Ampère's law [Eq. (23)]. For this purpose, we begin by noting that the only pattern of field lines consistent with the symmetry of the current distribution and consistent with the requirement that the field lines do not stop or start anywhere in the solenoid is a pattern of lines parallel to the axis. We can then apply Ampère's law to the rectangular path shown in Fig. 9.9. The integral $\int \mathbf{B}\cdot d\mathbf{l}$ receives no contribution from the short sides, which are perpendicular to the magnetic field, or from the long side exterior to the solenoid; the contribution from the long side interior to the solenoid is

Figure 9·9

A solenoid of infinite length and the path used for Ampère's law.

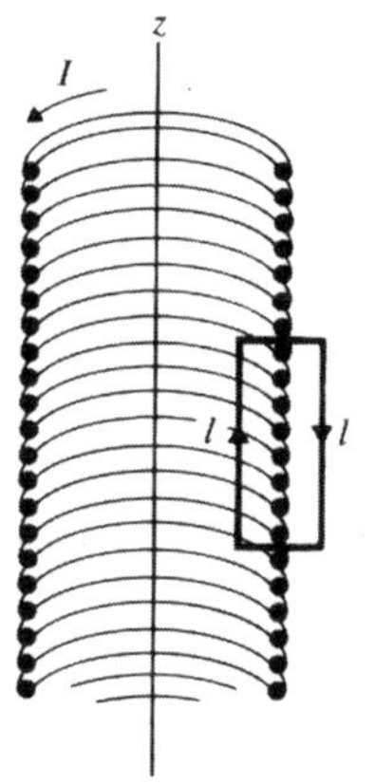

$B_z \Delta l$. The current intercepted by the rectangular area is $nI\Delta l$. Hence Ampère's law becomes

$$B_z \Delta l = \frac{4\pi}{c} nI\Delta l \qquad (50)$$

which is equivalent to Eq. (49). One important advantage of this derivation based on Ampère's law is that it does not hinge on the cylindrical symmetry of the solenoid; thus any solenoid of arbitrary, uniform cross section—for example, a solenoid of square cross section—produces a uniform magnetic field of the strength given by Eq. (49).

■ ***Exercise 4.*** Use Ampère's law to show that the magnetic field between two large, parallel conducting plates carrying uniform surface current densities of k statamp per centimeter in opposite directions is

$$B = \frac{4\pi k}{c} \qquad (51)$$

■

9·4 The Magnetic Scalar Potential; the Magnetic Dipole

In any region free of currents, Eq. (20) reduces to $\nabla \times \mathbf{B} = 0$. We can therefore apply the theorem stated in Section 1.5, according to which the field $\mathbf{B}$ can be expressed as the gradient of a potential function. Let us restate this theorem, and its restrictions, as follows:

• ***Theorem:*** Suppose that a region V is simply connected and free of currents. Then, within this region, there exists a function $\Phi_M(\mathbf{x})$ such that

$$\mathbf{B}(\mathbf{x}) = -\nabla\Phi_M(\mathbf{x}) \tag{52}$$

The function $\Phi_M(\mathbf{x})$ is called the *magnetic scalar potential.* In view of Eq. (21), this magnetic potential obeys Laplace's equation,

$$\nabla^2\Phi_M(\mathbf{x}) = 0 \tag{53}$$

This means that we can apply the mathematical methods that we developed in Chapter 3 to the solution of problems in magnetostatics. However, we must always keep in mind the crucial restriction imposed on the magnetic potential: It is well defined only in simply connected regions free of current. For instance, it is impossible to find a magnetic potential for the magnetic field of the ring of current (discussed in the preceding section) valid in all of space. We can find a potential for the region $z > 0$ and a separate magnetic potential for the region $z < 0$, but these two potentials do not join continuously across the plane $z = 0$.

EXAMPLE: *magnetic potential of a ring of current*

To find the magnetic potential for a ring with a current I, we begin with the observation that the magnetic field along the z axis,

$$B_z = \frac{I}{c}\frac{2\pi R^2}{(R^2+z^2)^{3/2}} \tag{54}$$

can be expressed as a gradient,

$$B_z = \frac{\partial}{\partial z}\left[\frac{2\pi I}{c}\frac{z}{(R^2+z^2)^{1/2}}\right] \tag{55}$$

Hence the magnetic potential along the z axis may be taken as

$$\Phi_M = -\frac{2\pi I}{c}\frac{z}{(R^2+z^2)^{1/2}} \tag{56}$$

To find the general expression for the magnetic potential, we must choose a simply connected region that does not include the ring of current. Let us choose the region $r > R$. In this region, the expression (56) has the following power-series expansion:

$$\Phi_M = -\frac{2\pi I}{c}\left(1 - \frac{R^2}{2z^2} + \frac{3R^4}{8z^4} - \frac{5R^6}{16z^6} + \ldots\right) \tag{57}$$

From Section 3.4, we know that the general solution of the Laplace equation with azimuthal symmetry is

$$\Phi_{\mathrm{M}} = \sum_{l=0}^{\infty}[A_l r^l P_l(\cos\theta) + B_l r^{-l-1} P_l(\cos\theta)] \tag{58}$$

Comparing this with Eq. (57), with $z = r$ along the z axis, we see that all the coefficients A_l are zero except A_0,

$$A_0 = -\frac{2\pi I}{c} \tag{59}$$

and that the coefficients B_l are

$$B_0 = 0 \qquad B_1 = \frac{2\pi I}{c}\frac{R^2}{2} \qquad B_2 = 0$$

$$B_3 = -\frac{2\pi I}{c}\frac{3R^4}{8} \qquad B_4 = 0 \qquad B_5 = \frac{2\pi I}{c}\frac{5R^6}{16} \tag{60}$$

and so on. Therefore, the magnetic potential is

$$\Phi_{\mathrm{M}} = \frac{2\pi I}{c}\left[-1 + \frac{R^2}{2r^2}\cos\theta - \frac{3R^4}{16r^4}(5\cos^3\theta - 3\cos\theta) + \ldots\right] \tag{61}$$

From this, the components of the magnetic field can be calculated:

$$B_r = -\frac{\partial\Phi_{\mathrm{M}}}{\partial r} = \frac{2\pi I}{c}\left[\frac{R^2}{r^3}\cos\theta - \frac{3R^4}{4r^5}(5\cos^3\theta - 3\cos\theta) + \ldots\right]$$

$$B_\theta = -\frac{1}{r}\frac{\partial\Phi_{\mathrm{M}}}{\partial\theta} = \frac{2\pi I}{c}\left[\frac{R^2}{2r^3}\sin\theta - \frac{9R^4}{16r^5}(5\cos^2\theta - 1)\sin\theta + \ldots\right]$$

$$B_\phi = -\frac{1}{r\sin\theta}\frac{\partial\Phi_{\mathrm{M}}}{\partial\phi} = 0 \tag{62}$$

The calculation of the magnetic field in the simply-connected region $r < R$ can be carried out by a similar method, but we will not do this here. Figure 9.10 shows the lines of magnetic field surrounding the ring.

For $r \gg R$, we can approximate the magnetic field as

$$B_r = \frac{2\pi IR^2}{c}\frac{\cos\theta}{r^3} \tag{63}$$

Figure 9·10

Magnetic field lines around a ring of current. The ring is perpendicular to the plane of the page.

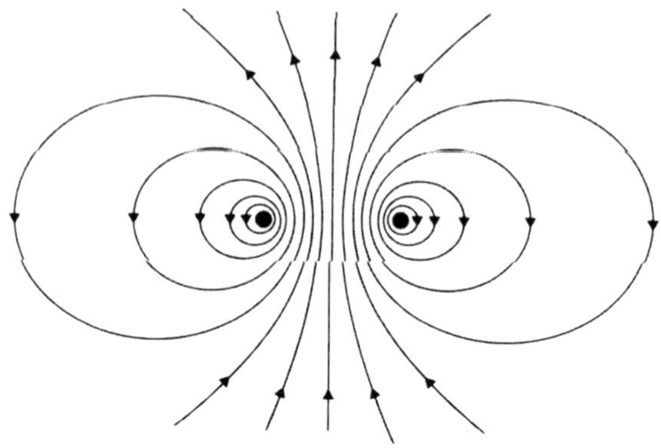

$$B_\theta = \frac{\pi I R^2}{c}\frac{\sin\theta}{r^3} \qquad (64)$$

EXAMPLE: *magnetic field of a pure dipole*

This is the magnetic field of an *ideal magnetic dipole*, or *pure magnetic dipole.* We will define the *magnetic dipole moment* as $1/c$ times the product of area and current:

$$m = \frac{\pi R^2 I}{c} \qquad (65)$$

so Eqs. (63) and (64) become

$$B_r = 2m\frac{\cos\theta}{r^3} \qquad (66)$$

$$B_\theta = m\frac{\sin\theta}{r^3} \qquad (67)$$

Evidently, the magnetic field of a pure magnetic dipole is similar to the electric field of a pure electric dipole.

Any arbitrary closed current loop will produce a magnetic dipole field at a large distance. This can be readily seen from the Biot-Savart law, Eq. (36):

$$\mathbf{B}(\mathbf{x}) = \frac{I}{c}\int d\mathbf{x}' \times \frac{\mathbf{x}-\mathbf{x}'}{|\mathbf{x}-\mathbf{x}'|^3} \qquad (68)$$

Figure 9·11

A loop of current of arbitrary shape. The point **x** is at a large distance from this loop.

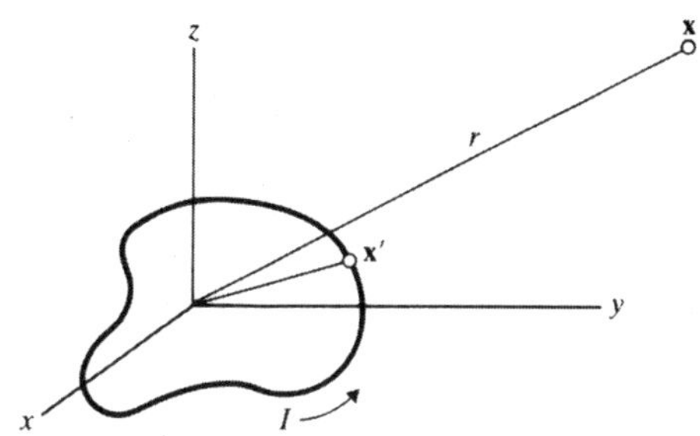

Suppose that the loop is near the origin of coordinates (see Fig. 9.11) and that the point **x** is at a large distance from the loop. Then we can use the approximation [see Eq. (3.114)]

$$\frac{1}{|\mathbf{x}-\mathbf{x}'|} \simeq \frac{1}{r} + \frac{\mathbf{x}\bullet\mathbf{x}'}{r^3} + \ldots \tag{69}$$

or

$$\frac{1}{|\mathbf{x}-\mathbf{x}'|^3} \simeq \frac{1}{r^3} + \frac{3\mathbf{x}\bullet\mathbf{x}'}{r^5} + \ldots \tag{70}$$

Inserting this in Eq. (68), we obtain four terms:

$$\mathbf{B}(\mathbf{x}) = \frac{I}{c}\left(\int \frac{1}{r^3} d\mathbf{x}' \times \mathbf{x} + \int \frac{3\mathbf{x}\bullet\mathbf{x}'}{r^5} d\mathbf{x}' \times \mathbf{x} - \int \frac{1}{r^3} d\mathbf{x}' \times \mathbf{x}' - \int \frac{3\mathbf{x}\bullet\mathbf{x}'}{r^5} d\mathbf{x}' \times \mathbf{x}' \right) + \ldots \tag{71}$$

The first term is (constant) $\times \int d\mathbf{x}'$, which vanishes for a closed loop. The last term has only one power of **x** in the numerator and is therefore of order $1/r^4$; this can be neglected. The remaining two terms (second and third) can be put in the form

$$\mathbf{B}(\mathbf{x}) = \frac{3(\mathbf{m}\bullet\mathbf{x})\mathbf{x}}{r^5} - \frac{\mathbf{m}}{r^3} \tag{72}$$

where the magnetic moment vector **m** is defined as

$$\mathbf{m} = \frac{I}{2c}\int \mathbf{x}' \times d\mathbf{x}' \tag{73}$$

■ ***Exercise 5.*** Verify Eq. (72). [*Hint:* It is best to work backward, from Eq. (72) to Eq. (71). Begin by substituting

$$\frac{(\mathbf{m}\cdot\mathbf{x})\mathbf{x}}{r^2} = \mathbf{m} + \frac{(\mathbf{m}\times\mathbf{x})\times\mathbf{x}}{r^2}$$

into Eq. (72). Since $\mathbf{m}$ is the integral of a cross product, $\mathbf{m}\times\mathbf{x}$ is the integral of a triple cross product; expand this triple cross product by means of Eq. (1.79), and then use integration by parts to show that the two terms of this expansion are equal.] ■

If the z axis is along the direction of $\mathbf{m}$, then the components of Eq. (72) are

$$B_r = 2m\frac{\cos\theta}{r^3}$$

$$B_\theta = m\frac{\sin\theta}{r^3} \qquad (74)$$

This agrees with Eqs. (66) and (67) and shows that $\mathbf{m}$ is the dipole moment of the current loop, expressed as a vector. Note that in the integral (73) for the dipole-moment vector, the direction of integration coincides with the direction of the current. For a plane loop, the dipole-moment vector is related to the direction of the current by a right-hand rule: If the fingers lie along the current, the thumb gives the direction of $\mathbf{m}$. For a three-dimensional loop, the direction of $\mathbf{m}$ must be determined by evaluating the integral (73), and the right-hand rule can at most give a rough indication.

■ ***Exercise 6.*** Show that the magnetic field (72) can be obtained from the magnetic potential

$$\Phi_{\mathrm{M}}(\mathbf{x}) = \frac{\mathbf{m}\cdot\mathbf{x}}{r^3} \qquad (75)$$

■

9·5 Motion of Particles in Magnetic Fields

The equation of motion of a charged particle in electric and magnetic fields is

$$\frac{d\mathbf{p}}{dt} = q\mathbf{E} + q\frac{\mathbf{u}}{c}\times\mathbf{B} \qquad (76)$$

First, we will discuss the motion of a particle in a purely magnetic field ($\mathbf{E} = 0$). In this case, Eq. (76) tells us that $\mathbf{u}\cdot d\mathbf{p}/dt = 0$. Since

$$\mathbf{u}\cdot\frac{d\mathbf{p}}{dt} = \mathbf{u}\cdot\frac{d}{dt}\frac{m\mathbf{u}}{\sqrt{1-u^2/c^2}}$$

$$= \frac{m}{(1-u^2/c^2)^{1/2}} \mathbf{u} \cdot \frac{d\mathbf{u}}{dt} + \frac{1}{2} \frac{mu^2/c^2}{(1-u^2/c^2)^{3/2}} \frac{du^2}{dt} \propto \frac{du^2}{dt} \tag{77}$$

we can immediately conclude that u^2 must remain constant. This establishes the well-known result that the magnetic force cannot change the speed or the energy of the particle. The rate of change of momentum in a magnetic field is therefore entirely due to the change in the direction of the velocity:

$$\frac{d\mathbf{p}}{dt} = \frac{m}{\sqrt{1-u^2/c^2}} \frac{d\mathbf{u}}{dt} \tag{78}$$

and Eq. (76) reduces to

$$\frac{d\mathbf{u}}{dt} = \frac{q}{mc} \sqrt{1 - \frac{u^2}{c^2}} \mathbf{u} \times \mathbf{B} \tag{79}$$

EXAMPLE: *motion of a charged particle in a constant magnetic field*

If the magnetic field is some function of space, Eq. (79) yields rather complicated orbits. For instance, Fig. 9.12 shows the motion of a cosmic-ray particle moving in the magnetic field of the Earth (a dipole field). Here we will deal only with motion in a constant magnetic field. According to Eq. (79), the velocity vector always changes perpendicular to itself and to the magnetic field. If $\mathbf{B}$ is constant, then $\mathbf{u}$ will precess uniformly around $\mathbf{B}$ (see Fig. 9.13); the component $\mathbf{u}_\parallel$ parallel to $\mathbf{B}$ remains constant, and the component $\mathbf{u}_\perp$ perpendicular to $\mathbf{B}$ oscillates with uniform circular motion. The frequency of this uniform circular motion is

Figure 9·12

Example of a trajectory of a cosmic-ray particle in the magnetic field of the Earth. The trajectory is three-dimensional. The diagram shows the projection of the trajectory into the (magnetic) equatorial plane.

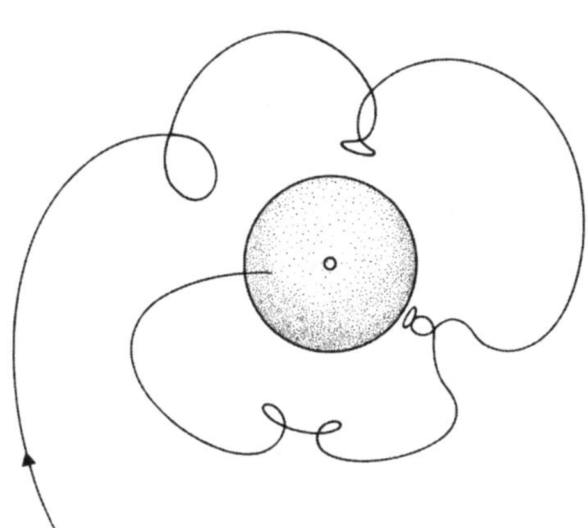

Figure 9·13

The velocity vector **u** precesses around **B**. The angle θ between **u** and **B** remains constant.

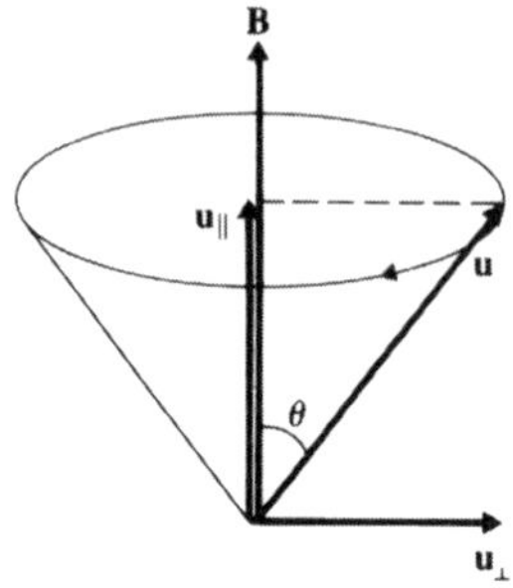

$$\omega = \frac{qB}{mc}\sqrt{1-\frac{u^2}{c^2}} \tag{80}$$

This equation emerges immediately if we compare the magnitude of Eq. (79),

$$\left|\frac{d\mathbf{u}_\perp}{dt}\right| = u_\perp \frac{qB}{mc}\sqrt{1-\frac{u^2}{c^2}} \tag{81}$$

with the standard expression for the centripetal acceleration in uniform circular motion,

$$\text{(centripetal acceleration)} = \omega \times \text{(speed in plane of motion)} \tag{82}$$

The frequency (80) is called the *cyclotron frequency*. For low speeds ($u/c \ll 1$), the cyclotron frequency is qB/mc, independent of the speed; but for relativistic speeds, it is smaller than qB/mc.

The radius of the circular motion is

$$r = \frac{u_\perp}{\omega} = \frac{u_\perp mc}{qB\sqrt{1-u^2/c^2}} \tag{83}$$

or

$$r = \frac{cp_\perp}{qB} \tag{84}$$

Thus, for a given magnetic field, the radius of the circular motion is directly proportional to the perpendicular momentum. High-energy physicists routinely exploit this relation to determine the momentum of a particle from the measured radius of the track that the particle makes in a bubble chamber in a magnetic field.

The combination of circular motion perpendicular to the magnetic field and translational motion parallel to the magnetic field gives the orbit the shape of a helix (see Fig. 9.14).

EXAMPLE: *motion of a charged particle in constant magnetic and electric fields*

Next, we want to discuss the motion of a particle under the simultaneous influence of magnetic and electric fields. For the sake of simplicity, we will suppose that the **B** and **E** fields are constant and at right angles to each other ("crossed fields"). Figure 9.15 shows a constant magnetic field along the x direction and a constant electric field along the z direction. Under these conditions, the x component of the velocity remains constant. Let us ignore this x component of the motion and concentrate on the y and z components.

Figure 9·14

In a uniform magnetic field, the orbit of a particle is a helix, with axis parallel to **B**.

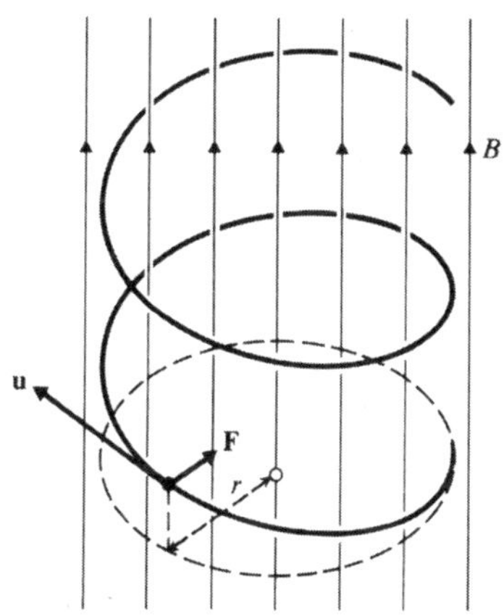

Figure 9·15

Constant, crossed electric and magnetic fields. The magnetic field is in the x direction; the electric field is in the z direction.

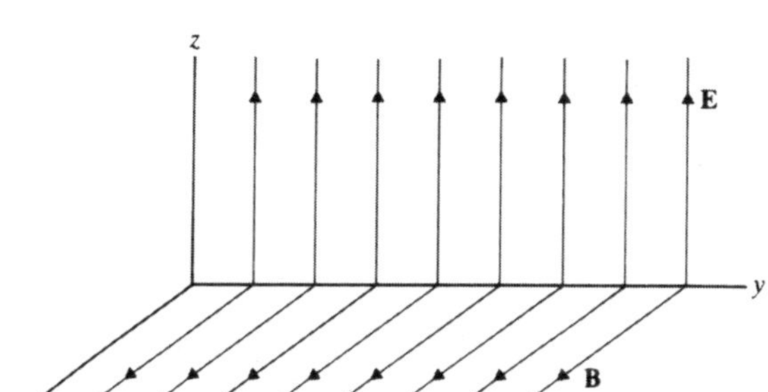

To analyze the motion in the y-z plane, we can take advantage of a clever trick: we go to a reference frame $x'y'z't'$ in which either $\mathbf{E}'$ or $\mathbf{B}'$ is zero. The Lorentz-transformation equations for the fields [Eqs. (8.43)–(8.44)] tell us that

$$\mathbf{E}'_{\parallel} = \mathbf{E}_{\parallel} \tag{85}$$

$$\mathbf{E}'_{\perp} = \gamma\left(\mathbf{E}_{\perp} + \frac{\mathbf{V}}{c}\times\mathbf{B}\right) \tag{86}$$

$$\mathbf{B}'_{\parallel} = \mathbf{B}_{\parallel} \tag{87}$$

$$\mathbf{B}'_{\perp} = \gamma\left(\mathbf{B}_{\perp} - \frac{\mathbf{V}}{c}\times\mathbf{E}\right) \tag{88}$$

where $\gamma = 1/\sqrt{1 - V^2/c^2}$. It turns out that if $B > E$ in our original reference frame, then we can make $\mathbf{E}'$ zero; if $E > B$, then we can make $\mathbf{B}'$ zero.

Let us consider the first case and assign the following value to $\mathbf{V}$:

$$\mathbf{V} = \frac{\mathbf{E}\times\mathbf{B}}{B^2}c \tag{89}$$

This is a permissible velocity, because $V = (EB/B^2)c = (E/B)c < c$. The velocity $\mathbf{V}$ is perpendicular to both $\mathbf{E}$ and $\mathbf{B}$ (in Fig. 9.15, the velocity is along the y axis); this makes the parallel components in Eqs. (85) and (87) zero. The electric and magnetic fields are therefore equal to their perpendicular components. Equations (86) and (88) give

$$\mathbf{E}' = \gamma\left(\mathbf{E} + \frac{\mathbf{E}\times\mathbf{B}}{B^2}\times\mathbf{B}\right) = \gamma(\mathbf{E} - \mathbf{E}) = 0 \tag{90}$$

and

$$\mathbf{B}' = \gamma\left(\mathbf{B} - \frac{\mathbf{E}\times\mathbf{B}}{B^2}\times\mathbf{E}\right) = \gamma\left(\mathbf{B} - \frac{E^2}{B^2}\mathbf{B}\right) = \mathbf{B}\sqrt{1 - \frac{E^2}{B^2}} \tag{91}$$

Thus, in the $x'y'z't'$ reference frame, we have a purely magnetic field. In this reference frame, the y-z motion of a charged particle is uniform circular motion.

What does this motion look like in the laboratory frame? It is circular motion combined with a translational motion of velocity $\mathbf{V} = (\mathbf{E}\times\mathbf{B}/B^2)c$. This gives a cycloidal motion, or, rather, a quasi-cycloidal motion (see Fig. 9.16). The orbits are not exactly cycloids because the relativity of simultaneity and the length contraction destroy the

Figure 9·16

Motion of positive and negative charged particles in constant, crossed electric and magnetic fields with $B > E$. The drift velocity is perpendicular to **E** and to **B**.

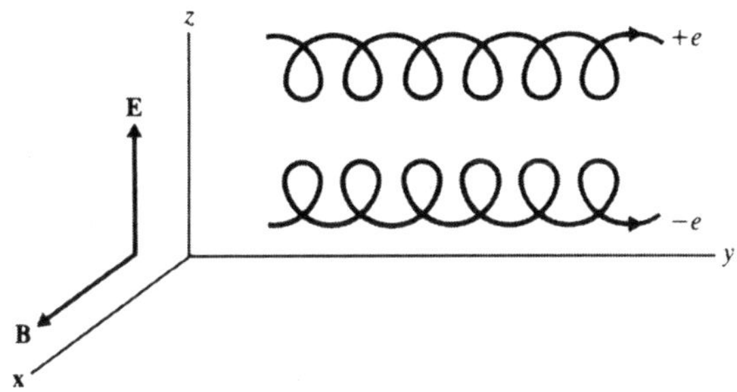

uniform character of the circular motion. If we ignore the gyrations of the particle, we see that the average motion (or the "secular" motion) is a gradual drift in the direction of $\mathbf{E}\times\mathbf{B}$, called the $\mathbf{E}\times\mathbf{B}$ *drift*. Note that the direction of this drift is independent of the sign of the charge. From the point of view of the laboratory frame, the drift results from the interplay of magnetic and electric forces. Qualitatively, what happens to, say, a positively charged particle is shown in Fig. 9.16: At the lowest point of its orbit, the particle has a small velocity, and the magnetic field forces it into a tight arc of small radius of curvature; as the particle moves upward, it gains energy from the electric field; the magnetic field can then not hold the particle in such a tight arc, and the radius of curvature increases; once the particle passes the highest point of its orbit and begins to move downward, it loses energy to the electric field, and the orbit again becomes a tight arc; and so on. The net result is that the particle has an excess motion to the right.

Note that if the velocity of the particle coincides with the velocity of the $x'y'z't'$ frame $[\mathbf{u} = \mathbf{V} = (\mathbf{E} \times \mathbf{B}/B^2)c]$, then the particle remains permanently at rest in this reference frame. In the laboratory frame, such a particle moves on a straight line with a velocity equal to the drift velocity.

■ ***Exercise 7.*** Consider the case $E > B$. What value must you assign to **V** to make $\mathbf{B}' = 0$? Described the motion of the particle qualitatively. ■

9·6 Magnetic Forces on Currents

When a wire carrying a current is immersed in a magnetic field, it will experience a force. Since the current consists of moving charges, we can calculate the force on the wire by integrating the forces acting on these moving charges. Consider a small segment of the wire; the current in this segment consists of charge dq moving with velocity $\mathbf{u}$. The magnetic force on this charge is

$$d\mathbf{F} = dq\frac{\mathbf{u}}{c}\times\mathbf{B} = \frac{1}{c}dq\frac{d\mathbf{x}}{dt}\times\mathbf{B} = \frac{1}{c}\frac{dq}{dt}d\mathbf{x}\times\mathbf{B} \tag{92}$$

Hence the magnetic force on the segment $d\mathbf{x}$ can be expressed as

$$d\mathbf{F} = \frac{I}{c}d\mathbf{x}\times\mathbf{B} \tag{93}$$

The net force on the wire is the integral of this equation:

$$\mathbf{F}_{\text{on}\,I} = \frac{I}{c}\int d\mathbf{x}\times\mathbf{B} \tag{94}$$

For a closed loop, $\int d\mathbf{x} = 0$; hence Eq. (94) tells us that the net force on a closed loop in a constant magnetic field is zero. However, such a loop will experience a torque. We can calculate the torque on a small segment from Eq. (93):

$$d\boldsymbol{\tau} = \mathbf{x}\times d\mathbf{F} = \frac{I}{c}\mathbf{x}\times(d\mathbf{x}\times\mathbf{B}) \tag{95}$$

so the net torque is

$$\boldsymbol{\tau} = \frac{I}{c}\int \mathbf{x}\times(d\mathbf{x}\times\mathbf{B}) \tag{96}$$

For a constant magnetic field, this integral is equal to

$$\boldsymbol{\tau} = \frac{I}{2c}\int (\mathbf{x}\times d\mathbf{x})\times\mathbf{B} \tag{97}$$

To show that the integrals in Eqs. (96) and (97) are indeed the same, we expand the triple cross product in each:

$$\int \mathbf{x}\times(d\mathbf{x}\times\mathbf{B}) = \int \mathbf{x}\cdot\mathbf{B}\,d\mathbf{x} - \int \mathbf{B}\,\mathbf{x}\cdot d\mathbf{x} \tag{98}$$

and

$$\frac{1}{2}\int (\mathbf{x}\times d\mathbf{x})\times\mathbf{B} = \frac{1}{2}\int \mathbf{x}\cdot\mathbf{B}\,d\mathbf{x} - \frac{1}{2}\int \mathbf{x}\,\mathbf{B}\cdot d\mathbf{x} \tag{99}$$

In Eq. (98), the second term on the right side involves the integration of a perfect differential; this gives zero for a closed loop:

$$\int \mathbf{x}\cdot d\mathbf{x} = \frac{1}{2}\int d(\mathbf{x}\cdot\mathbf{x}) = 0 \tag{100}$$

In Eq. (99), the second term on the right side can be integrated by parts; we then recognize that the second term is the same as the first. Thus the right sides of Eqs. (98) and (99) are the same—both are $\int \mathbf{x}\cdot\mathbf{B}d\mathbf{x}$. This establishes that Eqs. (96) and (97) are also the same.

Equation (97) states that the torque on the loop depends on the magnetic moment:

$$\tau = \mathbf{m} \times \mathbf{B} \tag{101}$$

This torque tends to align the magnetic moment with the magnetic field. Note that Eq. (101) is valid only for a constant magnetic field. If the magnetic field is not constant, we must return to Eq. (96).

If the magnetic field **B** that produces the force on a wire is generated by a second wire carrying a current I', then we can combine Eq. (94) with the Biot-Savart law to obtain an expression for the force on the wire carrying the current I:

$$\mathbf{F}_{\text{on } I} = \frac{II'}{c^2}\iint d\mathbf{x}\times\left(d\mathbf{x}'\times\frac{\mathbf{x}-\mathbf{x}'}{|\mathbf{x}-\mathbf{x}'|^3}\right) \tag{102}$$

By a similar argument, we obtain the force on the wire carrying the current I':

$$\mathbf{F}_{\text{on } I'} = \frac{I'I}{c^2}\iint d\mathbf{x}'\times\left(d\mathbf{x}\times\frac{\mathbf{x}'-\mathbf{x}}{|\mathbf{x}'-\mathbf{x}|^3}\right) \tag{103}$$

In Section 9.1, we found that the mutual magnetic forces of moving charges do not obey Newton's Third Law. We blamed this on a transfer of momentum to the electric and magnetic fields. What about the mutual magnetic forces of current loops? A superficial inspection of Eqs. (102) and (103) suggests that these forces, again, do not obey Newton's Third Law. But some clever manipulation reveals that for closed loops, the integrals in these equations can be expressed in the form

$$-\iint \frac{\mathbf{x}-\mathbf{x}'}{|\mathbf{x}-\mathbf{x}'|^3}d\mathbf{x}\cdot d\mathbf{x}' \quad \text{and} \quad -\iint \frac{\mathbf{x}'-\mathbf{x}}{|\mathbf{x}'-\mathbf{x}|^3}d\mathbf{x}'\cdot d\mathbf{x} \tag{104}$$

from which it is obvious that $\mathbf{F}_{\text{on } I} = -\mathbf{F}_{\text{on } I'}$. This result is of some importance because, in the case of steady currents flowing in closed loops, we cannot blame a failure of the Third Law on the fields (as we could in Exercise 1)—the fields are *constant*, and their momentum, if any, is also constant.

To change the integral in Eq. (102) into the form (104), we begin by expanding the triple vector product:

$$\iint d\mathbf{x}\times\left(d\mathbf{x}'\times\frac{\mathbf{x}-\mathbf{x}'}{|\mathbf{x}-\mathbf{x}'|^3}\right)=\iint d\mathbf{x}'\frac{(\mathbf{x}-\mathbf{x}')\cdot d\mathbf{x}}{|\mathbf{x}-\mathbf{x}'|^3}-\iint\frac{\mathbf{x}-\mathbf{x}'}{|\mathbf{x}-\mathbf{x}'|^3}d\mathbf{x}\cdot d\mathbf{x}' \qquad (105)$$

Since $(\mathbf{x}-\mathbf{x}')/|\mathbf{x}-\mathbf{x}'|^3=-\nabla(1/|\mathbf{x}-\mathbf{x}'|)$, the first term on the right side of this equation is

$$-\iint d\mathbf{x}'\nabla\frac{1}{|\mathbf{x}-\mathbf{x}'|}\cdot d\mathbf{x} \qquad (106)$$

By Stokes' theorem, the integral around the loop I can be expressed as a surface integral over the area bounded by this loop. Hence (106) becomes

$$-\iint d\mathbf{x}'\left(\nabla\times\nabla\frac{1}{|\mathbf{x}-\mathbf{x}'|}\right)\cdot d\mathbf{S}$$

But this is zero because the curl of any gradient is identically zero. The only term remaining on the right side of Eq. (105) is then the second term. This term is what we have in Eq. (104).

9·7 Magnetic Monopoles

A magnetic monopole is the hypothetical magnetic analog of an electric charge—the monopole is a source of magnetic field lines just as the electric charge is a source of electric field lines. Figure 9.17 shows the magnetic field lines of a monopole. If we designate the magnetic charge by g, then the magnetic field of a monopole located at the origin is

$$\mathbf{B}=g\frac{\mathbf{x}}{|\mathbf{x}|^3} \qquad (107)$$

This is the obvious magnetic analog of the familiar formula $\mathbf{E}=q\mathbf{x}/|\mathbf{x}|^3$ for the Coulomb field.

Magnetic monopoles were first proposed by P. A. M. Dirac in 1931.[1] Of late, there has been a resurgence of interest in magnetic monopoles because they are a collateral feature of modern grand unified theories (GUTs) of the weak, electromagnetic, and strong interactions; these theories lie within the realm of quantum physics, and we cannot discuss them here. Within the realm of classical physics, the motivation

Figure 9·17

Magnetic field lines of a positive magnetic monopole.

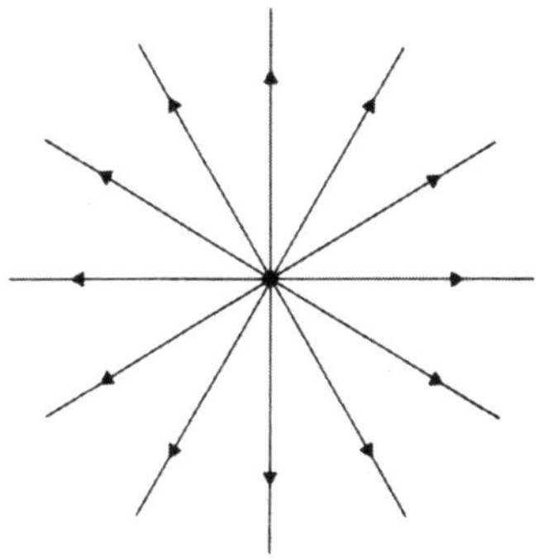

for introducing monopoles into electromagnetic theory arises from a lack of symmetry in Maxwell's equations [Eqs. (8.61)–(8.64)]. Inspection of these equations shows that the electric and magnetic fields enter almost symmetrically—time-dependent magnetic fields induce electric fields, and time-dependent electric fields induce magnetic fields. But the symmetry between these fields is destroyed by the presence of electric charge and current densities in the equations, or, rather, by the *absence* of magnetic charge and current densities. This suggests that we should restore the symmetry between the electric and magnetic fields by including a magnetic charge density ρ_m and current density $\mathbf{j}_m$ in Maxwell's equations:

$$\nabla \cdot \mathbf{E} = 4\pi\rho \tag{108}$$

$$\nabla \times \mathbf{B} = \frac{4\pi}{c}\mathbf{j} + \frac{1}{c}\frac{\partial \mathbf{E}}{\partial t} \tag{109}$$

$$\nabla \cdot \mathbf{B} = 4\pi\rho_m \tag{110}$$

$$\nabla \times \mathbf{E} = -\frac{4\pi}{c}\mathbf{j}_m - \frac{1}{c}\frac{\partial \mathbf{B}}{\partial t} \tag{111}$$

Now, the equations are symmetric between **E** and **B**. Mathematically, we can state this symmetry as follows: The equations are invariant under the substitution

$$\mathbf{E} \to \mathbf{B} \qquad \mathbf{j} \to \mathbf{j}_m \qquad \rho \to \rho_m \tag{112}$$

$$\mathbf{B} \to -\mathbf{E} \qquad \mathbf{j}_m \to -\mathbf{j} \qquad \rho_m \to -\rho \tag{113}$$

This substitution is called a *duality transformation.* Because of the invariance of the field equations under this transformation, there is no physical distinction between a system containing purely electric charges and a system containing purely magnetic charges—the only difference between such systems is what we call "electric" and what we call "magnetic."

Although the above generalization of Maxwell's equations seems perfectly straightforward, it contains some hidden obstacles that emerge when we try to express the magnetic field as the curl of the vector potential according to

$$\mathbf{B} = \nabla \times \mathbf{A} \tag{114}$$

To see why this leads to trouble, we integrate Eq. (110) over the volume of a sphere within which there is a magnetic monopole g. According to Gauss' theorem, we obtain

$$\int \mathbf{B} \cdot d\mathbf{S} = 4\pi g \tag{115}$$

or

$$\int \nabla \times \mathbf{A} \cdot d\mathbf{S} = 4\pi g \tag{116}$$

Now consider the left side of this equation. We can evaluate the integral of $\nabla \times \mathbf{A}$ over the closed spherical surface by first evaluating it over an open surface consisting of the spherical surface minus a small circle of radius δ (see Fig. 9.18); and we then let δ approach zero. According to Stokes' theorem, the integral of $\nabla \times \mathbf{A}$ over this "opened" spherical surface equals the integral of $\mathbf{A}$ around the small circle,

Figure 9·18

Spherical surface of integration for Eq. (116). A small circle of radius δ is missing from the spherical surface.

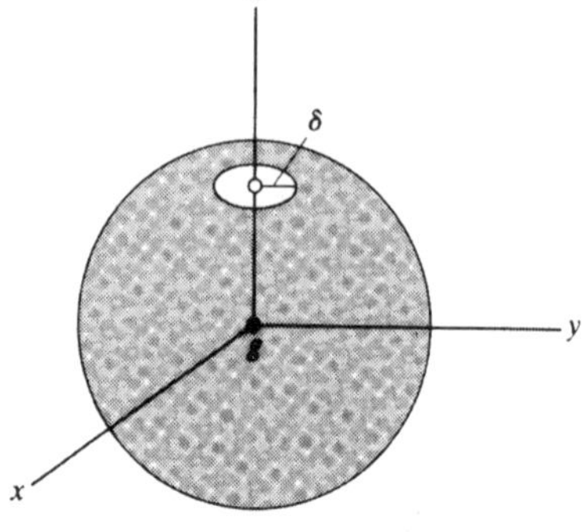

$$\int \nabla \times \mathbf{A} \cdot d\mathbf{S} = \oint_{\delta \to 0} \mathbf{A} \cdot d\mathbf{l} \qquad (117)$$

If **A** is free of singularities, then **A** approaches some constant as δ approaches zero. The right side of Eq. (117) then reduces to $\mathbf{A} \cdot \oint d\mathbf{l}$, which is zero; that is,

$$\int \nabla \times \mathbf{A} \cdot d\mathbf{S} = 0 \qquad (118)$$

But this contradicts Eq. (116)!

To avoid this trouble, we must admit the existence of singularities in **A**(**x**): on any closed surface surrounding a magnetic monopole, there must be at least one point at which **A** is singular in such a way that the path integral of **A** around the singularity gives $4\pi g$ as $\delta \to 0$. If we imagine that the magnetic monopole is surrounded by a sequence of concentric spherical surfaces of increasing radius, we recognize that the singular points on these surfaces form a line extending from the monopole to infinity (see Fig. 9.19); this line of singularities is called the *Dirac string* of the monopole.

■ ***Exercise 8.*** Show that for $0 < \theta \leq \pi$, the vector potential

$$\mathbf{A} = g \frac{1 + \cos\theta}{r \sin\theta} \hat{\boldsymbol{\phi}} \qquad (119)$$

yields the magnetic field (107) for a monopole. Show that with this choice of **A**, the Dirac string lies along the positive z axis. (There are other choices of **A** giving the same **B**.) ■

Incidentally: The vector potential (119) is that of an infinitesimally thin solenoid lying along the positive z axis and extending from the origin to

Figure 9·19

Dirac string of a monopole.

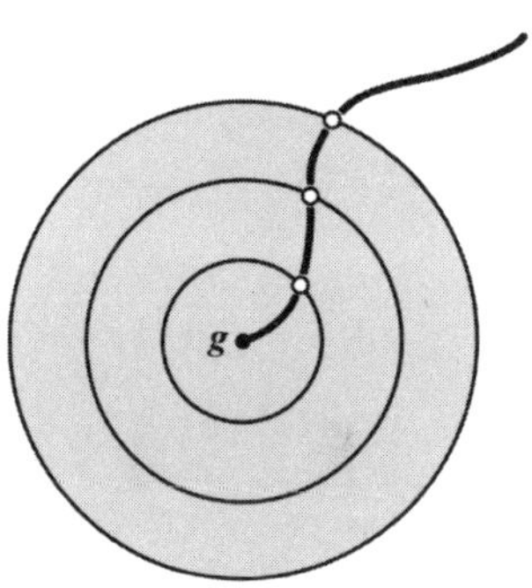

infinity. For such an infinitesimally thin solenoid, the field lines emerging at the lower end [see Figs. 9.20(*a*) and 9.20(*b*)] are radial—they mimic a magnetic monopole. This agreement between the fields of the infinitesimally thin solenoid and the magnetic monopole breaks down on the positive z axis; there, the field of the solenoid is infinite, whereas the monopole field remains finite. The Dirac string is the locus of the points of disagreement; its position is the same as that of the solenoid.

It is worth emphasizing that the magnetic field on the Dirac string is well behaved; only the vector potential is singular. In the context of classical electrodynamics, only the magnetic field is observable, and therefore the mathematical singularity in the vector potential has no physical implications and can be ignored. However, in quantum theory, the vector potential is observable—it affects the phases of quantum-mechanical wave functions and produces interference effects in these waves—and the singularity must be dealt with. It turns out that if the electric and monopole charges are quantized (see below), then the phase difference contributed by the string is always a multiple of 2π, which is unobservable.[2]

EXAMPLE: *fields of a magnetic monopole and an electric charge*

Perhaps the most fascinating consequence of the theory of magnetic monopoles is the quantization condition for electric charge, which emerges when we examine the angular momentum in the electromagnetic fields. Consider a magnetic monopole g and an electric charge q some distance apart (see Fig. 9.21). According to Eq. (8.106), the density of momentum in the fields is $\mathbf{E} \times \mathbf{B}/4\pi c$. This vector is perpendicular to $\mathbf{E}$ and $\mathbf{B}$; hence, in Fig. 9.21, the momentum flows in circles around the z axis, giving rise to an angular momentum along the z axis. This angular momentum is

Figure 9·20

(*a*) Field lines emerging from the end of a thin solenoid. (*b*) Limiting case of a solenoid of infinitesimal thickness.

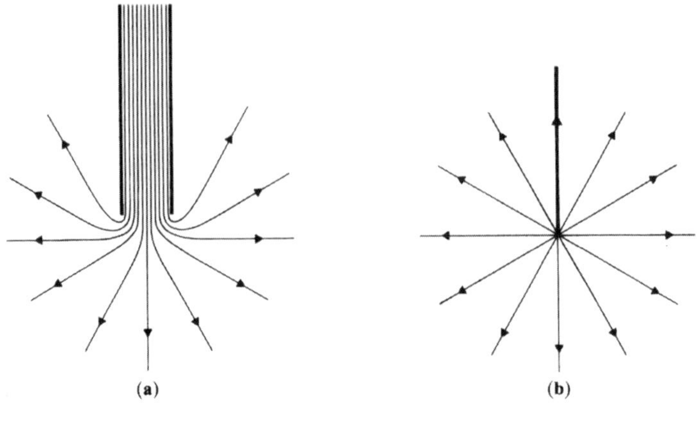

Figure 9·21

A magnetic monopole g at the origin and an electric charge q at some distance above.

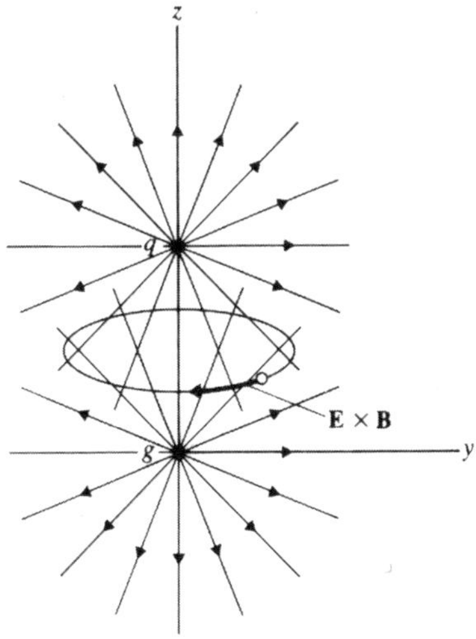

$$\mathbf{J}=\int \mathbf{r}\times\left(\frac{\mathbf{E}\times\mathbf{B}}{4\pi c}\right)dV=\frac{1}{4\pi c}\int(\mathbf{E}\,\mathbf{r}\cdot\mathbf{B}-\mathbf{B}\,\mathbf{r}\cdot\mathbf{E})\,dV \qquad (120)$$

Expressing this in component notation and inserting $B^k = gx^k/|\mathbf{x}|^3$, we obtain

$$J^k=\frac{g}{4\pi c}\int\left(E^k\frac{1}{|\mathbf{x}|}-\frac{x^k x^n}{|\mathbf{x}|^3}E^n\right)dV=\frac{g}{4\pi c}\int E^n\frac{\partial}{\partial x^n}\frac{x^k}{|\mathbf{x}|}dV \qquad (121)$$

Upon integration by parts, this becomes

$$J^k=-\frac{g}{4\pi c}\int\frac{x^k}{|\mathbf{x}|}\frac{\partial}{\partial x^n}E^n\,dV \qquad (122)$$

But $\partial E^n/\partial x^n = \nabla\cdot\mathbf{E} = 0$ at all points except at the position of the electric charge. Thus the integration in Eq. (122) can be restricted to a very small sphere surrounding the electric charge. Within this sphere, $x^k/|\mathbf{x}|$ is nearly constant, and its only nonvanishing component is $z/|\mathbf{x}| = z/z = 1$. We then obtain

$$J_z=-\frac{g}{4\pi c}\int\nabla\cdot\mathbf{E}\,dV \qquad (123)$$

which, by Gauss' law, equals

$$J_z = -\frac{gq}{c} \tag{124}$$

Thus the angular momentum in the fields is independent of the distance between the charges.

To obtain a quantization condition, we must trespass into the realm of quantum mechanics. There, we come under the sway of the well-known rule of quantization of angular momentum: the z component of angular momentum is either an integral multiple of $\hbar$ or, more generally, a half-integral multiple of $\hbar$. Since the angular momentum (124) is not an orbital angular momentum, it seems reasonable to adopt half-integral multiples:

$$J_z = \tfrac{1}{2} n\hbar \qquad n = 0, \pm 1, \pm 2, \pm 3, \ldots \tag{125}$$

or

$$\frac{gq}{c} = \frac{1}{2} n\hbar \tag{126}$$

This result is called the *Dirac quantization condition.* We have obtained it by a rather simple-minded, nonrigorous argument, but it can be confirmed by a more thorough quantum-mechanical analysis.[2]

From Eq. (126), we can draw the remarkable conclusion that both the electric charge and the magnetic charge are quantized: for a given g, q is proportional to the integer n; and for a given q, g is proportional to the integer n. Thus the magnetic monopole serves as an explanation of the quantization of electric charge. In the recently developed grand unified theories of the strong, electromagnetic, and weak interactions, the explanation of the quantization of electric charge relies on a somewhat more sophisticated version of our naive argument. The quantum field equations of these theories permit monopole solutions, and the consistency of the equations then demands Eq. (126); however, in contrast to our naive argument, these theories do not require the actual existence of monopoles in our universe.

If we substitute $q = e$ and $n = 1$ into Eq. (126), we find the fundamental unit of magnetic charge:

$$g = \frac{\hbar c}{2e} \tag{127}$$

From this, we see that for a given distance, the repulsive magnetic force between two magnetic monopoles is stronger than the repulsive electric force between two electrons by the factor

$$\frac{g^2}{e^2} = \frac{1}{4}\left(\frac{\hbar c}{e^2}\right)^2 \simeq \frac{1}{4}(137)^2 \simeq 5\times 10^3 \qquad (128)$$

The strength of the monopole forces is much larger than that of the electric forces! We can also recognize the larger strength of the monopole force by comparing the force on a monopole in a given magnetic field with force on an electron in an electric field of the same magnitude ($E = B$); the force on the monopole is then larger by a factor of 137/2.

Experimental searches for magnetic monopoles have focused on naturally occurring monopoles because the prospects of producing monopoles in the laboratory are very poor—the mass of a monopole is probably in excess of 10^{16} GeV/c^2, much larger than anything we can produce even with the most powerful of our available accelerators. Theoretical estimates suggest that a substantial number of monopoles may have been created during the very early stages of the Big Bang; furthermore, the impact of energetic cosmic rays on the Earth can presumably create monopoles. The magnetic field of the Earth would pull the monopoles down to the surface, and it seems likely that they would accumulate at the bottoms of the oceans, preferentially attaching themselves to ferromagnetic materials. Most of the experiments sought to extract monopoles from samples of materials by the application of a strong magnetic field. This would not only pull the monopole out of the sample but would also give it a substantial kinetic energy. The monopole could then be detected by the ionization trails it leaves when passing through a photographic emulsion or a plastic sheet. Unfortunately, none of the experimental searches have found any monopoles. If they exist, they are either very rare or very well hidden. The upper limit set by such experiments for monopole abundance in iron ores is 2×10^{-7} monopole per gram of ore.[3]

Searches for monopoles in cosmic rays have also been unsuccessful. Such searches have set a tight limit on the flux of magnetic monopoles reaching the Earth: no more than 10^{-16} monopole per cm^2 per second per unit solid angle.[4] A similar limit can be deduced from the persistence of the Galactic magnetic field. If there were an abundance of free monopoles in interstellar space, then they would proceed to cancel the magnetic field, in the same way that the free electric charges in a conductor proceed to cancel the electric field. The persistence of the Galactic magnetic field implies that the monopole flux is less than about 10^{-16} per cm^2 per second per unit solid angle.[5] A tighter, but less reliable, limit on monopole abundance can be obtained from theoretical predictions of the reactions of monopoles with neutrons in neutron stars. According to grand unified theory, the collision of a monopole with a proton or a neutron is likely to catalyze the decay of the proton or neutron into leptons, with the release of a large amount of energy. This means that monopoles captured by neutron stars would induce a high luminosity in these stars, mainly in the form of X rays. The observational limit on the X-ray luminosity in the sky therefore leads to a limit on the monopole flux, roughly 10^{-23} per cm^2 per second per unit solid angle.[6]

References

1. P. A. M. Dirac, *Proc. R. Soc. London, Ser.* A **133**, 60 (1931).
2. P. Goddard and D. I. Olive, *Rep. Prog. Phys., Phys. Soc. London* **41**, 1357 (1978).
3. T. Ebisu and T. Watanabe, *Phys. Rev.* D **36**, 3350 (1987).
4. M. Ambrosio et al., *Eur. Phys. J.* C **25**, 511 (2002).
5. F. C. Adams et al., *Phys. Rev. Lett.* **70**, 2511 (1993).
6. E. W. Kolb and M. S. Turner, *Astrophys. J.* **286**, 702 (1984).

Problems

1. Show that, for an extremely relativistic particle ($v \simeq c$), the magnetic field [Eq. (3)] and the electric field [Eq. (8.46)] are of nearly equal magnitudes. Show that the directions of these fields are nearly perpendicular and that $\mathbf{E} \times \mathbf{B}$ is in the direction of motion. How does this compare with the electric and magnetic fields of an electromagnetic wave?

2. Two electrons have equal instantaneous speeds of 3×10^8 cm/s. One electron is moving outward along the x axis; its instantaneous position is $x = 4\times10^{-9}$ cm. The other electron is moving outward along the y axis; its instantaneous position is $y = 0$. What is the magnetic force of the first electron on the second? Of the second electron on the first? What is the instantaneous rate of change of the momentum of the two-electron system?

3. Consider the electron in the smallest Bohr orbit, of radius 0.0529 nm, in a hydrogen atom. What is the magnitude of the magnetic field that this electron produces at the position of the nucleus? Pretend that the motion of the electron is classical.

4. Show that the magnetic force that a uniformly moving point charge q exerts on a point charge q' is never larger than the electric force. Use the exact, relativistic expression for the magnetic field [Eq. (3)] and the corresponding expression for the electric field.

5. In cylindrical coordinates, the electric potential inside and outside the cylindrical conductor shown in Fig. 9.3 is

$$\Phi = -E_0 z \qquad \text{for} \quad \rho \le R$$

$$\Phi = -E_0 z \frac{\ln(\rho / R')}{\ln(R / R')} \qquad \text{for} \quad \rho \ge R$$

where R is the radius of the cylindrical conductor and R' is the radius of the concentric shell at zero potential.

(a) Show that this potential is a solution of the Laplace equation, and show that it satisfies the boundary conditions at $\rho = R$ and at $\rho = R'$.

(b) Find the electric field just outside of the surface $\rho = R$ of the cylinder.

(c) Find the surface charge density on the surface of the cylinder.

6. A current I flows around a square loop of dimension $L \times L$. Find the magnetic field at a distance z from the center of the loop in a direction perpendicular to the plane of the loop.

7. A current I flows around a square loop of dimension $L \times L$. The sides of the loop are vertical or horizontal. Find the vector potential, and find the magnetic field in the plane inside the loop at a horizontal distance x and a vertical distance y from the center.

8. Two very large, thin sheets of conducting material intersect at right angles. The sheets lie in the x-z plane and the y-z plane, respectively. On each sheet, a surface current of k statamps/cm flows in the z direction. Find the magnetic field in each of the four quadrants between the sheets.

9. A thick solenoid has an inner radius a, an outer radius b, and a finite length L. The total current circulating around the solenoid is I; this current is uniformly distributed over the volume. Find the magnetic field on the axis of the solenoid, inside the solenoid

10. Consider the two parallel, current-carrying plates described in Exercise 4. What is the force per unit area, or the pressure, that the magnetic field of each plate exerts on the other? Show that this pressure can be expressed as $B^2/8\pi$.

11. A current I flows in a long straight wire of circular cross section, length Δl, with a voltage $\Delta\Phi$ and electric field $E = \Delta\Phi / \Delta l$. From this electric field and the magnetic field (25), calculate the Poynting vector at the surface of the wire and show that the rate at which electromagnetic energy flows into the wire per unit length equals the rate of Joule heating per unit length.

12. (a) From the equations $\nabla \times \mathbf{E} = 0$ and $\nabla \times \mathbf{B} = 0$ satisfied by the static electric and magnetic fields in empty space, show that the Poynting vector for such fields has zero divergence, $\nabla \cdot \mathbf{S} = 0$. This means the energy flow in such fields can be represented graphically by flow lines.

(b) Consider the electric and magnetic fields surrounding the central cylindrical conductor in Fig. 9.3. The field lines of the electric field are displayed in this figure, and the field lines of the magnetic field are concentric circles around the conductor. On a copy of Fig. 9.3, roughly sketch the flow lines of the energy flow in these electric and magnetic fields outside the conductor, and sketch the direction of the energy flow inside the conductor (**S** cannot be represented by flow lines inside the conductor, because $\nabla \cdot \mathbf{S} \neq 0$ there, but the direction of **S** does give the direction of the energy flow).

13. A toroid consists of many coils of wire tightly wound around the surface of a torus (see Fig. 9.22). The current in the coils is I and total number of coils is N. Show that the magnetic field within the toroid is

Figure 9·22

A toroid.

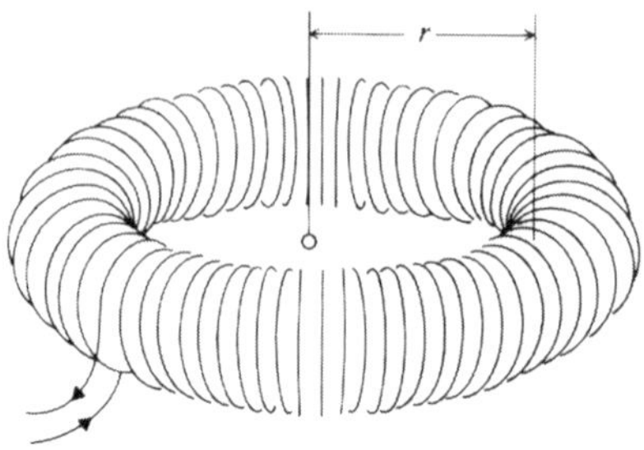

$$B = \frac{2NI}{cr}$$

Does this result depend on the shape of the cross section of the toroid?

14. A long cylindrical conductor of radius R has a cylindrical cavity of radius R' cut out of its volume; the axis of the cavity is parallel to the central axis but is displaced a distance h from the center (see Fig. 9.23). The conductor carries a current I, uniformly distributed over its remaining volume. Find the magnetic field in the cavity.

15. A long solenoid of rectangular cross section lies along the z axis. The magnitude of the magnetic field within the solenoid is B_0.

(a) Find all components of the Maxwell stress tensor [see Eq. (8.94)].

(b) From the stress tensor, deduce the magnetic force per unit area acting on the walls of the solenoid.

16. Show that the magnetic scalar potential for a closed loop of wire carrying a current I can be expressed as

$$\Phi_M(\mathbf{x}) = \frac{I\Omega}{c}$$

where Ω is the net solid angle subtended by the closed loop as seen from the point x. [*Hint:* A large current loop can be regarded as made of many small adjacent loops

Figure 9·23

A cylindrical conductor with a cavity.

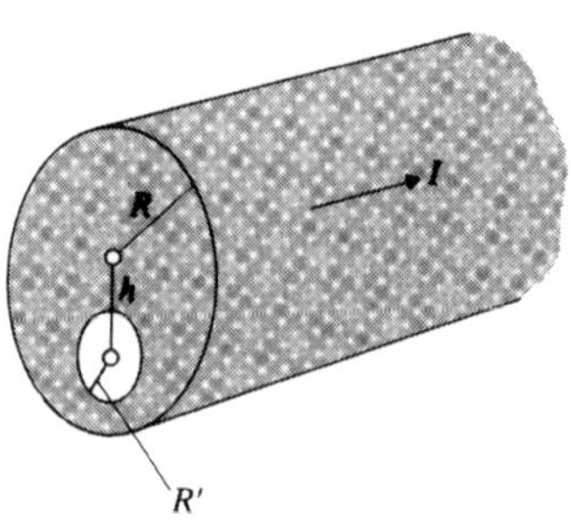

filling its area. For a small loop of radius R, Eq. (61) tells us that the magnetic potential is approximately, except for an irrelevant additive constant,

$$d\Phi_{\mathrm{M}}(\mathbf{x}) = \frac{2\pi I}{c}\frac{R^2}{2r^2}\cos\theta$$

Express this in terms of the solid angle $d\Omega$ subtended by the small loop.]

17. A pair of *Helmholtz coils* are often used in laboratories to produce fairly uniform magnetic fields. The coils consist of two rings of radius R on a common axis separated by a distance R (see Fig. 9.24). The rings carry equal currents I in the same direction. Suppose that the coils are parallel to the x-y plane and that their centers are at $z = \pm R/2$.

(a) Find the magnetic field B_z on the z axis.

(b) Show that at $z = 0$, the first, the second, and the third derivatives dB_z/dz, d^2B_z/dz^2, and d^3B_z/dz^3 vanish. Show that the fourth derivative d^4B_z/dz^4 does not vanish.

18. Find the magnetic dipole moment of a disk of paper of radius R with charge Q uniformly distributed over the surface spinning at angular velocity ω.

19. A sphere of radius R has a charge Q uniformly distributed over its volume. The sphere spins about a diameter at angular velocity ω.

(a) Calculate the magnetic field at the center of the sphere.

(b) Calculate the magnetic field on the axis of rotation at a large distance from the sphere. What is the magnetic dipole moment?

20. Show that the vector potential for an ideal magnetic dipole is $\mathbf{A} = \mathbf{m}\times\mathbf{x}/|\mathbf{x}|^3$.

21. Prove that if an ideal magnetic dipole of moment $\mathbf{m}$ moves with a uniform, nonrelativistic velocity $\mathbf{V}$, the associated electric field is that of an electric dipole of moment

$$\mathbf{p} = \frac{1}{c}\mathbf{V}\times\mathbf{m}$$

Figure 9·24

Helmholtz coils.

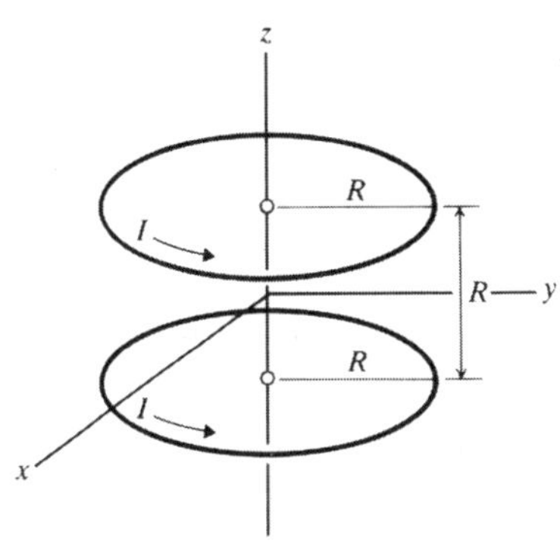

22. The magnetic field of the Earth is a dipole field. At the magnetic equator at the surface of the Earth, the field has a magnitude of 0.31 gauss.
   (a) Find the value of the magnetic moment.
   (b) What are the horizontal and vertical components of the magnetic field at Austin, Texas? Assume that the radius of the Earth is 6.4×10^3 km and that the magnetic latitude of Austin is 40°.

23. Consider the magnetic field of a magnetic dipole. Show that the equation of a field line that passes through the equatorial plane at $r = r_0$ is $r = r_0 \sin^2\theta$, where θ is the usual polar angle measured from the axis of the dipole. [*Hint:* Along a field line, $rd\theta / dr = B_\theta / B_r$. Why?]

24. A cylinder of radius R and length L spins about its axis with angular velocity ω. The cylinder has a charge Q uniformly distributed over its volume.
   (a) Find the magnetic field on the axis of the cylinder, outside the cylinder. Assume that the center of the cylinder is at the origin of coordinates.
   (b) Find the first two terms in a series expansion for the magnetic field at points off the axis; assume $r > R + L/2$.

25. Consider the ring of current described in Section 9.4. Find the magnetic scalar potential and the magnetic field in the region $r < R$.

26. A disk of paper of radius R has a charge Q uniformly distributed over its surface. The disk spins about its axis at angular velocity ω.
   (a) Find the magnetic field on the axis of the disk.
   (b) Find the first two terms in a series expansion for the magnetic field at points off the axis; assume $r > R$.

27. A sphere of radius R carries charge Q uniformly distributed over its surface. Find the magnetic field in the region inside the sphere if the sphere is spinning with angular velocity ω.

28. The magnetic field of the Earth is a dipole field. At the magnetic equator of the Earth, the field points north and has a magnitude of $B = 0.31$ gauss.
   (a) What is the magnetic field at an altitude of 3×10^3 km above the equator?
   (b) What must be the momentum of a proton so that it can remain in a circular orbit about the center of Earth at an altitude 3×10^3 km above the equator? Express your answer in GeV/c.

29. A nonrelativistic particle of charge q and mass m is in a uniform magnetic field $\mathbf{B} = B_0\hat{\mathbf{z}}$. The initial position of the particle is $x = 0$, $y = 0$, $z = 0$; the magnitude of the momentum of the particle is p_0. What must be the direction of the momentum if the particle is to reach the point $x = a < 2cp_0/qB_0$, $y = 0$, $z = 0$ in the least possible time? What is this time?

30. An electron moves in uniform, parallel electric and magnetic fields, $\mathbf{E} = E_0\hat{\mathbf{z}}$ and $\mathbf{B} = B_0\hat{\mathbf{z}}$. Find the position of the electron as a function of time, given the initial conditions $x = a$, $y = 0$, $z = 0$, $v_x = 0$, $v_y = aeB_0/m_e c$, $v_z = 0$. Assume that the motion is nonrelativistic.

31. Prove the following theorem known as *Bleakney's theorem:* If a (nonrelativistic) ion of mass M and initial velocity zero proceeds along some trajectory in given electric and magnetic fields **E** and **B**, then an ion of mass kM and the same charge will proceed along the same trajectory in electric and magnetic fields $\mathbf{E}/k$ and **B**. [*Hint:* Try changing the time scale in the equation of motion for the second ion.]

32. A compass needle is a thin rod of mass 0.5 g and length 2 cm. The magnetic moment of the needle is 9 gauss·cm^3. The needle is free to rotate around a vertical axis through its center. What is the frequency of small oscillations of this compass needle about the north-south direction when it is placed in a horizontal magnetic field of 0.2 gauss?

33. A thin, flexible wire carrying a current I hangs in a uniform magnetic field (see Fig. 9.25). A weight attached to one end of the wire provides a tension T.
(a) Show that within the magnetic field, the wire will adopt the shape of an arc of circle of radius $r = cT/BI$.
(b) Show that this arc of circle coincides with the orbit of a charged particle of momentum $p = qT/I$. Thus the shape of the wire mimics the shape of the orbit; this is useful for the experimental determination of the orbits of charged particles in magnets. Such a wire used for this purpose is called a *hodoscope.*

34. A strip of metal carries a current I at right angles to an external magnetic field **B**. Figure 9.26 shows the directions of the current and of the magnetic field; the flow of free electrons in the strip of metal is, of course, in a direction opposite to that of the current. Prove that under steady-state conditions, there must be a transverse electric field of magnitude vB/c in the metal and, consequently, a potential difference vBd/c across a strip of width d. This is the *Hall effect.* Which edge of the conductor is at the higher potential? Which edge of the conductor would be at the higher potential if the current were a flow of positive charges?

Figure 9·25

Flexible wire hanging in a uniform magnetic field.

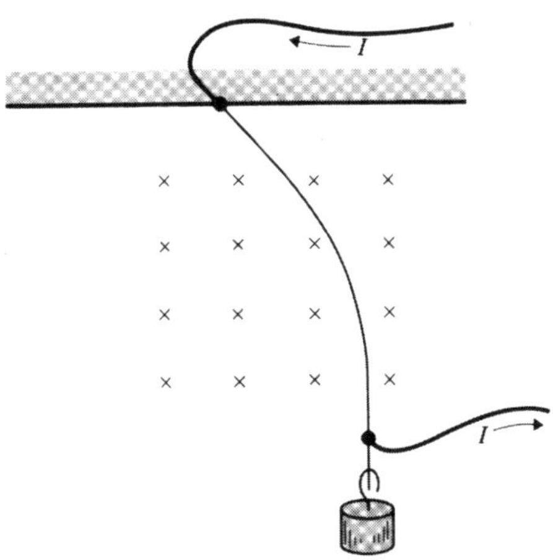

Figure 9·26

Metallic strip carrying a current perpendicular to a magnetic field.

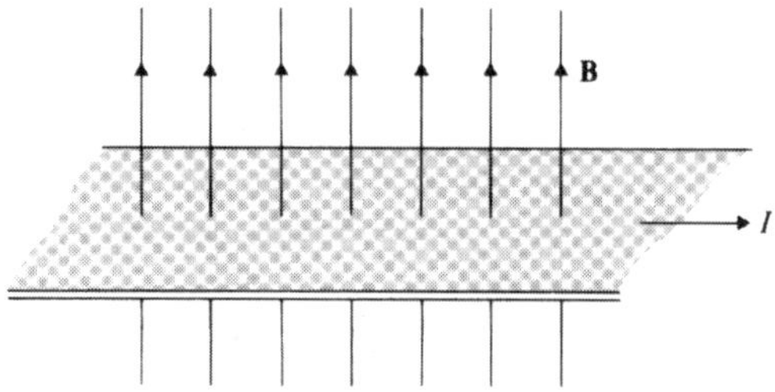

35. A particle of mass m and charge e, initially at rest, is acted upon by a uniform gravitational field (with gravitational acceleration g) and a uniform magnetic field, which is perpendicular to the gravitational field. Find the position of the particle as a function of time. Assume that $mg/e < B$, and assume that the motion is nonrelativistic.

36. The magnetic field of the Earth is approximately that of a magnetic dipole of moment m. If we use a coordinate system r, θ, ϕ centered on the Earth and with z axis in the magnetic north direction, the vector potential is

$$\mathbf{A} = -\frac{m \sin\theta}{r^2}\hat{\boldsymbol{\phi}}$$

Show that for a (relativistic) particle of charge q and momentum $|\mathbf{p}|$ moving in this field, the quantity

$$-r^2 \sin^2\theta \frac{d\phi / dt}{|\mathbf{u}|} + \frac{mq}{|\mathbf{p}|c}\frac{\sin^2\theta}{r}$$

is a constant of the motion. This is *Störmer's theorem.* Show that, for a cosmic ray approaching from large distance in the equatorial plane of the Earth $(\theta = 0)$, the value of this constant is ± the impact parameter. [*Hint:* The Lagrangian is $L = -Mc^2\sqrt{1-u^2/c^2} + q\mathbf{A}\cdot\mathbf{u}/c$. The angle ϕ is an ignorable coordinate.]

37. Show that in four-dimensional tensor notation, Eqs. (108)–(111) take the form

$$\partial_\mu F^{\mu\nu} = \frac{4\pi}{c} j^\nu \qquad \partial_\mu F^{*\mu\nu} = \frac{4\pi}{c} k^\nu$$

where $k^\nu = (c\rho_{\mathrm{m}}, j_{\mathrm{m},x}, j_{\mathrm{m},y}, j_{\mathrm{m},z})$ and $F^{*\mu\nu} = \frac{1}{2}\varepsilon^{\mu\nu\alpha\beta}F_{\alpha\beta}$ (the tensor $\varepsilon^{\mu\nu\alpha\beta}$ is defined in Problem 7.6.)

38. The duality transformation mentioned in Section 9.7 is a special case of the following general duality transformation:

$$\mathbf{E} \to \mathbf{E}\cos\theta + \mathbf{B}\sin\theta \qquad \mathbf{B} \to -\mathbf{E}\sin\theta + \mathbf{B}\cos\theta$$
$$\rho \to \rho\cos\theta + \rho_{\mathrm{m}}\sin\theta \qquad \rho_{\mathrm{m}} \to -\rho\sin\theta + \rho_{\mathrm{m}}\cos\theta$$
$$\mathbf{j} \to \mathbf{j}\cos\theta + \mathbf{j}_m\sin\theta \qquad \mathbf{j}_{\mathrm{m}} \to -\mathbf{j}\sin\theta + \mathbf{j}_{\mathrm{m}}\cos\theta$$

where θ is an arbitrary real parameter. Show that Eqs. (108)–(111) are invariant under this transformation. Because of this invariance, electromagnetic interactions do not make any physical distinction between a system of particles with combined electric and magnetic charges in a fixed (universal) ratio and a system of particles with purely electric charges.

39. By performing a Lorentz transformation on the field of a stationary magnetic monopole, find the magnetic and electric fields of a moving monopole, with a nonrelativistic speed. Describe the electric field lines qualitatively.

40. By manipulating the modified Maxwell equations (108)–(111), derive the following conservation law for the magnetic charge on monopoles:

$$\frac{\partial \rho_{\rm m}}{\partial t} + \nabla \boldsymbol{\cdot} \mathbf{j}_{\rm m} = 0$$

41. If an electric point charge q of mass m moving at a high speed v passes by a stationary magnetic monopole g, the effect of the magnetic force can be described as a transverse impulse, which produces a (small) transverse tangential deflection of the direction of motion. To a first approximation, we can ignore this deflection, and we can pretend that the point charge q moves approximately along a straight line, say the x axis. Assume that the stationary monopole g is on the y axis at $y = -b$. Then the transverse impulse given to q is

$$\int_{-\infty}^{\infty} F_z dt = \frac{q}{c}\int_{-\infty}^{\infty} B_y dx$$

where B_y is the magnetic field produced by g. Use Gauss' law to evaluate the integral $\int B_y dx$, and show that the transverse impulse is $2qg/cb$, in the z direction.

(a) From this, deduce that the point charge acquires an angular momentum L_x = $2qg/c$, and show that the quantization of charge and of angular momentum again leads to the Dirac quantization condition (126).

(b) Deduce that the point charge is deflected in the x direction through an angle $2qg/(mvcb)$. Evaluate this deflection for an alpha particle of 6.0 MeV passing by a magnetic monopole of one unit of magnetic charge with an impact parameter of 1.0×10^{-12} cm.

42. According to theoretical estimates, the mass of a magnetic monopole is around 10^{-8} g. What magnetic field is required to balance a monopole with one unit of magnetic charge against the gravitational force near the surface of the Earth?

43. A magnetic monopole of charge g and mass m, initially at rest, falls from infinity toward the surface of a planet. The planet has a mass M and a magnetic dipole moment m. If the monopole strikes the surface of the planet at a (magnetic) latitude θ, what is its impact speed? Evaluate numerically for the Earth; assume that $g = \hbar c/2e$ and $m = 1\times10^{-8}$ g, and ignore atmospheric friction. The magnetic dipole moment of the Earth is 8.1×10^{25} gauss·cm^3.

44. In consequence of Eq. (111), a monopole current generates an electric field that satisfies an "Ampère's law" of the form

$$\oint \mathbf{E}\boldsymbol{\cdot} d\mathbf{l} = -\frac{4\pi}{c} I_{\rm m}$$

Thus the monopole current generates an emf around any closed path encircling it; and if a wire loop lies along the closed path, the emf will drive an (electric) current around the loop. This phenomenon was exploited in one attempt at detecting monopoles. A sample of Moon rock suspected of containing monopoles was repeatedly propelled (in one direction) through a coil consisting of 1200 loops of wire, and the current in the coil was monitored. The sample was propelled through the coil about once every 4 s. If the sample had contained one monopole with $g = \hbar c / 2e$, what would have been the time-average emf in the coil?

45. Consider the motion of a particle of electric charge q_1 and magnetic charge g_1 in the fields of a particle of electric charge q_2 and magnetic charge g_2.

(a) Show that if the latter particle remains fixed at the origin, the (nonrelativistic) equation of motion of the former particle is

$$m_1 \frac{d\mathbf{v}}{dt} = (q_1 q_2 + g_1 g_2)\frac{\mathbf{x}}{|\mathbf{x}|^3} + (q_1 g_2 - q_2 g_1)\frac{\mathbf{v}}{c}\times\frac{\mathbf{x}}{|\mathbf{x}|^3}$$

(b) Take the cross product of the equation of motion with $\mathbf{x}$ and thereby show that the vector quantity

$$\mathbf{J} = \mathbf{x}\times m_1\mathbf{v} - \frac{1}{c}(q_1 g_2 - q_2 g_1)\frac{\mathbf{x}}{|\mathbf{x}|}$$

is a constant of the motion. This constant of the motion is the angular momentum of the system.

(c) Suppose that the first particle moves with high speed along a nearly straight line, passing by the second particle with some impact parameter. Show that the quantization of the changes in the component of the angular momentum along the line of motion leads to

$$(q_2 g_1 - q_1 g_2) = \tfrac{1}{2} n\hbar c$$

and show that this general quantization condition includes Eq. (126) as a special case.

46. Consider a pair of capacitor plates of arbitrary shape and separation immersed in a homogeneous medium of conductivity σ and dielectric constant ε. If the capacitor plates are charged, a current will flow from the positive plate to the negative. Show that the resistance that the medium offers to the current is related to the capacitance by

$$RC = \frac{\varepsilon}{4\pi\sigma}$$

[*Hint*: Evaluate $\int \mathbf{E}\cdot d\mathbf{S}$ along a surface passing between the plates and extending sufficiently far out to capture all the electric field lines. By Gauss' law, this integral is $4\pi Q/\varepsilon$, but in terms of the current, it is $(1/\sigma)\int \mathbf{j}\cdot d\mathbf{S} = I/\sigma$; this gives a proportionality between the charge on the capacitor and the current in the medium.] The above relation between resistance and capacitance can be used to calculate one of these from an experimentally measured value of the other. Note that it is not necessary to use the same medium as dielectric and as current carrier. If we first immerse the capacitor plates in a medium with dielectric constant and afterward immerse them in a medium with conductivity, the result is still valid because it does not depend on the actual strength of the electric field, only on the similar distribution of electric field lines [which is ensured by the arguments given in connection with Eq. (15).]

CHAPTER

Electromagnetic Induction 10

Besides the electric fields generated by static charges, there also exist electric fields generated by time-dependent magnetic fields; these electric fields are called *induced fields*. In this chapter, we will derive the laws governing induction from Maxwell's equations, and we will work out some examples involving applications of these laws.

By allowing the fields a time dependence, we leave the narrow stage of electrostatics and magnetostatics, and we enter the wide arena of electrodynamics. Not only do time-dependent magnetic fields induce electric fields, but, conversely, time-dependent electric fields induce magnetic fields. These mutual induction effects play a crucial role in electromagnetic waves and in electromagnetic cavity oscillations, topics we will cover in Chapters 12 and 15.

10·1 Faraday's Law of Induction

From Chapter 2, we know that the path integral of an electrostatic field around any closed path is zero. In contrast, the path integral of an induced electric field around a closed path is not zero. This integral is called the *electromotive force,* or emf, for the closed path. Faraday's law relates the emf for a closed path to the rate of change of magnetic flux intercepted by the area bounded by the path.

For the sake of simplicity, we will first deal with the special case of a stationary closed path. Figure 10.1 shows such a path immersed in an arbitrary, time-dependent magnetic field. The induced emf is defined as

$$\mathcal{E} = \oint \mathbf{E}\cdot d\mathbf{l} \qquad (1)$$

If the path coincides with a conducting wire, then this emf represents a driving voltage that forces a flow of current around the wire. However, the definition (1) is valid for any arbitrary closed path, regardless of whether it coincides with a conducting wire. To derive Faraday's law, we begin with the Maxwell equation (8.64),

Figure 10·1

A stationary closed path in a time-dependent magnetic field.

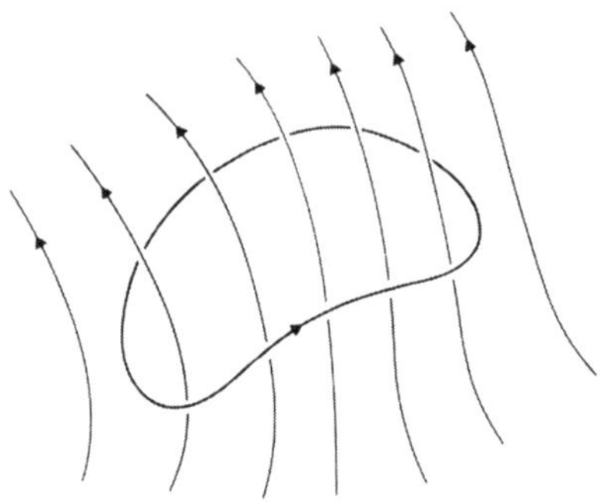

$$\nabla \times \mathbf{E} = -\frac{1}{c}\frac{\partial \mathbf{B}}{\partial t}$$

Integrating both sides of this equation over the area S bounded by the path, we obtain

$$\int \nabla \times \mathbf{E} \cdot d\mathbf{S} = -\frac{1}{c}\int_S \frac{\partial \mathbf{B}}{\partial t} \cdot d\mathbf{S} \tag{2}$$

On the left side, we can use Stokes' theorem to convert the surface integral into a path integral. On the right side, since the surface is fixed and the only time dependence occurs in $\mathbf{B}$, we can take the time derivative out of the integral. This yields

$$\oint \mathbf{E} \cdot d\mathbf{l} = -\frac{1}{c}\frac{d}{dt}\int_S \mathbf{B} \cdot d\mathbf{S} \tag{3}$$

The integral $\int \mathbf{B} \cdot d\mathbf{S}$ is called the *magnetic flux* intercepted by the surface,

$$\Phi_B = \int_S \mathbf{B} \cdot d\mathbf{S} \tag{4}$$

Hence the induced emf is

$$\mathcal{E} = -\frac{1}{c}\frac{d\Phi_B}{dt} \tag{5}$$

This is *Faraday's law:* The induced emf is proportional to the rate of change of the magnetic flux.

Note that in Eq. (3), the direction of integration for the path integral is related to the direction of the normal vector for the surface integral by a *right-hand rule*, as required by Stokes' theorem: If the fingers curl around the path in the direction of $d\mathbf{l}$, then the thumb lies along the direction of $d\mathbf{S}$. This rule is one version of *Lenz' law*; it determines the signs of the induced emf and of the magnetic flux in Eq. (5). The vulgar version of Lenz' law asserts that the induced current (if any) produces a magnetic field whose flux opposes the original change in flux. As we will see in the following examples, these versions of Lenz' law are equivalent.

Next, we will deal with the general case of a moving closed path. The motion of the path may be either a rigid translational motion, or a deformation, or a combination of both. The definition (1) of the emf for such a path must be modified. The emf must now be defined in terms of the electric field $\mathbf{E}'$ measured in the instantaneous rest frame of the segment $d\mathbf{l}$ of the path,

$$\mathcal{E} = \oint \mathbf{E}' \cdot d\mathbf{l} \tag{6}$$

If the velocity of the segment $d\mathbf{l}$ is $\mathbf{v}$, then the Lorentz-transformation equations for the electric field permit us to express the field $\mathbf{E}'$ in the rest frame as a function of the fields $\mathbf{E}$ and $\mathbf{B}$ in the laboratory frame. With approximation $\gamma \simeq 1$, the equations (8.43) for $\mathbf{E}'_{\parallel}$ and $\mathbf{E}'_{\perp}$ can be combined into

$$\mathbf{E}' \simeq \mathbf{E} + \frac{\mathbf{v}}{c} \times \mathbf{B} \tag{7}$$

The approximation $\gamma \simeq 1$ is satisfactory because, in most practical applications, the moving path coincides with a wire or some other conductor moving with a nonrelativistic velocity. We can then write the emf as

$$\varepsilon = \oint \left(\mathbf{E} + \frac{\mathbf{v}}{c} \times \mathbf{B} \right) \cdot d\mathbf{l} \tag{8}$$

To derive Faraday's law, we return to Eq. (2), which is valid regardless of whether the path is moving or stationary. On the left side of this equation, we again use Stokes' theorem to convert the surface integral into a path integral; but on the right side, we cannot take the time derivative out of the integral because the limits of integration, defined by the path, are time dependent. Our equation then looks as follows:

$$\oint \mathbf{E} \cdot d\mathbf{l} = -\frac{1}{c} \int_S \frac{\partial \mathbf{B}}{\partial t} \cdot d\mathbf{S} \tag{9}$$

If we add $\oint \mathbf{v}/c \times \mathbf{B} \cdot d\mathbf{l}$ to both sides of this equation, the left side becomes the emf as defined in Eq. (8). The right side becomes

$$-\frac{1}{c}\left(\int_S \frac{\partial \mathbf{B}}{\partial t} \cdot d\mathbf{S} - \oint \mathbf{v} \times \mathbf{B} \cdot d\mathbf{l} \right) \tag{10}$$

In this expression, the first term in parentheses represents the rate of change of the magnetic flux caused by the time dependence of the magnetic field, ignoring the motion of the path. The second term represents the rate of change of the magnetic flux caused by the motion of the path. We can recognize this easily by rewriting the second term in the parentheses in Eq. (10) as

$$+\oint \mathbf{B} \cdot \mathbf{v} \times d\mathbf{l} \tag{11}$$

Here $\mathbf{v} \times d\mathbf{l}$ is the area swept out per unit of time by the segment $d\mathbf{l}$ of the path (see Fig. 10.2); hence $\mathbf{B} \cdot \mathbf{v} \times d\mathbf{l}$ is the flux intercepted by the moving segment $d\mathbf{l}$ per unit time, and the integral (11) is the flux intercepted by the motion of the entire path per unit time. The sum of both terms in the expression (10) is then the net rate of change of flux. Thus Eq. (9) becomes

$$\oint \mathbf{E}' \cdot d\mathbf{l} = -\frac{1}{c} \frac{d\Phi_B}{dt}$$

Figure 10·2

A small segment $d\mathbf{l}$ of a moving path. The velocity of this segment is $\mathbf{v}$. The cross product $\mathbf{v} \times d\mathbf{l}$ is the area swept out by the segment in one unit of time.

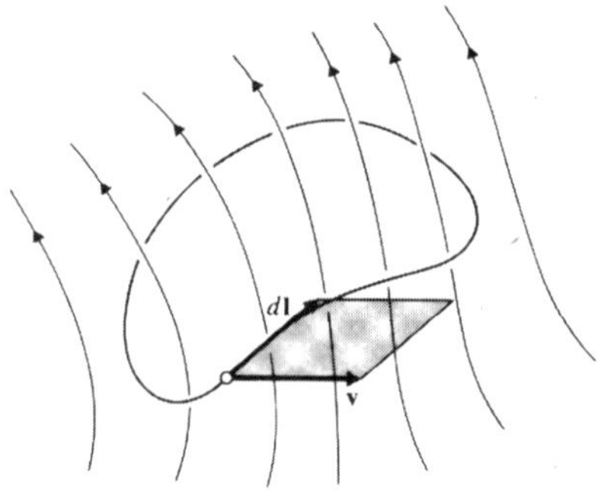

or

$$\mathcal{E} = -\frac{1}{c}\frac{d\Phi_B}{dt} \tag{12}$$

exactly as in the case of the stationary path. This shows that Faraday's law is valid for both stationary and moving paths.

The following examples are a few applications of Faraday's law.

EXAMPLE: *induced electric field in coaxial cable*

Coaxial cable. A coaxial cable consists of a thin conducting wire surrounded by a concentric cylindrical conducting shell of radius R. Suppose that the wire and the shell carry time-dependent currents I in opposite directions (see Fig. 10.3). We want to find the induced electric field in the space within the shell. For this purpose, we need the magnetic field; a simple calculation with Ampère's law tells us that the magnetic field has an azimuthal component

$$B_\phi(\rho) = \frac{2I}{c\rho} \quad \text{for} \quad \rho < R$$
$$B_\phi(\rho) = 0 \quad \text{for} \quad \rho > R \tag{13}$$

■ ***Exercise 1.*** Derive these expressions for the magnetic field. ■

By symmetry, the field lines of the induced electric field must be radial lines, concentric circles around the wire, or parallel lines in the direction of the wire. Radial lines would imply the presence of a net charge on the

Figure 10·3

A very long, straight coaxial cable consisting of a thin wire surrounded by a concentric shell of radius R. The dashed line shows a path of integration for Faraday's law.

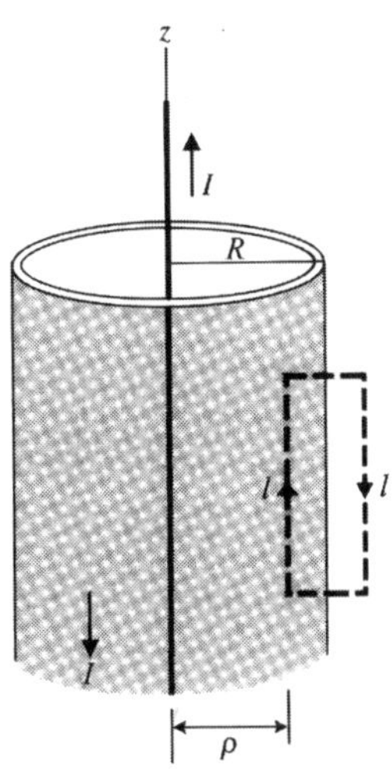

wire, which would be inconsistent with neutrality of the wire. Concentric circles would be inconsistent with Faraday's law, because the magnetic field (13) generates no flux through any circle perpendicular to the wire. Thus we conclude that the induced electric field must be parallel to the wire. To find the magnitude of this field, we consider the closed path shown in Fig. 10.3 For this path, the emf receives a contribution only from the segment of length l within the shell; this contribution is $lE_z(\rho)$. For the area bounded by the path, the rate of change of magnetic flux is

$$\frac{d\Phi_B}{dt} = \frac{d}{dt}\int_\rho^R \frac{2I}{c\rho'} l\,d\rho' = \frac{2l}{c}\frac{dI}{dt}\ln\frac{R}{\rho} \tag{14}$$

Faraday's law then implies

$$lE_z(\rho) = -\frac{2l}{c^2}\frac{dI}{dt}\ln\frac{R}{\rho} \tag{15}$$

or

$$E_z(\rho) = -\frac{2}{c^2}\frac{dI}{dt}\ln\frac{R}{\rho} \quad \text{for} \quad \rho < R \tag{16}$$

If dI/dt is positive, then E_z is negative. This direction for the induced electric field has emerged automatically from our calculation because we assigned the correct direction for the path of integration, according to the right-hand rule. The direction of the induced electric field agrees with the vulgar version of Lenz' law: if there were a conducting substance in the coaxial cable, the induced electric field would generate an induced current in the negative z direction whose magnetic field would oppose the increase of magnetic field of the original current.

Alternatively, we could have found this induced electric field from the vector potential [see Eq. (8.36)],

$$\mathbf{E} = -\frac{1}{c}\frac{\partial \mathbf{A}}{\partial t} \tag{17}$$

In the region $\rho < R$, the vector potential for our coaxial cable is the same as that for a very long, straight wire [see Eq. (9.42)]:

$$A_z(\rho) = -\frac{2I}{c}\ln\rho + \text{ constant} \tag{18}$$

In Section 9.3, we blithely ignored the additive constant in Eq. (18). Here, we must carefully take this constant into account—this "constant" is time dependent and makes a contribution to Eq. (17). To evaluate the "constant," we note that at points outside the coaxial cable ($\rho > R$), the vector potential is zero.

■ ***Exercise 2.*** Show that $A_z(\rho) = 0$ for $\rho > R$. [*Hint:* Since the wire and the concentric shell carry opposite currents, the vector potential must be constant for $\rho > R$; determine this constant by looking at Eq. (9.37), using $\rho \gg R$.] ■

We must therefore impose the boundary condition $A_z(R) = 0$ on the vector potential. This fixes the additive "constant" in Eq. (18):

$$A_z(\rho) = -\frac{2I}{c}\ln\rho + \frac{2I}{c}\ln R = -\frac{2I}{c}\ln\frac{\rho}{R} \tag{19}$$

Our induced electric field is then

$$E_z(\rho) = -\frac{1}{c}\frac{\partial A_z}{\partial t} = -\frac{1}{c}\frac{\partial}{\partial t}\left(-\frac{2I}{c}\ln\frac{\rho}{R}\right) \tag{20}$$

$$= -\frac{2}{c^2}\frac{dI}{dt}\ln\frac{R}{\rho} \tag{21}$$

which is the same as before.

Note that our calculation relied on the tacit assumption that the magnetic field is quasi-static, so the simple expression (13) gives the correct magnetic field even when the current is time dependent. This is not exactly true, but it is a good approximation if the current is changing sufficiently slowly, as in the case of an AC current with a frequency of 60 Hz. If the current is changing very quickly, as in the case of a radio-frequency current, then the quasi-static magnetic field is supplemented by an induced magnetic field caused by the time-dependent electric field (see Section 10.5). The calculation of the complete mutual induction effects of electric and magnetic fields is quite complicated; it involves a coupled system of wave equations, which we will examine in Chapter 15. Furthermore, at high frequencies the magnetic and electric fields cannot penetrate conductors (the "skin effect;" see Section 12.8), and the fields must then satisfy a corresponding boundary condition at the surface of the conductor. At high frequency, the resulting induced electric field in a coaxial cable is drastically different from the longitudinal electric field calculated above. In fact, the high-frequency induced electric field in a coaxial cable is mostly or completely *transverse* (that is, radial) rather than longitudinal.

EXAMPLE: *induced electric field in betatron*

Betatron. If the magnetic field in an electromagnet with cylindrical pole pieces is made to increase in time, an induced electric field with circular field lines will be generated. In the betatron, such an increasing magnetic field holds electrons in an orbit of fixed radius, and concurrently, the induced electric field accelerates these electrons to higher energy. Consider an orbit of radius R (see Fig. 10.4). By symmetry, the tangential component of the electric field at this radius is constant, and Faraday's law gives us

$$2\pi R E_\phi(R) = \frac{1}{c}\frac{d\Phi_B}{dt} \tag{22}$$

or

$$E_\phi(R) = \frac{1}{2\pi Rc}\frac{d\Phi_B}{dt} \tag{23}$$

The rate of increase in the magnitude of the momentum of an electron is then

$$\frac{dp}{dt} = eE_\phi(R) = \frac{e}{2\pi Rc}\frac{d\Phi_B}{dt} \tag{24}$$

For a circular orbit in a magnetic field, $p = eRB(R)/c$, so Eq. (24) becomes

$$\frac{eR}{c}\frac{dB(R)}{dt} = \frac{e}{2\pi Rc}\frac{d\Phi_B}{dt} \tag{25}$$

Integrating this, we obtain

$$\Phi_B = 2\pi R^2 B(R) \tag{26}$$

This constraint between the flux within the orbit and the magnetic field at the orbit must be satisfied at all times if the electron is to remain in its

Figure 10·4

Orbit of an electron (clockwise circle) in the magnetic field of a betatron. The lines of the increasing magnetic field, indicated by crosses, are directed into the plane of the page. The lines of the induced electric field are concentric counterclockwise circles in the plane of the page.

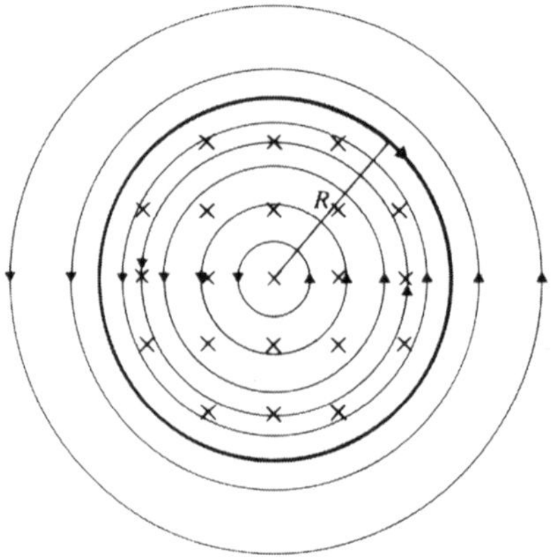

orbit of fixed radius in spite of the increase of energy. Note that for a uniform magnetic field, the flux would be $\pi R^2 B$, which is not enough to satisfy Eq. (26); thus, in the betatron, the magnetic field must be stronger inside the orbit than at the orbit.

EXAMPLE: *induced emf of Faraday disk*

Faraday disk. Consider a circular disk of conducting material spinning about its axis; the disk is immersed in a uniform and constant magnetic field parallel to this axis (see Fig. 10.5). Sliding contacts connect the axis and the rim of the disk to an external circuit. Under these conditions, the disk will generate an emf, which will drive a current through the circuit. We can calculate the emf from Faraday's law, but we must take care in the selection of the path of integration. Faraday's law is valid for any path, even for some path through empty space that does not coincide with any of the conductors shown in Fig. 10.5. But what we want is the emf delivered to the current flowing through the external circuit connected to the disk, and we must therefore choose a path of integration that follows the motion of the free charges around the circuit. In the disk, the free charges are almost at rest (the drift velocity of the conduction electrons is very low, and the random velocity does not count). Hence the relevant path must be at rest relative to the disk; for instance, we can choose the path shown in Fig. 10.5. As the disk turns, the radial segment of this path sweeps out area at the rate

$$\frac{dA}{dt} = \frac{\omega}{2\pi} \times \pi R^2$$

where ω is the angular frequency of rotation and R is the radius of the disk. The rate of change of magnetic flux is then $BdA/dt = \omega R^2 B/2$, and the induced emf is

Figure 10·5

A conducting disk rotating in a uniform and constant magnetic field. The dashed line shows the path of integration; the radial segment and the trailing tangential segment are at rest in the disk.

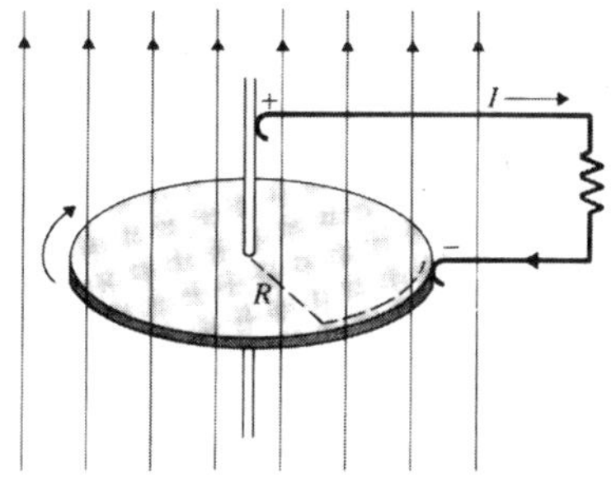

$$\mathcal{E} = \frac{\omega R^2 B}{2c}$$

With the direction of rotation and the direction of the magnetic field as in Fig. 10.5, the rim of the disk acts as the negative terminal of a source of emf and the axis as the positive terminal. Disk machines of this kind—called homopolar generators—are employed in practical applications, such as electroplating, requiring large currents but only modest voltages.

EXAMPLE: *induced emf of MHD generator*

Magnetohydrodynamic generator. The magnetohydrodynamic, or MHD, generator consists of a rectangular channel filled with a plasma flowing at high speed. The plasma is a very hot, ionized gas, which contains an abundance of free electrons and free ions and therefore is a good conductor. The channel is placed in a strong magnetic field (see Fig. 10.6). The top and the bottom of the channel are made of an insulator, whereas the opposite sides are made of a conductor; these opposite sides are connected to an external circuit. The calculation of the induced emf is much the same as in the preceding example. The column of plasma is a good conductor, and our path of integration must be at rest in this conductor. A possible path is shown in Fig. 10.6. If the velocity of flow of the plasma is $\mathbf{v}$, the transverse segment l of this path sweeps out area at the rate vl and sweeps out flux at the rate vlB. Faraday's law then leads to an induced emf

$$\mathcal{E} = \frac{vlB}{c} \tag{27}$$

We could also have derived this result by a simple Lorentz transformation of the magnetic field. If we use the Lorentz-transformation equation (7) with $\mathbf{E} = 0$, we find that the electric field in the rest frame of the plasma is

Figure 10·6

A magnetohydrodynamic generator. The magnetic field is perpendicular to the flow of plasma in the channel. The dashed line shows the path of integration, with its transverse segment at rest in the plasma.

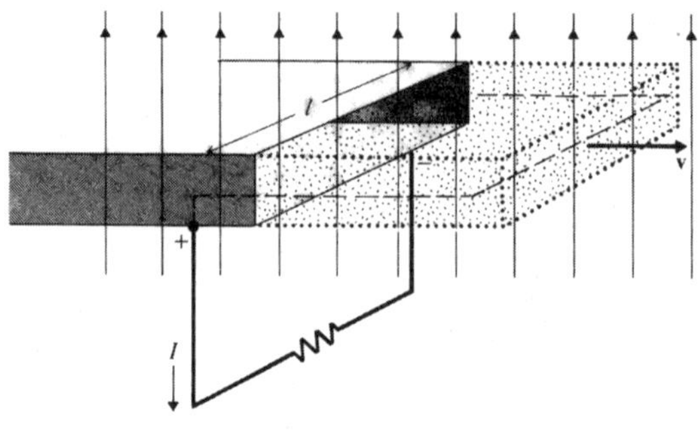

$$\mathbf{E}' = \frac{\mathbf{v}}{c} \times \mathbf{B}$$

Hence the emf, or the potential difference, between opposite sides of the plasma column is

$$E'l = \frac{vlB}{c}$$

This agrees with the above result—since Faraday's law incorporates the Lorentz-transformation equation (7), the agreement comes as no surprise.

The emf (27) is often called a *motional emf* because it is generated by the motion of a body in a magnetic field. Any kind of body moving through a magnetic field will develop such an emf. For instance, our result also applies to an electrolyte flowing in a channel and to a bar of metal sliding between stationary contact brushes.

In the above calculation of the induced emf, we ignored the electric field contributed by the positive and negative charges that may have accumulated on the sides of the channel enclosing the plasma. Since the electric field of these (stationary) charges is an electrostatic, conservative field, it does not contribute to the emf reckoned around a closed path. However, if we want to calculate the electric current that flows in the plasma from one side of the channel to the other, then we must take this field into account. Designating this field in the laboratory frame by **E**, we find from Eq. (7) that the net electric field in the rest frame of the plasma is

$$\mathbf{E}' = \mathbf{E} + \frac{\mathbf{v}}{c} \times \mathbf{B}$$

For calculating the current in the plasma, we can use Ohm's law, but we must keep in mind that this law is valid only in the *rest frame* of the conductor (this restriction is inherent in the assumptions underlying the derivation of the law). Thus the current is proportional to the net electric field in the rest frame of the plasma:

$$\mathbf{J} = \sigma \mathbf{E}' = \sigma \left(\mathbf{E} + \frac{\mathbf{v}}{c} \times \mathbf{B} \right) \tag{28}$$

This equation simply means that the current is driven by both the electrostatic field **E** and the induced electric field $\mathbf{v} \times \mathbf{B}/c$.

EXAMPLE: *plasma in a magnetic field*

Frozen field lines in plasma. At extreme temperatures, the ionization of a plasma is very high, with a great abundance of free electrons and free ions. The conductivity is then also very high; for instance, hydrogen plasma at 10^8 K has a conductivity roughly 10,000 times that of copper. For many purposes, we can pretend that the conductivity is infinite.

Figure 10·7

Plasma of infinite conductivity moving through a magnetic field carries the field lines along.

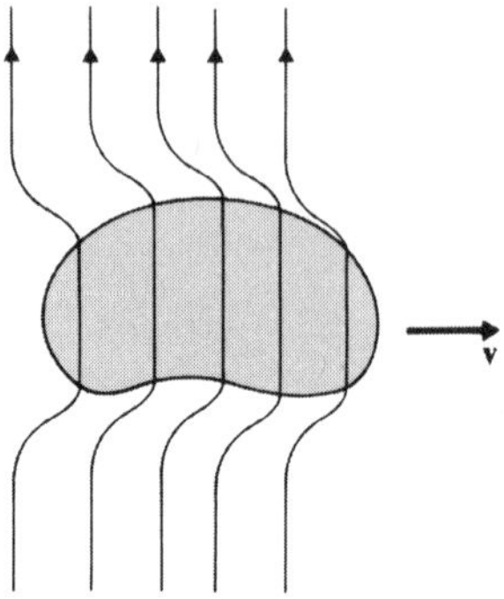

Since the current density must remain finite, it follows from Eq. (28) that the electric field in the instantaneous rest frame of the plasma must be zero, that is, $\mathbf{E}' = 0$. In terms of the electric and magnetic fields in the laboratory, we consequently find

$$\mathbf{E} + \frac{\mathbf{v}}{c} \times \mathbf{B} = 0 \tag{29}$$

This can be regarded as a constraint on the velocity of the plasma. If $\mathbf{E} = 0$, it requires $\mathbf{v} \times \mathbf{B} = 0$, which implies that the plasma can flow only along the direction of the magnetic field lines.* If we attempt to push the plasma in some other direction with a strong external force, it will drag the magnetic field lines along. We can also deduce that from Faraday's law: Since $\mathbf{E}' = 0$ and $\mathcal{E} = 0$, Eq. (12) yields $d\Phi_B / dt = 0$ for any path fixed in the plasma; this means that the number of field lines linked by the path has to remain constant. The field lines are carried along by the plasma—they are "frozen" in the plasma (see Fig. 10.7).

10·2 Inductance

In many practical applications, we need to calculate the emf that one circuit induces in another, nearby circuit. Such mutual induction effects will occur whenever the magnetic field lines produced by one circuit link the

*Note that the plasma in the MHD generator does *not* obey this constraint. This plasma does not have infinite conductivity, and it is forcibly made to flow across the magnetic field lines.

other circuit, so a change of current in one generates a change of flux in the other (see Fig. 10.8).

For a general system consisting of a number of separate circuits, with currents $I_1, I_2, I_3, \ldots, I_m$, the flux in any one of these must be linearly related to the currents. This is so because the magnetic fields are linearly related to the currents that produce them [see Eq. (9.35)], and the separate magnetic fields of the separate currents obey the principle of superposition, giving a net magnetic field and net flux that are sums of the separate contributions. Hence the flux in circuit n can be expressed as

$$\Phi_{B,n} = c\sum_{k=1}^{m} M_{nk} I_k \tag{30}$$

where the constants M_{nk} are called *coefficients of inductance.* Note that this equation rests on the implicit assumption that the circuits are in vacuum or in a linear material. The equation is not valid for circuits in the presence of ferromagnetic materials. The CGS electrostatic unit of inductance is s^2/cm.*

The coefficients of inductance depend on the shapes, sizes, and positions of the circuits. We can derive an explicit formula for these coefficients as follows: The term $cM_{nk}I_k$ in Eq. (30) is the magnetic flux that circuit k produces in circuit n; that is,

$$cM_{nk}I_k = \int_{S_n} \mathbf{B}_k \cdot d\mathbf{S}_n \tag{31}$$

Figure 10·8

Two circuits. Magnetic field lines produced by one circuit link the other circuit.

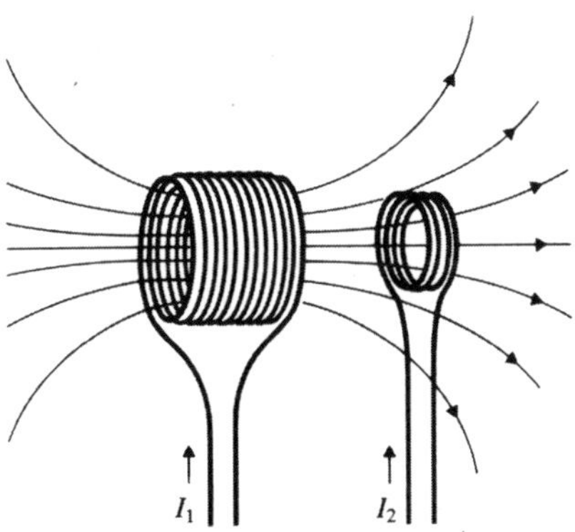

*This unit (which has no special name) is related to the SI unit (the henry) by the correspondence 1 henry $\Leftrightarrow 1.113\times10^{-12}$ s^2/cm.

where $\mathbf{B}_k$ is the magnetic field produced by circuit k, and $d\mathbf{S}_n$ is the area element for circuit n. With $\mathbf{B}_k = \nabla \times \mathbf{A}_k$ and Stokes' theorem, Eq. (31) becomes

$$cM_{nk}I_k = \int_{S_n} \nabla \times \mathbf{A}_k \bullet d\mathbf{S}_n = \oint \mathbf{A}_k \bullet d\mathbf{l}_n$$

Since the vector potential of circuit k is [compare Eq. (9.32)]

$$\mathbf{A}_k = \frac{1}{c}\oint \frac{I_k}{|\mathbf{x}_k - \mathbf{x}_n|} d\mathbf{l}_k$$

we readily obtain

$$M_{nk} = \frac{1}{c^2}\oint\oint \frac{1}{|\mathbf{x}_k - \mathbf{x}_n|} d\mathbf{l}_k \bullet d\mathbf{l}_n \qquad (32)$$

This expression is called the *Neumann formula*. From this formula, we immediately recognize that the coefficients of inductance obey a symmetry relation:

$$M_{nk} = M_{kn} \qquad (33)$$

which means that whenever the currents in circuits n and k are equal, the fluxes that these circuits produce through each other are also equal.

If the circuits are rigid and stationary, then the coefficients of inductance are time independent. The induced emf in circuit n can then be written

$$\mathcal{E}_n = -\frac{1}{c}\frac{d\Phi_{B,n}}{dt} = -\sum_{k=1}^{m} M_{nk}\frac{dI_k}{dt} \qquad (34)$$

The coefficients with $n = k$ are called the coefficients of *self-inductance*, usually designated by L_n,

$$L_n = M_{nn} \qquad (35)$$

These coefficients tell us what emf the circuit n induces in itself when its own current is changing. This is called the *self-emf* or the *back-emf* of the circuit. Incidentally: For $n = k$, the Neumann formula fails—it gives a divergent result because the magnetic field of an infinitesimally thin wire is infinite at the position of the wire. If we want to calculate the self-inductance of a circuit, we must take the thickness of the wire into account.

In terms of the coefficients of inductance, the magnetic energy in the system of circuits can be written as follows:

$$U = \tfrac{1}{2}\sum_{n=1}^{m}\sum_{k=1}^{m} M_{nk} I_n I_k \tag{36}$$

To derive this formula, we begin with the observation that whenever we increase any of the currents in the circuits, we must supply power from some external source of emf. The power absorbed by circuit n is, according to Eq. (34),

$$-I_n\mathcal{E}_n = \sum_{k=1}^{m} M_{nk} I_n \frac{dI_k}{dt}$$

and the total power absorbed by all the circuits is

$$\frac{dU}{dt} = -\sum_{n=1}^{m} I_n\mathcal{E}_n = \sum_{n=1}^{m}\sum_{k=1}^{m} M_{nk} I_n \frac{dI_k}{dt}$$

Because of the symmetry relation (33), we can also express this as

$$\frac{dU}{dt} = \tfrac{1}{2}\sum_{n=1}^{m}\sum_{k=1}^{m} M_{nk} \frac{d}{dt}(I_n I_k) \tag{37}$$

The total energy stored in the system of circuits as we gradually build up the currents from their initial value (zero) to their final values can be obtained by integrating Eq. (37). This immediately leads to Eq. (36).

In the special case of a single circuit of self-inductance L, this general formula reduces to the well-known expression

$$U = \tfrac{1}{2}LI^2 \tag{38}$$

for the magnetic energy in an inductor.

EXAMPLE: *mutual inductance of coaxial rings*

We end this section with two examples of calculations of inductance. First, we will calculate (approximately) the mutual inductance of two coaxial rings of wire of radius R separated by a distance d (see Fig. 10.9). With this geometry, the Neumann formula (32) is rather difficult to integrate (it leads to elliptic integrals), and we find it more convenient to return to Eq. (30). If a current I_1 in the first ring produces a flux $\Phi_{B,2}$ in the second ring, then Eq. (30) implies

$$M_{21} = \frac{1}{c}\frac{\Phi_{B,2}}{I_1}$$

For the sake of simplicity, we will assume $d > r$ so that we can use the power series (9.62) for the magnetic field of the first ring:

$$B_r = \frac{2\pi I_1}{c}\left[\frac{R^2\cos\theta}{r^3} - \frac{3}{4}\frac{R^4}{r^5}(5\cos^3\theta - 3\cos\theta) + \ldots\right]$$

Figure 10·9

Two coaxial rings. A mathematical spherical cap of radius $r=\sqrt{R^2+d^2}$ closes off one of the rings.

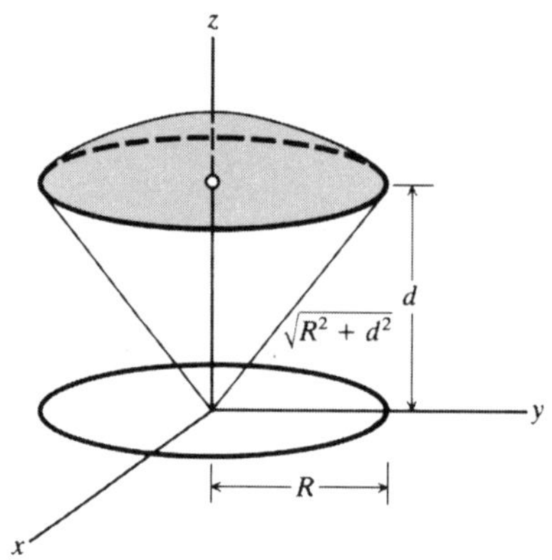

The evaluation of the flux through the second ring is best done by integrating B_r over a spherical cap of radius $r=\sqrt{R^2+d^2}$ closing off this ring (see Fig. 10.9), so

$$\mathbf{B}\cdot d\mathbf{S}=B_r dS=B_r\times r^2\sin\theta\, d\theta\, d\phi$$

and

$$\Phi_{B,2}=\iint B_r(R^2+d^2)\sin\theta\, d\theta\, d\phi$$

With $\mu=\cos\theta$, this becomes

$$\Phi_{B,2}=2\pi\int_0^{d/\sqrt{R^2+d^2}}\frac{2\pi I_1}{c}\times\left[\frac{R^2\mu}{(R^2+d^2)^{1/2}}-\frac{3}{4}\frac{R^4}{(R^2+d^2)^{3/2}}(5\mu^3-3\mu)+\dots\right]d\mu$$

$$=2\pi^2\frac{I_1}{c}\frac{R^2d^2}{(R^2+d^2)^{3/2}}\left[1+\frac{3}{8}\frac{R^2(d^2+6R^2)}{(R^2+d^2)^2}+\dots\right]$$

From this, we see that the mutual inductance is

$$M_{21}=\frac{1}{c}\frac{\Phi_{B,2}}{I_1}=\frac{2\pi^2R^2d^2}{c^2(R^2+d^2)^{3/2}}\left[1+\frac{3}{8}\frac{R^2(d^2+6R^2)}{(R^2+d^2)^2}+\dots\right]$$

Next, we want to calculate the self-inductance per unit length of a coaxial transmission line consisting of a long, cylindrical conducting

EXAMPLE: *self-inductance of coaxial transmission line*

shell of radius R_1 surrounded by a concentric cylindrical shell of radius R_2 (see Fig. 10.10). We will assume that both of these shells are thin (if one or both of these shells had a finite thickness, or if the inner shell were replaced by a solid cylinder, the calculation would be rather more complicated; see Section 11.5). The inner conductor carries a current I toward the right; the outer conductor carries the return current to the left. The magnetic field is then as given in Eq. (13). The rectangle in Fig. 10.10 shows a typical "streamline" for the current; obviously, all of the magnetic field lines between the two conductors are intercepted by the rectangular area. The flux Φ_B in a length l of the coaxial cable is then

$$\Phi_B = l\int_{R_1}^{R_2} B_\phi \, d\rho = l\int_{R_1}^{R_2} \frac{2I}{c}\frac{1}{\rho} d\rho = \frac{2Il}{c}\ln\frac{R_2}{R_1}$$

Consequently, the self-inductance is

$$L = \frac{1}{c}\frac{\Phi_B}{I} = \frac{2l}{c^2}\ln\frac{R_2}{R_1} \tag{39}$$

and the self-inductance per unit length is

$$\frac{L}{l} = \frac{2}{c^2}\ln\frac{R_2}{R_1}$$

If we compare this with the capacitance per unit length of this transmission line [see Eq. (5.55)], we find that

$$\frac{C}{l}\times\frac{L}{l} = \frac{\varepsilon}{c^2}$$

This relationship between the capacitance per unit length and the inductance per unit length is valid for any transmission line consisting of parallel conducting cylinders of arbitrary cross section, concentric or not.

Figure 10·10

A coaxial transmission line consisting of two thin conducting shells. The dashed rectangle shows a typical "streamline" for the current I moving toward the right on the inner conductor and toward the left on the outer.

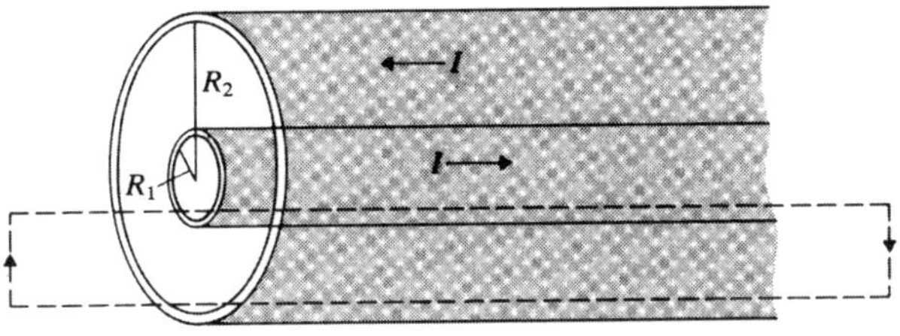

10·3 The LCR Circuit

The self-inductance gives rise to important "inertial" effects in circuits with oscillating or otherwise time-dependent currents. Here, we will discuss only the simple case of a single circuit with an inductor, a capacitor, a resistor, and an external source of oscillating emf, all connected in series (see Fig. 10.11). The equation of motion for such a driven LCR circuit follows from *Kirchhoff's rule*: The sum of all the emfs and voltage drops around the circuit must be zero. This rule merely expresses energy conservation: when a charge moves around the circuit, the sum of all the energy gains and losses must be zero. For our circuit,

$$L\frac{dI}{dt}+\frac{Q}{C}+RI-V_0\cos\omega t=0 \tag{40}$$

Here LdI/dt is the self-induced emf in the inductor, Q/C is the voltage drop across the capacitor, RI is the voltage drop across the resistor, and $V_0 \cos \omega t$ is the emf of the external source. If we differentiate Eq. (40) with respect to time and substitute I for dQ/dt, we obtain

$$L\frac{d^2I}{dt^2}+\frac{1}{C}I+R\frac{dI}{dt}=-\omega V_0\sin\omega t \tag{41}$$

Comparison with the standard equation

$$m\frac{d^2x}{dt^2}+kx+b\frac{dx}{dt}=F_0\sin\omega t$$

for a mechanical simple harmonic oscillator with friction shows that L is analogous to the mass of the mechanical oscillator, $1/C$ to the spring constant, and R to the frictional damping constant.

To obtain the steady-state solution for the circuit oscillations, it is best to regard Eq. (41) as the real part of a complex equation

Figure 10·11

An LCR circuit driven by an external source of emf.

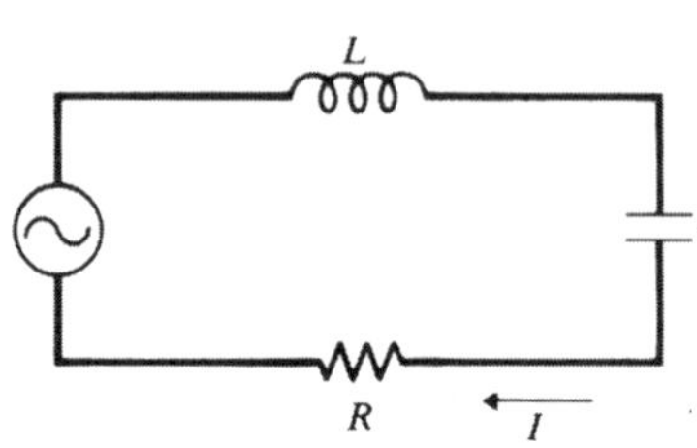

$$L\frac{d^2I}{dt^2}+\frac{1}{C}I+R\frac{dI}{dt}=-i\omega V_0 e^{-i\omega t} \tag{42}$$

After solving this complex equation, we will take the real part of the solution for our physical current. With the usual *Ansatz*

$$I = Ae^{-i\omega t} \tag{43}$$

our complex equation reduces to

$$-\omega^2 LA+\frac{1}{C}A-i\omega RA=-i\omega V_0 \tag{44}$$

from which

$$A=\frac{-i\omega V_0}{-\omega^2 L-i\omega R+1/C} \tag{45}$$

The complex current is then

$$I=\frac{-i\omega V_0 e^{-i\omega t}}{-\omega^2 L-i\omega R+1/C} \tag{46}$$

or

$$I=\frac{V_0 e^{-i\omega t}}{Z} \tag{47}$$

with

$$Z=-i\omega L+R+\frac{i}{\omega C} \tag{48}$$

This quantity Z is called the *impedance.* For a circuit without inductor and without capacitor ($L = 0$, $C=\infty$), Z reduces to R, and $I = V/Z = V/R$. In general, Z may be regarded as a complex "resistance" of the circuit.

The physical current is the real part of Eq. (46),

$$I=\frac{V_0\cos(\omega t-\phi)}{\sqrt{R^2+(\omega L-1/\omega C)^2}} \tag{49}$$

with

$$\phi=\tan^{-1}\frac{1}{R}\left(\omega L-\frac{1}{\omega C}\right) \tag{50}$$

Figure 10.12 shows the amplitude of the current as a function of the driving frequency. The current has a maximum at the frequency

Figure 10·12

Amplitude of the current as a function of driving frequency.

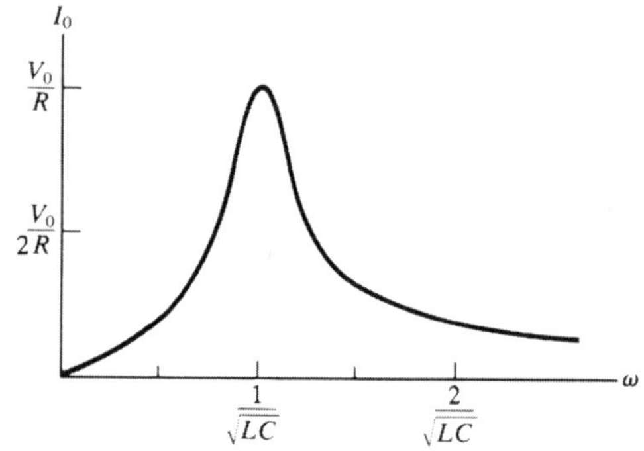

$$\omega_0 = \frac{1}{\sqrt{LC}} \tag{51}$$

This is the *resonant frequency* for the circuit. Figure 10.13 shows the phase angle ϕ as a function of the driving frequency. At resonance, the phase angle is zero—that is, the current is in phase with the driving voltage, and so

$$I = \frac{V}{R} = \frac{V_0 \cos \omega t}{R} \tag{52}$$

For $\omega < \omega_0$, the current "leads" the voltage ($\phi < 0$); for $\omega > \omega_0$, the current "lags" the voltage ($\phi > 0$).

Figure 10·13

Phase angle as a function of driving frequency.

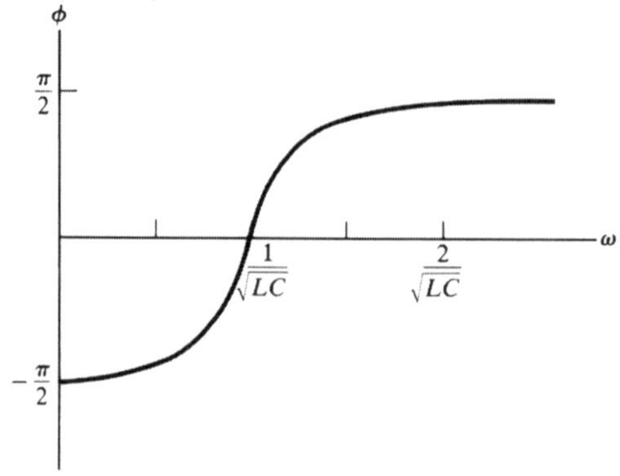

10·4 Motion in a Slowly Varying Magnetic Field

In Section 9.5, we found that the orbit of a charged particle in a uniform and constant magnetic field is a helix of radius

$$r = \frac{cp_\perp}{qB} \tag{53}$$

where $p_\perp$ is the component of the momentum perpendicular to the magnetic field. In this section, we want to inquire what happens to the orbit if we gradually increase or decrease the strength of the magnetic field. It is obvious that if the magnetic field changes only slowly, the orbit at each instant of time is (approximately) a circle or a helix, but the radius of this orbit will change in some way because the induced electric field increases or decreases the energy of the particle. We will prove that the radius of the orbit changes in such a way that

$$Br^2 = \text{constant} \tag{54}$$

This means that the flux through the orbit remains constant.

To prove this result, let us first consider a particle in a circular orbit, leaving the more general case of a helical orbit for later. Figure 10.14 shows one complete orbit of the particle; since the magnetic field varies only slowly, the change in the radius of the orbit is barely noticeable in this figure. By Faraday's law, the induced emf around the orbit is

$$\mathcal{E} = \frac{\pi r^2}{c}\frac{dB}{dt} \tag{55}$$

When multiplied by the charge of the particle, this gives us the increment of kinetic energy of the particle in one orbit. In terms of the increment of momentum, the increment of kinetic energy is $v\Delta p$, so

$$v\Delta p = q\frac{\pi r^2}{c}\frac{dB}{dt} \tag{56}$$

If we divide by the time Δt for one orbit and cancel the speed v against $2\pi r/\Delta t$, the equation reduces to

$$\frac{\Delta p}{\Delta t} = \frac{qr}{2c}\frac{dB}{dt} \tag{57}$$

For our circular orbit,

Figure 10·14

Orbit of a charged particle in an increasing magnetic field. The orbit is almost a closed circle.

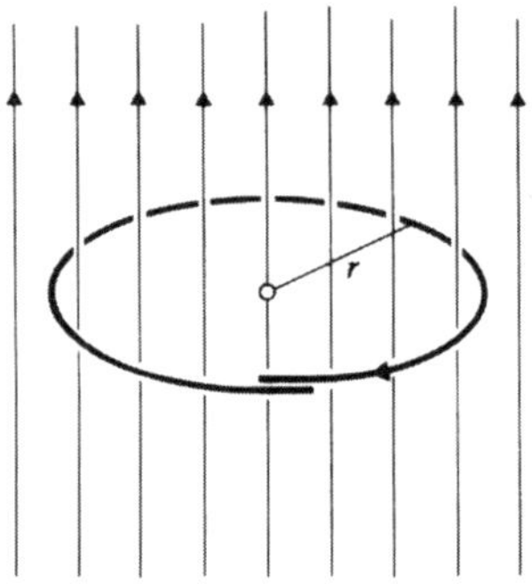

$$r = \frac{cp}{qB} \tag{58}$$

so

$$\frac{\Delta p}{\Delta t} = \frac{p}{2B}\frac{dB}{dt} \tag{59}$$

Integrating this, we find

$$\frac{B}{p^2} = \text{constant} \tag{60}$$

In view of Eq. (58), this is equivalent to

$$Br^2 = \text{constant} \tag{61}$$

which establishes the desired result for the special case of a circular orbit.

To complete the proof, we must treat the general case of a helical orbit. This we can do with a trick: If the orbit is helical in the reference frame of the laboratory, we go to a new reference frame in which the particle has no velocity parallel to the magnetic field, so the orbit is circular. Since the velocity of this new reference frame is parallel to the magnetic field, the Lorentz-transformation equation (8.44)* tells us that the magnetic fields in the new reference frame and in the laboratory frame are the same. Furthermore, the Lorentz-transformation equation for transverse distances tells us that the orbital radii in the two reference frames

*The electric field in Eq. (8.44) may be taken as zero (we ignore the induced electric field at this stage of the calculation).

are the same. Given that Eq. (61) is valid for the circular orbit in the new reference frame, it follows immediately that this equation is also valid for the helical orbit in the laboratory reference frame.

In mechanics, a quantity that remains constant while some parameter of the motion changes slowly is called an *adiabatic invariant.* Thus the flux Br^2 is an adiabatic invariant for a particle moving in a slowly changing magnetic field. This is true not only for a magnetic field changing in time but also for a magnetic field changing in space—from the point of view of a particle spiraling in a magnetic field, a slow variation in space is equivalent to a slow variation in time. For example, Fig. 10.15 shows a particle spiraling in a magnetic field that gradually increases from bottom to top. As the particle drifts upward, it experiences a magnetic field that increases with time. In such a magnetic field, the radius of the orbit will shrink, in accordance with Eq. (54).

This shrinking of the orbit in an increasing magnetic field can cause a reflection of the particle by a region of high magnetic field. To see how this comes about, we use $r = cp_\perp / qB$ to rewrite Eq. (54) as

$$\frac{p_\perp^2}{B} = \text{constant} \qquad (62)$$

This shows that when the particle drifts into a region of high magnetic field, $p\perp$ must increase. But $p\perp$ cannot increase indefinitely—in a static magnetic field, the kinetic energy is constant, and therefore, the perpendicular and the parallel components of the momentum are subject to the constraint

Figure 10·15

Orbit of a charged particle in an inhomogeneous magnetic field.

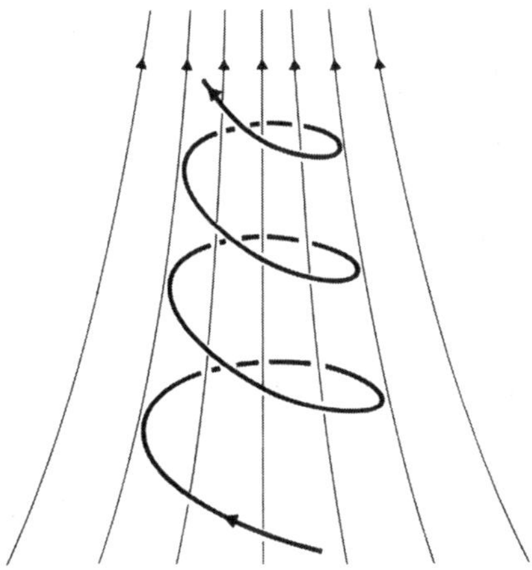

$$p_\perp^2 + p_\parallel^2 = \text{constant} \tag{63}$$

When $p_\perp$ increases, $p_\parallel$ must decrease. The maximum value of $p_\perp$ will be reached when $p_\parallel = 0$; at this point, the drift motion of the particle has to reverse (see Fig. 10.16). Note that the direction of the circular motion remains the same throughout this reflection process—only the drift motion reverses.

EXAMPLE: *motion of charged particles in magnetic field of Earth*

The Van Allen belts surrounding the Earth consist of electrons and protons trapped in the Earth's magnetic field by such a reflection process. The particles spiral around the magnetic field lines (see Fig. 10.17), and they bounce back and forth between the north and south polar regions, where the convergent magnetic field lines reflect them.

A similar reflection effect is exploited in some confinement schemes for plasma fusion experiments. The plasma is held in a "magnetic bottle" with "magnetic mirrors" at the ends. Some plasma machines of this kind use several mirrors in tandem at each end, with convergent field lines arranged in a very complicated pattern, to improve the reflection efficiency and inhibit the escape of even those particles with large components of motion along the axis of the bottle.

Figure 10·16

Reflection of a charged particle by a region of strong magnetic field.

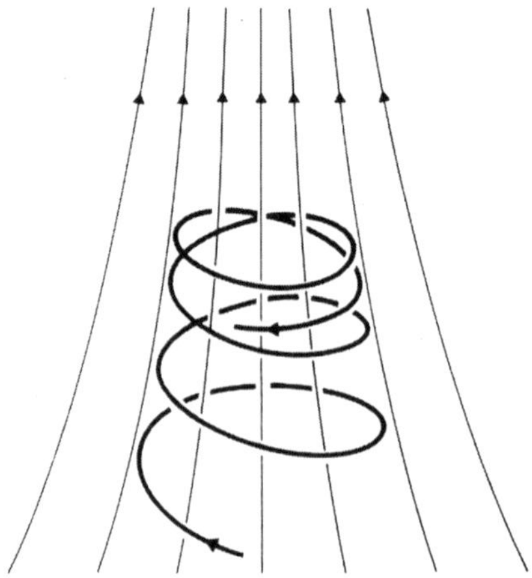

Figure 10·17

Orbit of an electron in one of the Van Allen belts.

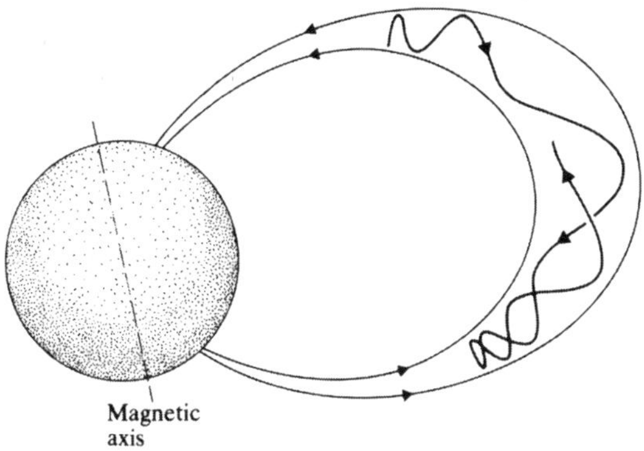

10·5 The Displacement Current

If we compare the Maxwell equations for $\nabla\times\mathbf{E}$ and $\nabla\times\mathbf{B}$,

$$\nabla\times\mathbf{E}=-\frac{1}{c}\frac{\partial\mathbf{B}}{\partial t} \tag{64}$$

$$\nabla\times\mathbf{B}=\frac{4\pi}{c}\mathbf{j}+\frac{1}{c}\frac{\partial\mathbf{E}}{\partial t} \tag{65}$$

we see that, except for the term $(4\pi/c)\mathbf{j}$, the equations are mathematically quite similar. The former equation tells us that a changing magnetic field induces an electric field; the latter equation tells us that, conversely, a changing electric field induces a magnetic field. By integrating Eq. (65) over an open surface S bounded by some arbitrary stationary path, we find

$$\oint\mathbf{B}\cdot d\mathbf{l}=\frac{4\pi}{c}\int_S\mathbf{j}\cdot d\mathbf{S}+\frac{1}{c}\int_S\frac{\partial\mathbf{E}}{\partial t}\cdot d\mathbf{S}$$

or

$$\oint\mathbf{B}\cdot d\mathbf{l}=\frac{4\pi}{c}\int_S\mathbf{j}\cdot d\mathbf{S}+\frac{1}{c}\frac{d\Phi_E}{dt} \tag{66}$$

where $\Phi_E=\int\mathbf{E}\cdot d\mathbf{S}$ is the electric flux.

■ ***Exercise 3.*** Show that, for a moving path, Eq. (66) can be written

$$\oint \mathbf{B}'\bullet d\mathbf{l} = \frac{4\pi}{c}\int_S \mathbf{j}\bullet d\mathbf{S} + \frac{1}{c}\frac{d\Phi_E}{dt} \tag{67}$$

where $\mathbf{B}'$ is the magnetic field measured in the instantaneous rest frame of the segment $d\mathbf{l}$ of the path, and $d\Phi_E/dt$ is the net change of flux arising from the time dependence of the electric field and the motion of the path. ■

This law of induction for magnetic fields is a generalized version of Ampère's law. It differs from Ampère's law by the extra term $1/c\int \partial\mathbf{E}/\partial t\bullet d\mathbf{S}$. Obviously, $\mathbf{j}$ and $\partial\mathbf{E}/\partial t$ have the same effect on the magnetic fields—both act as sources of magnetic fields. To emphasize this similarity, we can define a *displacement-current density*

$$\mathbf{j}_{\text{D}} = \frac{1}{4\pi}\frac{\partial\mathbf{E}}{\partial t} \tag{68}$$

in terms of which Eq. (66) becomes

$$\oint \mathbf{B}\bullet d\mathbf{l} = \frac{4\pi}{c}\int_S (\mathbf{j}+\mathbf{j}_{\text{D}})\bullet d\mathbf{S} \tag{69}$$

This displacement current can be regarded as a continuation of the real current "by other means." We can readily show that the sum $\mathbf{j} + \mathbf{j}_{\text{D}}$ has no sources or sinks; that is, it has zero divergence:

$$\begin{aligned}\nabla\bullet(\mathbf{j}+\mathbf{j}_{\text{D}}) &= \nabla\bullet\mathbf{j} + \frac{1}{4\pi}\frac{\partial}{\partial t}\nabla\bullet\mathbf{E}\\ &= \nabla\bullet\mathbf{j} + \frac{1}{4\pi}\frac{\partial}{\partial t}(4\pi\rho)\end{aligned} \tag{70}$$

The right side vanishes by the continuity equation.

A classic example of how the displacement current acts as the continuation of the true current is illustrated in Fig. 10.18. Two long straight wires carry a constant current into a capacitor. We want to evaluate $\oint \mathbf{B}\bullet d\mathbf{l}$ for a closed path around the wire. For the sake of simplicity, we will assume that the path is so far from the capacitor that the fringing electric field of the capacitor reaches neither the path nor its neighborhood. Figures 10.18(a) and 10.18(b) show two alternative surfaces that have the path as boundary. In Fig. 10.18(a), the surface of integration intercepts real current but no displacement current; we therefore obtain

$$\oint \mathbf{B}\bullet d\mathbf{l} = \frac{4\pi}{c}I \tag{71}$$

Figure 10·18

Two long, straight wires connected to a capacitor. (*a*) A possible surface of integration that has the given path as boundary. This surface intercepts the current of the wire. (*b*) Another possible surface of integration that has the same path as boundary. This surface passes between the capacitor plates.

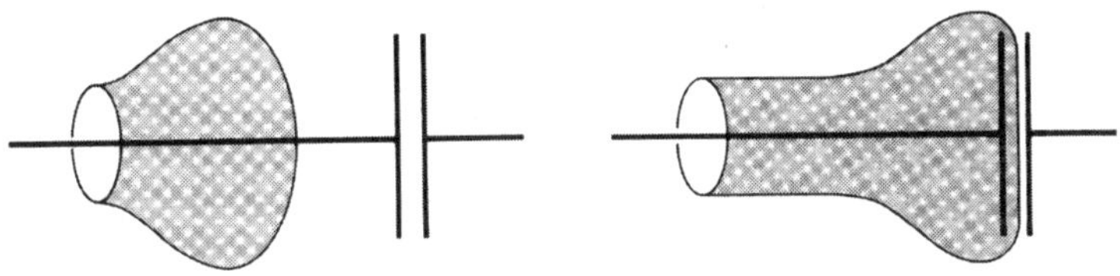

In Fig. 10.18(*b*), the alternative surface of integration intercepts displacement current but no real current; we then obtain

$$\oint \mathbf{B}\cdot d\mathbf{l} = \frac{4\pi}{c}\int_S \mathbf{j}_D \cdot d\mathbf{S} = \frac{1}{c}\frac{d}{dt}\int_S \mathbf{E}\cdot d\mathbf{S} \tag{72}$$

Since all the electric field lines that start on one of the plates must cross our surface of integration, the integral $\int \mathbf{E}\cdot d\mathbf{S}$ equals 4π times the charge on one plate. Equation (72) then reduces to

$$\oint \mathbf{B}\cdot d\mathbf{l} = \frac{4\pi}{c}\frac{dQ}{dt} \tag{73}$$

which is the same as (71). Thus the displacement current and the true current have the same magnitudes, and they produce the same magnetic field.

Problems

1. An ideal cylindrical solenoid of radius R and of infinite length has a uniform magnetic field $\mathbf{B}$ in its interior. The solenoid moves perpendicularly to its axis with velocity $\mathbf{V}$.

(a) From Faraday's law of induction, show that in consequence of this motion, the solenoid will have a uniform electric field $\mathbf{E} \simeq (\mathbf{V}/c)\times\mathbf{B}$ in its interior; ignore terms of order V^2/c^2.

(b) Is this electric field a conservative field?

(c) The field lines of $\mathbf{E}$ begin and end on the walls of the solenoid. Find the surface charge density on the walls.

(d) Suppose that this surface charge density were placed on a *stationary* cylindrical tube of paper of the same radius as the solenoid. Qualitatively, what would be the electrostatic field inside and outside the tube? How would the field lines of this conservative field differ from those of the field described in parts (a)–(c)?

2. A very long, straight wire carries a current I. A rod of length l oriented parallel to the wire is moving toward the wire at speed v. As a function of the distance from the wire to the rod, what is the motional emf induced between the ends of this rod? What is the induced electric field in the rest frame of the rod?

3. A square loop of dimension $l \times l$ is oriented with two of its sides parallel to a very long, straight wire, which is in the plane of the loop. The wire carries a current I and is moving away from the loop at speed v. If the instantaneous distance between the wire and the nearest side of the loop is h, what is the emf induced around the loop?

4. A *flip coil* is used to measure the strength of the magnetic field. The coil is a small wire loop of N turns and of area A; the resistance of the wire is R. The loop is placed face-on to the magnetic field and suddenly flipped over. During this procedure, the induced current gives rise to a net flow Q of charge around the loop. Express the magnitude of the magnetic field in terms of Q, R, A, and N.

5. A large metallic plate slides between the poles of a magnet. The magnet provides a uniform magnetic field **B** perpendicular to the plate and penetrating through a relatively small region $l \times l$ of the plate. The thickness of the plate is h, the conductivity of the metal is σ, and the (instantaneous) speed of the plate is v. Show that the induced currents, or eddy currents, in the plate give rise to a drag force of approximately $v\sigma h l^2 B^2/c^2$ opposing the motion of the plate. [*Hint*: Calculate the resistance for the current flowing from one edge of the area $l \times l$ to the opposite edge, and then make the crude approximation that the return current flowing through the (large) region surrounding this area has roughly the same resistance. Why?]

6. (a) Show that if the current on an infinitely long, straight wire changes at a finite rate, an infinite electric field will be induced parallel to the wire.
   (b) Estimate the induced electric field for a wire of finite length L. [*Hint*: Estimate how far the magnetic field extends from a finite wire; then use Eq. (16).]

7. Two infinitely long, straight wires separated by a distance d carry equal currents in opposite directions. If the currents change at the rate dI/dt, find the induced electric field at an arbitrary point in the plane of the two wires.

8. A coaxial transmission line consists of two thin, concentric, cylindrical conducting shells. The radius of the inner shell is 1 mm, and that of the outer shell is 3 mm. The two shells carry equal AC currents of amplitude 2.0×10^{10} statamps and frequency 2000 Hz in opposite directions; the currents are uniformly distributed over the shells. What is the maximum magnitude of the magnetic field generated by the current? What is the maximum magnitude of the electric field induced by the magnetic field?

9. A circular ring of mass m, radius r, and made of a piece of wire of electric resistance R, is free to turn about an axis coinciding with the vertical diameter. The ring is placed in a uniform, horizontal magnetic field **B**. Find the equation of motion for the rotational motion of the ring.

10. A uniformly magnetized conducting sphere of radius R produces a constant magnetic field

$$\mathbf{B} = B_0 \hat{\mathbf{z}} \qquad \text{for} \qquad r < R$$

in its interior and produces a dipole field

$$\mathbf{B} = \frac{1}{2}B_0R^3\left(\frac{2\cos\theta}{r^3}\hat{\mathbf{r}} + \frac{\sin\theta}{r^3}\hat{\boldsymbol{\theta}}\right) \quad \text{for} \quad r > R$$

in its exterior. Suppose that this sphere is made to rotate with an angular velocity ω about the z axis.

(a) The free charges at rest in the rotating conductor experience a magnetic force that produces a charge separation and a charge distribution in and on the sphere. When equilibrium is reached, the electric force exerted by the electric field arising from this charge distribution must compensate the magnetic force. Show that this implies that in the laboratory frame, the electric field in the interior of the sphere is

$$\mathbf{E} = -\frac{B_0\omega r\sin\theta}{c}\hat{\boldsymbol{\phi}} \times \hat{\mathbf{z}}$$

and show that the electric potential on the surface of the sphere is of the form

$$\Phi(R,\theta) = \frac{B_0\omega}{2c}R^2\sin^2\theta + \text{constant}$$

(b) From the given potential $\Phi(R,\theta)$ on the surface of the sphere, find the potential $\Phi(r,\theta)$ in the exterior of the sphere, and show that the electric field in the exterior is

$$\mathbf{E} = -\frac{B_0\omega R^5}{2c}\left(\frac{3\cos^2\theta - 1}{r^4}\hat{\mathbf{r}} + \frac{2\cos\theta\sin\theta}{r^4}\hat{\boldsymbol{\theta}}\right)$$

This is the electric field of a quadrupole.

11. A square loop of wire of dimension $l \times l$ has two of its sides parallel to a long, straight wire that lies in the plane of the loop at a distance h from the nearest side of the loop. Assume that the long, straight wire closes on itself at infinity in a direction away from the loop, so the wire and the loop form two separated circuits. Calculate the mutual inductance by means of the Neumann formula.

12. Two very long, thin strips of conductor are placed parallel and face to face. The width of each strip is h, and the separation between the strips is d. The strips carry equal currents in opposite directions and can be regarded as forming a single circuit. Assume that the currents are uniformly distributed over the strips, and assume that $h \gg d$ so that edge effects are small and the magnetic field between the strips can be treated as uniform. Find the self-inductance per unit length for the pair of strips.

13. Several inductors with self-inductances L_1, L_2, L_3, ... are connected in series [see Fig. 10.19(*a*)] or in parallel [see Fig. 10.19(*b*)]; their mutual inductances are negligible. Show that the effective net inductance of the series combination is

$$L = L_1 + L_2 + L_3 + \ldots$$

and that of the parallel combination is

$$L = \frac{1}{1/L_1 + 1/L_2 + 1/L_3 + \ldots}$$

Figure 10·19

(a) Inductors in series. *(b)* Inductors in parallel.

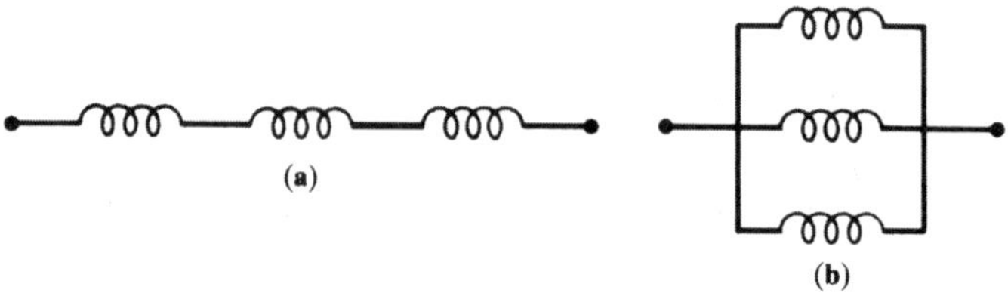

14. Two inductors with self-inductances L_1, L_2 and with mutual inductance M are connected in series. Show that the effective inductance of the combination is $L_1 + L_2 \pm 2M$, where the sign of the last term depends on whether the windings of the inductors are in the same direction or in opposite directions.

15. Two inductors with self-inductances L_1, L_2 and with mutual inductance M are connected in parallel. Show that the effective inductance of the combination is

$$L = \frac{L_1 L_2 - M^2}{L_1 + L_2 \mp 2M}$$

where the sign in the denominator depends on whether the windings of the inductors are in the same direction or in opposite directions.

16. Figure 10.20 shows a pair of very long, parallel wires with equal currents in opposite directions. A rectangular loop of wire is located in the plane of these wires. Calculate the emf induced around this loop if the currents on the wires are changing at the rate dI/dt. What is the mutual inductance between the loop and the pair of wires?

17. (a) A very long, cylindrical solenoid has radius R and n turns per unit length. What is the self-inductance per unit length?
(b) Suppose that the above solenoid is surrounded by another, concentric solenoid of radius $2R$ and n' turns per unit length. What is the mutual inductance?

18. A toroidal solenoid of square cross section has an inner radius R_1 and an outer radius R_2 (see Fig. 10.21). The total number of turns in the winding of the solenoid is N; assume N is very large. Find the self-inductance.

Figure 10·20

Two long, parallel wires carrying currents I and a rectangular loop of wire.

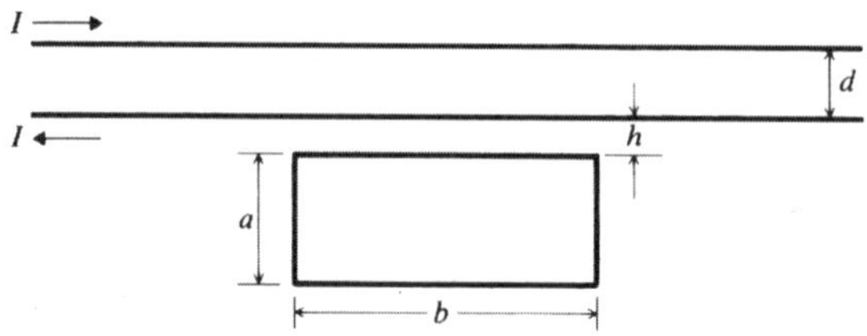

Figure 10·21

A toroidal solenoid of square cross section.

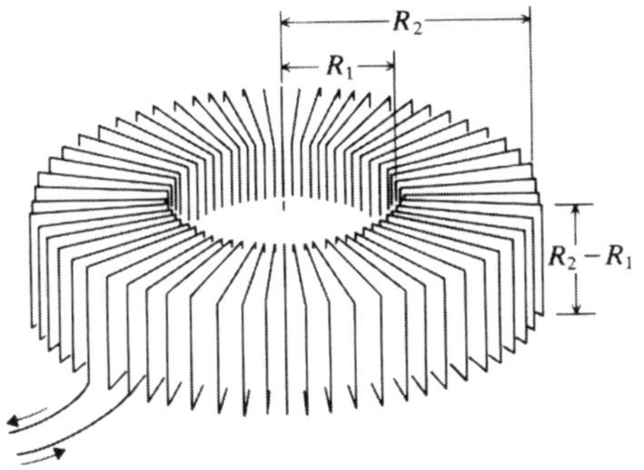

19. Figure 10.22 shows two circular rings of thin wire of radius R; their centers are separated by a distance r, and the planes of the rings are parallel. Calculate the mutual inductance as a function of the angle θ. Assume that $r \gg R$ so that the magnetic field of one ring is approximately uniform over the surface of the other. At what angle is the mutual inductance maximum?

20. An *electromagnetic launcher,* or rail gun, consists of two parallel conducting rails connected to a source of high voltage (usually a large capacitor) at one end. The projectile, a conducting mass, slides along the rails and completes the circuit (see Fig. 10.23). The large current in this circuit generates a strong magnetic field that interacts with the current flowing through the projectile and accelerates it along the rails.

Figure 10·22

Two parallel circular rings of wire.

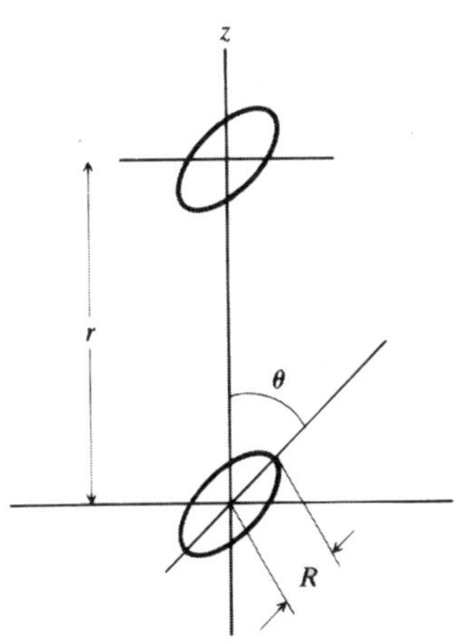

Figure 10·23

An electromagnetic launcher with a capacitor as source of emf.

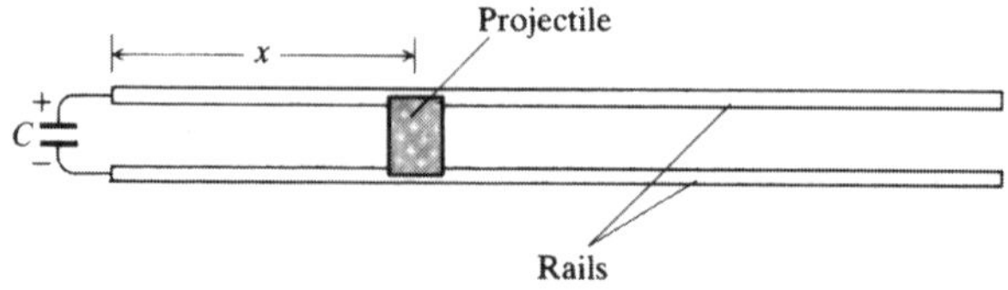

(a) Show that if the voltage source maintains a constant current in the rail circuit, the force on the projectile is directly proportional to the product of the square of the current and the self-inductance per unit length of the rails. [*Hint*: Neglect the resistance and show that the energy balance for the system can be written as $dW + \frac{1}{2}I^2 dL = I^2 dL$, where dW is the work done by the magnetic force on the projectile, $\frac{1}{2}I^2 dL$ is the increase of stored magnetic energy, and $I^2 dL$ is the work done by the voltage source to maintain the constant current.]

(b) Suppose that the inductance per unit length is $2/c^2$. Estimate the current required to give a projectile of 200 g a muzzle speed of 3000 m/s in a gun of length 10 m.

21. Three very long, concentric solenoids have radii $R_1 < R_2 < R_3$ and turns per unit length n_1, n_2, n_3. Calculate all the coefficients of inductance (per unit length); write your answer as a 3×3 matrix.

22. A small ring of thin wire of radius R is surrounded by a concentric large ring of radius R' lying in the same plane. Find an approximate expression for the mutual inductance. Your answer should be a power series in R/R'.

23. Find the self-inductance per unit length of a coaxial transmission line consisting of a long, solid, cylindrical rod of radius R_1 surrounded by a concentric cylindrical shell of radius R_2. [*Hint*: The current is uniformly distributed over the rod, which can be treated as an assembly of concentric shells.]

24. Consider a concentric arrangement of four long, conducting, cylindrical shells of radii R, $2R$, $3R$, and $4R$, respectively. A current I_1 flows along the shell of radius R and returns along the shell of radius $3R$; a current I_2 flows along the shell of radius $2R$ and returns along the shell of radius $4R$ (that is, the two pairs of shells form two separate circuits). Calculate the coefficients of inductance, per unit length, for this two-circuit system.

25. The differential equation for a freely oscillating LCR circuit (no external source of emf) is

$$L\frac{dI}{dt} + \frac{Q}{C} + RI = 0$$

(a) Solve this equation with the initial conditions $I = I_0$, $Q = Q_0$.

(b) Show that the energy in the circuit decreases exponentially, and find the time constant for the exponential decay.

26. From Eq. (49), deduce the instantaneous power that the external source of emf delivers to the circuit. Deduce the time-average power. Show that the time-average power is zero if $R = 0$.

27. Consider a driven LR circuit consisting of an inductor and a resistor connected in series to an external source of oscillating emf. Find the instantaneous steady-state current, and find the impedance. Show that your answers agree with Eqs. (47) and (48) in the limiting case $C \to \infty$.

28. A parallel LCR circuit consists of an inductor, a capacitor, a resistor, and an oscillating source of emf with $\mathcal{E} = V_0 \cos \omega t$, all connected in parallel (see Fig. 10.24). What is the (complex) impedance of this circuit? For what value of the frequency ω is the current maximum? Minimum?

29. Show that for a charged particle moving in a slowly varying magnetic field, the magnetic moment associated with the orbital motion is constant.

30. A proton of energy 10 MeV is initially in the magnetic equatorial plane of the Earth, at a distance of $1.5R_e$ from the center of the Earth (R_e is the Earth's radius). The magnetic field of the Earth is as described in Problem 9.22.

(a) What is the momentum of the proton in MeV/c? What is the magnitude of its velocity? If the velocity of the proton is in the equatorial plane, what will be the radius of its orbit?

(b) Suppose the initial velocity makes an angle α with the equatorial plane. At what latitude will the proton be reflected by the converging magnetic field lines? [*Hint:* The proton spirals around a field line whose equation is $r = 1.5R_e \sin^2\theta$ (see Problem 9.23).]

(c) Plot the latitude for reflection versus α. If α is too large, the proton will hit the surface of the Earth before it is reflected. What is the critical value of α?

31. A large, cylindrical solenoid has a uniform magnetic field in its interior. Under the influence of this magnetic field, a particle of charge q, momentum p is moving in a circular, planar orbit perpendicular to **B**, with center on the axis of the solenoid (see Fig. 10.25).

(a) At one instant, the particle is at point P. At this instant, the magnetic field is suddenly (almost instantaneously) switched off. Sketch the path of the particle after this instant. What is the new momentum of the particle?

(b) Suppose the magnetic field is not switched off suddenly but is decreased very gradually instead. Sketch the orbit for this case. What is the value of the momentum in the limit $B \to 0$? You may assume that the motion of the particle is nonrelativistic.

32. A coaxial transmission line consists of two concentric, cylindrical, thin conducting shells. The radius of the inner shell is a and that of the outer shell is b. The two shells carry AC currents of magnitude $I_0 \cos\omega t$ in opposite directions. Calculate the electric field induced between the shells, and then calculate the magnetic field induced by this electric field.

Figure 10·24

A parallel LCR circuit.

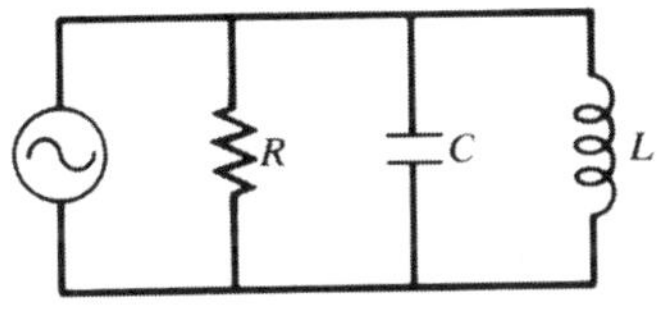

Figure 10·25

Solenoid with uniform magnetic field. The point P represents the initial instantaneous position of the particle.

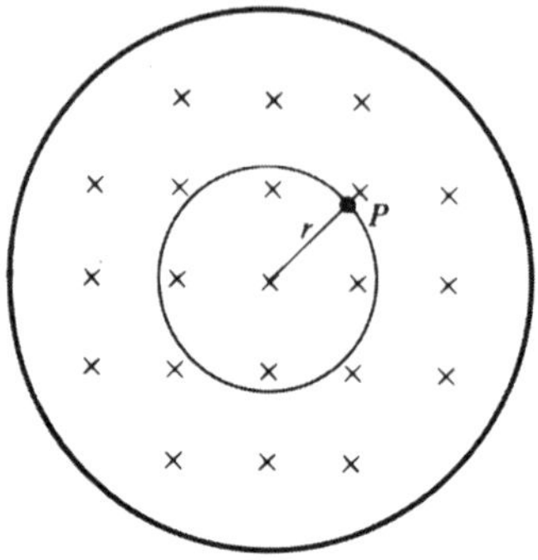

*33.** A parallel-plate capacitor with large circular plates is connected to an AC power supply. The charges on the plates then produce, in first approximation, a uniform electric field of magnitude

$$E_{(1)} = E_0 \sin \omega t$$

(a) As a function of the radial distance ρ from the axis, calculate the magnetic field $B_{(1)}$ induced by the electric field $E_{(1)}$ in the region between the plates (ignore fringing effects).
(b) Calculate the electric field $E_{(2)}$ induced by the magnetic field $B_{(1)}$. Assume that $E_{(2)}$ is zero on the axis of the capacitor.
(c) Calculate the magnetic field $B_{(2)}$ induced by the electric field $E_{(2)}$.
(d) This calculation can be continued; successive iterations yield a power series in powers of ρ. Stop the calculation at this stage, and plot the net electric field $E_{(1)} + E_{(2)}$ and the net magnetic field $B_{(1)} + B_{(2)}$ as a function of radial distance in the interval $\rho = 0$ to $\rho = 2c/\omega$.

34. A very long, cylindrical solenoid of radius R has an AC current in its windings. Within the solenoid, this current generates a uniform magnetic field

$$B_{(1)} = B_0 \cos \omega t$$

(a) As a function of the radial distance ρ from the axis of the solenoid, calculate the electric field $E_{(1)}$ induced by the magnetic field $B_{(1)}$.
(b) Calculate the magnetic field $B_{(2)}$ induced by the electric field $E_{(1)}$.
(c) Calculate the electric field $E_{(2)}$ induced by the magnetic field $B_{(2)}$.
(d) If we continue this calculation, successive iterations yield a power series in powers of ρ. Stop the calculation at this stage, and plot the net magnetic field $B_{(1)} + B_{(2)}$ and the net electric field $E_{(1)} + E_{(2)}$ as a function of ρ.

---

*Based on R. P. Feynman, R. B. Leighton, and M. Sands, *The Feynman Lectures on Physics.*

35. Show that the magnetic field of a uniformly moving point charge [see Eq. (9.4)] can be regarded as an induced magnetic field associated with the changing electric field in the laboratory frame. [*Hint:* Consider a *stationary* circular path centered on the line of motion of the point charge; calculate the displacement current intercepted by the area of the circle; then, use Eq. (66).]

36. Geomagnetic "storms" occur when clouds of plasma ejected from the Sun strike the Earth's magnetic field. The impact deforms the magnetic field, and this can induce electric fields along the surface of the Earth. The emf induced in a large grid of power lines stretched along the surface can be quite large, and the resulting surge of current can trigger safety switches and cause blackouts. Estimate the maximum emf that might be induced in a grid $1000\text{ km} \times 1000\text{ km}$ when the Earth's magnetic field changes by 10^{-2} gauss in a time of about 1 minute.

37. Since the displacement current has the same effect as a real current, we expect that it is possible to write a generalization of the Biot-Savart law (9.35) that includes a contribution from the displacement current:

$$\mathbf{B}(\mathbf{x}) = \frac{1}{c}\int \left(\mathbf{j}(\mathbf{x}') + \frac{1}{4\pi}\frac{\partial \mathbf{E}}{\partial t} \right) \times \frac{\mathbf{x}-\mathbf{x}'}{|\mathbf{x}-\mathbf{x}'|^3}\, dV'$$

An analogous expression for the electric field, including contributions from electric charges and from induction, is

$$\mathbf{E}(\mathbf{x}) = \int \left(\rho(\mathbf{x}') - \frac{1}{4\pi}\frac{\partial \mathbf{B}}{\partial t} \times \right) \frac{\mathbf{x}-\mathbf{x}'}{|\mathbf{x}-\mathbf{x}'|^3}\, dV'$$

These are a coupled system of integral equations for **E** and **B**. Show that these two equations are, indeed, valid by substituting them into Maxwell's equations and verifying that the results are identically true. [*Hint*: For evaluating, say, $\nabla \times \mathbf{B}$, use the double cross identity $\nabla \times [\mathbf{j}(\mathbf{x}') \times (\mathbf{x}-\mathbf{x}')] = \mathbf{j}(\mathbf{x}')\nabla \cdot (\mathbf{x}-\mathbf{x}') - \mathbf{j}(\mathbf{x}') \cdot \nabla(\mathbf{x}-\mathbf{x}')$ and then replace ∇ by $-\nabla'$; use Gauss' theorem in the first term to reduce the volume integral to a surface integral over a small sphere around the point $\mathbf{x}' = \mathbf{x}$; and then use integration by parts to change the second term into $\nabla' \cdot \mathbf{j}(\mathbf{x}')$.]

CHAPTER

11

Magnetic Dipoles and Magnetic Materials

Within the atoms of any piece of material, all the electrons are forever in motion, both orbital motion and spin motion. These circulating charges constitute currents, which give the atom a magnetic moment generating a magnetic field surrounding the atom. When we place the material in an external magnetic field, the atomic magnetic moments suffer readjustments, and the magnetic fields of these moments then modify the original external magnetic field. The magnitude of this modification depends on the kind of magnetic material. In paramagnetic and ferromagnetic materials, the main readjustment within the magnetic material is an alignment of the atomic magnetic moments with the external field. This enhances the magnetic field—in a ferromagnetic material, this enhancement can amount to a factor of several thousand. In diamagnetic materials, the main readjustment arises from currents and magnetic moments induced in the atoms by the external magnetic field. As we will see, this reduces the magnetic field.

11·1 Magnetic Dipole in an External Field

According to the results of Section 9.6, the torque that an external magnetic field **B** exerts on a closed loop carrying a steady current I is

$$\boldsymbol{\tau} = \mathbf{m} \times \mathbf{B} \tag{1}$$

where

$$\mathbf{m} = \frac{I}{2c}\int \mathbf{x}' \times d\mathbf{l} \tag{2}$$

is the magnetic moment of the loop (see Fig. 11.1).

Equation (1) is exact for a constant magnetic field, but it also is a good approximation for a small loop in an arbitrary magnetic field. In the latter case, the magnetic field in Eq. (1) must be evaluated in the vicinity of the loop. If the loop is at or near the origin (see Fig. 11.1), then we may evaluate **B** at the origin:

Figure 11·1

Current loop in a magnetic field.

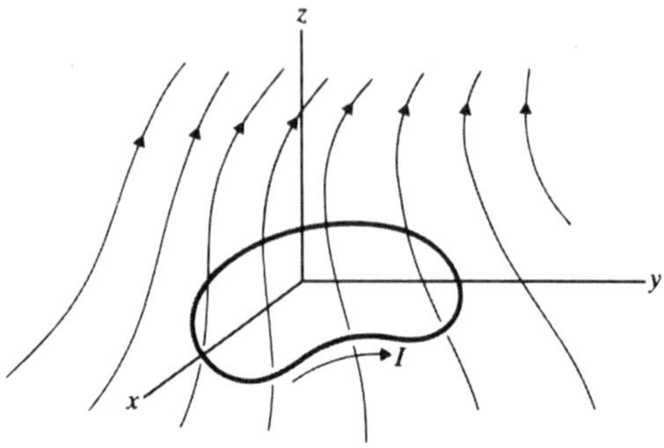

$$\boldsymbol{\tau} = \mathbf{m} \times \mathbf{B}(0) \tag{3}$$

If we define a potential energy

$$U = -\mathbf{m} \cdot \mathbf{B}(0) \tag{4}$$

then the above torque equals the negative derivative of the potential energy with respect to the angle of orientation of the dipole relative to the magnetic field.

■ ***Exercise 1.*** Verify this. [Hint: Write Eq. (4) as $U = -mB\cos\alpha$.] ■

When immersed in a magnetic field that depends on position, the loop will experience not only a torque but also a translational force. We can prove that this force equals the negative gradient of the potential energy defined above,

$$\mathbf{F} = -\nabla U = \nabla[\mathbf{m} \cdot \mathbf{B}(\mathbf{x})] \tag{5}$$

A simple proof of this expression for the force is as follows: Let us pretend that the loop is built of many small, square, plane loops (see Fig. 11.2), all carrying the same current. Wherever the adjacent small loops touch, their currents effectively cancel; the only remaining current is that flowing along the boundary of the original loop. The net magnetic effect of this assembly of small loops is then equivalent to that of the original loop, and it will therefore be sufficient to establish Eq. (5) for the small loops. Figure 11.3 shows one of the small loops, conveniently placed near the origin in the x-y plane. The force on this loop is

Figure 11·2

The current loop can be regarded as consisting of many small, square loops.

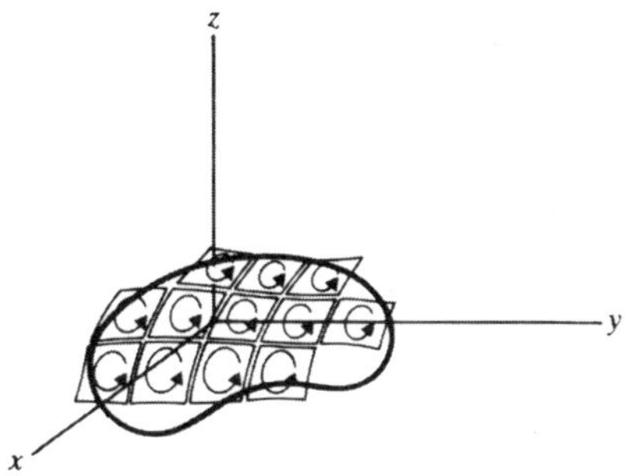

Figure 11·3

One of the small, square loops, located in the x-y plane.

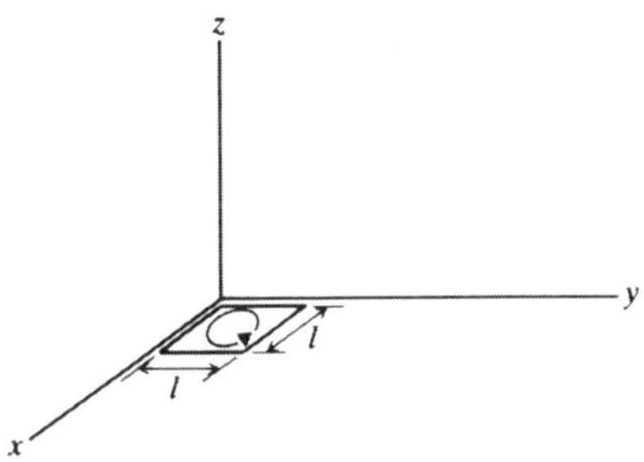

$$\mathbf{F} = \frac{I}{c}\int d\mathbf{x}' \times \mathbf{B} \tag{6}$$

Let us look at the z component of this force,

$$F_z = \frac{I}{c}\int (dx' B_y - dy' B_x) \tag{7}$$

For a small loop, it is a good approximation to use a Taylor-series expansion for the magnetic field; this gives us the following expression for the x component of the magnetic field (and similar expressions for the other components):

$$B_x(\mathbf{x}') = B_x(0) + x'\frac{\partial B_x}{\partial x} + y'\frac{\partial B_x}{\partial y} + z'\frac{\partial B_x}{\partial z} + \ldots \tag{8}$$

where all the derivatives are evaluated at the origin. The integral of the constant term $B_x(0)$ cancels upon integration in Eq. (7). Hence we can ignore this and other constant terms in the other components of the magnetic field. Integrating around the four sides of the loop, we obtain

$$F_z = \frac{I}{c}\left[\int_0^l x'\frac{\partial B_y}{\partial x}dx' - \int_0^l\left(l\frac{\partial B_x}{\partial x} + y'\frac{\partial B_x}{\partial y}\right)dy'\right]$$

$$+\frac{I}{c}\left[\int_l^0\left(l\frac{\partial B_y}{\partial y} + x'\frac{\partial B_y}{\partial x}\right)dx' - \int_l^0 y'\frac{\partial B_x}{\partial y}dy'\right] \qquad (9)$$

$$= -\frac{I}{c}l^2\left(\frac{\partial B_x}{\partial x} + \frac{\partial B_y}{\partial y}\right) \qquad (10)$$

Since $\nabla\cdot\mathbf{B} = \partial B_x/\partial x + \partial B_y/\partial y + \partial B_z/\partial z = 0$, this is the same as

$$F_z = \frac{Il^2}{c}\frac{\partial B_z}{\partial z} \qquad (11)$$

The magnetic moment of the small loop is $\mathbf{m} = \hat{\mathbf{z}}Il^2/c$, in terms of which Eq. (11) becomes

$$F_z = \frac{\partial}{\partial z}(\mathbf{m}\cdot\mathbf{B}) \qquad (12)$$

This is in agreement with Eq. (5). It is easy to check that F_y and F_x are also in agreement with this equation. This completes the derivation.

■ ***Exercise 2.*** Check F_y and F_z. ■

Note that, according to Eq. (5), a magnetic dipole will be attracted by a region of strong magnetic field if **m** is parallel to **B**, and it will be repelled if **m** is antiparallel to **B**. This gives an alternative explanation of the reflection effect for a particle spiraling in a magnetic field (see Section 10.4). The magnetic moment associated with the orbital motion of such a particle is always antiparallel to **B** (see Fig. 11.4); hence the force on the magnetic moment pushes the particle away from any region of high magnetic field.

Although the function U given by Eq. (4) is an effective potential energy for the rotational and translational motion of a dipole, it does not take into account all of the energy exchanges between the dipole and the magnetic field. When a current loop moves in a magnetic field, the induced emfs will do work on the current; this will increase or decrease the current (and the magnetic moment) unless some internal source of emf acts to keep the current constant. Equation (4) does not include this work,

Figure 11·4

Positively charged particle in a magnetic field. The magnetic moment associated with the circular component of the orbital motion is antiparallel to the magnetic field.

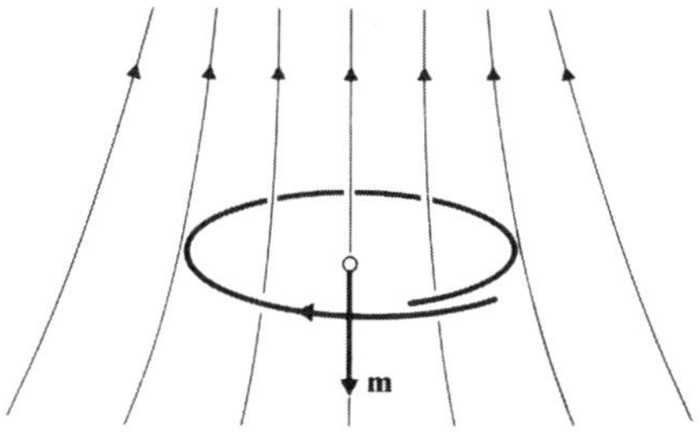

nor does it include the "internal" magnetic and other energy of the current loop.

The above expressions for the torque and the force are also valid for a continuous distribution of steady currents confined to some bounded volume. For such a distribution of currents, the definition of magnetic moment is

$$\mathbf{m} = \frac{1}{2c}\int \mathbf{x}'\times\mathbf{j}(\mathbf{x}')\,dV \qquad (13)$$

We could derive Eq. (13) by a generalization of the argument that led us to Eq. (2). But we will not bother because Eq. (13) is rather self-evident: In a continuous, steady current distribution, the charges move around closed loops (see Fig. 11.5). Hence this distribution can be regarded as a combination of many current loops, and the magnetic moment is the sum of the magnetic moments of these current loops. The two extra integrations in Eq. (13), as compared with Eq. (2), accomplish this summation over current loops.

A slight modification of Eq. (13) permits us to calculate the magnetic moment contributed by the orbital motion of the electrons in an atom. In this case, the integral must be replaced by a discrete sum over the individual electrons:

$$\mathbf{m}_{\text{orb}} = \frac{1}{2c}\sum_i (-e)(\mathbf{x}'_i\times\mathbf{v}_i) \qquad (14)$$

where $\mathbf{x}'_i$ and $\mathbf{v}_i$ are the positions and the velocities of the electrons. Since the orbital angular momenta of the electrons are $\mathbf{L}_i = m_{\text{e}}(\mathbf{x}'_i\times\mathbf{v}_i)$, Eq. (14) becomes

$$\mathbf{m}_{\text{orb}} = \frac{-e}{2m_e c}\sum_i \mathbf{L}_i = \frac{-e}{2m_e c}\mathbf{L} \qquad (15)$$

Figure 11·5

An example of a continuous current distribution: a rigidly rotating sphere carrying a uniform charge density. All the charges move along circular paths, which form closed current loops.

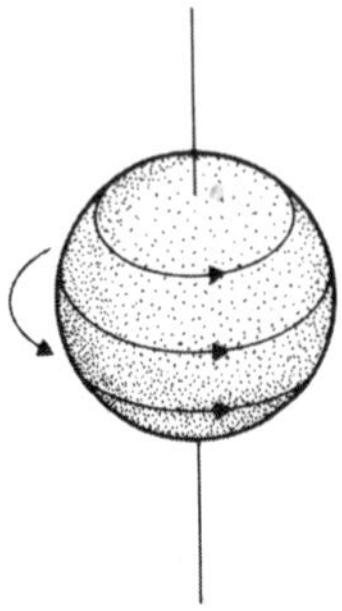

Thus the orbital contribution to the magnetic moment is directly proportional to the total angular momentum of the atom. This result is confirmed by a more sophisticated calculation based on quantum mechanics.

On the other hand, the spin contribution to the magnetic moment is not given by this formula. Instead, the spin angular momentum **s** of one or several electrons contributes a magnetic moment

$$\mathbf{m}_{\text{spin}} = -\frac{e}{2m_{\text{e}}c} g\mathbf{s} \tag{16}$$

where g is the gyromagnetic factor of the electron, $g = 2.0023$. For a single electron whose spin is directed along the z axis, $s_z = \hbar/2$ and

$$m_{\text{spin},z} = -\frac{e\hbar}{2m_{\text{e}}c}\frac{g}{2} \simeq -\frac{e\hbar}{2m_{\text{e}}c} \tag{17}$$

The spins of the other particles also contribute magnetic moments. For instance, the proton and the neutron with $s_z = \hbar/2$ have, respectively,

$$m_{\text{spin},z} = \frac{e\hbar}{2m_{\text{p}}c} \times 2.7928 \tag{18}$$

and

$$m_{\text{spin},z} = -\frac{e\hbar}{2m_{\text{p}}c} \times 1.9130 \tag{19}$$

where m_p is the mass of the proton. These magnetic moments are much smaller than that of the electron, and they usually can be neglected in the computation of the magnetic moment of an atom.

From a classical point of view, the discrepancy between Eqs. (15) and (16) indicates that the charge and mass within the rotating "body" of the electron are not equally distributed; to obtain $g > 1$, we need to concentrate the mass near the center and the charge near the surface. It is not worthwhile to investigate such mechanical models of the electron in detail, because the electron is a quantum-mechanical *point particle*, and its spin arises from relativistic quantum effects in the wave field of the electron rather than from rigid-body rotation. The spin is analogous to the angular momentum carried by a circularly polarized electromagnetic wave; as we will see in Section 12.3, such an angular momentum arises from the distribution of momentum density in the wave.

EXAMPLE: *precession of spin in a magnetic field*

However, regardless of the mechanics underlying the spin, we can investigate the motion of an atom or subatomic particle, with some given intrinsic angular momentum and magnetic moment, when placed in a uniform magnetic field. We know from Eq. (1) that on such a body there is a torque $\mathbf{m}\times\mathbf{B}$; this must equal the rate of change of spin:

$$\frac{d\mathbf{s}}{dt} = \mathbf{m}_{\text{spin}} \times \mathbf{B} \tag{20}$$

Thus, for an electron,

$$\frac{d\mathbf{s}}{dt} = -g\frac{e}{2m_e c}\mathbf{s}\times\mathbf{B} \tag{21}$$

This equation shows that the change of **s** is always perpendicular to **s**, and therefore **s** precesses around **B** (see Fig. 11.6). The precession frequency is

$$\omega = g\frac{eB}{2m_e c} \tag{22}$$

A quantum-mechanical calculation confirms this result for the frequency; however, the frequency must be interpreted as a transition frequency for the spin flip of the electron from the state with spin parallel to **B** ("spin up") to the state with spin antiparallel to **B** ("spin down").

The corresponding precession frequency for a proton is

$$\omega = 2.27928\frac{eB}{m_p c} \tag{23}$$

This frequency plays a crucial role in magnetic resonance imaging (MRI), where protons in the hydrogen atoms of a human body are subjected simultaneously to a strong static magnetic field and an extra oscil-

Figure 11·6

A magnetic field causes a precession of the electron spin around the direction of the field.

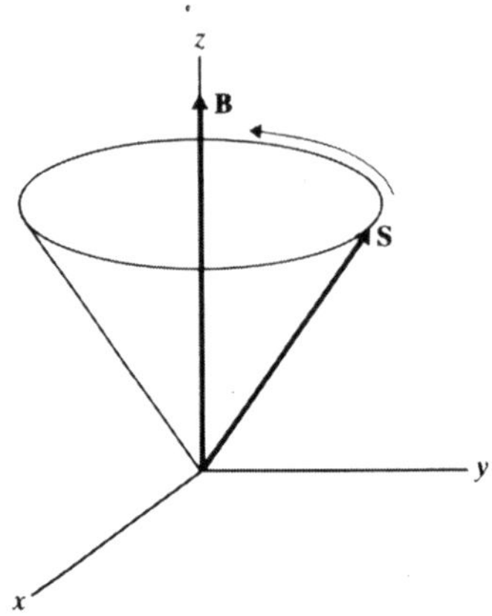

lating magnetic field that triggers resonant transitions at the precession frequency. The locations at which these transitions occur can be identified by means of a spatial modulation of the static field, and this information permits the construction of images of cross sections of the body.

Although the focus of this section has been the interaction of magnetic dipoles with an *external* magnetic field, it is well to keep in mind that the magnetic moment generates a magnetic field of its own. For instance, an electron at rest at the origin with its spin aligned with the z axis has a magnetic moment

$$m_{\text{spin},z} = -\frac{e\hbar}{2m_e c}\frac{g}{2}$$

and generates a magnetic field [see Eqs. (9.66) and (9.67)]

$$\mathbf{B} = -\frac{e\hbar}{2m_e c}\frac{g}{2}\left(\frac{2\cos\theta}{r^3}\hat{\mathbf{r}} + \frac{\sin\theta}{r^3}\hat{\boldsymbol{\theta}}\right) \qquad (24)$$

Thus the electron is always surrounded by both electric and magnetic fields (see Fig. 11.7). The magnetic fields generated by electrons control the behavior of magnetic materials.

Note that the formula (24) for the magnetic field of an individual electron (and also the formula for the electric field of an individual electron) must be used with some discretion. The trouble is that these fields are not directly measurable—the quantum uncertainties in the measurements of the fields of an individual electron are larger than the fields themselves. This is true both at large distances (where the fields are

Figure 11·7

Electric field (solid lines) and magnetic field (dashed lines) of an electron. The magnetic field is that of a dipole. The blank region near the center is occupied by the electron (more precisely, by its wave packet).

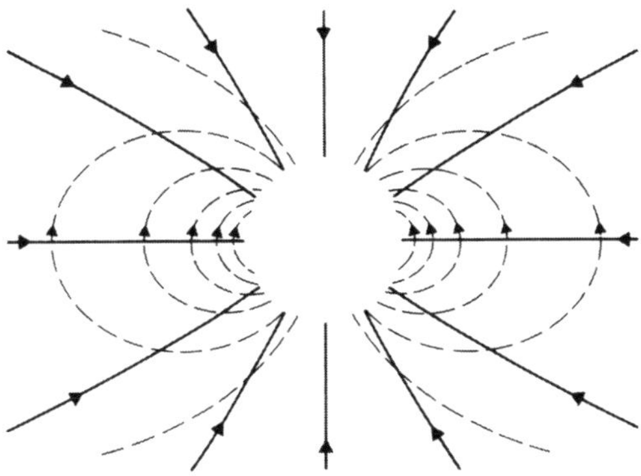

weak) and at short distances (where the fields are stronger, but the quantum uncertainties in the measurements are increased; see Problem 6). However, the magnetic and electric fields of electrons become measurable when the individual fields of a large number of electrons are combined. For instance, the combined magnetic fields of the aligned electrons in a ferromagnet are readily measurable, and so are the combined electric fields of the electrons accumulated on, say, the negative plate of a capacitor. Thus, although formulas such as (24) deal with quantities that are not directly measurable, they can be used to calculate the combined fields of a large number of electrons, and these combined fields are measurable.

11·2 The Magnetization

Because of the orbital motion and the spin of the electrons, most atoms have a permanent magnetic dipole moment. When we place a piece of material in an external magnetic field, the torques on these permanent dipole moments tend to align them with the field. The aligned dipole moments then produce a magnetic field of their own, which, as we will see, increases the strength of the original field. In *paramagnetic* materials, the increase of the original field is modest. But in *ferromagnetic* materials, the increase of the original field can be enormous—an increase by a factor of several thousand is not uncommon.

If the atoms of the material do not have any permanent magnetic dipole moment, then the magnetic effects of the material hinge on induced dipole moments. By Lenz' law, the direction of the induced dipole moment is opposite to that of the magnetic field. Such induced dipole moments therefore produce a magnetic field that reduces the strength of the original field. This is the mechanism underlying the behavior of *diamagnetic materials*. Incidentally: Since the atomic currents circulate in loops of finite size, they produce not only dipole fields but also higher multipoles. In the following calculations, we will neglect these higher multipoles because they are small.

Let us designate the *average* magnetic dipole moment of one atom or molecule by **m**. In the presence of an external magnetic field, this average magnetic dipole moment will be different from zero, and it will lie along the direction of the magnetic field. We can then define the magnetic moment per unit volume as

$$\mathbf{M} = n\mathbf{m} \tag{25}$$

with n being the number of atoms per unit volume. The vector **M** is called the *magnetization*. If the material, or the magnetic field, or both are inhomogeneous, then **M** will be a function of position. We want to express the magnetic field contributed by the dipoles in the material as an integral over **M**. To do this, let us begin by writing the expression for the magnetic field of one single atomic dipole **m** located at the position $\mathbf{x}'$. According to Eq. (9.75), the magnetic potential for such a dipole is

$$\Phi_{\mathrm{M}}(\mathbf{x}) = \frac{\mathbf{m}\bullet(\mathbf{x}-\mathbf{x}')}{|\mathbf{x}-\mathbf{x}'|^3} \tag{26}$$

$$= -\mathbf{m}\bullet\nabla\frac{1}{|\mathbf{x}-\mathbf{x}'|} \tag{27}$$

Hence

$$\mathbf{B}(\mathbf{x}) = -\nabla\Phi_{\mathrm{M}}(\mathbf{x}) = \nabla\left(\mathbf{m}\bullet\nabla\frac{1}{|\mathbf{x}-\mathbf{x}'|}\right) \tag{28}$$

By means of the operator identity

$$\nabla\times(\mathbf{m}\times\nabla) = \mathbf{m}\nabla\bullet\nabla - \nabla(\mathbf{m}\bullet\nabla) \tag{29}$$

we can write this as

$$\mathbf{B}(\mathbf{x}) = \mathbf{m}\nabla^2\frac{1}{|\mathbf{x}-\mathbf{x}'|} - \nabla\times\left(\mathbf{m}\times\nabla\frac{1}{|\mathbf{x}-\mathbf{x}'|}\right) \tag{30}$$

But $\nabla^2(1/|\mathbf{x}-\mathbf{x}'|) = 0$, so

$$\mathbf{B}(\mathbf{x}) = -\nabla\times\left(\mathbf{m}\times\nabla\frac{1}{|\mathbf{x}-\mathbf{x}'|}\right) \tag{31}$$

Since $\mathbf{B} = \nabla \times \mathbf{A}$, this equation tells us that the vector potential of a single dipole may be taken as

$$\mathbf{A}(\mathbf{x}) = -\mathbf{m}\times\nabla\frac{1}{|\mathbf{x}-\mathbf{x}'|} \tag{32}$$

We are now ready to calculate the magnetic field contributed by a piece of magnetized material. The dipole moment associated with a small volume dV' is $\mathbf{M}(\mathbf{x}')dV'$, and according to Eq. (32), this gives the following contribution to the vector potential:

$$d\mathbf{A}(\mathbf{x}) = -\mathbf{M}(\mathbf{x}')\times\nabla\frac{1}{|\mathbf{x}-\mathbf{x}'|}dV' \tag{33}$$

Next we must integrate this over the volume of the magnetized material, which yields

$$\mathbf{A}(\mathbf{x}) = -\int \mathbf{M}(\mathbf{x}')\times\nabla\frac{1}{|\mathbf{x}-\mathbf{x}'|}dV' \tag{34}$$

$$= +\int \mathbf{M}(\mathbf{x}')\times\nabla'\frac{1}{|\mathbf{x}-\mathbf{x}'|}dV' \tag{35}$$

$$= -\int \nabla'\times\frac{\mathbf{M}(\mathbf{x}')}{|\mathbf{x}-\mathbf{x}'|}dV' + \int \frac{\nabla'\times\mathbf{M}(\mathbf{x}')}{|\mathbf{x}-\mathbf{x}'|}dV' \tag{36}$$

The first of these volume integrals can be transformed into a surface integral by means of the identity

$$\int \nabla'\times\mathbf{F}(\mathbf{x}')dV' = -\int_{S'} \mathbf{F}(\mathbf{x}')\times d\mathbf{S}' \tag{37}$$

which is valid for any vector function $\mathbf{F}(\mathbf{x})$.

■ ***Exercise 3.*** Prove this vector identity. [*Hint:* Do this first for a cubical volume, treating the components of Eq. (37) separately.] ■

With this transformation, Eq. (36) becomes

$$\mathbf{A}(\mathbf{x}) = \int_{S'} \frac{\mathbf{M}(\mathbf{x}')}{|\mathbf{x}-\mathbf{x}'|} \times d\mathbf{S}' + \int \frac{\nabla' \times \mathbf{M}(\mathbf{x}')}{|\mathbf{x}-\mathbf{x}'|} dV' \tag{38}$$

where the first integration is over the surface of the piece of magnetized material. To this expression, we must still add the vector potential contributed by all the conduction currents arising from the motion of free charges. Designating the real current density by $\mathbf{j}_F$, we then obtain the net result [see Eq. (9.32)]

$$\mathbf{A}(\mathbf{x}) = \frac{1}{c}\int \frac{\mathbf{j}_F(\mathbf{x}')}{|\mathbf{x}-\mathbf{x}'|} dV' + \int_{S'} \frac{\mathbf{M}(\mathbf{x}')}{|\mathbf{x}-\mathbf{x}'|} \times d\mathbf{S}' + \int \frac{\nabla' \times \mathbf{M}(\mathbf{x}')}{|\mathbf{x}-\mathbf{x}'|} dV' \tag{39}$$

Note that this equation is valid regardless of whether the point $\mathbf{x}$ is outside or inside the magnetized material [see the discussion given in connection with Eq. (4.4)].

If we compare the first term in this equation with the other two, we see that the magnetized material acts as though it had a current density

$$\mathbf{j}_M(\mathbf{x}') = c\nabla' \times \mathbf{M}(\mathbf{x}') \tag{40}$$

distributed over its volume, and a surface current density

$$\mathbf{k}_M(\mathbf{x}') = c\mathbf{M}(\mathbf{x}') \times \hat{\mathbf{n}}' \tag{41}$$

distributed over its surface. These are called the *magnetization current densities*.

The magnetization currents result from the flow of charges within the atoms or molecules. We can understand how these currents come about by the following simple argument: Figure 11.8 shows a piece of magnetized material with a uniform distribution of aligned magnetic

Figure 11·8

A piece of uniformly magnetized material containing small current loops.

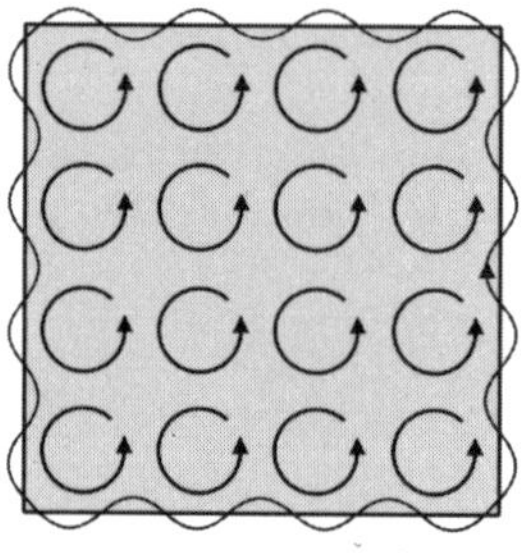

Figure 11·9

A piece of material with a nonuniform magnetization. The magnetization is larger on the left, where there is a higher density of the small current loops.

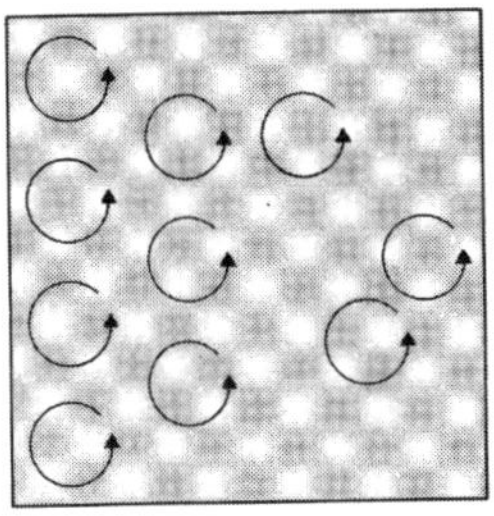

dipoles. The currents of the small current loops then cancel everywhere within the material, leaving an effective current only along the surface of the material—this is the surface current density of Eq. (41). If we are to obtain a volume current density, then we need a material with a nonuniform magnetization. For example, Fig. 11.9 shows a material within a nonuniform density of aligned dipoles. Here the adjacent currents of the small loops fail to cancel, and the net result is an effective current in the upward direction in the volume of the material. This is the volume current density of Eq. (40).

11·3 The H Field

Consider a magnetic material in which there are both conduction currents, or currents of free charges, and magnetization currents. The effects of the magnetic dipoles in the material are entirely determined by the magnetization currents they produce, and we can write Eq. (9.20) for the static magnetic field at any point in the interior of the material as follows:

$$\nabla \times \mathbf{B} = \frac{4\pi}{c}(\mathbf{j}_F + \mathbf{j}_M) \tag{42}$$

$$= \frac{4\pi}{c}\mathbf{j}_F + 4\pi \nabla \times \mathbf{M} \tag{43}$$

where $\mathbf{j}_F$ is the conduction current density, or free-charge current density, and $\mathbf{j}_M = c\nabla \times \mathbf{M}$ is the magnetization current density. Equation (43) is equivalent to

$$\nabla \times (\mathbf{B} - 4\pi\mathbf{M}) = \frac{4\pi}{c}\mathbf{j}_F \tag{44}$$

Here we can define a new field quantity

$$\mathbf{H} \equiv \mathbf{B} - 4\pi\mathbf{M} \tag{45}$$

in terms of which

$$\nabla \times \mathbf{H} = \frac{4\pi}{c}\mathbf{j}_F \tag{46}$$

We will simply call this new field the **H** field. Most textbooks follow an old tradition, according to which this **H** field is called the "magnetic field," whereas the **B** field is called the "magnetic induction." That is a very confusing terminology, which we will take pains to avoid. As Chapters 8–10 make quite clear, **B** is the fundamental magnetic field—it is the magnetic analog of the electric field **E**. On the other hand, **H** is merely an auxiliary mathematical artifact designed to simplify Eq. (44)—it is the analog of the displacement field **D**. Historically, the unfortunate confusion in terminology arose from the inability of nineteenth-century electricians to draw a clear distinction between the vacuum and material media;* they regarded the vacuum as a material medium consisting of "ether," and they failed to appreciate the fundamental significance of Maxwell's equations in vacuum.

Note that the magnetization vector **M**(**x**), like the polarization vector **P**(**x**) of Chapter 4, is defined only in an average sense: we must take the average of the atomic dipole moments over a small volume surrounding the point **x**, a volume small by macroscopic standards but large enough to contain many atoms. For the sake of the consistency of Eq. (45), we will then also have to regard the magnetic **B** field and the **H** field appearing in the equations of this section as average, macroscopic fields.

For the solution of problems in magnetostatics in the presence of magnetic materials, besides Eq. (46) we must also take into account the other Maxwell equation for the magnetic field:

$$\nabla \cdot \mathbf{B} = 0 \tag{47}$$

These two equations are sufficient provided we know the relationship between **B** and **H** or, equivalently, the relationship between **M** and **H**. In *linear, isotropic* magnetic materials, the magnetization is directly proportional to the **H** field,

$$\mathbf{M} = \chi\mathbf{H} \tag{48}$$

The constant χ is called the magnetic *susceptibility*.† This yields

---

*The SI system of units perpetuates this misunderstanding by assigning to the vacuum a permittivity and permeability, as though it were a material medium.

†Note that, by analogy with Eq. (4.19), it would have made more sense to write $\mathbf{M} = \chi\mathbf{B}$. But Eq. (48) is sanctified by tradition.

$$\mathbf{B} = \mathbf{H} + 4\pi\mathbf{M} = (1 + 4\pi\chi)\mathbf{H} = \mu\mathbf{H} \tag{49}$$

where the constant is called the magnetic *permittivity*.

The linear relation (48) is a good approximation in paramagnetic and in diamagnetic materials. Paramagnets have $\chi > 0$ and $\mu > 1$, so the magnetic field in the material is increased beyond the value it would have if the material were absent. Diamagnets have $\chi < 0$ and $\mu < 1$, so the magnetic field in the material is decreased. As mentioned in Section 11.2, this decrease of the original magnetic field results from the production of induced dipole moments in the material, which are antiparallel to the magnetic field. Since all kinds of atoms exhibit this induction effect when placed in a magnetic field, all kinds of materials have some degree of diamagnetism. But in most materials, the diamagnetic effects are overshadowed by paramagnetic effects.* Table 11.1 lists the values of the magnetic susceptibility for a few paramagnetic and diamagnetic materials.

The linear relation (48) fails in ferromagnetic materials. In these materials, **B** and **M** are complicated nonlinear functions of **H**. Figure 11.10 shows a plot of B as a function of H for commercial iron. The function $B = B(H)$ is nonlinear and multiple-valued. It consists of several branches that depend on the previous history of the material; this dependence on previous history is called *hysteresis*. For instance, if the iron sample is initially annealed (heat-treated), then $B = 0$ for $H = 0$. Suppose that we now place the sample in an electromagnet and gradually increase H. This will lead to a sharp increase of B, until B reaches about 10^4 gauss (see Fig. 11.10). Beyond this point, B increases only slowly, in proportion to the increase of H (the slope of the curve tends toward 1). What has happened is that at 10^4 gauss, all of the available magnetic dipoles in the iron have achieved perfect alignment; thus M has reached its maximum value (saturation), and any further increase of B must come directly from H [see Eq. (45)]. If we begin to decrease H, B will decrease according to the upper branch of Fig. 11.10. When H reaches zero, B will not be zero; this means that the sample of iron has become a permanent magnet. To bring B back to zero, we need to make H negative. As we continue to vary H, B will vary according to the outer loop of Fig. 11.10; this is a *hysteresis loop*. The size of the hysteresis loop depends on the initial increment in H, starting with the annealed sample ($H = 0$, $B = 0$). If we begin to decrease H before the material reaches the saturation point, then B will vary according to a smaller hysteresis loop.

---

* According to a theorem of classical statistical mechanics (van Leeuwen's theorem), the diamagnetic contribution to the magnetization exactly cancels the paramagnetic contribution in any system of pointlike (spinless) classical particles in thermal equilibrium. This theorem is no more than an irrelevant curiosity, because the particles within atoms do not obey the laws of classical physics.

Table 11·1 Susceptibilities of some materials*

Paramagnetic Materials	$4\pi\chi$
Air	30.4×10^{-5}
Oxygen	133×10^{-5}
Oxygen (liquid, –190°C)	327×10^{-5}
Nickel monoxide	67.5×10^{-5}
Manganese chloride	134×10^{-5}
Manganese	12.4×10^{-5}
Platinum	1.38×10^{-5}
Aluminum	0.817×10^{-5}
Diamagnetic Materials	$4\pi\chi$
Bismuth	-1.9×10^{-5}
Beryllium	-1.3×10^{-5}
Methane	-3.1×10^{-5}
Ethylene	-2.0×10^{-5}
Ammonia	-1.4×10^{-5}
Carbon dioxide	-0.53×10^{-5}
Glass (heavy flint)	-1.5×10^{-5}

*At room temperature (20°C) and 1 atm, unless otherwise noted.

Figure 11·10

B as a function of H for commercial iron. Note that the two axes are plotted on very different scales. (From R. M. Bozorth, *Ferromagnetism*.)

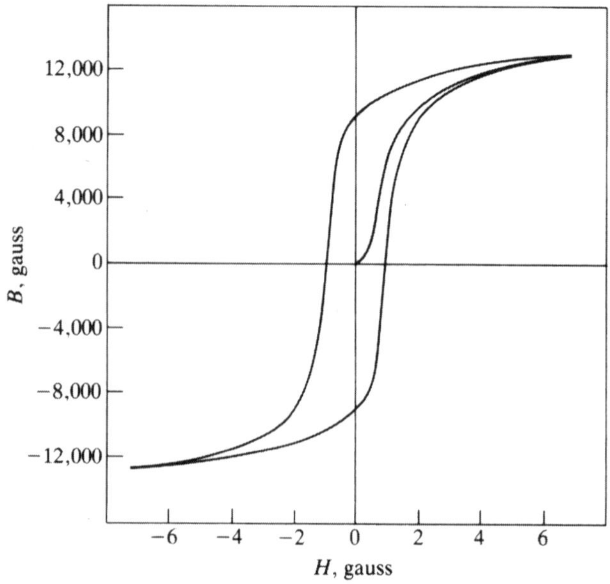

The magnetization in ferromagnetic materials is due to the spin magnetic moments of electrons. The magnetization is a very complicated function of H because the electron spins in ferromagnetic materials tend to align spontaneously, even in the absence of any external magnetic field. These spontaneous alignments are caused by a quantum-mechanical force acting between neighboring spins. On a small scale, a ferromagnetic material is always magnetized; it consists of a large number of small domains, of the order of a millimeter across, within each of which the spins of the valence electrons are perfectly aligned. However, on a large scale, this magnetization does not show up because the domains are oriented at random. When we place the material in an external magnetic field, the domains tend to orient along the magnetic field, leading to a large-scale magnetization.

The magnitude of the magnetization attained by the domains is a decreasing function of temperature, because the alignment of the spins is inhibited by the random thermal disturbances. At a critical temperature, called the *Curie temperature*, the thermal disturbances break the quantum-mechanical bonds that hold neighboring spins in alignment. Beyond this temperature, the ferromagnet behaves as a simple paramagnet. For iron, the Curie temperature is 1040 K.

11·4 The Magnetic Potential and the Boundary-Value Problem

We have already discovered that a magnetic potential can be useful in the solution of magnetic field problems in a region without currents. According to the theorem of Section 9.4, the existence of a magnetic potential for **B** in some region hinges on the condition $\nabla \times \mathbf{B} = 0$. This condition is satisfied in the absence of currents, and it ensures that in a simply-connected region the magnetic field **B** can be expressed as the gradient of a potential. Since magnetic materials carry magnetization currents, in the presence of which $\nabla \times \mathbf{B} \neq 0$, the theorem is not valid in such materials. However, the magnetic **H** field satisfies the condition $\nabla \times \mathbf{H} = 0$ wherever the conduction currents are zero, regardless of the presence of magnetization currents. Hence the theorem of Section 9.4 is valid for the magnetic **H** field: in any simply-connected region without conduction currents, this field can be expressed as the gradient of a magnetic potential,

$$\mathbf{H} = -\nabla\Phi_M \tag{50}$$

The Poisson equation for Φ_M follows from $\nabla\cdot\mathbf{B} = 0$:

$$\nabla\cdot\mathbf{B} = \nabla\cdot(\mathbf{H} + 4\pi\mathbf{M}) = -\nabla^2\Phi_M + 4\pi\nabla\cdot\mathbf{M} = 0 \tag{51}$$

or

$$\nabla^2 \Phi_M = 4\pi \nabla \cdot \mathbf{M} \tag{52}$$

If we define a virtual *magnetic pole density*

$$\rho_M = -\nabla \cdot \mathbf{M} \tag{53}$$

then the Poisson equation for the magnetic potential takes the familiar form

$$\nabla^2 \Phi_M = -4\pi \rho_M \tag{54}$$

In connection with Eq. (54), we must keep in mind that the magnetic pole or charge density ρ_M is nothing but a mathematical artifact; Eq. (54) does not imply the existence of real magnetic charges.

From our study of electrostatics, we know that the potential of a continuous charge distribution is simply an integral over the charge distribution; thus

$$\Phi_M(\mathbf{x}) = \int \frac{\rho_M(\mathbf{x}')}{|\mathbf{x} - \mathbf{x}'|} dV' = -\int \frac{\nabla' \cdot \mathbf{M}(\mathbf{x}')}{|\mathbf{x} - \mathbf{x}'|} dV' \tag{55}$$

Here, the volume of integration must enclose all of the magnetized body (including its surface). Since the magnetization has a discontinuity at the surface of the body (it is zero outside, nonzero inside), the derivatives in $\nabla' \cdot \mathbf{M}(\mathbf{x}')$ have singularities at the surface. It is best to isolate these singularities by separating Eq. (55) into a volume integral over the interior of the magnetized body (excluding the surface) plus a separate integral over the surface. Figure 11.11 shows the volume of integration for Eq. (55); this volume consists of a volume entirely within the body plus

Figure 11·11

The volume of integration encloses the magnetized body (shaded), including its surface. This volume consists of one portion entirely within the body plus a second portion in the shape of a thin skin bracketing the surface of the body. Note that the normal vector $d\mathbf{S}'$ at the inside of the skin has been assigned the direction opposite to what is customary for Gauss' theorem; hence the difference between the signs of the two terms in Eq. (57).

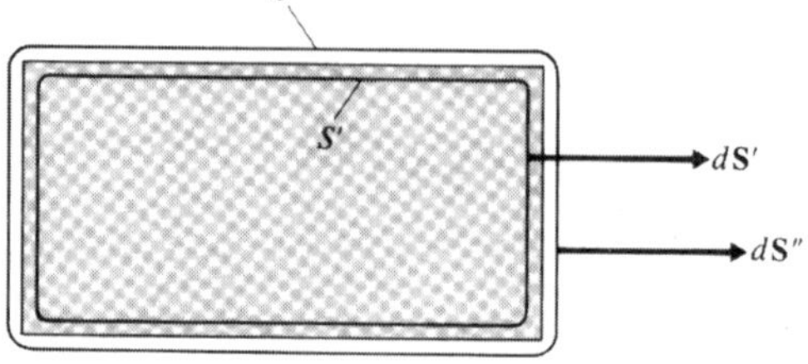

the volume of a thin skin bracketing the surface:

$$-\int_{\text{skin}} \frac{\nabla'\cdot\mathbf{M}}{|\mathbf{x}-\mathbf{x}'|}dV' = -\int_{\text{skin}} \nabla'\cdot\left(\frac{\mathbf{M}}{|\mathbf{x}-\mathbf{x}'|}\right)dV' + \int_{\text{skin}} \mathbf{M}\cdot\nabla'\frac{1}{|\mathbf{x}-\mathbf{x}'|}dV' \tag{56}$$

The first integral on the right side can be transformed by Gauss' theorem into a surface integral over the surfaces S' and S'' shown in Fig. 11.11. The second integral has no significant singularities in its integrand; hence, if we proceed to the limit of infinitesimal skin thickness, this integral will vanish. We are then left with

$$+\int_{S'} \frac{\mathbf{M}}{|\mathbf{x}-\mathbf{x}'|}\cdot d\mathbf{S}' - \int_{S''} \frac{\mathbf{M}}{|\mathbf{x}-\mathbf{x}'|}\cdot d\mathbf{S}'' \tag{57}$$

On the surface S'', which is just outside the magnetized body, $\mathbf{M} = 0$. On the surface S', which is just inside the body, $\mathbf{M} \neq 0$. Hence the first term in Eq. (57) represents the entire contribution due to the thin skin. With this, our expression for Φ_M becomes

$$\Phi_M(\mathbf{x}) = -\int \frac{\nabla'\cdot\mathbf{M}(\mathbf{x}')}{|\mathbf{x}-\mathbf{x}'|}dV' + \int_{S'} \frac{\mathbf{M}(\mathbf{x}')}{|\mathbf{x}-\mathbf{x}'|}\cdot d\mathbf{S}' \tag{58}$$

where it is now understood that the volume integration excludes the surface of the body. Obviously, this equation ascribes to the surface of the body a virtual magnetic pole density

$$\sigma_M = \mathbf{M}\cdot\hat{\mathbf{n}} \tag{59}$$

If the magnetization $\mathbf{M}(\mathbf{x}')$ is prescribed, Eq. (58) gives an explicit solution for the magnetic potential. However, it is often more convenient to return to the Poisson equation (54) and treat its solution as a boundary-value problem. We will work out an example of this procedure (see below), but we must first lay down the boundary conditions for the magnetic fields. Since $\nabla \times \mathbf{H} = 0$ in a region free of conduction currents, and since $\nabla\cdot\mathbf{B} = 0$, the arguments given in Section 4.3 [see Eqs. (4.26)–(4.30)] tell us that the tangential component of $\mathbf{H}$ and the normal component of $\mathbf{B}$ must be continuous across the interface of two magnetic media:

$$\mathbf{H}_{1\parallel} = \mathbf{H}_{2\parallel} \tag{60}$$

$$\mathbf{B}_{1\perp} = \mathbf{B}_{2\perp} \tag{61}$$

■ ***Exercise 4.*** Derive the boundary conditions in (60) and (61). ■

■ ***Exercise 5.*** Show that the discontinuity of the perpendicular component of **H** is directly proportional to the magnetic pole density at the interface:

$$\mathbf{H}_{2\perp} - \mathbf{H}_{1\perp} = 4\pi\sigma_{\mathrm{M}} \tag{62}$$

■

EXAMPLE: *magnetic potential and field of magnetized sphere*

We are now ready for an example. Figure 11.12 shows a sphere of radius R made of uniformly magnetized material. The magnetization **M** is along the z direction. We want to find the magnetic fields in the regions $r > R$ and $r < R$. In this problem, the conduction current is zero everywhere—the magnetic effects are entirely due to the permanent magnetism of the sphere. The magnetic potential is therefore well defined throughout the space. With a uniform magnetization $\mathbf{M} = M\hat{\mathbf{z}}$, the magnetic pole density is zero in the interior volume of the sphere:

$$\rho_{\mathrm{M}} = -\nabla \cdot \mathbf{M} = 0 \tag{63}$$

and it is

$$\sigma_{\mathrm{M}} = \mathbf{M} \cdot \hat{\mathbf{n}} = M\cos\theta \tag{64}$$

along the surface of the sphere.

Our problem then reduces to finding the potential generated by the surface charge distribution (64). We could obtain the answer by evaluating the surface integral in Eq. (58), but we will choose to deal with Eq. (54). With $\rho_{\mathrm{M}} = 0$, this is the Laplace equation. From Chapter 3, we know that the general solution of this equation is

$$\Phi_{\mathrm{M1}} = \sum_{l=1}^{\infty} A_l \frac{P_l(\cos\theta)}{r^{l+1}} \qquad \text{for} \qquad r > R \tag{65}$$

$$\Phi_{\mathrm{M2}} = \sum_{l=0}^{\infty} B_l r^l P_l(\cos\theta) \qquad \text{for} \qquad r < R \tag{66}$$

Here the $l = 0$ term has been omitted from Eq. (65); this is required by the absence of a net "magnetic charge" (σ_{M} is positive on the upper

Figure 11·12

A uniformly magnetized sphere. The magnetization **M** is in the z direction.

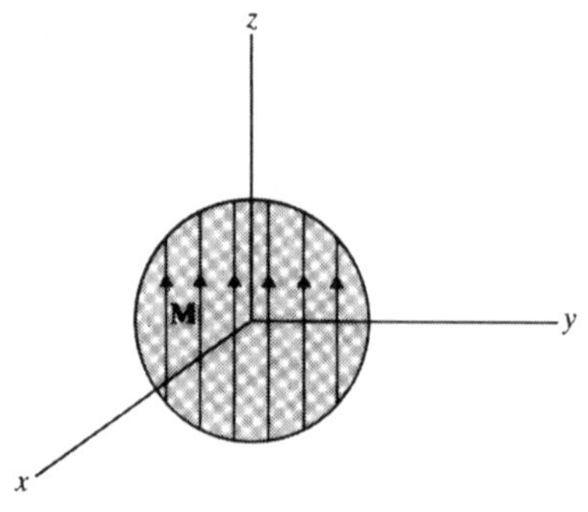

hemisphere but negative on the lower). Our boundary conditions at $r = R$ are the continuity of $\mathbf{H}_{\parallel}$,

$$\frac{\partial \Phi_{M1}}{\partial \theta} = \frac{\partial \Phi_{M2}}{\partial \theta} \tag{67}$$

and the continuity of $\mathbf{B}_{\perp}$, or of $\mathbf{H}_{\perp} + 4\pi \mathbf{M}_{\perp}$,

$$-\frac{\partial \Phi_{M1}}{\partial r} = -\frac{\partial \Phi_{M2}}{\partial r} + 4\pi M \cos\theta \tag{68}$$

As in the solution of the problem of the dielectric sphere (see Section 4.3), we can replace Eq. (67) by the equivalent condition

$$\Phi_{M1} = \Phi_{M2} \tag{69}$$

With our formulas for Φ_{M1} and Φ_{M2}, these boundary conditions become

$$\sum_{l=1}^{\infty} A_l \frac{P_l(\cos\theta)}{R^{l+1}} = \sum_{l=0}^{\infty} B_l R^l P_l(\cos\theta) \tag{70}$$

and

$$\sum_{l=1}^{\infty} (l+1) A_l \frac{P_l(\cos\theta)}{R^{l+2}} = -\sum_{l=1}^{\infty} l B_l R^{l-1} P_l(\cos\theta) + 4\pi M \cos\theta \tag{71}$$

■ ***Exercise 6.*** Show that the solution of these equations yields

$$A_1 = \frac{4\pi}{3} M R^3 \tag{72}$$

$$B_1 = \frac{4\pi}{3} M \tag{73}$$

with all other A_l and B_l equal to zero. ■

The magnetic potential is then

$$\Phi_{M1} = \frac{4\pi}{3} \frac{MR^3}{r^2} \cos\theta \qquad \text{for} \qquad r > R \tag{74}$$

and

$$\Phi_{M2} = \frac{4\pi}{3} M r \cos\theta = \frac{4\pi}{3} M z \qquad \text{for} \qquad r < R \tag{75}$$

From this, the **H** field outside the sphere is found to be

$$\mathbf{H} = -\nabla\Phi_{\mathrm{M1}} = \frac{4\pi}{3} MR^3 \left(\frac{2\cos\theta}{r^3}\hat{\mathbf{r}} + \frac{\sin\theta}{r^3}\hat{\boldsymbol{\theta}} \right) \tag{76}$$

whereas inside the sphere it is found to be

$$\mathbf{H} = -\nabla\Phi_{\mathrm{M2}} = -\frac{4\pi}{3} M\hat{\mathbf{z}} = -\frac{4\pi}{3}\mathbf{M} \tag{77}$$

The magnetic field **B** outside the sphere is the same as the **H** field. The magnetic field **B** inside the sphere is

$$\mathbf{B} = \mathbf{H} + 4\pi\mathbf{M} = -\frac{4\pi}{3}\mathbf{M} + 4\pi\mathbf{M} = \frac{8\pi}{3}\mathbf{M} \tag{78}$$

Figures 11.13 and 11.14 show the field lines for **H** and for **B**. Inside the sphere, the fields are uniform. Outside the sphere, the fields are those of an ideal magnetic dipole of moment $(4\pi/3)\mathbf{M}R^3$ (this is the product of the volume and the dipole moment per unit volume, as we might have expected).

Figure 11·13

The **H** field of the uniformly magnetized sphere. Note that the surface of the sphere acts as a source of field lines of **H**.

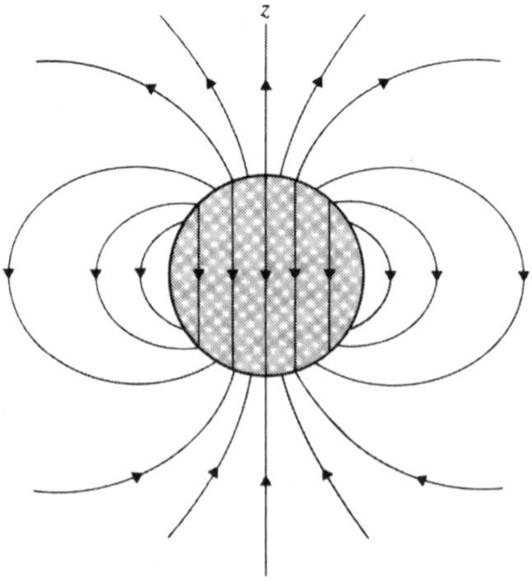

Figure 11·14

The magnetic field **B** of the uniformly magnetized sphere. The lines of **B** have no sources or sinks.

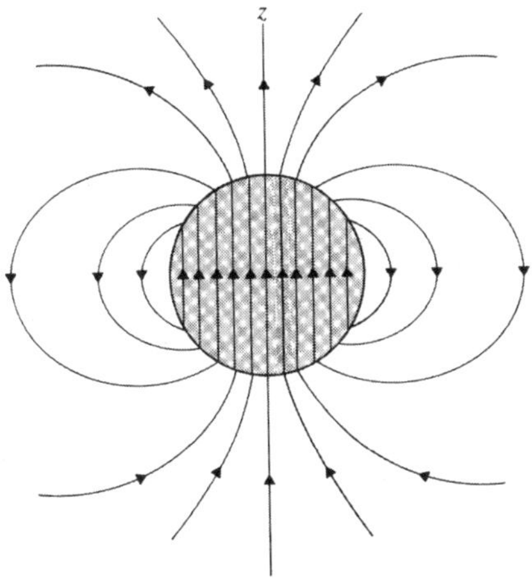

11·5 Energy in the Magnetic Field

Our study of the energy momentum tensor of the electromagnetic fields (see Section 8.4) revealed that the energy density in a magnetic field **B** in vacuum is

$$T^{00} = \frac{1}{8\pi} B^2 \tag{79}$$

so the total energy in the magnetic field is

$$U = \int \frac{1}{8\pi} B^2 \, dV \tag{80}$$

This integral for the field energy must, of course, be evaluated with the exact, microscopic field **B**(**x**).

Although Eq. (80) remains valid in the presence of magnetic materials, it ceases to be useful—in a magnetic material we do not know the exact microscopic field, and even if we did, we do not really care about the field energy but only about the amount of work that we have to supply to change the microscopic fields in some way or another. This work includes changes in the field energy and also changes in the potential and kinetic energies of the atoms brought about by the forces exerted by the fields.

We can find the work required to produce a small change of the magnetic field by the following argument: Suppose that the current density that generates the magnetic field is $\mathbf{j}(\mathbf{x})$. If we increase the current slightly and generate an increase $\delta\mathbf{B}$ in the magnetic field in some time δt, we will also generate an induced electric field $\mathbf{E}$. By Lenz' law, this electric field opposes the increase of current—it does work on the current and removes energy at the rate

$$\int \mathbf{E}\cdot\mathbf{j}\,dV \tag{81}$$

The work we do to produce to change $\delta\mathbf{B}$ must balance the work done by the induced electric field. Thus the power we have to supply is

$$\frac{\delta W}{\delta t} = -\int \mathbf{E}\cdot\mathbf{j}\,dV \tag{82}$$

In terms of the vector potential, the induced electric field is [see Eq. (10.17)],

$$\mathbf{E} = -\frac{1}{c}\frac{\delta\mathbf{A}}{\delta t}$$

so

$$\frac{\delta W}{\delta t} = \frac{1}{c}\int \frac{\delta\mathbf{A}}{\delta t}\cdot\mathbf{j}\,dV \tag{83}$$

or

$$\delta W = \frac{1}{c}\int \delta\mathbf{A}\cdot\mathbf{j}\,dV \tag{84}$$

We can transform this expression by using the equation $\nabla\times\mathbf{H} = (4\pi/c)\mathbf{j}$, which gives

$$\delta W = \frac{1}{4\pi}\int \delta\mathbf{A}\cdot\nabla\times\mathbf{H}\,dV \tag{85}$$

In view of the vector identity

$$\mathbf{F}\cdot\nabla\times\mathbf{G} = \nabla\cdot(\mathbf{G}\times\mathbf{F}) + \mathbf{G}\cdot\nabla\times\mathbf{F} \tag{86}$$

this is the same as

$$\delta W = \frac{1}{4\pi}\int \mathbf{H}\cdot\nabla\times\delta\mathbf{A}\,dV + \frac{1}{4\pi}\int \nabla\cdot(\mathbf{H}\times\delta\mathbf{A})\,dV \tag{87}$$

By Gauss' theorem, the second of these integrals equals a surface integral over a surface enclosing the entire volume of integration, that is, a surface integral over a sphere of infinite radius. For any finite current distribution, **H** is zero at infinity, and this surface integral can be discarded. We are then left with

$$\delta W = \frac{1}{4\pi}\int \mathbf{H}\cdot\nabla\times\delta\mathbf{A}\,dV$$

or

$$\delta W = \frac{1}{4\pi}\int \mathbf{H}\cdot\delta\mathbf{B}\,dV \tag{88}$$

To find the finite amount of work for a finite change in the magnetic field, we must integrate the small amounts of work δW. This we can do only if we know the relationship between **H** and **B**. In a linear magnetic material, **H** is proportional to **B**, and consequently,

$$\mathbf{H}\cdot\delta\mathbf{B} = \tfrac{1}{2}\,\delta(\mathbf{H}\cdot\mathbf{B}) \tag{89}$$

so

$$\delta W = \frac{1}{8\pi}\int \delta(\mathbf{H}\cdot\mathbf{B})\,dV \tag{90}$$

and

$$W = \frac{1}{8\pi}\int \mathbf{H}\cdot\mathbf{B}\,dV \tag{91}$$

We can regard this work as the potential energy stored in the macroscopic fields in the magnetic material. Accordingly, we can interpret

$$u = \frac{1}{8\pi}\mathbf{H}\cdot\mathbf{B} \tag{92}$$

as the energy density in the magnetic material. When using Eq. (91) or (92), we must not forget that in the present context, the fields are the macroscopic fields.

In a nonlinear magnetic material, Eqs. (89)–(92) are, of course, not valid. However, Eq. (88) remains valid, because it relies on no special assumptions about the material. If we want to calculate finite energy changes, we must return to the latter equation and express **H** in terms of **B** by means of the hysteresis curve, making the appropriate choice among the branches of this curve.

EXAMPLE: *magnetic energy in coaxial transmission line*

As an example of the use of these formulas, we will calculate the magnetic energy per unit length in a long coaxial transmission line consisting of a solid cylindrical conductor of radius R_1 surrounded by a thin cylindrical shell of radius R_2 (see Fig. 11.15). The inner conductor carries a current I toward the right, and the outer conductor carries a current I toward the left; both of these currents are uniformly distributed over the conducting material. Assuming that the conducting material is nonmag-

Figure 11·15

A coaxial transmission line consisting of a solid cylindrical conductor surrounded by a thin cylindrical shell.

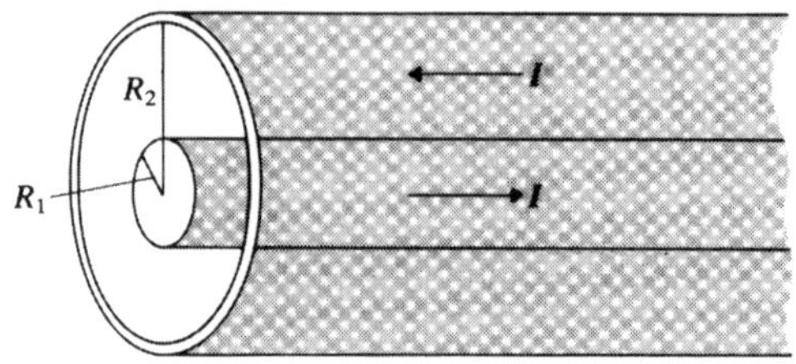

netic, we find from Ampère's law that the fields in the regions $\rho < R_1$ and $R_1 < \rho < R_2$ are

$$B_\phi(\rho) = \frac{2I\rho}{cR_1^2} \qquad \text{for} \qquad \rho < R_1$$

$$B_\phi(\rho) = \frac{2I}{c\rho} \qquad \text{for} \qquad R_1 < \rho < R_2 \tag{93}$$

The magnetic energy in a length l of the transmission line is then

$$U = \int \frac{B^2}{8\pi} dV = 2\pi l \int_0^{R_1} \frac{1}{8\pi}\left(\frac{2I\rho}{cR_1^2}\right)^2 \rho\, d\rho + 2\pi l \int_{R_1}^{R_2} \frac{1}{8\pi}\left(\frac{2I}{c\rho}\right)^2 \rho\, d\rho$$

$$= \frac{lI^2}{c^2}\left(\frac{1}{4} + \ln\frac{R_2}{R_1}\right) \tag{94}$$

Thus the energy per unit length is

$$\frac{U}{l} = \frac{I^2}{c^2}\left(\frac{1}{4} + \ln\frac{R_2}{R_1}\right) \tag{95}$$

Incidentally: From this expression for the energy, we can deduce the self-inductance per unit length. According to Eq. (10.38),

$$\frac{U}{l} = \frac{1}{2}\frac{L}{l}I^2 \tag{96}$$

Comparing Eqs. (95) and (96), we immediately recognize that

$$\frac{L}{l}=\frac{2}{c^2}\left(\frac{1}{4}+\ln\frac{R_2}{R_1}\right) \tag{97}$$

Note that this differs from the corresponding result for a transmission line with a hollow interior conductor [see Eq. (10.39)] by the extra $\frac{1}{4}$ in the term in parentheses; this extra $\frac{1}{4}$ represents of the contribution of the magnetic field and flux inside the inner conductor.

Problems

1. According to a naive classical model, the electron is a rigid, spherical body of radius R spinning about its axis with a spin angular momentum $\hbar/2$. Assume the R is ten times the Compton wavelength (that is, $R=10\times\hbar/m_e c$) and assume that the mass is uniformly distributed over the volume; find the angular velocity of rotation and the speed of rotation at the equator. Assume that the electric charge $-e$ is distributed over the surface; find a distribution that yields the correct magnetic moment for the electron. What theoretical or experimental evidence can you adduce against such a model?

2. Suppose that a proton is placed in a uniform magnetic field. Treating the proton as a classical top, show that the precession frequency of the spin of the proton is

$$\omega=2.7928\frac{e}{m_p c}B$$

This is the frequency of *nuclear-magnetic resonance* of the proton. What is the numerical value of this frequency in a magnetic field of 3000 gauss?

3. In its rest frame, the current loop shown in Fig. 11.3 has a magnetic dipole moment **m** but no electric dipole moment. By means of the Lorentz-transformation equations for charge and current [Eqs. (7.115) and (7.116)], prove that in a reference frame with respect to which the current loop is moving at velocity **v** in the x direction, the loop has an electric dipole moment **p** = **v** × **m**/c. (Problem 9.21 deduces the same result by a Lorentz transformation of the magnetic field of the current loop.)

4. A horizontal beam of electrons enters a magnet that has a vertical magnetic field whose strength increases in the vertical direction (that is, $\partial B_z/\partial z>0$). The spins of the electrons are either parallel to this magnetic field ("spin up") or antiparallel ("spin down"). The speed of the electrons is 3.0×10^8 cm/s, and the length of the magnet is 40 cm. What must be the magnitude of the field gradient $\partial B_z/\partial z$ if the spin-up and the spin-down electrons are to be separated by a vertical distance of 0.3 mm when they emerge from the magnet?

5. (a) What is the momentum density **E** × **B**/($4\pi c$) in the electric and magnetic fields [see Eq. (24)] of an electron?
   (b) What is the density of angular momentum? Integrate this density from $r = R$ to $r=\infty$ to find the (classical) electromagnetic angular momentum in the fields of the electron.

(c) To obtain an angular momentum equal to the spin of the electron, or $\hbar/2$, what radius R must you adopt?

(These calculations are no more than an amusing exercise, because the quantum uncertainties in the value of the electric field of an electron are much larger than the magnitude e/r^2 of the electron's Coulomb field; see the next problem.)

6. From the quantum theory of fields it is known that when measurements are performed within a length scale L, the uncertainty in the value of the electric field is of the order of $\Delta E \approx \sqrt{\hbar c}/L^2$. Since $e^2/\hbar c = 1/137$ is a dimensionless number, this can also be written as $\Delta E \approx \sqrt{137}e/L^2$, which is always larger than the Coulomb field e/r^2 of a single electron (or a single proton) and renders this Coulomb field undetectable.

(a) Consider a uranium nucleus (92 protons). Is the Coulomb field of this nucleus detectable?

(b) Consider a sodium nucleus (11 protons). Is the Coulomb field detectable?

(c) Consider a collection of N electrons, which jointly produce some electric field. What is the minimum number N that permits these electrons to have a detectable electric field?

7. According to the classical model, the electron is a small particle with a spin angular momentum $\mathbf{s}$ and a magnetic moment $\mathbf{m} = (e/m_e c)\mathbf{s}$. The magnitude of the spin is $\hbar/2$ and the magnitude of the magnetic moment is $e\hbar/2m_e c$. As established in Problem 3, an electron moving at speed $\mathbf{v}$ will then be endowed with an electric dipole moment $\mathbf{p} = \mathbf{v}\times\mathbf{m}/c$. Show that the dipole interaction energy $U' = -\mathbf{p}\cdot\mathbf{E}$ for an electron moving in the radial electric field of a nucleus can be expressed as follows:

$$U' = \frac{1}{m_e^2 c^2}\frac{1}{r}\frac{d\Phi}{dr}\mathbf{s}\cdot\mathbf{L}$$

where $\mathbf{L}$ is the orbital angular momentum of the electron and Φ the Coulomb potential of the nucleus. This interaction energy is called the spin-orbit interaction, and it accounts for the "fine-structure" splittings observed in spectral lines [the correct expression, obtained from relativistic quantum mechanics, is smaller by a factor of 2; classically, such a factor of 2 can be obtained by considering the relativistic "Thomas precession" of the reference frame of the accelerated electron; but since the classical model is good only for order-of-magnitude estimates, we will ignore this correction]. Use this equation to estimate the energy difference between two states in the hydrogen atom of the same orbital radius $r = 4a_0 = 4\times 0.0529$ nm, the same orbital angular momentum $L = \hbar$, but opposite directions of spin relative to the angular momentum.

8. Consider an electron in the smallest, circular Bohr orbit ($r = a_0 = \hbar^2/m_e e^2$) around a fixed proton.

(a) Treating the motion classically, show that if we immerse this orbiting electron in a uniform magnetic field B perpendicular to the plane of the orbit, the angular frequency of the orbital motion changes by approximately

$$\Delta\omega = \pm\frac{eB}{2m_e c}$$

and show that the radius of the orbit remains unchanged. [*Hint*: Show that the change of the electron's energy caused by the emf induced around the orbit during immersion in the magnetic field agrees with the change of kinetic energy implied by the change of frequency at fixed radius.]

(b) Show that in consequence, the change in the orbital contribution to the magnetic moment is approximately

$$\frac{e^2 B}{4m_e c^2} a_0^2$$

This change of magnetic moment gives rise to diamagnetism. Assuming that all hydrogen atoms in a sample of hydrogen gas at STP make a contribution of the above magnitude to the diamagnetic magnetization, calculate the value of the diamagnetic contribution to the susceptibility.

9. Since all the atomic magnetic moments are associated with angular momenta, a reversal of the magnetization of a magnetized body entails a change in the atomic angular momenta. If the body is free of external torques, then the change in the atomic angular momenta must be balanced by a change in the gross rotational state of the body, that is, the body must begin to rotate during the reversal of magnetization. This is the *Einstein–de Haas effect*. Suppose that an iron cylinder hangs vertically in a vertical magnetic field. The length of the cylinder is 20 cm, its radius is 0.30 cm, and its mass is 44 g. The cylinder is uniformly magnetized, with $M = 900$ gauss. If the suspension exerts no torque about the cylinder axis, predict the angular speed of rotation that the cylinder acquires when the magnetization is suddenly reversed. Explain how a measurement of the angular speed can be used to confirm experimentally that the magnetization is caused by aligned electron spins rather than by aligned orbital angular momenta.

10. Two very long, parallel wires carrying currents I are embedded in a paramagnetic material of permittivity μ. Find the magnitude of the force per unit length that one of the wires exerts on the other.

11. Figure 11.10 shows B versus H for a sample of commercial iron. From the numerical data in this figure, deduce the maximum value of the magnetization M in this iron. Under the assumption that the magnetization is entirely due to alignment of electron spins, how many electrons per atom contribute to this magnetization? The density of the iron is 7.8 g/cm^3.

12. (a) A long, thin (needlelike) cavity has been cut out of a magnetic material of permeability μ. The magnetic field in the material is $\mathbf{B}_0$ and is parallel to the needle. Find the **B** and **H** fields in the cavity.

(b) Repeat the calculation for a flat (disklike) cavity with the $\mathbf{B}_0$ field perpendicular to the face.

13. Is $\nabla \cdot \mathbf{H} = 0$ in a magnetic material? Prove your answer.

14. A very long solenoid is filled with transformer steel. The solenoid has 15 turns/cm and carries a current of 2.2×10^8 statamp. What are the magnetic B and H fields in the solenoid? What is the magnetization M in the steel? What is the magnetization current density along the surface of the steel? The following table gives the magnetic properties of transformer steel:

B	H
2,000 gauss	0.60 gauss
4,000	0.87
6,000	1.10
8,000	1.48
10,000	2.28
12,000	3.85
14,000	10.9
16,000	43.0
18,000	149

15. A long, straight wire of copper is surrounded by a concentric, thick shell of transformer steel of inner radius 2.0 cm and outer radius 4.0 cm. The wire carries a current of 1.7×10^{11} statamp. Find the magnetic field B at $r = 1.0$ cm, $r = 3.0$ cm, and $r = 5.0$ cm. See the table listing the magnetic properties of transformer steel in Problem 14.

16. An electromagnet is made of transformer steel in the shape shown in Fig. 11.16. The cross-sectional area of the steel is A; the winding has N turns and carries a current I. Find the approximate strength of the magnetic field in the gap if $NI = 16 \times 10^{12}$ statamps, $h = 0.5$ cm, $l = 100$ cm, and $A = 100$ cm^2. Neglect all fringing effects. The magnetic properties of transformer steel are listed in the table accompanying Problem 14. [*Hint:* Assume that the **B** field has a constant magnitude everywhere in the iron and in the gap, with field lines roughly as shown in Fig. 11.16. Integration along a path following a field line gives $\int \mathbf{H} \cdot d\mathbf{l} = (4\pi/c)\, NI$, or $(4l - h)H + hB = (4\pi/c)NI$. Combine this equation with the dependence of H on B summarized in the table, and solve numerically.]

17. A very long cylinder of radius R of magnetic material has a uniform magnetization **M** perpendicular to its axis. Find the magnetic field **B** outside and inside the cylinder.

18. A solid iron cylinder of length L and radius R has a uniform magnetization **M** in its volume; the direction of **M** is parallel to the axis of the cylinder.

(a) Find the magnetic field on the axis of the cylinder, outside the cylinder.

Figure 11·16

An electromagnet

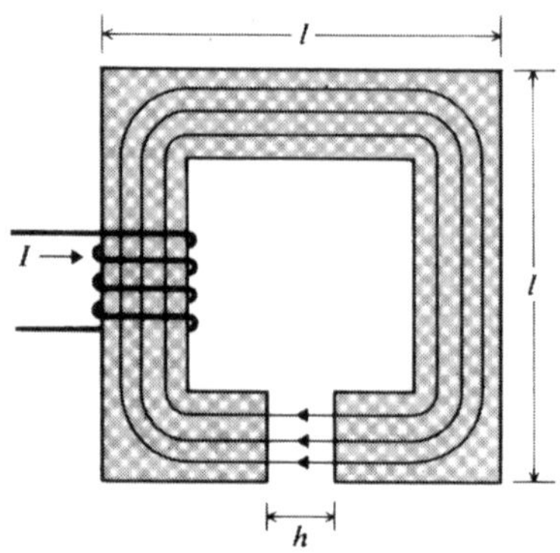

(b) Using the method of Section 9.4, find the first two terms in a power-series expansion (in powers of $1/r$) for the magnetic field in the space surrounding the cylinder.

19. A sphere of radius R made of a paramagnetic material of permittivity is placed in an originally uniform magnetic field $\mathbf{B}_0 = B_0\hat{\mathbf{z}}$. Find the magnetic field inside the sphere and outside the sphere.

20. A spherical shell of paramagnetic material of permittivity μ has an inner radius a and an outer radius b. The shell is placed in an initially uniform magnetic field $\mathbf{B}_0$. Find the magnetic potential and the fields $\mathbf{B}$ and $\mathbf{H}$ in the regions $r < a$, $a < r < b$, and $r > b$. If $\mu \gg 1$, by what factor is the magnetic field in the interior of the shell smaller than B_0? This explains how spheres or other containers made of an alloy with a large value of μ (μ-metal) can be used to shield magnetic fields.

21. A very long cylindrical shell has an inner radius a and an outer radius b. The shell is made of a uniformly magnetized material of magnetization $\mathbf{M}$, directed perpendicular to the axis of the pipe. Taking the z axis along the direction of $\mathbf{M}$, find the magnetic field $\mathbf{B}$ in all regions inside and outside the shell.

22. A very long cylindrical shell of paramagnetic material of permittivity μ has an inner radius a and an outer radius b. This shell is placed in an originally uniform vertical magnetic field $\mathbf{B}_0$ at right angles to the axis of the shell. Find the magnetic potential and the fields $\mathbf{B}$ and $\mathbf{H}$ in all regions inside and outside the shell. Verify that if $\mu >> 1$, the magnetic field in the space enclosed by the shell is much smaller than $\mathbf{B}_0$. (This reduction of the magnetic field has a practical application in the construction of magnetic shields of μ-metal, an alloy with a large value of μ.)

23. A spherical shell has an inner radius a and an outer radius b. The shell is made of a uniformly magnetized material of magnetization $\mathbf{M}$. Taking the z axis along the direction of $\mathbf{M}$, find the magnetic field $\mathbf{B}$ in all regions inside and outside the shell.

24. (a) By integrating Eq. (84), show that the magnetic energy in a linear magnetic material can be expressed as

$$W = \frac{1}{2c}\int \mathbf{A}\cdot\mathbf{j}\,dV$$

(b) Show that in a nonmagnetic material, this energy reduces to

$$W = \frac{1}{2c^2}\iint \frac{\mathbf{j}(\mathbf{x}')\cdot\mathbf{j}(\mathbf{x})}{|\mathbf{x}-\mathbf{x}'|}\,dV'\,dV$$

25. A coaxial transmission line consists of a solid conducting cylinder of radius a surrounded by a thick, conducting, cylindrical shell of inner radius b and outer radius c (see Fig. 11.17). The conductors carry equal currents I in opposite directions. Calculate the magnetic energy in a segment of length l of this transmission line, and deduce the self-inductance per unit length.

Figure 11·17

A coaxial transmission line.

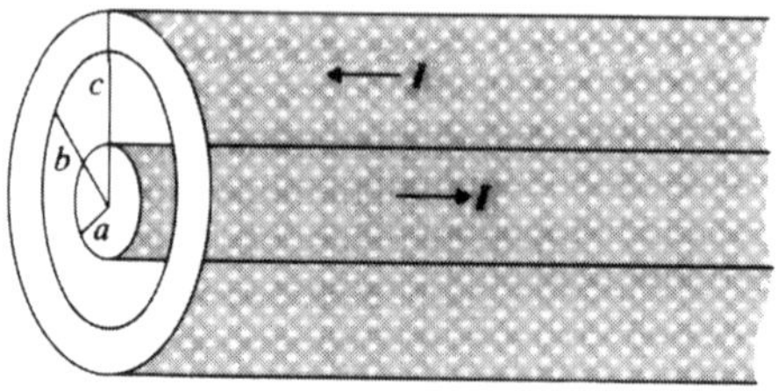

CHAPTER

12

Electromagnetic Waves

According to the laws of induction, changing magnetic fields induce electric fields, and conversely. This phenomenon of mutual induction of electric and magnetic fields leads to self-supporting electromagnetic oscillations in space. As we will see, these oscillations take the form of waves propagating with the speed of light. Of course, some initial disturbance is required to start the oscillations. For instance, we might take a point charge and shake it suddenly, starting a disturbance in its electric field; this will then generate an induced magnetic field, which will generate an induced electric field, and so on. The resulting electromagnetic pulse will spread outward from the point charge as a spherical wave. Far from the source of the disturbance, a small portion of such a spherical wave can be approximated as a plane wave. In this chapter, we will study the propagation of plane electromagnetic waves in vacuum and in homogeneous-material media, but we will leave for a later chapter the detailed analysis of how the waves are initiated by sudden changes in the motion of charges.

12·1 Plane Waves in Vacuum

If we describe the electromagnetic field by the four-vector potential, our fundamental field equation is Eq. (8.10),

$$\partial_\mu \partial^\mu A^\nu = \frac{4\pi}{c} j^\nu$$

In vacuum, this becomes

$$\partial_\mu \partial^\mu A^\nu = 0 \tag{1}$$

that is,

$$\left(\frac{1}{c^2}\frac{\partial^2}{\partial t^2} - \frac{\partial^2}{\partial x^2} - \frac{\partial^2}{\partial y^2} - \frac{\partial^2}{\partial z^2}\right) A^\nu = 0 \tag{2}$$

We must supplement this equation with the Lorentz condition, Eq. (8.12),

$$\partial_\nu A^\nu = 0 \tag{3}$$

Equation (2) is a wave equation. Its simplest solutions are plane waves with a harmonic dependence on space and time. For instance, if we assume that the direction of propagation is along the z axis, we can solve Eq. (2) with an expression of the form

$$A^\nu = \varepsilon^\nu \sin(\omega t - kz) \tag{4}$$

where ω and k are constants and ε^ν is a four-vector with constant components. The constants k and ω are, respectively, the wave number (or the magnitude of the wave vector) and the angular frequency of the wave. In terms of the wavelength and the frequency,

$$k = \frac{2\pi}{\lambda} \qquad \omega = 2\pi\nu \tag{5}$$

Substitution of Eq. (4) into Eq. (2) leads to

$$\left(-\frac{\omega^2}{c^2} + k^2\right)\varepsilon^\nu \sin(\omega t - kz) = 0 \tag{6}$$

This implies

$$k = \pm\frac{\omega}{c} \tag{7}$$

By convention, ω is always positive; hence, for a wave propagating in the $+z$ direction, k must also be positive,

$$k = \frac{\omega}{c} \tag{8}$$

The speed of the wave is

$$u = \frac{\omega}{k} = c \tag{9}$$

Thus the speed of the wave coincides with the speed of light. This coincidence is not accidental—light is an electromagnetic wave.

We have yet to take into account the Lorentz condition. Substitution of the expression (4) into Eq. (3) gives

$$\left(\frac{\omega}{c}\varepsilon^0 - k\varepsilon^3\right)\cos(\omega t - kz) = 0 \tag{10}$$

which implies

$$\varepsilon^0 = \varepsilon^3 \tag{11}$$

This means that only three of the four components of ε^ν are independent. The most general vector ε^ν must then necessarily be a superposition of

$$\varepsilon^\nu_{(1)} = (0,1,0,0) \qquad \varepsilon^\nu_{(2)} = (0,0,1,0) \qquad \varepsilon^\nu_{(3)} = (1,0,0,1) \tag{12}$$

■ ***Exercise 1.*** Show that any four-vector satisfying Eq. (11) can be expressed as a superposition of the three basis vectors given by Eq. (12). ■

EXAMPLE: *plane wave with linear polarization*

For the plane wave

$$A^\nu = A\varepsilon^\nu_{(1)} \sin(\omega t - kz) = A(0,1,0,0)\sin(\omega t - kz) \tag{13}$$

the electric and magnetic fields are

$$\begin{aligned} E_x &= F^{10} = \partial^1 A^0 - \partial^0 A^1 = -\partial^0 A^1 = -A\frac{\omega}{c}\cos(\omega t - kz) \\ B_y &= F^{13} = \partial^1 A^3 - \partial^3 A^1 = \partial_3 A^1 = -A\frac{\omega}{c}\cos(\omega t - kz) \\ E_y &= E_z = B_x = B_z = 0 \end{aligned} \tag{14}$$

This shows that the electric and magnetic fields are perpendicular to each other and to the direction of propagation, they have equal magnitudes,* and they oscillate in phase.

For the plane wave

$$A^\nu = A\varepsilon^\nu_{(2)} \sin(\omega t - kz) = A(0,0,1,0)\sin(\omega t - kz) \tag{15}$$

the fields are

$$\begin{aligned} E_y &= -A\frac{\omega}{c}\cos(\omega t - kz) \\ B_x &= A\frac{\omega}{c}\cos(\omega t - kz) \\ E_x &= E_z = B_y = B_z = 0 \end{aligned} \tag{16}$$

---

*This equality of the magnitudes of **E** and **B** in a wave (in vacuum) is another advantage of the CGS system of units.

Again, these electric and magnetic fields are perpendicular to each other and to the direction of propagation, they have equal magnitudes, and they oscillate in phase.

Finally, for the plane wave

$$A^{\nu} = A\varepsilon^{\nu}_{(3)} \sin(\omega t - kz) = A(1,0,0,1)\sin(\omega t - kz) \tag{17}$$

all components of the electric and magnetic fields vanish identically.

■ ***Exercise 2.*** Show that Eq. (17) yields $F^{\mu\nu} = 0$. ■

This third solution therefore does not represent a physical wave. The third solution is a "ghost": it carries no energy, it carries no momentum, and it never exerts any force on any charged particle. We can ignore it.

The fields given by Eqs. (14) and (16) represent electromagnetic waves with two independent *linear polarizations*. By definition, the plane of polarization of a wave is taken to be the plane of the electric field vector; thus the wave (14) is polarized in the x direction, whereas the wave (16) is polarized in the y direction. In three-dimensional vector notation, the fields of the former wave are

$$\mathbf{E}_{(1)} = E_0 \hat{\mathbf{x}} \cos(\omega t - kz) \tag{18}$$

$$\mathbf{B}_{(1)} = E_0 \hat{\mathbf{y}} \cos(\omega t - kz) \tag{19}$$

and those of the latter wave are

$$\mathbf{E}_{(2)} = E_0 \hat{\mathbf{y}} \cos(\omega t - kz) \tag{20}$$

$$\mathbf{B}_{(2)} = -E_0 \hat{\mathbf{x}} \cos(\omega t - kz) \tag{21}$$

where $E_0 = -A\omega / c$. Figures 12.1(a) and 12.1(b) show the fields of these two linearly polarized waves. Note that the directions of the **E** and **B** are related by the right-hand rule: if the fingers curl from **E** toward **B**, the thumb is along the direction of propagation.

It is often convenient to express wavefunctions in exponential notation. For instance, the electric fields (18) and (20) can be written as follows:

$$\mathbf{E}_{(1)} = E_0 \hat{\mathbf{x}} \mathrm{e}^{-i(\omega t - kz)} \tag{22}$$

$$\mathbf{E}_{(2)} = E_0 \hat{\mathbf{y}} \mathrm{e}^{-i(\omega t - kz)} \tag{23}$$

where it is understood that the *real* part of these expressions represents the physical electric field, whereas the imaginary part is to be discarded. These wavefunctions can be regarded as basis functions—any wave of

Figure 12·1

Electric and magnetic field lines for plane waves. (*a*) Polarized in the x direction. (*b*) Polarized in the y direction.

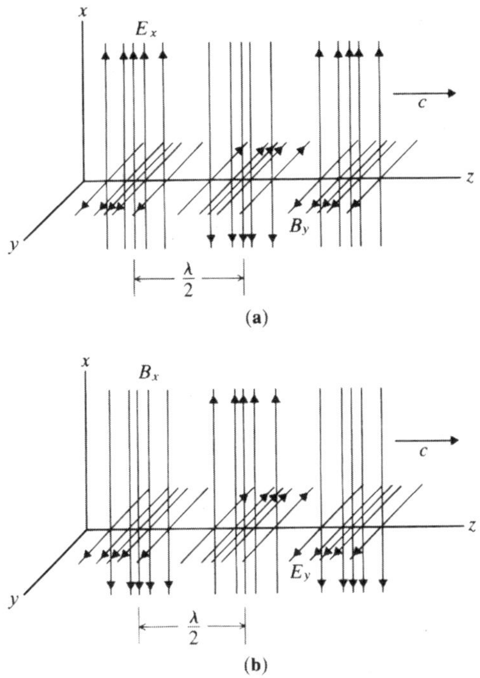

arbitrary direction of polarization is a superposition of the functions (22) and (23) (with their corresponding magnetic fields).

Instead of these linearly polarized basis functions, we can also use circularly polarized basis functions. The latter are defined by

$$\mathbf{E}_+ \equiv (\mathbf{E}_{(1)} + i\mathbf{E}_{(2)}) = E_0(\hat{\mathbf{x}} + i\hat{\mathbf{y}})e^{-i(\omega t - kz)} \tag{24}$$

$$\mathbf{E}_- \equiv (\mathbf{E}_{(1)} - i\mathbf{E}_{(2)}) = E_0(\hat{\mathbf{x}} - i\hat{\mathbf{y}})e^{-i(\omega t - kz)} \tag{25}$$

with corresponding equations for the magnetic fields.

■ ***Exercise 3.*** Express $E_{(1)}$ and $E_{(2)}$ in terms of E_+ and E_-. ■

The wave $\mathbf{E}_+$ is said to have *positive helicity*, the wave $\mathbf{E}_-$ *negative helicity*; they are also said to be right-handed and left-handed, respectively.*

* The terminology used here is that of quantum physics. In optics, *right* and *left* are commonly used the other way around.

Let us take the real part of Eq. (24) and examine the electric field:

$$\mathrm{Re}\mathbf{E}_+ = E_0\hat{\mathbf{x}}\cos(\omega t - kz) + E_0\hat{\mathbf{y}}\sin(\omega t - kz) \tag{26}$$

At fixed z, say $z = 0$, this is $E_0(\hat{\mathbf{x}}\cos\omega t + \hat{\mathbf{y}}\sin\omega t)$, which shows that the electric field rotates counterclockwise around the z axis, as seen from the positive z direction (direction of propagation). Note that the directions of propagation and rotation are related by the right-hand rule. In contrast, the real part of Eq. (25) gives an electric field that rotates clockwise.

■ ***Exercise 4.*** Examine the real part of Eq. (25) and check the direction of rotation of the electric field. ■

As we will see in a later section, electromagnetic waves carry energy and momentum. Furthermore, polarized waves carry angular momentum. For instance, the positive-helicity wave (24) carries angular momentum with a positive z component—the flux of angular momentum is $1/\omega$ times the flux of energy.

Depending on their frequency and wavelength, electromagnetic waves are called radio waves $(\nu \approx 10^4 \text{ to } 10^{11}\text{ Hz}, \lambda \approx 10^6 \text{ to } 10^{-1}\text{ cm})$, infrared waves $(10^{11} \text{ to } 4\times10^{14}\text{ Hz}, 10^{-1}\text{ cm to } 7\times10^{-5}\text{ cm})$, visible light $(4\times10^{14} \text{ to } 7\times10^{14}\text{ Hz}, 7\times10^{-5} \text{ to } 4\times10^{-5}\text{ cm})$, ultraviolet waves $(7\times 10^{14} \text{ to } 10^{16}\text{ Hz}, 4\times10^{-5} \text{ to } 10^{-6}\text{ cm})$, X rays $(10^{16} \text{ to } 10^{19}\text{ Hz}, 10^{-6} \text{ to } 10^{-9}\text{ cm})$, or gamma rays (more than 10^{19}Hz, less than 10^{-9} cm). The boundaries of the visible region of the electromagnetic spectrum are sharply defined, but the boundaries of the other regions overlap to some extent, because the labeling of an electromagnetic wave depends not only on its frequency or wavelength, but also on its mode of production. For instance, a wave of 10^{17} Hz will be called ultraviolet when produced by a thermal source, but it will be called an X ray when produced in an X-ray machine.

In the above derivation of the properties of electromagnetic waves, we took the relativistic wave equation as a starting point. Instead, we can obtain the same results by manipulating Maxwell's equations. This approach is somewhat more laborious, but it is instructive because it shows that the existence of electromagnetic waves hinges on the induction terms in Maxwell's equations. This means that the electric and magnetic fields of the wave mutually induce one another. In vacuum, the Maxwell equations (8.61)–(8.64) are

$$\nabla\cdot\mathbf{E} = 0 \tag{27}$$

$$\nabla\times\mathbf{B} = \frac{1}{c}\frac{\partial\mathbf{E}}{\partial t} \tag{28}$$

$$\nabla\cdot\mathbf{B} = 0 \tag{29}$$

$$\nabla\times\mathbf{E} = -\frac{1}{c}\frac{\partial\mathbf{B}}{\partial t} \tag{30}$$

To construct a wave equation for **E**, we take the curl of the last equation, which describes the induction of the electric field by a changing magnetic field:

$$\nabla\times(\nabla\times\mathbf{E}) = -\frac{1}{c}\frac{\partial}{\partial t}\nabla\times\mathbf{B} \tag{31}$$

On the right side, we can substitute Eq. (28), which describes the induction of the magnetic field by a changing electric field. On the left side, we can use the identity $\nabla\times(\nabla\times\mathbf{E}) = \nabla(\nabla\cdot\mathbf{E}) - \nabla^2\mathbf{E}$; with $\nabla\cdot\mathbf{E} = 0$, this reduces to $\nabla\times(\nabla\times\mathbf{E}) = -\nabla^2\mathbf{E}$. Equation (31) then becomes

$$-\nabla^2\mathbf{E} = -\frac{1}{c^2}\frac{\partial^2\mathbf{E}}{\partial t^2} \tag{32}$$

which is the wave equation for the electric field.

■ ***Exercise 5.*** Show that the magnetic field satisfies a similar wave equation:

$$-\nabla^2\mathbf{B} = -\frac{1}{c^2}\frac{\partial^2\mathbf{B}}{\partial t^2} \tag{33}$$

■

As before, we will look for plane-wave solutions with the direction of propagation along the z axis. Obviously, Eq. (32) has the solution

$$\mathbf{E} = \mathbf{E}_0\cos(\omega t - kz) \tag{34}$$

where $\mathbf{E}_0$ is any constant vector. With this expression for **E**, we find

$$\nabla\cdot\mathbf{E} = kE_{0,z}\sin(\omega t - kz) \tag{35}$$

The Maxwell equation $\nabla\cdot\mathbf{E} = 0$ then demands $E_{0,z} = 0$, that is, the vector $\mathbf{E}_0$ must be perpendicular to the z axis. Thus $\mathbf{E}_0$ is a superposition of $\hat{\mathbf{x}}$ and $\hat{\mathbf{y}}$, and the two independent linearly polarized waves are again the same as given by Eqs. (18) and (20).

The magnetic fields of these waves follow from the Maxwell equation (30). For instance, if

$$\mathbf{E} = E_0\hat{\mathbf{x}}\cos(\omega t - kz) \tag{36}$$

then

$$\frac{\partial \mathbf{B}}{\partial t} = -c\nabla \times \mathbf{E} = -ckE_0\hat{\mathbf{y}}\sin(\omega t - kz) \tag{37}$$

We can integrate this with respect to time and obtain

$$\mathbf{B} = \frac{ck}{\omega}E_0\hat{\mathbf{y}}\cos(\omega t - kz) = E_0\hat{\mathbf{y}}\cos(\omega t - kz) \tag{38}$$

which agrees with Eq. (19).

In the above discussion, we dealt only with harmonic waves of some given frequency (monochromatic waves). To deal with wave pulses of some arbitrary shape, we can take advantage of the Fourier theorem: an arbitrary wave (with some given direction of propagation) is a superposition of an infinite number of harmonic waves of many different frequencies. We will see an example of this kind of superposition when we examine the propagation of wave pulses in Section 12.5.

12·2 The Wave Four-Vector; Doppler Shift and Aberration

For a plane wave propagating in some direction different from the z direction, the wavefunction $\sin(\omega t - kz)$ in the equations of the preceding section must be replaced by

$$\sin(\omega t - \mathbf{k}\cdot\mathbf{x}) \tag{39}$$

where the *wave vector* **k** is in the direction of propagation. The wavefunction (39) is a solution of Eq. (2), provided that

$$k_x^2 + k_y^2 + k_z^2 = \frac{\omega^2}{c^2} \tag{40}$$

■ ***Exercise 6.*** Check that Eq. (2) implies Eq. (40). ■

In four-dimensional notation, we can write the argument of the sine function as

$$\omega t - \mathbf{k}\cdot\mathbf{x} = k^\mu x_\mu \tag{41}$$

where k^μ is the four-component quantity

$$k^\mu = (k^0, k^1, k^2, k^3) = \left(\frac{\omega}{c}, k_x, k_y, k_z\right) \tag{42}$$

This quantity is actually a four-vector, called the *wave four-vector.* To prove that the components of k^μ satisfy the Lorentz-transformation equations, we begin by considering the nodes of the wave, that is, the points at which **E** = **B** = 0. Since the fields are proportional to $\cos(k^\mu x_\mu)$, the condition for nodes is $k^\mu x_\mu = \pm\pi/2, \pm 3\pi/2, \pm 5\pi/2, \ldots$. Whenever **E** = **B** = 0 in the *xyzt* frame, then also **E**′ = **B**′ = 0 in the *x'y'z't'* frame. Hence $k'^\mu x'_\mu = \pm\pi/2, \pm 3\pi/2, \pm 5\pi/2, \ldots$, which shows that $k^\mu x_\mu$ and $k'^\mu x'_\mu$ have the same numerical values at the nodes. Since both $k^\mu x_\mu$ and $k'^\mu x'_\mu$ are linear functions of the spacetime coordinates, their agreement at two points implies that they also agree everywhere *between* these points; that is, in general

$$k^\mu x_\mu = k'^\mu x'_\mu \tag{43}$$

Differentiating this with respect to x'_α, we obtain

$$k'^\alpha = k^\mu \frac{\partial x_\mu}{\partial x'_\alpha} \tag{44}$$

Since $\partial x_\mu / \partial x'_\alpha = a^\alpha{}_\mu$ [see Eq. (7.26)], this is the Lorentz-transformation law of a four-vector. If the Lorentz transformation has velocity *V* along the *x* axis, then

$$k'^0 = \frac{k^0 - (V/c)k_x}{\sqrt{1 - V^2/c^2}} \tag{45}$$

$$k'_x = \frac{k_x - (V/c)k^0}{\sqrt{1 - V^2/c^2}} \tag{46}$$

$$k'_y = k_y \tag{47}$$

$$k'_z = k_z \tag{48}$$

EXAMPLE: *Doppler shift and aberration of light*

From this, we can obtain the Doppler shift and the aberration of light. Suppose that in our laboratory frame (that is, the *xyzt* frame), the wave vector makes an angle θ with the *x* axis, so

$$k_x = k\cos\theta = \frac{\omega}{c}\cos\theta \tag{49}$$

where $k = (k_x^2 + k_y^2 + k_z^2)^{1/2}$ is the magnitude of the (three-dimensional) wave vector. Then Eq. (45) yields

$$\omega' = \frac{\omega[1-(V/c)\cos\theta]}{\sqrt{1-V^2/c^2}} \tag{50}$$

Furthermore, Eq. (46) yields

$$\cos\theta' = \frac{k'_x}{k'} = \frac{k'_x}{\omega'/c} = \frac{(\omega/c)\cos\theta - V\omega/c^2}{\sqrt{1-V^2/c^2}} \frac{1}{\dfrac{\omega[1-(V/c)\cos\theta]}{c\sqrt{1-V^2/c^2}}}$$

or

$$\cos\theta' = \frac{\cos\theta - V/c}{1-(V/c)\cos\theta} \tag{51}$$

Equation (50) is the relativistic Doppler-shift formula—it gives the change of frequency of the wave. Equation (51) is the relativistic aberration formula—it gives the change of direction of propagation of the wave.

In the special case $\theta = 0$ (propagation along the x axis), we obtain the *longitudinal* Doppler shift:

$$\omega' = \omega\frac{1-V/c}{\sqrt{1-V^2/c^2}} = \omega\sqrt{\frac{1-V/c}{1+V/c}} \tag{52}$$

In the special case $\theta = \pi/2$, we obtain the *transverse* Doppler shift:

$$\omega' = \frac{\omega}{\sqrt{1-V^2/c^2}} \tag{53}$$

This implies that a wave emitted by a source in the $x'y'z't'$ frame and reaching our laboratory frame (or $xyzt$ frame) from a direction perpendicular to the velocity $\mathbf{V}$ will have a frequency ω that is lower by a factor $\sqrt{1-V^2/c^2}$ than the emission frequency ω'. This is essentially the time-dilation effect.

According to the aberration formula (51), the direction of propagation makes a larger angle with the x' axis than with the laboratory x axis. For instance, if the wave is to reach the laboratory at an angle $\theta = \pi/2$, then it must be emitted from the $x'y'z't'$ frame at an angle such that

$$\cos\theta' = -\frac{V}{c} \tag{54}$$

That is, the wave must be emitted in a backward direction ($\theta' > \pi/2$) in order to compensate for the effect of the motion of the $x'y'z't'$ frame on the direction of propagation.

12·3 Energy, Momentum, and Angular Momentum in a Plane Wave

The components of the energy-momentum tensor $T^{\mu\nu}$ (see Section 8.4) give us the energy density, energy flux, and momentum flux in the electromagnetic fields. Here we will look at these components for the fields of a plane wave—say, the fields specified in Eqs. (18) and (19).

EXAMPLE: *energy density and energy flux in a plane wave*

The energy density is T^{00}. By substituting **E** and **B** into Eq. (8.90), we immediately find

$$T^{00} = \frac{1}{8\pi}(E^2 + B^2) = \frac{1}{4\pi}E_0^2 \cos^2(\omega t - kz) \tag{55}$$

The components of the energy flux, or the energy flow per unit area and unit time, are cT^{0k}; these components constitute the Poynting vector **S**. From Eq. (8.96), we find

$$\mathbf{S} = \frac{c}{4\pi}\mathbf{E}\times\mathbf{B} = \hat{\mathbf{z}}\frac{c}{4\pi}E_0^2 \cos^2(\omega t - kz) \tag{56}$$

As expected, the direction of this energy flux coincides with the direction of propagation of the wave. Comparing Eqs. (55) and (56), we see that

$$(\text{energy flux}) = c \times (\text{energy density}) \tag{57}$$

This relation is intuitively obvious because the energy of the wave moves with the wave, at the speed of light.

The components of the momentum flux are T^{mn}. Of these, only T^{33} is nonzero. From Eq. (8.94), we find

$$T^{33} = \frac{1}{4\pi}\left[-E_z E_z - B_z B_z + \frac{1}{2}(E^2 + B^2)\right] = \frac{1}{4\pi}E_0^2 \cos^2(\omega t - kz) \tag{58}$$

This indicates that the wave carries z momentum in the z direction. We see that

$$(\text{energy flux}) = c \times (\text{momentum flux}) \tag{59}$$

This relation becomes obvious if we take a quantum-mechanical point of view: the wave is a stream of photons, each of momentum p and energy cp; and the relation (59) merely expresses this correspondence between the energy and momentum carried by the photons.

For practical purposes, we are often not interested in the instantaneous energy density, energy flux, or momentum flux but only in their time-average values. Taking an average over one or several periods of the wave, we obtain

$$\overline{T}^{00} = \frac{1}{8\pi} E_0^2 \tag{60}$$

$$\overline{\mathbf{S}} = \hat{\mathbf{z}} \frac{c}{8\pi} E_0^2 \tag{61}$$

$$\overline{T}^{33} = \frac{1}{8\pi} E_0^2 \tag{62}$$

EXAMPLE: *pressure of light*

The momentum flux in an electromagnetic wave is capable of exerting a pressure, called the *pressure of light*. Suppose that light is absorbed, without reflection. Then the amount of z momentum absorbed by an area A in unit time is, on the average,

$$\overline{T}^{33} A = \frac{1}{8\pi} E_0^2 A$$

and the force exerted on this area is

$$F = \frac{1}{8\pi} E_0^2 A$$

Hence there is a pressure

$$\frac{F}{A} = \frac{E_0^2}{8\pi} = \frac{S}{c} \tag{63}$$

A beautiful example of the effects of this pressure can be seen in comets. The curved tail of a comet is made of dust separated from the comet's head by the pressure of solar light. At the distance of the Earth's orbit, the value of the Poynting vector for solar light is S = 1.4×10^6 erg/cm$^2\cdot$s, and the corresponding pressure is quite small, $S/c = 4.7\times10^{-5}$ dyne/cm^2. However, for a dust grain of very small mass, this pressure can exceed the Sun's gravitational attraction, and it then drives the dust grain outward.

From the Lorentz-transformation equations for the energy-momentum tensor, we can deduce the transformation equations for the energy flux. The relevant equations are

$$T'^{0k} = a^0{}_\mu \, a^k{}_\nu \, T^{\mu\nu}$$

If the energy-momentum tensor is that of a plane wave propagating in the z direction [see Eqs. (55), (56), and (58)], then only T^{00}, T^{30}, T^{03}, and T^{33} are nonzero, which leads to

$$T'^{0k} = a^0{}_0 a^k{}_0 T^{00} + a^0{}_3 a^k{}_0 T^{30} + a^0{}_0 a^k{}_3 T^{03} + a^0{}_3 a^k{}_3 T^{33}$$

For a Lorentz transformation in the z direction with velocity V, this yields

$$T'^{03} = -\gamma^2 \frac{V}{c} T^{00} + \gamma^2 \left(\frac{V^2}{c^2} + 1 \right) T^{03} - \gamma^2 \frac{V}{c} T^{33}$$

or, since $T^{00} = T^{30} = T^{03} = T^{33}$ for our wave,

$$T'^{03} = \gamma^2 \left(\frac{V^2}{c^2} + 1 - 2\frac{V}{c} \right) T^{03} = \gamma^2 \left(1 - \frac{V}{c} \right)^2 T^{03}$$

Note that for $V/c \to 1$, $T'^{03} \to 0$; thus the energy flux of the wave tends to zero in a reference frame traveling at very high speed in the same direction as the wave. This means that for an observer in this reference frame, the wave (almost) disappears.

Besides energy and momentum, an electromagnetic wave can also carry angular momentum. We know from Section 8.4 that the density of momentum in the electromagnetic fields is $\mathbf{E} \times \mathbf{B}/(4\pi c)$; consequently, the density of angular momentum is $\mathbf{x} \times (\mathbf{E} \times \mathbf{B})/(4\pi c)$. We can regard this density of angular momentum as arising from the energy flow—the moving energy is equivalent to a moving mass, which contributes momentum and angular momentum. In an infinite plane wave, the **E** and **B** fields are everywhere perpendicular to the wave vector, and the energy flow is everywhere parallel to the wave vector. Hence the only angular momentum in such a wave is an "orbital" angular momentum, that is, an angular momentum associated with the translational motion of the wave.* However, in a wave of finite transverse extent, the **E** and **B** fields near the edges of the wave have a component parallel to the wave vector (the field lines are closed loops), and the energy flow has components perpendicular to the wave vector. For instance, Fig. 12.2 shows the time-average transverse energy flow in a circularly polarized wave propagating in the z direction; the wave has a finite extent in the x and y directions, and it has axial symmetry about the z axis. Besides the circulating energy flow shown in Fig. 12.2, the wave has a translational flow in the z direction; thus the net energy flow is helical. The circulating energy flow in the wave implies the existence of an extra, nonorbital an-

---

*If the wave is axially symmetric and the axis interests the origin, then the net orbital angular momentum is, of course, zero.

Figure 12·2

This pattern of circular flow lines represents the time-average energy flow in a circularly polarized wave propagating in the z direction. On a given wave front, say $z = 0$, the time-average fields are assumed to be constant within a circular area of radius a and to decrease toward zero between radius a and radius b. The picture shows only the flow in the transverse directions. The flow in the direction of propagation is much larger; the net flow is helical.

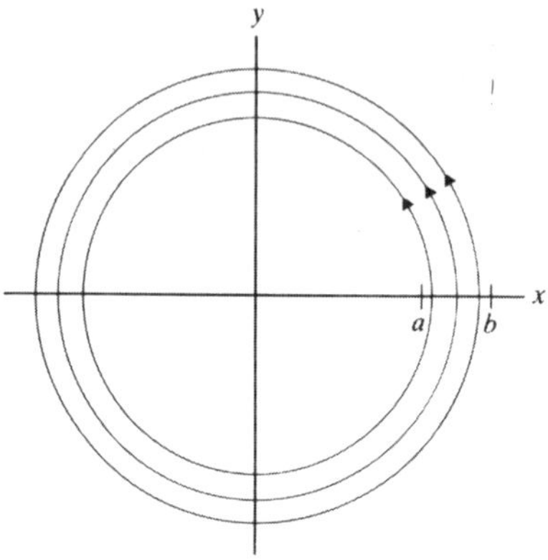

gular momentum, with a direction parallel or antiparallel to the direction of propagation. This angular momentum is what in quantum mechanics we call the *spin* of the wave.

To obtain an expression for the net angular momentum in an arbitrary wave of finite extent (a wave packet), we express the momentum density and the angular-momentum density in terms of the vector potential:

$$\frac{1}{4\pi c}\mathbf{E}\times\mathbf{B} = \frac{1}{4\pi c}\mathbf{E}\times(\nabla\times\mathbf{A}) = \frac{1}{4\pi c}[E^n\nabla A^n - (\mathbf{E}\cdot\nabla)\mathbf{A}] \tag{64}$$

and

$$\frac{1}{4\pi c}\mathbf{x}\times(\mathbf{E}\times\mathbf{B}) = \frac{1}{4\pi c}\mathbf{x}\times[E^n\nabla A^n - (\mathbf{E}\cdot\nabla)\mathbf{A}]$$

Integrating this angular-momentum density over the entire volume of the wave, we find the net angular momentum:

$$\mathbf{J} = \frac{1}{4\pi c}\int \mathbf{x}\times(E^n\nabla A^n)\,dV - \frac{1}{4\pi c}\int \mathbf{x}\times(\mathbf{E}\cdot\nabla)\mathbf{A}\,dV \tag{65}$$

In the second integral, we can use the identity

$$\mathbf{x}\times(\mathbf{E}\cdot\nabla)\mathbf{A} = \nabla^n(E^n\mathbf{x}\times\mathbf{A}) - \mathbf{x}\times\mathbf{A}\nabla\cdot\mathbf{E} + \mathbf{A}\times(\mathbf{E}\cdot\nabla)\mathbf{x}$$

Since $\nabla\cdot\mathbf{E} = 0$ and $(\mathbf{E}\cdot\nabla)\mathbf{x} = \mathbf{E}$, this reduces to

$$\mathbf{x}\times(\mathbf{E}\bullet\nabla)\mathbf{A} = \nabla^n(E^n\mathbf{x}\times\mathbf{A}) - \mathbf{E}\times\mathbf{A} \tag{66}$$

By the arguments familiar from the proof of Gauss' theorem (see Section 1.5), the volume integral of the divergence term can be converted into a surface integral over a closed surface surrounding the wave packet:

$$\int \nabla^n(E^n\mathbf{x}\times\mathbf{A})\,dV = \int_\sigma E^n\mathbf{x}\times\mathbf{A}\,d\sigma^n$$

This surface integral vanishes because the electric field is zero at all points outside the wave packet. We can therefore ignore the divergence term in Eq. (66), and upon substituting the remaining term into Eq. (65), we find

$$\mathbf{J} = \frac{1}{4\pi c}\int \mathbf{x}\times(E^n\nabla A^n)\,dV + \frac{1}{4\pi c}\int \mathbf{E}\times\mathbf{A}\,dV \tag{67}$$

This equation is mathematically equivalent to Eq. (65), but it is much more convenient for calculations. For instance, it permits a simple and direct evaluation of the spin of a quasi-plane wave, consisting of an inner region of large volume within which the amplitudes of the electric and magnetic fields are nearly uniform, surrounded by an outer region of small volume within which the amplitudes are decreasing and nonuniform. According to Eq. (65), the evaluation of the spin angular momentum for such a wave requires knowledge of the (very complicated) fields in the outer region, where the energy flow circulates around the wave. But as we will see below, according to Eq. (67), the spin angular momentum receives most of its contribution from the inner region, and the outer region can be neglected, vastly simplifying the calculation. In essence, the mathematical transformation leading from Eq. (65) to Eq. (67) has replaced the energy flow in the large loop around the outer edge of the wave by a virtually equivalent energy flow consisting of many small, adjacent loops filling the entire volume of the wave; the spin contributions of these small loops are then likewise distributed over the entire volume.

The first integral on the right side of Eq. (67) represents the orbital angular momentum, and the second integral represents the spin. To justify this interpretation, consider a circularly polarized wave of positive or of negative helicity with an electric field

$$\mathbf{E}_\pm = E_0\hat{\mathbf{x}}\cos(\omega t - kz) \pm E_0\hat{\mathbf{y}}\sin(\omega t - kz) \tag{68}$$

The vector potential for this wave is [see Eqs. (13) and (15)]

$$\mathbf{A}_\pm = -\frac{cE_0}{\omega}\hat{\mathbf{x}}\sin(\omega t - kz) \pm \frac{cE_0}{\omega}\hat{\mathbf{y}}\cos(\omega t - kz)$$

Of course, at the edge of the wave, this electric field and vector potential will have to be modified to fit the decreasing, nonuniform fields, but for a quasi-plane wave we can neglect this modification. With these expressions for **E** and **A**, the time-average values of the two integrals in Eq. (67) are

$$\mathbf{L} = \frac{1}{4\pi c}\int \mathbf{x}\times(E''\nabla A'')\,dV = \int \mathbf{x}\times\frac{\hat{\mathbf{z}}E_0^2}{4\pi c}\,dV$$

and

$$\mathbf{s} = \frac{1}{4\pi c}\int \mathbf{E}\times\mathbf{A}\,dV = \pm\frac{1}{4\pi}\int \hat{\mathbf{z}}\frac{E_0^2}{\omega}\,dV$$

■ ***Exercise 7.*** Obtain these results. ■

The quantity $\hat{\mathbf{z}}E_0^2/4\pi c$ in the integrand of the expression for **L** is simply the time-average density of translational momentum in our circularly polarized wave. Hence **L** is indeed the orbital angular momentum, and **s** must therefore be the extra, nonorbital angular momentum—that is, it must be the spin. The direction of the spin depends on the state of polarization: for a wave of positive helicity, the spin is parallel to the direction of propagation ($s_z > 0$); for a wave of negative helicity, it is antiparallel ($s_z < 0$). Note that spin is a characteristic feature of circularly polarized waves; plane-polarized waves do not have spin.

■ ***Exercise 8.*** Show that **s** = 0 for a linearly polarized wave. ■

EXAMPLE: *spin of a light wave*

It is instructive to compare the energy and the spin in a circularly polarized wave. The time-average energy density in the circularly polarized wave (68) is $E_0^2/4\pi$. Hence the ratio of spin to energy is

$$\frac{s_z}{U} = \frac{\pm 1/4\pi\int E_0^2/\omega\,dV}{1/4\pi\int E_0^2\,dV} = \pm\frac{1}{\omega} \tag{69}$$

From the point of view of quantum theory, this is simply the ratio of the spin of a photon to its energy, $\pm\hbar/\hbar\omega$. The wave of positive helicity consists of photons of spin $+\hbar$ ("spin up," or spin parallel to the direction of propagation), and the wave of negative helicity consists of photons of spin $-\hbar$ ("spin down," or spin antiparallel to the direction of motion).

The physical picture of the spin that emerges from the above calculations is that of an angular momentum arising from a circulating flow of energy in the wave field. It can be established that this picture is of general validity; it applies to gravitational waves, and it applies to all kinds of quantum-mechanical waves. For instance, the spin of the electron arises from a circulating energy flow in the quantum-mechanical electron wave field. Thus, in contrast to a common misconception, the electron spin does *not* arise from an internal rotating structure of the electron—rather, it arises from the rotating structure of the electron wave field.[1]

12·4 Plane Waves in a Nonconducting Medium

The crucial difference between the field equations in vacuum and in a material medium is that in the latter, the charge and the current densities are not zero. If the medium is nonconducting, then the charge density and the current density of *free* charges are zero. However, if the medium is capable of polarization and magnetization, then there will still exist a charge density and a current density of *bound* charges, and these must be taken into account in our field equations. From Section 11.2, we already know that the magnetization of the material contributes a magnetization current

$$\mathbf{j}_{\mathrm{M}} = c\nabla\times\mathbf{M} \tag{70}$$

We will now show that the polarization of the material contributes a polarization current

$$\mathbf{j}_{\mathrm{P}} = \frac{\partial\mathbf{P}}{\partial t} \tag{71}$$

We recall that wherever the polarization is inhomogeneous, there exists a density of bound charges [see Eq. (4.8)]

$$\rho_{\mathrm{P}} = -\nabla\bullet\mathbf{P} \tag{72}$$

The current $\mathbf{j}_{\mathrm{P}}$ represents the flow of bound charge that must occur wherever the polarization changes in time. By definition, the polarization is the dipole moment per unit volume. This means that the polarization can be expressed in terms of the sum of all the dipole moments of the bound charges in some small volume ∇V:

$$\mathbf{P} = \frac{1}{\Delta V}\sum_i q_i\mathbf{x}_i \tag{73}$$

If the charges move, then*

$$\frac{\partial\mathbf{P}}{\partial t} = \frac{1}{\Delta V}\sum_i q_i\frac{d\mathbf{x}_i}{dt} \tag{74}$$

---

*Note that the bound charges move only within their atoms, which remain fixed. Hence the volume ∇V remains constant in time.

Here the sum on the right side is the current of bound charges, and when multiplied by $1/\Delta V$, it becomes the current density, as was to be shown. Incidentally: The polarization charge density and current density satisfy the continuity equation

$$\frac{\partial \rho_P}{\partial t} + \nabla \cdot \mathbf{j}_P = 0 \tag{75}$$

This is an immediate consequence of Eqs. (71) and (72).

In vacuum, we found that we could derive the properties of waves equally well from the equations for the four-vector potential or from Maxwell's equations. In a material medium, the equations for the four-vector potential are a bit awkward to handle, and we will give preference to Maxwell's equations. Taking into account the polarization charge and current, the magnetization current, and the free charge and current, we find that the Maxwell equations in a material medium are

$$\nabla \cdot \mathbf{E} = 4\pi\rho_F - 4\pi\,\nabla \cdot \mathbf{P} \tag{76}$$

$$\nabla \times \mathbf{B} = \frac{1}{c}\frac{\partial \mathbf{E}}{\partial t} + \frac{4\pi}{c}\left(\mathbf{j}_F + c\nabla \times \mathbf{M} + \frac{\partial \mathbf{P}}{\partial t}\right) \tag{77}$$

$$\nabla \cdot \mathbf{B} = 0 \tag{78}$$

$$\nabla \times \mathbf{E} = -\frac{1}{c}\frac{\partial \mathbf{B}}{\partial t} \tag{79}$$

In terms of the electric displacement field ($\mathbf{D} = \mathbf{E} + 4\pi\mathbf{P}$) and the **H** field ($\mathbf{H} = \mathbf{B} - 4\pi\mathbf{M}$), these equations become

$$\nabla \cdot \mathbf{D} = 4\pi\rho_F \tag{80}$$

$$\nabla \times \mathbf{H} = \frac{1}{c}\frac{\partial \mathbf{D}}{\partial t} + \frac{4\pi}{c}\mathbf{j}_F \tag{81}$$

$$\nabla \cdot \mathbf{B} = 0 \tag{82}$$

$$\nabla \times \mathbf{E} = -\frac{1}{c}\frac{\partial \mathbf{B}}{\partial t} \tag{83}$$

These equations have no ostensible dependence on the density and current of the bound charges—these contributions from the bound charges are hidden within **D** and **H**. But to solve these equations, we need to know how **D** and **H** depend on **E** and **B**, and this dependence involves the bound charges and currents. And there is one other, subtle point: the microscopic Maxwell equations are exact, but the macroscopic Maxwell equations (80)–(83) are only approximations—they rest on the assumption that the bound charges and currents contribute dipoles, but that the quadrupoles and higher multipoles can be neglected.

In this section, we want to study waves in a nonconducting medium, with $\rho_F = 0$ and $\mathbf{j}_F = 0$. Furthermore, we will assume that the medium is linear, isotropic, and homogeneous, so $\mathbf{D} = \varepsilon\mathbf{E}$ and $\mathbf{B} = \mu\mathbf{H}$, with ε and μ constant in space and time. This permits us to write our equations as

$$\nabla \cdot \mathbf{E} = 0 \tag{84}$$

$$\frac{1}{\mu}\nabla \times \mathbf{B} = \frac{\varepsilon}{c}\frac{\partial \mathbf{E}}{\partial t} \tag{85}$$

$$\nabla \cdot \mathbf{B} = 0 \tag{86}$$

$$\nabla \times \mathbf{E} = -\frac{1}{c}\frac{\partial \mathbf{B}}{\partial t} \tag{87}$$

These equations differ from the vacuum equations only by the factors μ and ε. The derivation of the wave equation for **E** can therefore be carried out by the same steps as in Section 12.1; this leads to

$$-\nabla^2 \mathbf{E} = -\frac{\mu\varepsilon}{c^2}\frac{\partial^2 \mathbf{E}}{\partial t^2} \tag{88}$$

Obviously, the plane-wave solutions of this wave equation with direction of propagation along the z axis are again of the form given in Eq. (34), but the relationship between k and ω is now

$$k = \sqrt{\mu\varepsilon}\,\frac{\omega}{c} \tag{89}$$

and the speed of the wave is

$$u_p = \frac{\omega}{k} = \frac{c}{\sqrt{\mu\varepsilon}} \tag{90}$$

A calculation similar to that of Section 12.1 shows that the electric field of the wave is again perpendicular to the direction of propagation; thus there exist two independent, linearly polarized waves. The magnetic field in these waves is perpendicular to the electric field; however, the magnitude of the magnetic field differs from that of the electric field by a factor $\sqrt{\mu\varepsilon}$. The two independent, linearly polarized waves are

$$\mathbf{E}_{(1)} = E_0\hat{\mathbf{x}}\cos(\omega t - kz)$$
$$\mathbf{B}_{(1)} = \sqrt{\mu\varepsilon}\,E_0\hat{\mathbf{y}}\cos(\omega t - kz) \qquad (91)$$

and

$$\mathbf{E}_{(2)} = E_0\hat{\mathbf{y}}\cos(\omega t - kz)$$
$$\mathbf{B}_{(2)} = -\sqrt{\mu\varepsilon}\,E_0\hat{\mathbf{x}}\cos(\omega t - kz) \qquad (92)$$

■ ***Exercise 9.*** Derive the formulas for the magnetic fields from those for the electric fields. ■

In a linear, isotropic medium with dielectric constant ε and permittivity constant μ, the energy density in the electromagnetic fields is [see Eqs. (5.26) and (11.92)]

$$T^{00} = \frac{1}{8\pi}\mathbf{E}\cdot\mathbf{D} + \frac{1}{8\pi}\mathbf{H}\cdot\mathbf{B} = \frac{1}{8\pi}(\varepsilon E^2 + \frac{1}{\mu}B^2)$$

Hence the energy density in the above plane waves is

$$T^{00} = \frac{\varepsilon}{4\pi}E_0^2\cos^2(\omega t - kz)$$

Since the expression for the energy density in a material medium differs from that in vacuum, we expect that the expression for the Poynting vector must also be modified. To discover the required modification, let us take the time derivative of the energy contained in some given volume:

$$\frac{d}{dt}\int T^{00}\,dV = \int \frac{1}{4\pi}\left(\varepsilon\mathbf{E}\cdot\frac{\partial\mathbf{E}}{\partial t} + \frac{1}{\mu}\mathbf{B}\cdot\frac{\partial\mathbf{B}}{\partial t}\right)dV \qquad (93)$$

With the Maxwell equations (85) and (87), the right side of this equation becomes

$$\int \frac{c}{4\pi}(\mathbf{E}\cdot\nabla\times\mathbf{H} - \mathbf{H}\cdot\nabla\times\mathbf{E})\,dV$$

In consequence of the identity

$$\nabla\cdot(\mathbf{E}\times\mathbf{H}) = \nabla^n(\varepsilon^{nkl}E^kH^l) = E^k\varepsilon^{nkl}\nabla^nH^l + H^l\varepsilon^{nkl}\nabla^nE^k$$
$$= -\mathbf{E}\cdot\nabla\times\mathbf{H} + \mathbf{H}\cdot\nabla\times\mathbf{E}$$

this is equal to

$$-\int \frac{c}{4\pi}\nabla\cdot(\mathbf{E}\times\mathbf{H})\,dV$$

If we apply Gauss' theorem, we can transform this volume integral into a surface integral over the surface σ bounding the volume:

$$-\int_{\sigma} \frac{c}{4\pi}(\mathbf{E}\times\mathbf{H})\cdot d\boldsymbol{\sigma}$$

Substituting this expression into Eq. (93), we find

$$\frac{d}{dt}\int T^{00}\, dV = -\int_{\sigma} \frac{c}{4\pi}(\mathbf{E}\times\mathbf{H})\cdot d\boldsymbol{\sigma}$$

Since the left side of this equation is the rate of change of the energy in the given volume, the right side must represent the rate at which energy flows through the surface bounding the volume—that is, $(c/4\pi)(\mathbf{E}\times\mathbf{H})\cdot d\boldsymbol{\sigma}$ must be the rate at which energy flows across the area $d\boldsymbol{\sigma}$. Thus the Poynting vector representing the energy flux in the material medium must be

$$\mathbf{S} = \frac{c}{4\pi}\mathbf{E}\times\mathbf{H} \tag{94}$$

For the plane waves described by Eqs. (91) and (92), this energy flux is

$$\mathbf{S} = \hat{\mathbf{z}}\frac{c}{4\pi}\sqrt{\frac{\varepsilon}{\mu}}E_0^2\cos^2(\omega t - kz)$$

12·5 Dispersion and the Oscillator Model

The velocity u_{p} defined by Eq. (90) is called the *phase velocity* because it describes the motion of the wave peaks or wave troughs, that is, the surfaces at which $\omega t - kz$ = constant. If we define the *index of refraction* as

$$n = \sqrt{\mu\varepsilon} \tag{95}$$

then

$$k = n\frac{\omega}{c} \tag{96}$$

and

$$u_{\mathrm{p}} = \frac{c}{n} \tag{97}$$

In calculating the phase velocity from this formula, we must keep in mind that the index of refraction is a function of frequency—the value of

ε for, say, a light wave propagating through some dielectric is very different from the static value used in Chapter 4. A medium with an index of refraction that depends on frequency is called a *dispersive* medium. When a beam of white light, containing a mixture of all colors, penetrates the surface of such a medium, the different colors will be refracted through different angles—that is, the colors will be dispersed.

To understand how such a frequency dependence comes about, we will examine a very simple very simple model for the behavior of the bound electrons in a dielectric. This model postulates that inside the atoms, the electrons are held near their equilibrium position by elastic forces (like the forces of springs). The electric field of a light wave will then cause oscillations of the electrons about their equilibrium positions, and the resulting oscillating dipole moments will affect the propagation of the light wave.

Suppose that the light wave has an electric field (in complex notation)

$$\mathbf{E} = E_0 \hat{\mathbf{x}} \mathrm{e}^{-i(\omega t - kz)}$$

Consider an electron with equilibrium position $x = y = z = 0$. The equation of motion of this electron under the influence of the wave is then*

$$m_e \frac{d^2x}{dt^2} + m_e \omega_0^2 x = -eE_0 \mathrm{e}^{-i\omega t} \tag{98}$$

where $m_e\omega_0^2$ is the spring constant associated with the elastic force holding the electron within the atom. The steady-state solution of this equation is

$$x = \frac{(-e/m_e)E_0 \mathrm{e}^{-i\omega t}}{\omega_0^2 - \omega^2}$$

The dipole moment that this electron contributes is then

$$p = -ex = \frac{(e^2/m_e)E_0 \mathrm{e}^{-i\omega t}}{\omega_0^2 - \omega^2} \tag{99}$$

This gives a dipole moment per unit volume

$$P = Np = \frac{Ne^2/m_e}{\omega_0^2 - \omega^2} E_0 \mathrm{e}^{-i\omega t} = \frac{Ne^2/m_e}{\omega_0^2 - \omega^2} E \tag{100}$$

---

*We neglect the magnetic force of the light wave on the electron. This is a good approximation whenever the motion is nonrelativistic.

where N is the number of electrons per unit volume. Note that P has the same sign as E for $\omega < \omega_0$, but the sign opposite to that of E for $\omega > \omega_0$ (at low frequencies, the response of the oscillator is quasistatic; the electron moves in a direction opposite to that of the electric field, and its dipole moment is in the *same* direction as the electric field).

From Eq. (100), we conclude that the electric susceptibility is

$$\chi = \frac{Ne^2 / m_e}{\omega_0^2 - \omega^2} \tag{101}$$

and from Eq. (4.22), we then find that the dielectric constant is

$$\varepsilon = 1 + 4\pi\chi = 1 + \frac{4\pi Ne^2 / m_e}{\omega_0^2 - \omega^2} \tag{102}$$

If we assume $\mu \simeq 1$, which is a good approximation for most materials, then $n^2 = \varepsilon$, or

$$n^2 = 1 + \frac{4\pi Ne^2 / m_e}{\omega_0^2 - \omega^2} \tag{103}$$

This simple formula—based on a rather crude model—can be brought into good agreement with measurements by a slight refinement. Instead of assigning the same spring constant to all the electrons, let us assign spring constants $m_e\omega_j^2$ to a fraction f_j of the electrons. This leads to

$$n^2 = 1 + \frac{4\pi Ne^2}{m_e} \sum_j \frac{f_j}{\omega_j^2 - \omega^2} \tag{104}$$

This is *Sellmeier's equation*. Although we have here derived this equation from classical mechanics, it is also valid in quantum mechanics.[2] The fractions f_j are called *oscillator strengths* in quantum mechanics; they are proportional to probabilities for atomic transitions with transition frequencies ω_j. With suitable values for the constants f_j and ω_j, Eq. (104) gives a good fit to the experimental data for the index of refraction of gases.

Note that Eq. (104) rests on the implicit assumption that the polarization P in Eq. (100) is small compared with the electric field, so the local electric field E_{local} acting on the electron is approximately equal to the macroscopic electric field E. If this is not the case, then the relation (100) between P and E must be modified, as discussed in Section 4.5, and, correspondingly, the expression (100) for the index of refraction must also be modified (see Problem 12.34). These modifications become important whenever the index of refraction differs from 1 by more than a few percent, for instance, in gases at high density or in glasses.

Figure 12.3 shows a plot of n^2 versus ω according to the simple formula (104) for a material with two different resonant frequencies ω_j. At these resonant frequencies, our formula breaks down; it gives a divergent value of n^2 at the resonant frequencies, with a discontinuous jump from $+\infty$ to $-\infty$.

To remedy this divergence, we need to take into account the damping forces (or frictional forces) that act on the electrons and prevent the amplitude of oscillation from becoming infinite at resonance. The following exercise deals with this improvement of the formula.

■ ***Exercise 10.*** The equation of motion of an electron including a frictional force proportional to the velocity (with a damping constant Γ) is

$$m_e \frac{d^2x}{dt^2} + m_e\Gamma\frac{dx}{dt} + m_e\omega_0^2 x = -eE_0 e^{-i\omega t}$$

Show that the solution of this equation leads to an electric susceptibility

$$\chi = \frac{Ne^2/m_e}{\omega_0^2 - \omega^2 - i\Gamma\omega}$$

If different fractions of the electrons have different spring constants and different damping constants, this becomes

$$\chi = \frac{Ne^2}{m_e}\sum_j \frac{f_j}{\omega_j^2 - \omega^2 - i\Gamma_j\omega} \qquad (105)$$

This equation leads to an index of refraction

Figure 12·3

Square of the index of refraction as a function of frequency according to Eq. (104).

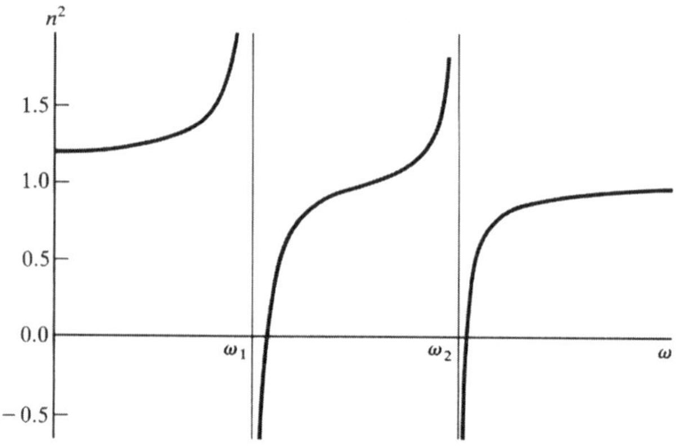

$$n^2 = 1 + \frac{4\pi N e^2}{m_e} \sum_j \frac{f_j}{\omega_j^2 - \omega^2 - i\Gamma_j \omega}$$

Show that the right side of this equation can be expressed as a sum of real and imaginary parts, as follows:

$$n^2 = 1 + \frac{4\pi N e^2}{m_e} \sum_j \frac{(\omega_j^2 - \omega^2) f_j}{(\omega_j^2 - \omega^2)^2 + \Gamma_j^2 \omega^2} + i \frac{4\pi N e^2}{m_e} \sum_j \frac{\omega \Gamma_j f_j}{(\omega_j^2 - \omega^2)^2 + \Gamma_j^2 \omega^2} \tag{106}$$

■

According to Eq. (106), n^2 is a complex quantity, with real and imaginary parts. Figure 12.4 is a plot of the real part of n^2, for a case with two resonant frequencies. The value of n^2 is now everywhere finite.

Note that since the imaginary part of n^2 is positive, the imaginary part of n must also be positive [for any complex number z, $\text{Im}(z^2) > 0$ implies $\text{Im}(z) > 0$, although the converse is not true]. The imaginary parts of n^2 and of n are small, except when the frequency of the wave is near a resonant frequency, where the imaginary parts become large. In general, a positive imaginary part of the index of refraction indicates damping of the wave—the electrons absorb the energy of the wave, and the amplitude of the wave decreases. We can see this from Eq. (96), where a positive imaginary part in n leads to a positive imaginary part in k. When inserted into the wavefunction $e^{-i\omega t + ikz}$, such a positive imaginary part in k produces an exponentially decreasing function of z (we will later explore a mathematically similar instance of such damping in our discussion of waves in a conducting medium, in Section 12.8). Thus a wave with a frequency near resonance does not propagate far into the medium; instead, it is quickly absorbed.

The dispersion in the medium is said to be *normal* if the real part of n increases with frequency (as in Fig. 12.3 in each of the three intervals $\omega < \omega_1$, $\omega_1 < \omega < \omega_2$, and $\omega > \omega_2$). It is said to be *anomalous* if the real part of n decreases with frequency (as in Fig. 12.4 where the real part of n decreases sharply in the narrow intervals near ω_1 and ω_2). The intervals of anomalous dispersion, where the real part of n decreases, coincide with the intervals where the imaginary part of n is large; hence a wave with anomalous dispersion coincidentally also experiences strong absorption. Note that if the index of refraction is complex, then the phase velocity (97) must be calculated from the real part of n,

$$u_p = \frac{c}{\text{Re}(n)}$$

In optics, it is conventional to reserve the name "index of refraction" and the symbol n for the real part of the complex quantity that we designated by n in Eq. (106). With this change of notation, the phase velocity can always be expressed in the simple form $u_p = c/n$. We will hereafter

Figure 12·4

Real part of the square of the index of refraction as a function of frequency, including the effects of frictional damping, according to Eq. (106). The index of refraction is finite for all values of the frequency, but it decreases sharply in the vicinity of ω_1 and ω_2.

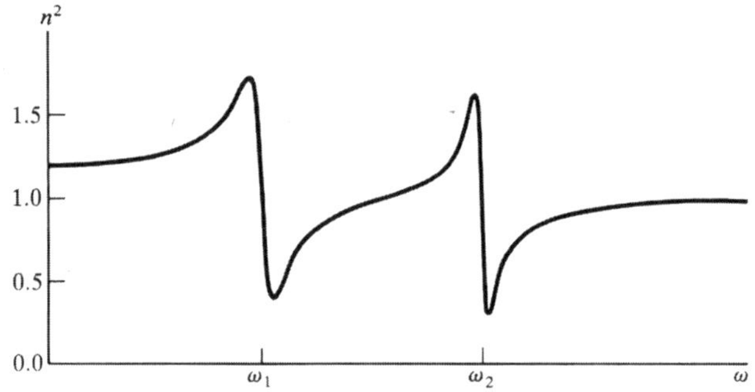

adopt this change of notation for the treatment of waves in nonconducting media, so any n that appears in the remainder of this section and in the next two sections will be real.*

So far, our discussion has been restricted to the propagation of harmonic waves of a single frequency. Now we will briefly examine the propagation of wave pulses. As mentioned in Section 12.1, an arbitrary wave pulse is a superposition of an infinite number of harmonic waves of different frequencies. If each of these component waves has a different velocity of propagation, these waves will not keep up with one another, and the wave pulse will suffer a gradual distortion. For example, consider a wave pulse in a medium with the index of refraction (103), and suppose that for all the component waves $\omega < \omega_0$. The high-frequency components of this wave pulse then have a smaller speed than the low-frequency components. The latter get ahead, while the former fall behind—the wave pulse spreads out, growing broader as time passes.

A wave pulse consisting of a superposition of harmonic waves with a fairly narrow range of frequencies is called a wave group. The velocity of the peak of such a wave group is called the group velocity. We will show that this velocity is

$$u_g = \frac{d\omega}{dk} \tag{107}$$

*This convention in optics differs from the convention used in the study of waves in a conducting medium (see Section 12.8); there, the symbol n represents the complex index of refraction, including both its real and its imaginary parts. Unfortunately, this difference in conventions has become a tradition.

where it is understood that the derivative on the right side of this equation is to be evaluated at the typical frequency of the waves in the group.*

EXAMPLE: *a simple wave group*

For the sake of mathematical simplicity, we will treat only the case of a wave group consisting of the superposition of just two harmonic waves of equal amplitudes whose frequencies and wavelengths differ by only a small amount. If the direction of propagation is along the z axis, we can write this superposition as

$$\mathbf{E} = E_0\hat{\mathbf{x}}\cos(\omega t - kz) + E_0\hat{\mathbf{x}}\cos[(\omega + \Delta\omega)t - (k + \Delta k)z] \tag{108}$$

According to the identity $\cos\alpha + \cos\beta = 2\cos\frac{1}{2}(\alpha - \beta)\cos\frac{1}{2}(\alpha + \beta)$, this is the same as

$$\mathbf{E} = 2E_0\hat{\mathbf{x}}\cos(\tfrac{1}{2}\Delta\omega t - \tfrac{1}{2}\Delta kz)\cos[(\omega + \tfrac{1}{2}\Delta\omega)t - (k + \tfrac{1}{2}\Delta k)z] \tag{109}$$

Figure 12.5 shows this wave at one instant of time. The wave has an average wave vector $k + \frac{1}{2}\,\Delta k$, and its amplitude is modulated, with an envelope that displays beats. If the phase velocity of both waves in Eq. (109) were the same (nondispersive medium), then the waves would always remain in step. The entire pattern in Fig. 12.5 would then move to the right rigidly, and the peak of the envelope would move with the same velocity as that of the individual waves. But if the phase velocities are

Figure 12·5

Superposition of two waves of slightly different frequencies and wavelengths. The envelope of the combined wave displays beats.

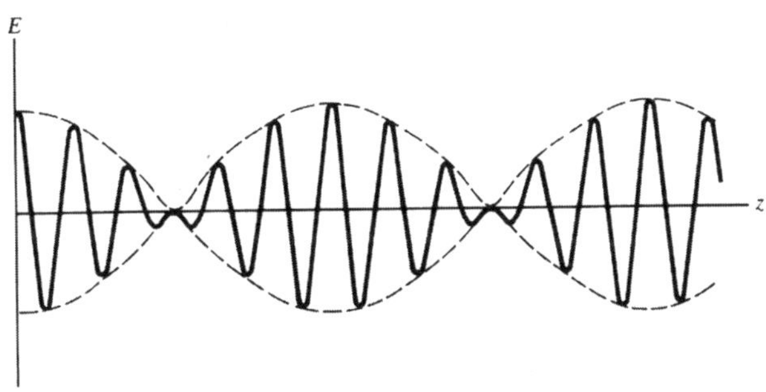

*Since all the waves under consideration lie within a narrow range of frequencies, the result is not sensitive to the precise definition of *typical* frequency. Note that for a general wave pulse, consisting of a superposition of harmonic waves with a *wide* range of frequencies, the group velocity is not well defined. Such a wave pulse will suffer quick and drastic changes of shape as it propagates, and we will not be able to identify a distinctive peak whose velocity we could measure.

different (dispersive medium), then the peak of the envelope moves relative to the individual waves. To obtain the velocity of the peak, we note that the first factor in Eq. (109) represents the envelope shown in Fig. 12.5. The peak then corresponds to $\frac{1}{2}\Delta\omega t - \frac{1}{2}\Delta\, kz = 0$, and consequently, the velocity of the peak is

$$u_g = \frac{\Delta\omega}{\Delta k} \tag{110}$$

Since $\Delta\omega$ and Δk are small, the ratio $\Delta\omega/\Delta k$ can be approximated by $d\omega/dk$, which yields Eq. (107).

The group velocity is the velocity of wave signals in the dispersive medium. Any wave signal must consist of some finite wave pulse. A signal cannot be sent with a purely harmonic wave—this wave lasts forever, it has no beginning and no end, and it conveys no information. To transmit information, we must switch the harmonic wave on and off (in the terminology of radio, we must modulate the carrier wave), and this amounts to the same thing as constructing a finite wave pulse. The velocity of such a wave signal will then be the group velocity.

For the evaluation of the group velocity, it is usually convenient to regard k as a function of ω, so

$$u_g = \frac{1}{dk/d\omega} = \frac{c}{d(\omega n)/d\omega} = \frac{c}{n + \omega dn/d\omega} \tag{111}$$

EXAMPLE: *calculation of group velocity*

For instance, if we take the value of n from Eq. (103), then

$$n(\omega) = \sqrt{1 + \frac{4\pi Ne^2/m_e}{\omega_0^2 - \omega^2}} \tag{112}$$

which gives

$$u_g = \frac{c}{n + \omega dn/d\omega} = \frac{c\sqrt{1 + \dfrac{4\pi Ne^2/m_e}{\omega_0^2 - \omega^2}}}{1 + \dfrac{4\pi Ne^2\omega_0^2/m_e}{(\omega_0^2 - \omega^2)^2}} \tag{113}$$

It is easy to check that $u_g < c$ for all values of the frequency ω, except values near the resonant frequency ω_0. Thus wave signals satisfy the requirement of relativity that signals cannot exceed the standard speed of light. For values of the frequency near ω_0, the group velocity calculated from Eq. (113) can exceed the standard speed of light. But in this frequency interval, the wave is strongly absorbed and does not propagate; hence the group velocity loses its physical significance.

■ **Exercise 11.** Verify that, according to the above equation, $u_g < c$ whenever $|\omega_0^2 - \omega^2| \gg 4\pi Ne^2 / m_e$. ■

Recently, some spectacular experiments have exhibited propagation of light pulses with extremely low group velocities, 10 m/s and even less. Such a drastic reduction of the group velocity hinges on a peculiarity of Eq. (106) that emerges from quantum mechanics: for atoms in excited states, the oscillator strengths f_j for downward transitions are negative.* In a medium that contains a preponderance of excited atoms, the behavior of the index of refraction near the transition frequency for the excited state is then the opposite of what is shown in Fig. 12.4: instead of a sharp decrease of the index of refraction near the transition frequency, we now have a sharp *increase*. This implies a very large, positive value of $dn/d\omega$, and Eq. (111) then gives a very small value for the group velocity. The abundance of excited atoms required for these experiments can be attained in an atomic vapor cell, in which the atoms are pre-excited by a laser beam. Some experiments have even succeeded in bringing a pulse of light to a complete stop ("frozen light"). But that involves some extra quantum-mechanical tricks with superpositions of atomic states, for which there is no analog in classical electrodynamics.

12·6 Refraction

As is well known, a light wave suffers refraction when penetrating from one medium (with index of refraction n) into a second medium (with index of refraction n'). The change of direction of propagation of the wave at the interface is given by *Snell's law of refraction*:

$$n \sin\theta = n' \sin\theta' \qquad (114)$$

where θ and θ' are the angles between the directions of propagation and the normal to the interface (see Fig. 12.6). In this section, we will use Maxwell's equations to derive this law of refraction and the formulas for the intensities of the refracted and the reflected waves.

Figure 12.6 shows the wave vectors of the incident, refracted, and reflected waves. The electric and magnetic fields of these waves are, respectively, as follows:

*These negative oscillator strengths are also crucial for the operation of lasers; see Section 15.7.

Figure 12·6

Wave vectors of the incident, refracted, and reflected waves, and angles of incidence, refraction, and reflection.

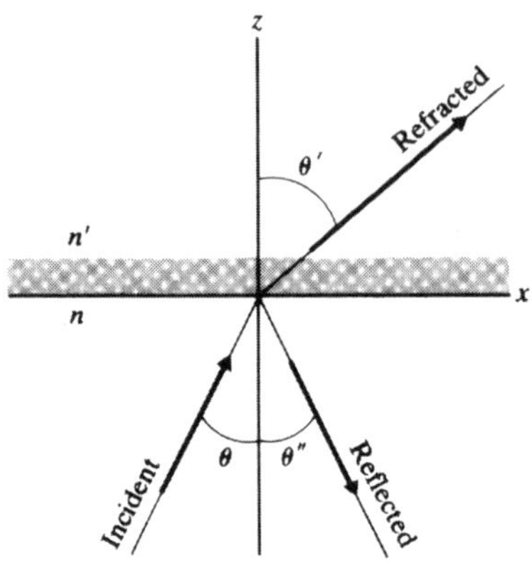

Incident:

$$\mathbf{E} = \mathbf{E}_0 e^{-i(\omega t - \mathbf{k}\cdot\mathbf{x})} \tag{115}$$

$$\mathbf{B} = \frac{c}{\omega}\mathbf{k}\times\mathbf{E}_0 e^{-i(\omega t - \mathbf{k}\cdot\mathbf{x})} \tag{116}$$

Refracted:

$$\mathbf{E}' = \mathbf{E}'_0 e^{-i(\omega t - \mathbf{k}'\cdot\mathbf{x})} \tag{117}$$

$$\mathbf{B}' = \frac{c}{\omega}\mathbf{k}'\times\mathbf{E}'_0 e^{-i(\omega t - \mathbf{k}'\cdot\mathbf{x})} \tag{118}$$

Reflected:

$$\mathbf{E}'' = \mathbf{E}''_0 e^{-i(\omega t - \mathbf{k}''\cdot\mathbf{x})} \tag{119}$$

$$\mathbf{B}'' = \frac{c}{\omega}\mathbf{k}''\times\mathbf{E}''_0 e^{-i(\omega t - \mathbf{k}''\cdot\mathbf{x})} \tag{120}$$

Since the magnitudes of the wave vectors are $k = n\omega/c$, $k' = n'\omega/c$, and $k'' = k = n\omega/c$, the magnitude of each magnetic field equals the index of refraction times the magnitude of the electric field, as it should.

In the above equations, we have taken for granted that the frequencies of all waves are the same. If this were not so, the boundary conditions could not be satisfied at all times—for waves of different frequencies, the boundary conditions can be satisfied at one initial instant of time, but after that the waves would get out of step, and the boundary conditions would break down. This rule concerning the equality of the frequencies of all the waves generated by one incident wave holds in general for any linear medium under steady-state conditions, regardless of any dependence of the index of refraction on position and regardless of the shape of the boundaries. Thus the waves produced by multiple reflections or refractions or by diffraction all have the same frequency as the incident wave, but their wavelength varies according to the index of refraction. The frequency can vary only if the index of refraction depends on time. Throughout all our calculations, we will assume that this is not the case.

At the interface $z = 0$, the fields must satisfy the boundary conditions (4.27), (4.30), (11.60), and (11.61), which express the continuity of the appropriate electric and magnetic field components parallel or perpendicular to the interface:

$$\mathbf{E}_{\parallel} + \mathbf{E}''_{\parallel} = \mathbf{E}'_{\parallel} \tag{121}$$

$$\mathbf{D}_{\perp} + \mathbf{D}''_{\perp} = \mathbf{D}'_{\perp} \tag{122}$$

$$\mathbf{H}_{\parallel} + \mathbf{H}''_{\parallel} = \mathbf{H}'_{\parallel} \tag{123}$$

$$\mathbf{B}_{\perp} + \mathbf{B}''_{\perp} = \mathbf{B}'_{\perp} \tag{124}$$

To deal with these boundary conditions, we find it convenient to assume that the incident wave has one or another of two linear polarizations: either the electric field is parallel to the plane of incidence (z-x plane) or it is perpendicular to the plane of incidence. Any other polarization can be regarded as a superposition of these two basic polarizations. Figure 12.7 shows the directions of the fields for the former case; the electric field is parallel to the plane of incidence, and the magnetic field is perpendicular. Here $\mathbf{E}_{\parallel}$ coincides with E_x, $\mathbf{D}_{\perp}$ with D_z, $\mathbf{H}_{\parallel}$ with H_y, and $\mathbf{B}_{\perp}$ with B_z. At $z = 0$, the boundary conditions (121) and (123) are then

EXAMPLE: *refraction and reflection, E_0 parallel to plane of incidence*

$$-E_0 \cos\theta\, e^{-i(\omega t - kx\sin\theta)} + E_0'' \cos\theta''\, e^{-i(\omega t - kx\sin\theta'')} = -E_0' \cos\theta'\, e^{-i(\omega t - k'x\sin\theta')} \tag{125}$$

and

$$\frac{n}{\mu} E_0 e^{-i(\omega t - kx\sin\theta)} + \frac{n}{\mu} E_0'' e^{-i(\omega t - kx\sin\theta'')} = \frac{n'}{\mu'} E_0' e^{-i(\omega t - k'x\sin\theta')} \tag{126}$$

Without much loss of generality, we could omit the factors μ and μ' in this last equation, because the materials that permit the propagation of light waves or of microwaves usually have negligible magnetic susceptibilities. Nevertheless, we will keep the factors μ and μ' because it costs us little to do so.

Since the time dependence in these equations is contained in the overall factor $e^{-i\omega t}$, the equations will be satisfied at all times if they are satisfied at one initial time, say $t = 0$. [This is where the rule of equal frequency for all waves enters our calculations; if we had not already as-

Figure 12·7

Electric and magnetic fields of the incident, refracted, and reflected waves. The wave is incident from below. The electric fields are parallel to the plane of incidence (x-z plane), and the magnetic fields are perpendicular to it. The dots indicate the tips of the arrows representing the magnetic fields. The arrows fix the sign conventions for the amplitudes E_0 , E'_0 , and E''_0. The amplitude is positive if the electric field has the direction of the arrow; it is negative if the electric field has the opposite direction.

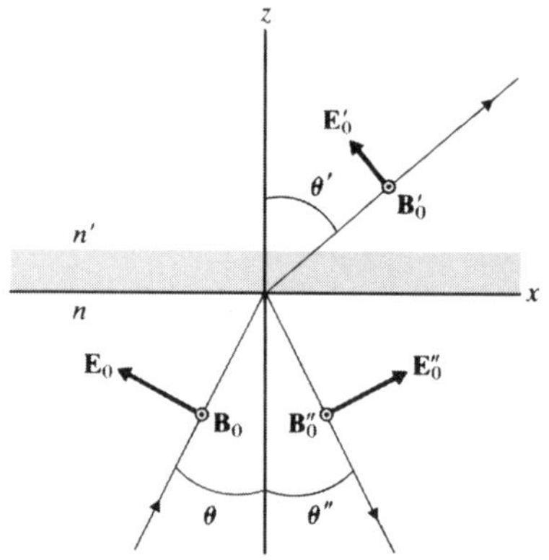

sumed the equality of frequencies for all our waves, then Eqs. (125) and (126) would have forced us to accept this equality.] We can now ignore the time dependence and cancel the overall factor $e^{-i\omega t}$.

The boundary conditions must hold everywhere along the interface—that is, they must hold for all values of x. This requires that the factors $e^{ikx\sin\theta}$,$e^{ikx\sin\theta''}$, and $e^{ik'x\sin\theta'}$ be equal. Accordingly,

$$k\sin\theta = k\sin\theta'' \quad \text{and} \quad k\sin\theta = k'\sin\theta' \tag{127}$$

The first of these equations implies that the angles of incidence and of reflection are equal,

$$\theta = \theta'' \tag{128}$$

This is the law of reflection. With $k = n\omega/c$, $k' = n'\omega/c$, the second equation implies that

$$n\sin\theta = n'\sin\theta' \tag{129}$$

This is Snell's law of refraction. Note that both of these laws are independent of the details of the boundary conditions; they hinge only on the existence of a boundary condition of some kind—Eq. (125) leads to these laws, and so does Eq. (126) and any other equation expressing the equality of some wave fields at the interface.

We can now cancel all of the exponential factors in Eqs. (125) and (126):

$$-E_0 \cos\theta + E_0'' \cos\theta = -E_0' \cos\theta' \qquad (130)$$

$$\frac{n}{\mu} E_0 + \frac{n}{\mu} E_0'' = \frac{n'}{\mu'} E_0' \qquad (131)$$

Solving these for E_0'' and E_0' in terms of E_0, we obtain the following results:

For $\mathbf{E}_0$ *parallel to the plane of incidence,*

$$E_0' = \frac{2(n/\mu)\cos\theta}{(n'/\mu')\cos\theta + (n/\mu)\cos\theta'} E_0 \qquad (132)$$

$$E_0'' = \frac{(n'/\mu')\cos\theta - (n/\mu)\cos\theta'}{(n'/\mu')\cos\theta + (n/\mu)\cos\theta'} E_0 \qquad (133)$$

These are the *Fresnel equations*. If the incident wave is normally incident on the interface ($\theta = \theta' = \theta'' = 0$), these equations reduce to

$$E_0' = \frac{2(n/\mu)}{(n'/\mu') + (n/\mu)} E_0 \qquad (134)$$

$$E_0'' = \frac{(n'/\mu') - (n/\mu)}{(n'/\mu') + (n/\mu)} E_0 \qquad (135)$$

Figure 12.8 gives plots of the ratios E_0'/E_0 and E_0''/E_0 as a function of the angle of incidence. For the purposes of these plots, we have assumed $n = 1$, $n' = 1.5$, and $\mu = \mu' = 1$; these numbers correspond to a light wave traveling from air into crown glass. Near $\theta = \pi/2$, Fig. 12.8 shows that $E_0' \to 0$ and $E_0'' \to -1$; this explains why plates of glass and other materials give strong reflections if the angle of incidence is very shallow. At $\theta = 0$, both E_0' and E_0'' are nonzero; at this angle, the reflected amplitude is 20% of the incident amplitude. For a range of angles around $\theta = 0$, Fig. 12. 8 shows that $E_0''/E_0 > 0$. This means that during reflection, the wave suffers a phase change by 180°—the direction of the electric field of the reflected wave is opposite to that of the incident wave, that is, the directions of the fields are as drawn in Fig. 12.7. Note that the phase change for angles near $\theta = 0$ is a consequence of $n' > n$. This phase change would be absent if $n' < n$, but there would then be a phase change for a range of angles near $\pi/2$.

At an intermediate angle, Fig. 12.8 shows $E_0'' = 0$. At this angle, we have zero reflection. According to Eq. (133), the critical angle corresponds to

Figure 12·8

Amplitudes of the refracted and reflected waves as a function of the angle of incidence. ($\mathbf{E}_0$ is parallel to the plane of incidence.)

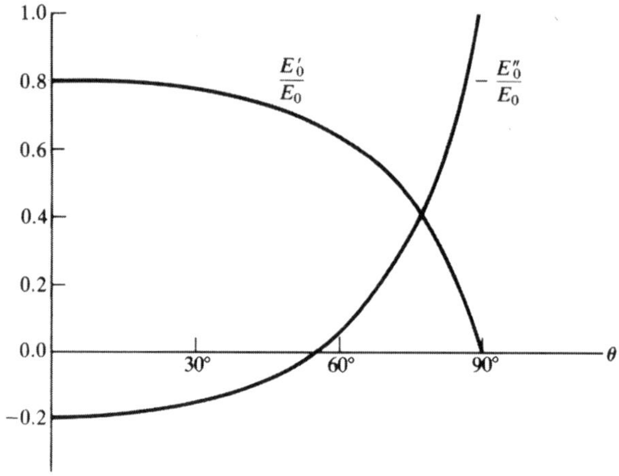

$$\frac{n'}{\mu'}\cos\theta = \frac{n}{\mu}\cos\theta' \tag{136}$$

If $\mu = \mu' = 1$, we can simplify this expression by multiplying it into Snell's law; this yields

$$\sin\theta\cos\theta = \sin\theta'\cos\theta' \tag{137}$$

or

$$½\sin 2\theta = ½\sin 2\theta' \tag{138}$$

The relevant solution of this equation is

$$2\theta = \pi - 2\theta' \tag{139}$$

or

$$\theta + \theta' = \pi/2 \tag{140}$$

As we can see from Fig. 12.9, this condition makes the refracted wave vector perpendicular to the reflected wave vector. Using Snell's law once more, we can express the condition for zero reflection in terms of the angle of incidence:

$$\frac{n'}{n} = \frac{\sin\theta}{\sin\theta'} = \frac{\sin\theta}{\sin(\pi/2-\theta)} = \tan\theta \tag{141}$$

Figure 12·9

When $\theta + \theta' = \pi / 2$, the reflected wave has amplitude zero; its wave vector is perpendicular to the wave vector of the refracted wave.

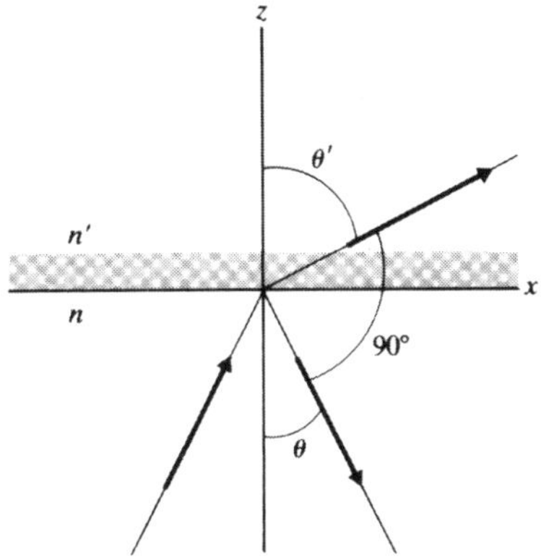

The angle $\tan^{-1}(n'/n)$ is called the *Brewster angle.*

EXAMPLE: *refraction and reflection, $\mathbf{E}_0$ perpendicular to plane of incidence*

Next we must deal with the second case: a wave polarized perpendicular to the plane of incidence. Figure 12.10 shows the direction of the fields for this case; the electric field is perpendicular to the plane of incidence, and the magnetic field is parallel. The boundary conditions then yield the following Fresnel equations:

Figure 12·10

Electric and magnetic fields of the incident, refracted, and reflected waves. The wave is incident from below. The magnetic fields are parallel to the plane of incidence (x-z plane), and the electric fields are perpendicular to it. The crosses indicate the tails of the arrows representing the electric fields. As in Fig. 12.7, the arrows fix the sign conventions for the amplitudes E_0, E_0', E_0''.

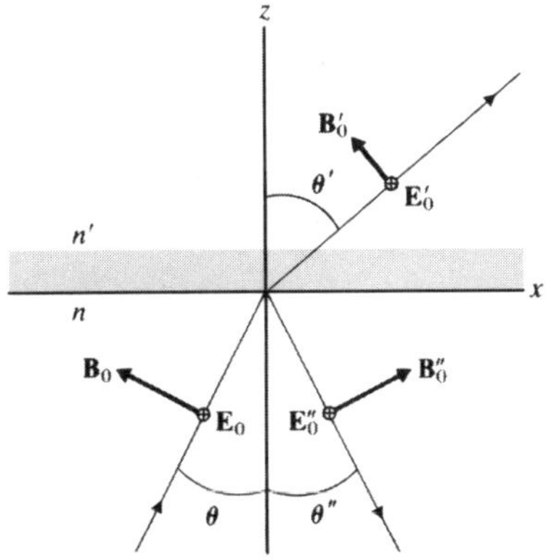

For $\mathbf{E}_0$ *perpendicular to the plane of incidence,*

$$E_0' = \frac{2(n/\mu)\cos\theta}{(n/\mu)\cos\theta + (n'/\mu')\cos\theta'} E_0 \tag{142}$$

$$E_0'' = \frac{(n/\mu)\cos\theta - (n'/\mu')\cos\theta'}{(n/\mu)\cos\theta + (n'/\mu')\cos\theta'} E_0 \tag{143}$$

For normal incidence ($\theta = \theta' = \theta'' = 0$), these equations reduce to the same pair of equations as in the case of $\mathbf{E}_0$ parallel to the plane of incidence. [Eq. (142) yields Eq. (134), and Eq. (143) yields Eq. (135) with a reversed sign; this reversed sign is due to the difference between the conventions for the directions of $\mathbf{E}_0''$ and $\mathbf{E}_0$ established in Figs. 12.7 and 12.10.] This congruity can be understood by noting that if θ = 0, the plane of incidence is undefined—there is then no distinction between the cases $\mathbf{E}_0$ parallel to and $\mathbf{E}_0$ perpendicular to the plane of incidence.

■ ***Exercise 12.*** Derive the Fresnel equations (142) and (143). ■

Figure 12.11 shows plots of E_0'/E_0 and E_0''/E_0. Again, we have assumed $n = 1$, $n' = 1.5$, and $\mu = \mu' = 1$. We see that $E_0''/E_0 < 0$ for all values of θ. This means that the reflected wave has suffered a phase change by 180°—the direction of the electric field of the reflected wave is opposite to that of the incident wave, that is, it is *opposite* to the direction drawn in Fig. 12.10. Note that this phase change hinges on $n' > n$; there would be no phase change if $n' < n$.

One important difference between the cases of $\mathbf{E}_0$ parallel to and $\mathbf{E}_0$ perpendicular to the plane of incidence is that in the latter case, the

Figure 12·11

Amplitudes of the refracted and reflected waves as a function of the angle of incidence. ($\mathbf{E}_0$ is perpendicular to the plane of incidence.)

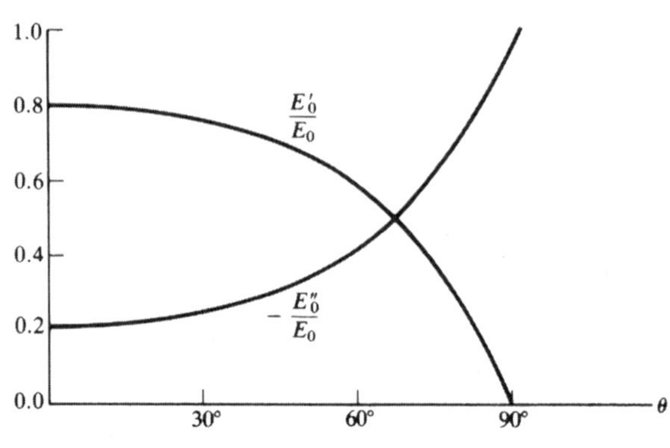

EXAMPLE: *polarization by reflection*

reflected wave is never zero and there is no Brewster angle.* This difference causes the phenomenon of *polarization by reflection.* A wave of some arbitrary direction of polarization can be regarded as a superposition of the two basic polarizations discussed above. If this wave strikes a plate of glass or some other surface at the Brewster angle, the emerging reflected wave will consist entirely of the component of the electric field perpendicular to the plane of incidence; that is, the emerging reflected wave will be polarized perpendicular to the plane of incidence. If the wave strikes the surface at an angle different from the Brewster angle, the component of the electric field parallel to the plane of incidence will not be completely suppressed; however, comparison of Figs. 12.8 and 12.11 shows that in the reflected wave, this component tends to be smaller than the component perpendicular to the plane of incidence. This explains why unpolarized sunlight—consisting of a mixture of waves of random polarizations—becomes partially polarized when reflected by a surface of water. This partial polarization can be readily detected by viewing the reflected image of the Sun with Polaroid sunglasses.

12·7 Total Internal Reflection

If the medium in which the incident wave travels has a higher index of refraction than the medium in which the refracted wave travels (that is, $n > n'$), then the angle of refraction will reach its maximum value 90° while the angle of incidence is smaller than 90°. The critical angle of incidence follows from Snell's law with $\sin\theta' = 1$:

$$\sin\theta = \frac{n'}{n} \tag{144}$$

EXAMPLE: *total internal reflection in glass*

For instance, for crown glass in air, $\sin\theta = 1/1.5$ and $\theta = 42°$. Any angle of incidence beyond that given by Eq. (144) will result in *total internal reflection*—there will be no refracted wave propagating into the second medium. However, as we will see, under conditions of total internal reflection, some of the incident wave does spill over into the second medium and propagates *along the interface.*

To discover the equations for the fields in the second medium, we start with the boundary conditions. Since these conditions must be satisfied regardless of the values of n, n', or θ, we must still have

$$\sin\theta' = \frac{n}{n'}\sin\theta \tag{145}$$

*If $\mathbf{E}_0$ is perpendicular to the plane of incidence, the reflected wave can be zero for $n = 1$ and $n' = 1.5$ if $\mu'/\mu > 1.5$. But this is of little practical interest.

Under conditions of total internal reflection, $\sin\theta' > 1$; this is impossible if θ' is real, but it is possible if θ' is complex. We will not need the value of θ' in our calculations, but we will need $\cos\theta'$:

$$\cos\theta' = \pm\sqrt{1-\sin^2\theta'} = \pm i\sqrt{\sin^2\theta' - 1}$$

$$= \pm i\sqrt{\left(\frac{n}{n'}\sin\theta\right)^2 - 1} \qquad (146)$$

If, as in Section 12.6, the interface is the *x-y* plane and the plane of incidence is the *x-z* plane, then the wave in the second medium is

$$\mathbf{E}' = \mathbf{E}'_0 \exp[-i\omega t + ik'(x\sin\theta' + z\cos\theta')] \qquad (147)$$

$$= \mathbf{E}'_0 \exp\left[-i\omega t + ik'x\frac{n}{n'}\sin\theta \mp k'z\sqrt{\left(\frac{n}{n'}\sin\theta\right)^2 - 1}\right] \qquad (148)$$

Of the two alternative signs in this last function, the – sign corresponds to the + sign in Eq. (146), and the + sign to the – sign in Eq. (146). Obviously, the + sign in the function (148) is unacceptable—it would lead to $\mathbf{E}' \to \infty$ as $z \to +\infty$. Thus we must choose the – sign in the function (148) and the + sign in Eq. (146); that is,

$$\cos\theta' = i\sqrt{\left(\frac{n}{n'}\sin\theta\right)^2 - 1} \qquad (149)$$

This expression for $\cos\theta'$ is to be substituted in our old equations for E'_0 and E''_0 [Eqs. (132), (133), and (142), (143)]. The imaginary value of $\cos\theta'$ makes E'_0 and E''_0 complex—the wave in the second medium and the reflected wave in the first medium have phase shifts relative to the incident wave.

Let us examine the wave in the second medium somewhat more carefully. With the correct choice of sign, Eq. (148) becomes

$$\mathbf{E}' = \mathbf{E}'_0 \exp\left[-i\omega t + ik'x\frac{n}{n'}\sin\theta - k'z\sqrt{\left(\frac{n}{n'}\sin\theta\right)^2 - 1}\right] \qquad (150)$$

This wave propagates in the x direction [with a wavelength $2\pi n'/(k'n\sin\theta)$], but it decays exponentially in the positive z direction without propagation. It is therefore a wave traveling parallel to the interface, with an amplitude that is small except in the immediate vicinity of the interface. If we set the argument of the decaying exponential equal to −1, we find the distance Δz within which the amplitude of the wave decreases by a factor e:

$$k'\Delta z\sqrt{\left(\frac{n}{n'}\sin\theta\right)^2-1}=1 \tag{151}$$

or

$$\Delta z=\frac{\lambda'}{2\pi}\frac{1}{\sqrt{(n/n'\sin\theta)^2-1}} \tag{152}$$

Roughly, we can regard this as the depth of penetration of the wave into the second medium. For instance, if $n = 1.5$, $n' = 1$, and $\theta = 45°$, the depth of penetration is

$$\Delta z = 0.45\lambda' \tag{153}$$

This shows that the wave that spills over into the second medium remains within a very thin layer parallel to the interface.

There is a curious and little-known effect associated with total internal reflection: when a narrow beam is reflected internally, the point of intersection of the incident beam with the interface between the two media does not exactly coincide with the point of intersection of the reflected beam—the intersection points of the two beams with the interface are separated by a distance of the order of one wavelength. This shift can be qualitatively understood as a consequence of the penetration of the incident wave into the second medium. The wave does not immediately turn around at the interface but penetrates a distance of the order of one wavelength [see, for instance, Eq. (153)]; hence the incident and reflected beams intersect at a distance of about one wavelength beyond the interface, rather than at the interface itself. The shift of the reflected beam relative to the incident beam has been experimentally demonstrated by performing multiple internal reflections, to enhance the magnitude of the effect.[3]

12·8 Plane Waves in a Conducting Medium

If the medium in which our wave is propagating is a conductor, the electric field will generate a conduction current

$$\mathbf{j}_F = \sigma\mathbf{E} \tag{154}$$

and this will affect the propagation of the wave. With the inclusion of the conduction current, Maxwell's equations (84)–(87) for a linear, isotropic, and homogeneous medium become

$$\nabla \cdot \mathbf{E} = 0 \tag{155}$$

$$\frac{1}{\mu}\nabla \times \mathbf{B} = \frac{\varepsilon}{c}\frac{\partial \mathbf{E}}{\partial t} + \frac{4\pi\sigma}{c}\mathbf{E} \tag{156}$$

$$\nabla \cdot \mathbf{B} = 0 \tag{157}$$

$$\nabla \times \mathbf{E} = -\frac{1}{c}\frac{\partial \mathbf{B}}{\partial t} \tag{158}$$

where we made the usual assumption that the medium carries no excess of free charges, so $\rho_F = 0$. For an oscillating electric field with the time dependence $e^{-i\omega t}$, we have

$$\frac{\partial \mathbf{E}}{\partial t} = -i\omega\mathbf{E} \tag{159}$$

or

$$\mathbf{E} = \frac{i}{\omega}\frac{\partial \mathbf{E}}{\partial t} \tag{160}$$

so Eq. (156) can be replaced by

$$\frac{1}{\mu}\nabla \times \mathbf{B} = \frac{1}{c}\left(\varepsilon + \frac{4\pi i\sigma}{\omega}\right)\frac{\partial \mathbf{E}}{\partial t} \tag{161}$$

This shows that the inclusion of the conduction current (154) in Maxwell's equations has the same effect as replacing the dielectric constant ε by $\varepsilon + 4\pi i\sigma/\omega$. We can therefore think of the conducting medium as having a complex index of refraction,

$$n = \sqrt{\left(\varepsilon + \frac{4\pi i\sigma}{\omega}\right)\mu} \tag{162}$$

With this new expression for the index of refraction, many of the equations of Sections 12.4 and 12.6 remain valid. For instance, the wave vector is

$$k = \frac{n\omega}{c} = \frac{\omega}{c}\sqrt{\left(\varepsilon + \frac{4\pi i\sigma}{\omega}\right)\mu} \tag{163}$$

and a plane wave traveling in the z direction with a polarization in, say, the x direction has the form

$$\mathbf{E} = E_0\hat{\mathbf{x}}e^{-i(\omega t - kz)} \tag{164}$$

$$= E_0\hat{\mathbf{x}}\exp\left[-i\omega t + \frac{i\omega}{c}\sqrt{\left(\varepsilon + \frac{4\pi i\sigma}{\omega}\right)\mu}\, z\right] \tag{165}$$

with a magnetic field

$$\mathbf{B} = nE_0\hat{\mathbf{y}}\exp\left[-i\omega t + \frac{i\omega}{c}\sqrt{\left(\varepsilon + \frac{4\pi i\sigma}{\omega}\right)\mu}\, z\right]$$

$$= \sqrt{\left(\varepsilon + \frac{4\pi i\sigma}{\omega}\right)\mu}\, E_0\hat{\mathbf{y}}\exp\left[-i\omega t + \frac{i\omega}{c}\sqrt{\left(\varepsilon + \frac{4\pi i\sigma}{\omega}\right)\mu}\, z\right] \tag{166}$$

Since the index of refraction is complex, there will be a phase difference between the electric and the magnetic fields.

It is instructive to examine two cases in detail: a poor conductor with $\varepsilon \gg 4\pi\sigma/\omega$ and a very good conductor with $\varepsilon \ll 4\pi\sigma/\omega$. In the former case,

$$k \simeq \frac{\omega}{c}\sqrt{\mu\varepsilon} + \frac{2\pi i\sigma}{c}\sqrt{\frac{\mu}{\varepsilon}} \tag{167}$$

so the wave (165) becomes

$$\mathbf{E} = E_0\hat{\mathbf{x}}\exp\left(-i\omega t + \frac{i\omega}{c}\sqrt{\varepsilon\mu}\, z - \frac{2\pi\sigma}{c}\sqrt{\frac{\mu}{\varepsilon}} z\right) \tag{168}$$

This is a damped wave—as it travels in the z direction, its amplitude decreases exponentially (see Fig. 12.12). The characteristic decay distance within which the wave amplitude decreases by a factor e is

$$\Delta z = \frac{c}{2\pi\sigma}\sqrt{\frac{\varepsilon}{\mu}} \tag{169}$$

In consequence of our assumption $\varepsilon \gg 4\pi\sigma/\omega$, this decay distance is much longer than a wavelength.

■ ***Exercise 13.*** Show that $\Delta z \gg \lambda$. ■

In the case of a very good conductor, we can neglect ε in Eq. (163). This gives

$$k \simeq \frac{\omega}{c}\sqrt{\frac{4\pi i\sigma\mu}{\omega}} \tag{170}$$

Figure 12·12

Electric field as a function of position in a wave propagating in a conducting medium of low conductivity. The wavelength of the wave is $\lambda = 2\pi c/\omega\sqrt{\mu\varepsilon}$.

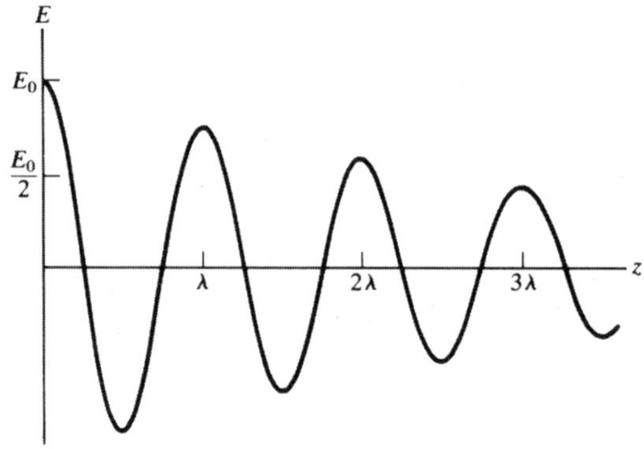

or, since $\sqrt{i} = (1+i)/\sqrt{2}$,

$$k \simeq (1+i)\frac{\sqrt{2\pi\sigma\mu\omega}}{c} \tag{171}$$

Our plane wave now becomes

$$\mathbf{E} = E_0\hat{\mathbf{x}}\exp\left(-i\omega t + \frac{i}{c}\sqrt{2\pi\sigma\mu\omega}\, z - \frac{1}{c}\sqrt{2\pi\sigma\mu\omega}\, z\right) \tag{172}$$

This is a strongly damped wave (see Fig. 12.13). It has a characteristic decay distance

$$\Delta z = \frac{c}{\sqrt{2\pi\sigma\mu\omega}} \tag{173}$$

EXAMPLE: *skin depth and skin effect*

This distance is called the *skin depth*, usually designated by δ,

$$\delta = \frac{c}{\sqrt{2\pi\sigma\mu\omega}} \tag{174}$$

Note that the skin depth depends on the frequency—apart from the explicit dependence on the factor $1/\sqrt{\omega}$, Eq. (174) may also have an implicit dependence, because in some conductors σ is a function of frequency. However, in metals σ is nearly constant, and therefore $\delta \propto 1/\sqrt{\omega}$. For instance, copper has $\delta = 6.3 \times 10^{-2}$ cm at 10^4 Hz, and $\delta = 2.0 \times 10^{-4}$ cm at 10^9 Hz. Thus, high-frequency waves in a conductor are

Figure 12·13

Electric field as a function of position for a wave in a conducting medium of high conductivity.

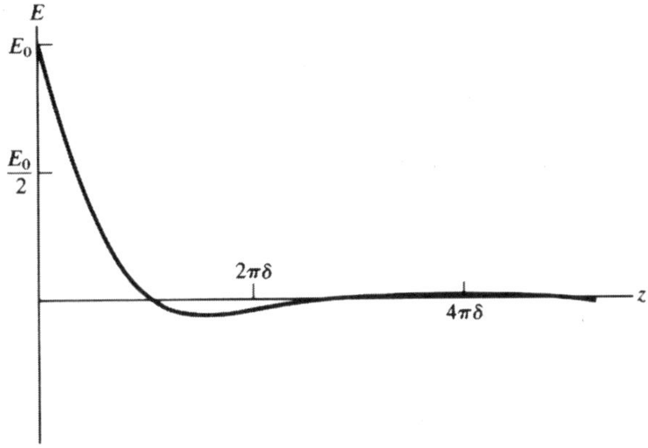

very quickly attenuated, their energy being dissipated by Joule heating. For a wave incident on a good conductor, the penetration is restricted to a thin layer; this phenomenon is called the *skin effect.* This inability of the oscillating electric and magnetic fields of a wave to penetrate a conductor is also true of the oscillating electric and magnetic fields generated quasi-statically by a charge and current distribution. For example, if a wire carries a high-frequency current, the electric field within the wire will not be uniform, as it would be for zero frequency (steady current); instead, the electric field is largest at the surface of the wire and decays exponentially with the skin depth given by Eq. (174). The reason why a high-frequency alternating electric field exhibits the same skin effect as a wave field is that by neglecting ε in Eq. (161), we are eliminating the displacement current; with this elimination, the equation for the wave field coincides with that for the alternating field.

To calculate the amplitudes of the refracted and the reflected waves for a wave incident on a conductor, we can simply use the Fresnel equations with the complex index of refraction specified by Eq. (162).

The equations developed in this section also apply to waves in a plasma, or ionized gas. The abundance of free electrons and ions gives the plasma a high conductivity. The free electrons make the main contribution to the conductivity because their small mass permits them to respond rapidly to an electric field. We can obtain a simple formula for the conductivity of plasma by examining the motion of an electron, as we did in Section 12.5. For a free electron under the influence of an electromagnetic wave polarized in the x direction, the equation of motion is simply Eq. (98) with $\omega_0 = 0$:

$$m_e \frac{d^2x}{dt^2} = -eE_0 e^{-i\omega t} \tag{175}$$

This has the steady-state solution

$$x = \frac{e}{m_e \omega^2} E_0 e^{-i\omega t} \tag{176}$$

If there are N electrons per unit volume, the current density is

$$j_x = -Ne\frac{dx}{dt} = \frac{iNe^2}{m_e \omega} E_0 e^{-i\omega t} \tag{177}$$

From this, we see that the conductivity of the plasma is

$$\sigma = \frac{iNe^2}{m_e \omega} \tag{178}$$

The imaginary factor in the conductivity merely means that the current and the electric field are out of phase; the current "lags" the electric field by 90°.

Substituting our expression for the conductivity into Eq. (163), we find

$$k = \frac{\omega}{c}\sqrt{\left(\varepsilon - \frac{4\pi Ne^2}{m_e \omega^2}\right)\mu} \tag{179}$$

We can write this as

$$k = \frac{\omega}{c}\sqrt{\varepsilon\mu}\sqrt{1 - \frac{\omega_p^2}{\omega^2}} \tag{180}$$

where

$$\omega_p = \sqrt{\frac{4\pi}{\varepsilon}\frac{Ne^2}{m_e}} \tag{181}$$

EXAMPLE: *plasma frequency and radio communication*

is called the *plasma frequency*. If ω is smaller than ω_p, then the wave number becomes purely imaginary, and the wave fails to propagate. Thus the plasma frequency is the cutoff frequency for the propagation of electromagnetic waves in the plasma. For example, in the ionosphere of the Earth, $N \simeq 10^5/\text{cm}^3$, and $\varepsilon \simeq 1$, for which Eq. (181) gives $\omega_p \simeq 2 \times 10^7$/s, that is, a frequency of 3 MHz. Radio waves below this frequency cannot penetrate through the ionosphere. Such low-frequency radio waves bounce back and forth between the ionosphere and the ground, and they can thereby follow the curvature of the Earth and reach distant locations out of the line of sight. This makes them useful for long-range radio communication. In contrast, radio waves transmitted via communication satellites must be of high frequency so that they can pass upward and downward through the ionosphere.

12·9 The Approach to Equilibrium in a Good Conductor

Induction effects play a crucial role in the approach to electro- and magnetostatic equilibrium in a good conductor. As we saw in Section 9.2, any free charge will be expelled from a good conductor very quickly, in a time of about 10^{-14} s. But the transient current in the conductor produces a time-dependent magnetic field, which induces an electric field, which drives an extra current, and so forth. A good conductor tends to retain these dynamic fields and currents for a long time. The gradual expulsion of the fields and currents is governed by the Maxwell equations (155)–(158). Since in the present context, the time derivatives are small whereas the conductivity is large, we can neglect $(\varepsilon/c)\partial\mathbf{E}/\partial t$ compared with $(4\pi\sigma/c)\mathbf{E}$; that is, we can neglect the displacement current compared with the real current. Equation (156) then reduces to

$$\frac{1}{\mu}\nabla\times\mathbf{B}=\frac{4\pi\sigma}{c}\mathbf{E} \tag{182}$$

As in the calculation at the end of Section 12.1, we can extract a second-order differential equation for **B** from the first-order Maxwell equations by taking the curl of Eq. (182) and by substituting Eqs. (157) and (158). We readily find

$$\nabla^2\mathbf{B}=\frac{4\pi\mu\sigma}{c^2}\frac{\partial\mathbf{B}}{\partial t} \tag{183}$$

Likewise, by taking the curl of Eq. (158), we find

$$\nabla^2\mathbf{E}=\frac{4\pi\mu\sigma}{c^2}\frac{\partial\mathbf{E}}{\partial t} \tag{184}$$

■ ***Exercise 14.*** Supply the steps for the derivations of Eqs. (183) and (184). ■

The differential equations (183) and (184) are *diffusion* equations, analogous to the equation of heat conduction. These equations show how the electric and magnetic fields "flow" out of the volume of the conductor. For a crude estimate of the relaxation time, we note that

$\nabla^2\mathbf{B}$ will be of the order of magnitude of B/l^2, where B is a typical value of the magnetic field and l is a characteristic length of the conductor; and $\partial \boldsymbol{B}/\partial t$ will be of the order of magnitude of B/t_0, where t_0 is the relaxation time. According to Eq. (183), we then have

$$\frac{B}{l^2} \simeq \frac{4\pi\mu\sigma}{c^2}\frac{B}{t_0} \tag{185}$$

Hence the time that a good conductor takes to achieve electro- and magnetostatic equilibrium is approximately

$$t_0 \simeq \frac{4\pi\mu\sigma}{c^2} l^2 \tag{186}$$

EXAMPLE: *relaxation time in good conductor*

This expression for the relaxation time has a reasonable dependence on the relevant parameters—the relaxation time increases with the conductivity, and it also increases with the length of the conductor. The expression also yields reasonable numbers. For a piece of copper 1 cm across, we find, with $\sigma \simeq 5\times10^{17}\ \text{s}^{-1}$ and $\mu \simeq 1$,

$$t_0 \simeq 10^{-2}\ \text{s} \tag{187}$$

Of course, this is only a rough estimate. To obtain a detailed description of the process of expulsion of the fields and currents from the volume of the conductor, we must solve the diffusion equations (183) and (184) with suitable initial conditions and boundary conditions. This is a quite formidable mathematical problem unless the initial configuration has a high degree of symmetry. Exact solutions have been obtained in a few cases with simple geometries.[4]

The diffusion of the magnetic field out of the volume of a conductor finds practical application in an elegant method for the measurement of the conductivity of samples of material.[5] Since the relaxation time depends on the conductivity [see Eq. (186)], a direct experimental measurement of the relaxation time permits the calculation of the conductivity, provided the exact expression for the relaxation time has been obtained from the exact mathematical solution of the diffusion equation for the geometry of the sample. The relaxation time of an initial magnetic field supplied to the conductor can be measured quite easily by surrounding the sample with a pickup coil and monitoring the emf induced during the diffusion and decay of the magnetic field. One important advantage of this method for the measurement of the conductivity is that no electrical connections need to be made to the sample; such connections are a major source of uncertainty in other, conventional methods. Furthermore, bulk samples can be handled very conveniently.

Another practical application of the diffusion of magnetic fields arises in connection with experiments that attempt to create extremely large magnetic fields (megagauss fields) by compressing an initial, smaller magnetic field in a chamber with a conducting liner. The strength of the field that can be achieved with such an arrangement is limited by the amount of magnetic field that is lost by diffusion into and across the conducting liner.

It may be worth pointing out that the expulsion of fields and of currents from the volume of the conductor is not quite the final step in the approach to electrostatic and magnetostatic equilibrium.[6] A conductor the interior of which is in equilibrium can still have oscillating currents on its surface and oscillating wave fields in the surrounding space. Complete equilibrium will be achieved only when these oscillations have been damped. We will analyze oscillating wave fields near a conductor (or in a cavity in a conductor) in Chapter 15. It turns out that the time scale for damping these fields is usually much shorter than the time scale for the expulsion of the fields from the volume [Eq. (186)]. Hence the damping of the wave fields makes a negligible contribution to the total time required to achieve equilibrium.

References

1. H. C. Ohanian, *Am. J. Phys.* **54**, 500 (1986). For a detailed exploration of spin in some quantum-mechanical wave packets, see K. Mita, *Am. J. Phys.* **68**, 259 (2000).
2. E. U. Condon and G. H. Shortley, *The Theory of Atomic Spectra* (Cambridge University Press, Cambridge, 1935), Section 4.9.
3. F. Goos and M. Hänchen, *Ann. Physik* (*Leipzig*) **1**, 333 (1947); for extra complications concerning circularly polarized waves, see D. G. Boulware, *Phys. Rev.* D **7**, 2375 (1973).
4. H. E. Wilhelm, J. Math. Phys. 23, 10 (1982).
5. C. P. Bean, R. W. DeBlois, and L. B. Nesbitt, *J. Applied Phys.* **30**, 1976 (1959).
6. H. C. Ohanian, *Am. J. Phys.* **51**, 1020 (1983).

Problems

1. A plane electromagnetic wave in vacuum has an electric field

$$\mathbf{E} = E_0\hat{\mathbf{x}}e^{-i(\omega t - kz)}$$

For this wave, compute the time average of all the components of the energy-momentum tensor $T^{\mu\nu}$.

2. In the laboratory frame, or $xyzt$ frame, a plane electromagnetic wave of frequency ω is traveling in the $+z$ direction. Suppose the $x'y'z't'$ frame has velocity V in the $+x$ direction. What is the frequency of the wave as seen in the $x'y'z't'$ frame? What is the direction of propagation of the wave in this frame?

3. In the laboratory, a plane electromagnetic wave is traveling in the $+z$ direction. At some given place, the instantaneous Poynting vector of the wave has a magnitude S and the energy density a magnitude S/c. Find the Poynting vector and the energy density in a reference frame moving at velocity V in the $+x$ direction.

4. The orbital velocity of the Earth around the Sun is 30 km/s.
   (a) As observed from the Earth, how much does the angular position of a distant star placed directly above the Earth's orbit shift in 6 months?
   (b) How much does the wavelength of visible (green) light from a distant star placed in the plane of the orbit shift in 6 months? Assume the star is initially on the line of motion of the Earth.

5. A pulse of radiation travels in the z direction. The pulse has infinite length in the z direction but has *finite width* in the x and y directions. The electric and magnetic fields are

$$\mathbf{E}(t,\mathbf{x}) = \mathbf{E}(x,y)e^{-i(\omega t - k_z z)}$$

$$\mathbf{B}(t,\mathbf{x}) = \mathbf{B}(x,y)e^{-i(\omega t - k_z z)}$$

(*Note*: This is not a plane wave, $k_z \neq \omega/c$.) Assume that $E_z = 0$; the wave is then called a "transverse electric," or TE, wave.
   (a) Show that $B_z(x,y)$ satisfies the equation

$$\left[\frac{\partial^2}{\partial x^2} + \frac{\partial^2}{\partial y^2} + \left(\frac{\omega^2}{c^2} - k_z^2\right)\right] B_z(x,y) = 0$$

   (b) Suppose a function $B_z(x,y)$ satisfying this equation is given. Use the Maxwell equations (28) and (30) to find expressions for E_x, E_y, B_x, B_y in terms of this $B_z(x,y)$.
   (c) Show that $\omega > ck_z$ for such a pulse of radiation.
   (d) Show that if $B_z(x,y) = 0$, then the fields $E_x(x,y)$, $E_y(x,y)$ must be constant and the pulse of radiation does *not* have finite width in the x and y directions.

6. A circularly polarized wave of positive helicity propagating in the z direction has an electric field $\mathbf{E}_+$ given by Eq. (24) and a corresponding magnetic field. The wave is incident from below on a conducting plate that lies in the x-y plane. What is the reflected wave? What is the helicity of the reflected wave? What is the rate at which the incident wave deposits angular momentum in the plate, per unit area and per unit time?

7.* A circularly polarized wave travels in the z direction. The wave has infinite extent in the z direction but finite extent in the x and y directions. Show that

$$\mathbf{E}(t,x,y,z) = \left[(\hat{\mathbf{x}} + i\hat{\mathbf{y}})E_0(x,y) + \frac{i}{k}\left(\frac{\partial E_0}{\partial x} + i\frac{\partial E_0}{\partial y}\right)\hat{\mathbf{z}}\right]e^{-i(\omega t - kz)}$$

$$\mathbf{B}(t,x,y,z) = -i\mathbf{E}(t,x,y,z)$$

---

*Based on J. D. Jackson, *Classical Electrodynamics*.

with $k = \omega/c$ is an approximate solution of the Maxwell equations (27)–(30). Here $E_0(x,y)$ is a slowly varying function that decreases to zero when $\sqrt{x^2+y^2}$ becomes larger than some finite distance (in this context, "slowly varying" means that the change in the function and the change in its derivative are small in a distance $2\pi/k$).

8. A grain of dust is spherical; it consists of material of density 2 g/cm^3. The grain floats in interplanetary space. The forces on the grain are the gravitational attraction of the Sun and the radiation pressure of sunlight; both of these obey an inverse-square law, but the gravitational force is proportional to the volume (or mass) or the grain, whereas the radiation-pressure force is proportional to the surface area. What are the critical radius and mass of the grain if these forces are to balance? Assume that the grain completely absorbs the sunlight. The luminosity of the Sun is 3.9×10^{33} erg/s.

9. Show that a circularly polarized wave remains circularly polarized under Lorentz transformations. [*Hint*: Show that under a Lorentz transformation, the amplitude of each of the two plane waves in the circularly polarized wave changes by a factor of $\gamma[1-(V/c)\cos\theta]$, where θ is the angle between the direction of propagation and the velocity **V**.]

10. According to Eq. (64), the momentum density in the electromagnetic fields can be expressed as a sum of two terms. Show that when the second of these terms is integrated over the volume of a wave pulse of finite extent, it gives zero:

$$-\frac{1}{4\pi c}\int (\mathbf{E}\cdot\nabla)\mathbf{A}\,dV = 0$$

Hence this term makes no contribution to the net momentum of the wave pulse.

11. An *elliptically polarized wave* has an electric field

$$\mathbf{E} = (a\hat{\mathbf{x}} + ib\hat{\mathbf{y}})e^{-i(\omega t - kz)}$$

where a and b are real constants. Find the ratio of the time-average spin to the time-average energy in this wave.

12. To deal with the Lorentz transformations of the fields **H** and **D** in a material medium, we can define a field tensor

$$H^{\mu\nu} = \begin{pmatrix} 0 & -D_x & -D_y & -D_z \\ D_x & 0 & -H_z & H_y \\ D_y & H_z & 0 & -H_x \\ D_z & -H_y & H_x & 0 \end{pmatrix}$$

and a dipole-moment tensor

$$M^{\mu\nu} = \begin{pmatrix} 0 & P_x & P_y & P_z \\ -P_x & 0 & -M_z & M_y \\ -P_y & M_z & 0 & -M_x \\ -P_z & -M_y & M_x & 0 \end{pmatrix}$$

(a) Show that $H^{\mu\nu} = F^{\mu\nu} - 4\pi M^{\mu\nu}$.

(b) Show that the Maxwell equations (80) and (81) can be written in the Lorentz-invariant form

$$\partial_\mu H^{\mu\nu} = \frac{4\pi}{c} j_{\mathrm{F}}^{\nu}$$

(c) Both $H^{\mu\nu}$ and $M^{\mu\nu}$ have the Lorentz-transformation properties of tensors of the second rank. Find the Lorentz-transformation equations analogous to Eqs. (8.43) and (8.44) for **H**, **D** and for **M**, **P**.

13. The index of refraction of water is a function of frequency, or of wavelength. Figure 12.14 shows the dependence of the index of refraction on wavelength.

(a) Roughly, what is the phase velocity and what is the group velocity for red light (700 nm)? For violet light (400 nm)?

(b) Suppose that a pulse of white light propagates through a layer of water 300 m thick. Roughly, by how much will the red and the violet portions of this pulse separate?

14. A ray of white light approaches a water surface at an angle of incidence of 5°. Find the angles of refraction for red (700 nm), green (550 nm), and violet light (400 nm). Figure 12.14 gives the index of refraction of water as a function of wavelength.

15. Rainbows are caused by the refraction of sunlight by raindrops. Figure 12.15 shows the path of a ray of light entering a spherical drop of water. The angle of incidence at A is θ and the angle of refraction is θ'.

(a) Prove that at B, the angle of incidence and the angle of reflection coincide with θ'. Prove that at C, the angle of incidence is θ' and the angle of refraction is θ.

(b) Prove that the net angular deflection of the ray by the raindrop is $\Delta = 2(\theta - \theta') + \pi - 2\theta'$.

Figure 12·14

Index of refraction of water as a function of wavelength.

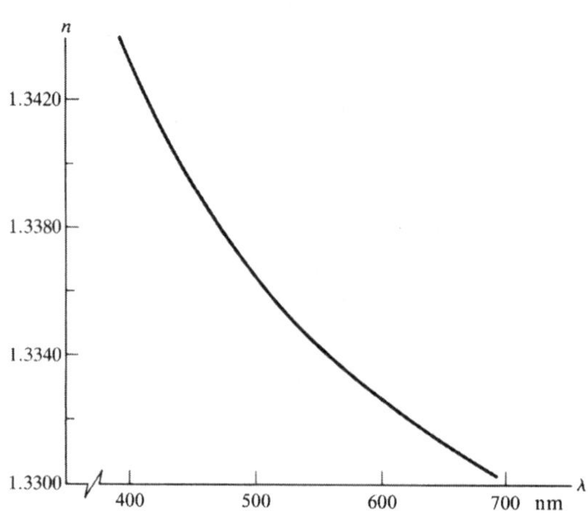

Figure 12·15

Path of a light ray in a raindrop.

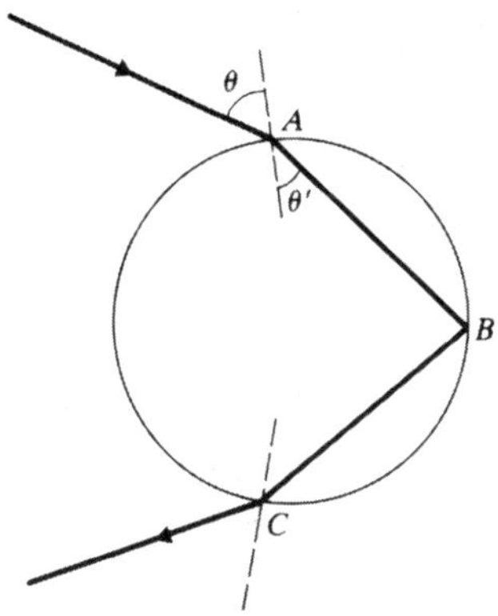

(c) The condition for formation of a rainbow is that the derivative of Δ with respect to θ is zero. Under these conditions, all the rays within a small range of angles of incidence (or a small range of impact points on the drop) emerge at nearly the same angle; the rays are then strongly concentrated (they form a caustic surface). Prove that the critical angle for rainbow formation is

$$\theta = \cos^{-1}\sqrt{(n^2-1)/3}$$

(d) Using Fig. 12.14, calculate this critical angle for red light (700 nm) and for violet light (400 nm). Calculate the angle between the direction of the Sun and the direction of the rainbow as seen by an observer standing on the ground. What is the angular width of the rainbow? What is the color of the upper edge of the rainbow?

16. A vertical meterstick is halfway submerged in the water of a pool.

(a) Assume that your eye is barely above the surface of the water. Plot the apparent length of the submerged half of the meterstick as a function of the horizontal distance from your eye.

(b) Assume that your eye is barely below the surface of the water. Plot the apparent length of the upper half of the meterstick as a function of the horizontal distance.

17. A light wave is incident on the interface between two dielectrics with $n' > n$. Assume that the electric field is parallel to the plane of incidence and that $\mu = \mu' = 1$. Show that the incident and reflected electric fields are in phase if the angle of incidence is larger than Brewster's angle and out of phase otherwise.

18. Suppose that we reverse the sign of t in Eqs. (115)–(133). Is the result a solution of Maxwell's equations? Does is satisfy all the boundary conditions at the interface? How could you contrive to generate such a "time-reversed" wave experimentally?

19. Show that the Fresnel equations can be put into the following alternative forms:

For $\mathbf{E}_0$ *parallel to the plane of incidence,*

$$E_0' = 2\frac{n}{n'}\frac{\mu'\sin 2\theta}{\mu'\sin 2\theta' + \mu\sin 2\theta}E_0$$

$$E_0'' = \frac{\mu\sin 2\theta - \mu'\sin 2\theta'}{\mu'\sin 2\theta' + \mu\sin 2\theta}E_0$$

For $\mathbf{E}_0$ *perpendicular to the plane of incidence,*

$$E_0' = \frac{2\mu'\tan\theta'}{\mu'\tan\theta' + \mu\tan\theta}E_0$$

$$E_0'' = \frac{\mu'\tan\theta' - \mu\tan\theta}{\mu'\tan\theta' + \mu\tan\theta}E_0$$

20. Consider Eq. (143) with $n' > n$. Prove the following:

(a) If $\mu = \mu'$, then Eq. (143) gives $E_0''/E_0 < 0$ for all values of θ between 0 and $\pi/2$.

(b) If $n/\mu > n'/\mu'$, then Eq. (143) gives $E_0'' = 0$ for some value of θ between 0 and $\pi/2$. Show that the equation for the critical angle is

$$\mu\tan\theta = \mu'\tan\left[\sin^{-1}\left(\frac{n}{n'}\sin\theta\right)\right]$$

21. For a plane wave incident on the interface between two nonconducting media, we can define reflection and transmission coefficients

$$R = \frac{|S_z''|}{|S_z|} \quad \text{and} \quad T = \frac{|S_z'|}{|S_z|}$$

where S_z, S_z', and S_z'' are the components of the Poynting vectors perpendicular to the interface for the incident, refracted, and reflected waves, respectively. These reflection and transmission coefficients represent the fractions of the energies reflected and transmitted at the interface.

(a) Show that if $\mu = \mu'$, the Fresnel equations yield the following results:

For $\mathbf{E}_0$ *parallel to the plane of incidence,*

$$R = \left(\frac{n'\cos\theta - n\cos\theta'}{n'\cos\theta + n\cos\theta'}\right)^2$$

$$T = \frac{4nn'\cos\theta\cos\theta'}{(n'\cos\theta + n\cos\theta')^2}$$

For $\mathbf{E}_0$ *perpendicular to the plane of incidence,*

$$R = \left(\frac{n\cos\theta - n'\cos\theta'}{n\cos\theta + n'\cos\theta'}\right)^2$$

$$T = \frac{4nn'\cos\theta\cos\theta'}{(n\cos\theta + n'\cos\theta')^2}$$

(b) Show that $R + T = 1$.

22. Consider an electromagnetic wave normally incident on the interface between two dielectrics. Show that the fraction of the wave energy reflected by the interface is independent of whether the wave is incident from one side or from the other.

23. Unpolarized light is incident on a thick plate of glass of index of refraction 1.5. The unpolarized light consists of a mixture of waves with polarization parallel and polarization perpendicular to the plane of incidence, with equal intensities for these two polarizations. The reflected light will then be partially polarized. The *degree of polarization* is defined as the difference between the parallel-polarized and the perpendicular-polarized intensities divided by the sum of these intensities. Calculate the degree of polarization of the reflected light for angles of incidence of 0°, 20°, 40°, 60°, and 80°.

24. For X rays of a wavelength 0.0708 nm, the index of refraction of calcite is $n = 1 - 1.85 \times 10^{-6}$. Describe how calcite may be used to achieve total reflection of an X-ray beam at a calcite-air interface. Find the angle of incidence that is needed.

25. When a plane wave suffers total internal reflection at a glass-vacuum interface, the reflected wave has the same magnitude as the incident wave but differs in phase. Show that for $\mu = 1$, the Fresnel equations for the reflected wave reduce to $E_0'' = e^{i\delta} E_0$, where

$$\tan\frac{\delta}{2} = -\frac{\sqrt{\sin^2\theta - (n'/n)^2}}{(n'/n)^2\cos\theta} \quad \text{for } \mathbf{E}_0 \text{ parallel to plane of incidence}$$

$$\tan\frac{\delta}{2} = -\frac{\sqrt{\sin^2\theta - (n'/n)^2}}{\cos\theta} \quad \text{for } \mathbf{E}_0 \text{ perpendicular to plane of incidence}$$

26. A plane wave suffers total internal reflection at a glass-vacuum interface. The electric field is parallel to the plane of incidence (the *x*-*z* plane).

(a) Show that the x and z components of the transmitted amplitude ($E'_{0x} = -E'_0\cos\theta'$ and $E'_{0z} = E'_0\sin\theta'$) are 90° out of phase. Describe the behavior of the electric field vector $\mathbf{E}'_0$ of the transmitted wave at a fixed point as a function of time.

(b) Assume $n = 1.5$, $\mu = 1$. Plot the magnitudes of E'_{0x}/E_0 and E'_{0z}/E_0 as a function of θ in the range $42°$ to $90°$.

27. Light suffers total internal reflection at a glass-vacuum interface. Assuming that the electric field is perpendicular to the plane of incidence, calculate the time-average Poynting vector in the vacuum region just outside the glass. Show that this vector is parallel to the interface.

28. A horizontal slab of dielectric with $n' > 1$, $\mu' = 1$ has a thickness *d*; the space above and below the dielectric is empty. A plane wave is perpendicularly incident on the slab. Find the amplitude of the net reflected wave and that of the net transmitted wave.

29. The U.S. Navy has deployed large antenna arrays in northern Michigan, capable of transmitting extremely low-frequency (ELF) radio waves for communication with submerged submarines. The frequency of these waves is 75 Hz. What is their skin depth in salt water, with a conductivity 5×10^{10}/statohm cm?

30. A plane wave is incident from vacuum on a very good conductor. The electric field is perpendicular to the plane of incidence. Using the index of refraction $n' \simeq \sqrt{4\pi i\sigma\mu'/\omega}$ in the Fresnel equations [Eqs. (142), (143)], find the complex amplitude of the reflected wave. Write this amplitude as the product of a magnitude and a phase. What fraction of the incident energy is reflected?

31. Show that for a plane wave in a good conductor, the magnitude of the magnetic field is $\sqrt{4\pi\sigma\mu/\omega}$ times that of the electric field, and the magnetic field is 45° out of phase with the electric field.

32. During reentry into the atmosphere, spacecraft suffer a blackout of radio communication because friction heats and ionizes the surrounding air. Given that radio waves of frequency 200 MHz fail to reach the spacecraft, what can you conclude about the density of free electrons in the vicinity of the spacecraft?

33. Figure 12.16 shows the experimental data obtained for a determination of the conductivity of an aluminum sample by the magnetic-diffusion method described at the end of Section 12.9. The sample, kept at a temperature of 4.2 K, was in the form of a bar of cross section $a \times b$ = 1.1 cm × 1.2 cm, for which the mathematical solution of the diffusion problem yields an exponential decrease of the magnetic field with a relaxation time

$$t_0 = 0.101 \frac{4\pi\sigma}{c^2} \frac{a^2b^2}{a^2+b^2}$$

What conductivity do the data imply?

Figure 12·16

Decay of the magnetic field in a sample of aluminum. (Courtesy of C. P. Bean, Rensselaer Polytechnic Institute.)

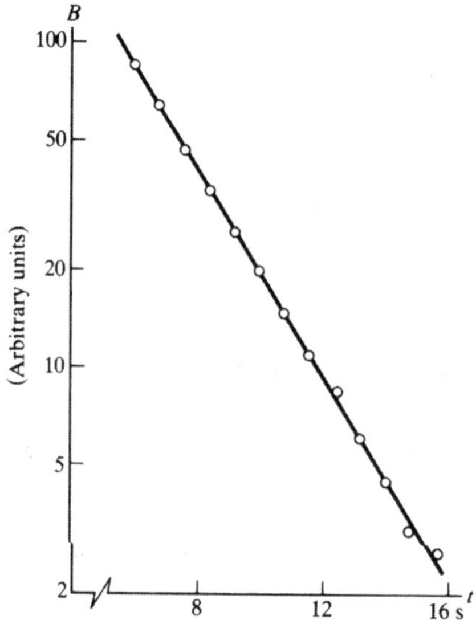

34. Suppose that the number N of electrons per unit volume is large, so the polarization P is large, and the local field acting on each electron is given by Eq. (4.74). Show that in this case, the formula (104) for the index of refraction is modified to

$$\frac{n^2-1}{n^2+2}=\frac{4\pi Ne^2}{3m_e}\sum_j \frac{f_j}{\omega_j^2-\omega^2}$$

35. For most gases, the resonant frequencies are in the ultraviolet. For visible light, this implies that $\omega^2 \ll \omega_j^2$, so $1/(\omega_j^2-\omega^2) \simeq 1/\omega_j^2 + \omega^2/\omega_j^4$. With this, and with the extra assumption $4\pi Ne^2/m_e\omega_j^2 \ll 1$, show that the formula (104) for the index of refraction can be expressed in the approximate form

$$n-1 \simeq A\left(1+\frac{B}{\lambda^2}\right)$$

where $\lambda = 2\pi c/\omega$ is the nominal wavelength of the light and A and B are constants. This is *Cauchy's formula*.

CHAPTER

13

Radiation by a Point Charge

A point charge at rest or in uniform motion is surrounded by static or quasi-static electric and magnetic fields. As long as the point charge remains in its state of rest or uniform motion, these fields move with the point charge as though they were rigidly attached to it. However, if the point charge is subjected to a sudden acceleration caused by some external force, then pieces of the electric and magnetic fields break away from the point charge and propagate outward as a self-supporting electromagnetic wave pulse.

Such a wave pulse is a disturbance of the static or quasi-static electric and magnetic fields. The disturbance arises from the need of the electric and magnetic fields to readjust from their initial configuration (before the acceleration) to their new configuration (after the acceleration). Figure 13.1 illustrates how a propagating disturbance accomplishes the readjustment of the electric field of an accelerated point charge. Initially, the point charge was at rest at the point P; it then received a sharp acceleration during the interval PP'; and it finally continued with uniform velocity during the interval $P'Q$. The initial electric field lines converged on the point P; the final electric field lines converge on the point Q. We know from relativity that the readjustment of the electric field cannot be accomplished instantaneously. Instead, it is accomplished by a disturbance, or kink, in the field lines that travels outward at the speed of light. In Fig. 13.1, this traveling disturbance constitutes a spherical wave pulse. Within the region of the disturbance, the electric field has both a longitudinal (radial) component and a transverse (tangential) component. The longitudinal component is the usual Coulomb field, which is proportional to the inverse square of the distance. The transverse component is the radiation field. As we will see in this chapter, the radiation field is inversely proportional to the distance rather than to the square of the distance. Thus the radiation field can reach much farther from the charge than the Coulomb field, and it carries energy away from the accelerated charge.

Figure 13·1

Electric field lines surrounding a point charge that has suffered an acceleration during the interval PP' and then continued at uniform velocity. The electric field lines acquired a kink during this acceleration interval. The kink propagates outward, readjusting the electric field.

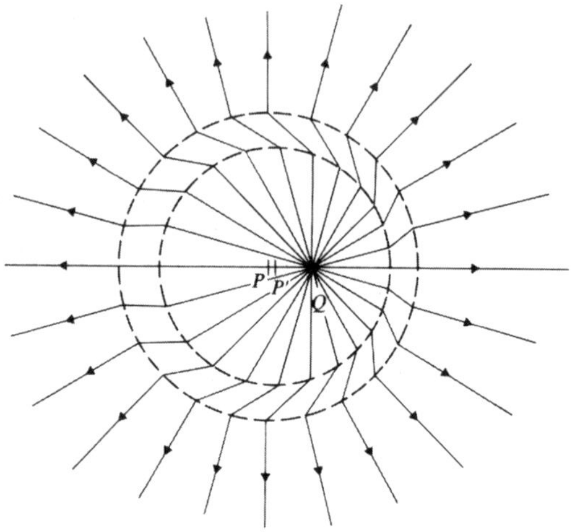

13·1 The Retarded Potential

The four-vector potential generated by some distribution of charge and current obeys the fundamental equation

$$\partial_\nu \partial^\nu A^\mu = \frac{4\pi}{c} j^\mu \tag{1}$$

or

$$\left(\frac{1}{c^2}\frac{\partial^2}{\partial t^2} - \nabla^2\right) A^\mu(t,\mathbf{x}) = \frac{4\pi}{c} j^\mu(t,\mathbf{x}) \tag{2}$$

A point charge is a special case of a distribution of charge and current, with all the charge and current concentrated at one point. Hence we can use Eq. (2) to find the potential of an accelerated point charge.

For a start, we will write the general solution of Eq. (2). This general solution will be useful for calculating not only the radiation from a point charge but also the radiation from a system of charges, such as the charges on an antenna (see Chapter 14). The solution is an integral over the distribution of charge and current:

$$A^{\mu}(t,\mathbf{x}) = \frac{1}{c}\int \frac{j^{\mu}(t-|\mathbf{x}-\mathbf{x}'|/c,\mathbf{x}')}{|\mathbf{x}-\mathbf{x}'|}dV' \tag{3}$$

This expression is called the *retarded four-potential.* By direct substitution into Eq. (2), we can verify that this expression is indeed a solution. But before doing that, let us take a careful look at the integral in Eq. (3). Obviously, if the charge and current distribution is time independent, then the integral reduces to our familiar results for the electrostatic potential and the vector potential [see Eqs. (2.40) and (9.32)], each small element of charge or current making a contribution inversely proportional to the distance. If j^{μ} is time dependent, then each small element of charge or current still makes a contribution inversely proportional to the distance; however, the contribution must be evaluated at the time $t' = t - |\mathbf{x} - \mathbf{x}'|/c$. This is called the *retarded time.* It is earlier than the time t by exactly the time that a light signal takes to travel from $\mathbf{x}'$ to $\mathbf{x}$. Thus Eq. (3) represents an electromagnetic disturbance that propagates outward from the charges and currents.* We can adopt the following intuitive picture: Each small element of charge or current generates a disturbance of potential or vector potential with an amplitude proportional to j^{μ} ; this disturbance spreads outward in all directions at the speed of light, its amplitude decreasing in inverse proportion to the distance. To find the potential at the point $\mathbf{x}$ at time t, we must add all the contributions that arrive at this point at this time. Hence very distant charges make contributions according to what they were doing a very long time ago. For instance, charges in the Andromeda nebula — at a distance of 2×10^6 light-years — make a contribution to the potential at the Earth today that depends on what those charges were doing 2×10^6 years ago.

When faced with the mathematical problem of evaluating the integral (3), we will find it helpful to think in terms of an *information-collector shell.* Suppose we imagine a thin (mathematical) spherical shell centered on the point $\mathbf{x}$, and suppose that this shell collapses toward the point $\mathbf{x}$ at the speed of light so as to arrive at the point $\mathbf{x}$ at the time t (see Fig. 13.2). The shell will then pass the point $\mathbf{x}'$ at the retarded time $t - |\mathbf{x} - \mathbf{x}'|/c$. To evaluate (3), we can then pretend that each small element of charge or current contributes according to the usual static formulas for potential, but the magnitude of the small element of charge or current must be reckoned at the time the information-collector shell sweeps over it.

---

*If we replace the retarded time $t - |\mathbf{x} - \mathbf{x}'|/c$ in Eq. (3) by the advanced time $t + |\mathbf{x} - \mathbf{x}'|/c$, we obtain the *advanced potential,* which is also a solution of Maxwell's equations. However, this solution is unphysical because it represents an electromagnetic disturbance that converges toward the charges and currents and is *exactly* absorbed by them (the advanced potential is simply the time reverse of the retarded potential). Such exact absorption is impossible unless the motions of the charges have been artificially preadjusted so that they correlate in a precise way with the converging waves coming toward them.

Figure 13·2

Information-collector shell shown at the instant $t' = t - |\mathbf{x} - \mathbf{x}'|/c$ when it sweeps over the point $\mathbf{x}'$.

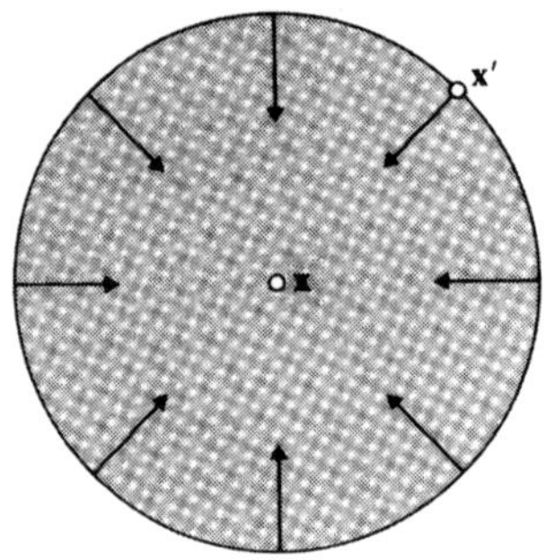

Now we will verify that the expression (3) is indeed a solution of the differential equation (2). For this, we need the derivatives $\partial^2 A^\mu / \partial t^2$ and $\nabla^2 A^\mu$. With

$$t' = t - \frac{|\mathbf{x}-\mathbf{x}'|}{c} \tag{4}$$

we have

$$\frac{\partial t'}{\partial t} = 1 \tag{5}$$

so

$$\frac{\partial^2 A^\mu}{\partial t^2} = \frac{\partial^2 A^\mu}{\partial t'^2} = \frac{1}{c}\int \frac{1}{|\mathbf{x}-\mathbf{x}'|}\frac{\partial^2}{\partial t'^2} j^\mu(t',\mathbf{x}')\,dV' \tag{6}$$

Next,

$$\nabla A^\mu = \frac{1}{c}\int\left[\frac{1}{|\mathbf{x}-\mathbf{x}'|}\frac{\partial}{\partial t'} j^\mu(t',\mathbf{x}')\nabla t' + j^\mu(t',\mathbf{x}')\nabla\frac{1}{|\mathbf{x}-\mathbf{x}'|}\right]dV' \tag{7}$$

and

$$\nabla^2 A^\mu = \frac{1}{c}\int\left[\frac{1}{|\mathbf{x}-\mathbf{x}'|}\frac{\partial^2}{\partial t'^2} j^\mu(t',\mathbf{x}')(\nabla t')^2 + \frac{1}{|\mathbf{x}-\mathbf{x}'|}\frac{\partial}{\partial t'} j^\mu(t',\mathbf{x}')\nabla^2 t' \right.$$
$$\left. + 2\frac{\partial}{\partial t'} j^\mu(t',\mathbf{x}')\nabla t'\bullet\nabla\frac{1}{|\mathbf{x}-\mathbf{x}'|} + j^\mu(t',\mathbf{x}')\nabla^2\frac{1}{|\mathbf{x}-\mathbf{x}'|}\right]dV' \tag{8}$$

Here we need to evaluate $\nabla t'$ and $\nabla^2 t'$.

■ ***Exercise 1.*** Show that

$$\nabla t' = -\frac{1}{c}\frac{\mathbf{x}-\mathbf{x}'}{|\mathbf{x}-\mathbf{x}'|} \tag{9}$$

$$\nabla^2 t' = -\frac{2}{c}\frac{1}{|\mathbf{x}-\mathbf{x}'|} \tag{10}$$

■

Substituting these expressions for $\nabla t'$ and $\nabla^2 t'$ into Eq. (8), we find that the second and third terms on the right side cancel. This leaves us with

$$\nabla^2 A^\mu = \frac{1}{c}\int\left[\frac{1}{c^2}\frac{1}{|\mathbf{x}-\mathbf{x}'|}\frac{\partial^2}{\partial t'^2}j^\mu(t',\mathbf{x}') + j^\mu(t',\mathbf{x}')\nabla^2\frac{1}{|\mathbf{x}-\mathbf{x}'|}\right]dV' \tag{11}$$

Combining this with Eq. (6), we find

$$\frac{1}{c^2}\frac{\partial^2 A^\mu}{\partial t^2} - \nabla^2 A^\mu = -\frac{1}{c}\int j^\mu(t',\mathbf{x}')\nabla^2\frac{1}{|\mathbf{x}-\mathbf{x}'|}dV' \tag{12}$$

This equation ought to coincide with Eq. (2). To see that it does, we must manipulate the right side slightly. We know that $\nabla^2(1/|\mathbf{x}-\mathbf{x}'|) = 0$ at all points $\mathbf{x}'$ except at $\mathbf{x}' = \mathbf{x}$. Hence the integral receives its contribution only from the immediate vicinity of the point $\mathbf{x}$. Provided the function $j^\mu(t', \mathbf{x}')$ is smooth near this point, we can ignore its dependence on $\mathbf{x}'$ and treat it as constant, so the right side of Eq. (12) equals

$$-\frac{1}{c}j^\mu(t,\mathbf{x})\int \nabla^2\frac{1}{|\mathbf{x}-\mathbf{x}'|}dV' \tag{13}$$

Since the integrand is zero except at $\mathbf{x} = \mathbf{x}'$, we can replace the integration over all volume by an integration over a small sphere centered on $\mathbf{x}$ (see Fig. 13.3). Noting that the integrand is a divergence, we can use Gauss' theorem:

$$\int \nabla^2\frac{1}{|\mathbf{x}-\mathbf{x}'|}dV' = \int \nabla\cdot\left(\nabla\frac{1}{|\mathbf{x}-\mathbf{x}'|}\right)dV' = \int_S \nabla\frac{1}{|\mathbf{x}-\mathbf{x}'|}\cdot d\mathbf{S}$$

If we write $|\mathbf{x}-\mathbf{x}'| = r$, then $\nabla(1/|\mathbf{x}-\mathbf{x}'|) = -\hat{\mathbf{r}}/r^2$ and $d\mathbf{S} = \hat{\mathbf{r}}dS$. Hence the last integral is simply -4π, and the expression (13), or the right side of Eq. (12), is $(4\pi/c)\, j^\mu(t, \mathbf{x})$.* This confirms that the retarded four-potential is a solution of Eq. (2).

---

*Readers familiar with the delta function will recognize that $\nabla^2(1/|\mathbf{x}-\mathbf{x}'|) = -4\pi\delta(\mathbf{x}-\mathbf{x}')$.

Figure 13·3

Small, spherical Gaussian surface centered on the point **x**. The radius of the sphere is r = $|\mathbf{x} - \mathbf{x}'|$.

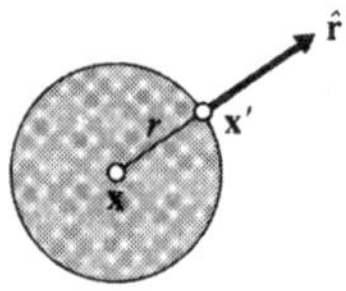

■ ***Exercise 2.*** Use the conservation equation for electric charge to prove that our expression for the retarded four-potential satisfies the Lorentz condition [Eq. (8.12)]

$$\partial_\mu A^\mu(t, \mathbf{x}) = 0 \tag{14}$$

■

13·2 The Liénard-Wiechert Potential

Now that we have at our disposal the solution for the four-potential generated by an arbitrary time-dependent distribution of charge and current, we can apply this solution to the case of a point charge. We will treat the point charge as the limiting case of a small, spherical ball of charge whose radius tends to zero. The components of the four-potential (3) are then integrals over the volume of the point charge. For instance, the A^0 component is

$$A^0(t, \mathbf{x}) = \Phi(t, \mathbf{x}) = \int \frac{1}{|\mathbf{x} - \mathbf{x}'|} \rho(t - |\mathbf{x} - \mathbf{x}'|/c, \mathbf{x}')\, dV' \tag{15}$$

Since the volume of integration is very small, the factor $1/|\mathbf{x} - \mathbf{x}'|$ in the integrand is essentially constant. Setting this factor aside for the moment, we are left with an integral over the charge density,

$$\int \rho(t - |\mathbf{x} - \mathbf{x}'|/c, \mathbf{x}')\, dV' \tag{16}$$

If the point charge is at rest, then ρ is time independent and

$$\rho(\mathbf{x}')\, dV' = dq \tag{17}$$

so the integral (16) merely equals the total charge of our point charge. Including the factor $1/|\mathbf{x} - \mathbf{x}'|$, we immediately obtain

$$\Phi(t,\mathbf{x}) = \frac{q}{|\mathbf{x}-\mathbf{x}'|} \qquad (18)$$

where we are now using $\mathbf{x}'$ to designate the position of the point charge (rather than the variable of integration).

However, if the point charge is not at rest, then

$$\rho(t-|\mathbf{x}-\mathbf{x}'|/c,\mathbf{x}')dV' \neq dq \qquad (19)$$

and the integral (16) is *not* equal to q. We can understand the source of this discrepancy by recalling that the integral for the retarded potential is evaluated by means of the information-collector shell described in Section 13.1. This shell takes some time dt to sweep over the volume dV'. During that time, some of the charge will leave the volume and therefore will not be intercepted by the collapsing shell. Figure 13.4 shows the information-collector shell sweeping over the volume dV' in the time dt. The volume element dV' has a face area dS and a thickness $c\,dt$. The direction of collapse of the shell is specified by the unit vector $\hat{\mathbf{n}}$ pointing from $\mathbf{x}'$ toward $\mathbf{x}$, that is, $\hat{\mathbf{n}} = (\mathbf{x}-\mathbf{x}')/|\mathbf{x}-\mathbf{x}'|$. If the velocity of the charges in the volume dV' is $\mathbf{v}'$, then the amount of charge that leaves the volume at the inner edge in the time dt is $\rho\hat{\mathbf{n}}\cdot\mathbf{v}'dt\,dS$ (this is positive if charge is leaving the volume, negative if entering). The actual amount of charge intercepted by the collapsing shell is then the initial amount of charge in dV' minus the amount of charge that leaves,*

Figure 13·4

In a time dt, the area dS of the information-collector shell sweeps out a volume $dV' = c\,dt\,dS$.

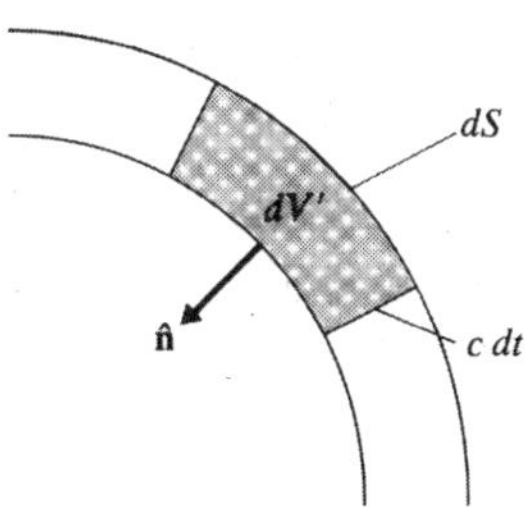

*Of course, under the assumption that the velocity $\mathbf{v}'$ is constant over the volume element, an equal amount of charge enters at the outer edge of this volume element. But this charge cannot catch up with the collapsing shell and therefore does not affect our calculation.

$$dq = \rho\, dV' - \rho \hat{\mathbf{n}} \cdot \mathbf{v}' dt\, dS \tag{20}$$

Since $dV' = c\, dt\, dS$, this is the same as

$$dq = \rho\, dV' - \rho \frac{\hat{\mathbf{n}} \cdot \mathbf{v}'}{c} dV' \tag{21}$$

Consequently,

$$\rho\left(t - |\mathbf{x} - \mathbf{x}'| / c, \mathbf{x}'\right) dV' = \frac{dq}{1 - \hat{\mathbf{n}} \cdot \mathbf{v}' / c} \tag{22}$$

With this new relation between $\rho\, dV'$ and dq, we see that the integral (16) equals $q/(1 - \hat{\mathbf{n}} \cdot \mathbf{v}'/c)$, and our expression for Φ equals

$$\Phi(t, \mathbf{x}) = \frac{q}{(1 - \hat{\mathbf{n}} \cdot \mathbf{v}'/c)|\mathbf{x} - \mathbf{x}'|} \tag{23}$$

The other components of the four-potential may be evaluated by a similar calculation, with the result

$$\mathbf{A}(t, \mathbf{x}) = \frac{1}{c} \frac{q\mathbf{v}'}{(1 - \hat{\mathbf{n}} \cdot \mathbf{v}'/c)|\mathbf{x} - \mathbf{x}'|} \tag{24}$$

Note that in all these expressions for $\Phi(t, \mathbf{x})$ and $\mathbf{A}(t, \mathbf{x})$, the vector $\mathbf{x}'$ represents the position of the point charge at the retarded time $t' = t - |\mathbf{x} - \mathbf{x}'|/c$. If the position of the point charge is some given function $\boldsymbol{\xi}(t)$, then $\mathbf{x}'$ is determined implicitly by the equation

$$\mathbf{x}' = \boldsymbol{\xi}(t - |\mathbf{x} - \mathbf{x}'|/c) \tag{25}$$

Here the retarded position $\mathbf{x}'$ appears on both the left side and the right side. In principle, this equation determines $\mathbf{x}'$ at any given t; but in practice, the equation is difficult to solve, except for the simple case of motion with constant velocity or uniform motion around a circle. Likewise, the velocity $\mathbf{v}'$ and the normal vector $\hat{\mathbf{n}}$ are to be evaluated at the retarded time. As a reminder of these retardations, we will write our equations with an extra pair of square brackets:

$$\Phi(t, \mathbf{x}) = \left[\frac{q}{(1 - \hat{\mathbf{n}} \cdot \mathbf{v}'/c)|\mathbf{x} - \mathbf{x}'|}\right] \tag{26}$$

$$\mathbf{A}(t, \mathbf{x}) = \frac{1}{c}\left[\frac{q\mathbf{v}'}{(1 - \hat{\mathbf{n}} \cdot \mathbf{v}'/c)|\mathbf{x} - \mathbf{x}'|}\right] \tag{27}$$

Any quantities enclosed in such square brackets are to be evaluated at the retarded time. The expressions (26) and (27) are called the *Liénard-Wiechert potentials*.

In four-dimensional notation, we can write these potentials as

$$A^{\mu}(t,\mathbf{x}) = \frac{1}{c}\left[\frac{q\sqrt{1-v'^2/c^2}\,u'^{\mu}}{(1-\hat{\mathbf{n}}\boldsymbol{\cdot}\mathbf{v}'/c)|\mathbf{x}-\mathbf{x}'|}\right] \tag{28}$$

which shows that the direction of A^{μ} in spacetime is parallel to the direction of the four-velocity at the retarded time. The Lorentz condition $\partial_{\mu}A^{\mu} = 0$ tells us that A^{μ} has a vanishing divergence in four dimensions; hence A^{μ} can be represented by field lines in spacetime. This leads us to the following graphical construction of A^{μ}: For a point charge initially at rest, the worldline is straight and the field lines for A^{μ} are also straight lines parallel to the worldline, that is, parallel to the x^0 axis (see Fig. 13.5). If the point charge is given an acceleration starting at time $x^0 = 0$, its worldline will thereafter be a curve. The field lines for A^{μ} will

Figure 13·5

Field lines of the four-potential A^{μ} in spacetime for a particle given an oscillatory motion starting at time $x^0 = 0$. These field lines are replicas of the worldline of the particle, displaced along the forward light cone.

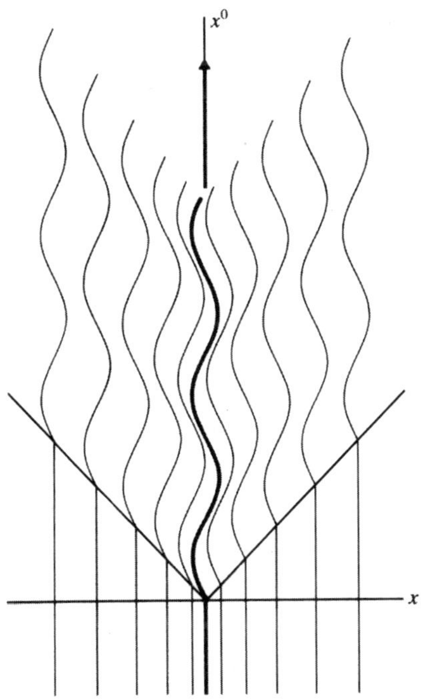

then be congruent curves, displaced forward in time by the appropriate retardation (see Fig. 13.5).*

13·3 The Radiation Fields

The electric and magnetic fields of an accelerated point charge can be calculated from the Liénard-Wiechert potentials in the usual way by taking derivatives:

$$\mathbf{E} = -\nabla\Phi - \frac{1}{c}\frac{\partial \mathbf{A}}{\partial t} \tag{29}$$

$$\mathbf{B} = \nabla \times \mathbf{A} \tag{30}$$

Unfortunately, the calculation of these derivatives is rather laborious because the dependence of the potentials on position and on time is quite complicated—the potentials are functions of the retarded time, and this is a function of position and of time. To ease the labor, we will content ourselves with a calculation of those terms in **E** and **B** that have the *least decrease* with distance. Those terms are the *radiation fields*; as we will see, they decrease in proportion to $1/|\mathbf{x} - \mathbf{x}'|$, and they depend on the acceleration of the point charge.

In our calculation, we will need the derivative of the retarded time with respect to time:

$$\frac{\partial t'}{\partial t} = \frac{\partial}{\partial t}\left(t - \frac{|\mathbf{x}-\mathbf{x}'|}{c}\right) = 1 - \frac{1}{c}\frac{\mathbf{x}-\mathbf{x}'}{|\mathbf{x}-\mathbf{x}'|}\cdot\frac{\partial}{\partial t}(-\mathbf{x}') = 1 - \frac{\hat{\mathbf{n}}}{c}\cdot\frac{\partial}{\partial t'}(-\mathbf{x}')\frac{\partial t'}{\partial t}$$

$$= 1 + \frac{\hat{\mathbf{n}}\cdot\mathbf{v}'}{c}\frac{\partial t'}{\partial t} \tag{31}$$

Solving this for $\partial t'/\partial t$, we find†

$$\frac{\partial t'}{\partial t} = \frac{1}{1 - \hat{\mathbf{n}}\cdot\mathbf{v}'/c} \tag{32}$$

We will also need $\nabla t'$:

---

*This beautiful construction was invented by R. Lichtenstein.

†This differs from Eq. (5) because here $\mathbf{x}'$ is the position of the point charge and depends on time, whereas in the calculation leading up to Eq. (5), $\mathbf{x}'$ was merely an integration variable that did not depend on time.

$$\nabla t' = \nabla\left(t - \frac{|\mathbf{x}-\mathbf{x}'|}{c}\right) = -\frac{1}{c}\nabla|\mathbf{x}-\mathbf{x}'|$$

$$= -\frac{\hat{\mathbf{n}}}{c} - \frac{1}{c}\frac{\partial}{\partial t'}|\mathbf{x}-\mathbf{x}'|\nabla t' \tag{33}$$

Here, the first term arises directly from differentiating the explicit $\mathbf{x}$ appearing in $|\mathbf{x}-\mathbf{x}'|$ while holding $\mathbf{x}'$ fixed; the second term arises from differentiating $\mathbf{x}'$ (which depends implicitly on $\mathbf{x}$ through t') while holding the explicit $\mathbf{x}$ fixed. Since $(\partial/\partial t')|\mathbf{x}-\mathbf{x}'| = -\hat{\mathbf{n}}\cdot\mathbf{v}'$, Eq. (33) becomes

$$\nabla t' = -\frac{\hat{\mathbf{n}}}{c} + \frac{\hat{\mathbf{n}}\cdot\mathbf{v}'}{c}\nabla t' \tag{34}$$

from which we find

$$\nabla t' = -\frac{1}{c}\frac{\hat{\mathbf{n}}}{1-\hat{\mathbf{n}}\cdot\mathbf{v}'/c} \tag{35}$$

With these preliminaries, we are ready to compute the derivatives of the potentials. We begin with $\nabla\Phi$. As in Eq. (33), there are terms that arise from the explicit $\mathbf{x}$ dependence including the $\mathbf{x}$ dependence in $\hat{\mathbf{n}} = (\mathbf{x}-\mathbf{x}')/|\mathbf{x}-\mathbf{x}'|$ and terms that arise from the implicit $\mathbf{x}$ dependence through the t' dependence of $\mathbf{x}'$. With

$$\nabla\hat{\mathbf{n}}\cdot\mathbf{v}' = \nabla\frac{(\mathbf{x}-\mathbf{x}')\cdot\mathbf{v}'}{|\mathbf{x}-\mathbf{x}'|} = \frac{\mathbf{v}'}{|\mathbf{x}-\mathbf{x}'|} - \frac{(\mathbf{x}-\mathbf{x}')\cdot\mathbf{v}'}{|\mathbf{x}-\mathbf{x}'|^3}(\mathbf{x}-\mathbf{x}')$$

$$= \frac{\mathbf{v}'}{|\mathbf{x}-\mathbf{x}'|} - \frac{\hat{\mathbf{n}}\cdot\mathbf{v}'}{|\mathbf{x}-\mathbf{x}'|}\hat{\mathbf{n}}$$

we find

$$\nabla\Phi = \nabla\left[\frac{q}{(1-\hat{\mathbf{n}}\cdot\mathbf{v}'/c)|\mathbf{x}-\mathbf{x}'|}\right]$$

$$= q\left[\frac{\mathbf{v}'/|\mathbf{x}-\mathbf{x}'|c - (\hat{\mathbf{n}}\cdot\mathbf{v}'\hat{\mathbf{n}})/|\mathbf{x}-\mathbf{x}'|c}{(1-\hat{\mathbf{n}}\cdot\mathbf{v}'/c)^2|\mathbf{x}-\mathbf{x}'|} - \frac{\hat{\mathbf{n}}}{(1-\hat{\mathbf{n}}\cdot\mathbf{v}'/c)|\mathbf{x}-\mathbf{x}'|^2}\right.$$

$$\left. + \frac{\partial}{\partial t'}\frac{1}{(1-\hat{\mathbf{n}}\cdot\mathbf{v}'/c)|\mathbf{x}-\mathbf{x}'|}\nabla t'\right] \tag{36}$$

Here the first term arises from the explicit $\mathbf{x}$ dependence in $1/(1-\hat{\mathbf{n}}\cdot\mathbf{v}'/c)$, the second term from the explicit $\mathbf{x}$ dependence in $1/|\mathbf{x}-\mathbf{x}'|$, and the last term from the implicit **x** dependence in $\mathbf{x}'$. Combining the first two terms and performing the differentiations in the last term, we obtain

$$\nabla\Phi = q\left[\frac{\mathbf{v}'/c-\hat{\mathbf{n}}}{(1-\hat{\mathbf{n}}\bullet\mathbf{v}'/c)^2|\mathbf{x}-\mathbf{x}'|^2}+\frac{\nabla t'}{(1-\hat{\mathbf{n}}\bullet\mathbf{v}'/c)^2|\mathbf{x}-\mathbf{x}'|}\left(\frac{\partial\hat{\mathbf{n}}}{\partial t'}\bullet\frac{\mathbf{v}'}{c}+\hat{\mathbf{n}}\bullet\frac{\mathbf{a}'}{c}\right)\right.$$
$$\left.+\frac{\nabla t'}{(1-\hat{\mathbf{n}}\bullet\mathbf{v}'/c)|\mathbf{x}-\mathbf{x}'|^3}(\mathbf{x}-\mathbf{x}')\bullet\mathbf{v}'\right] \quad (37)$$

Inspecting this expression, we see that all the terms vary with distance as $1/|\mathbf{x}-\mathbf{x}'|^2$, except the term proportional to the acceleration $\mathbf{a}'$, which varies as $1/|\mathbf{x}-\mathbf{x}'|$.* In the limit as $|\mathbf{x}-\mathbf{x}'|\to\infty$, we can ignore the former terms and retain only the latter term. This gives us the radiative part of $\nabla\Phi$:

$$\nabla\Phi = \frac{q}{c}\left[\frac{\nabla t'}{(1-\hat{\mathbf{n}}\bullet\mathbf{v}'/c)^2|\mathbf{x}-\mathbf{x}'|}\hat{\mathbf{n}}\bullet\mathbf{a}'\right]+\ldots$$
$$= -\frac{q}{c^2}\left[\frac{\hat{\mathbf{n}}\,\hat{\mathbf{n}}\bullet\mathbf{a}'}{(1-\hat{\mathbf{n}}\bullet\mathbf{v}'/c^2)^3|\mathbf{x}-\mathbf{x}'|}\right]+\ldots \quad (38)$$

Next, we want $\partial\mathbf{A}/\partial t$. The dependence **A** on t is due to the dependence on t':

$$\frac{\partial\mathbf{A}}{\partial t}=\frac{\partial\mathbf{A}}{\partial t'}\left[\frac{\partial t'}{\partial t}\right]=\frac{q}{c}\left[\frac{\partial}{\partial t'}\frac{\mathbf{v}'}{(1-\hat{\mathbf{n}}\bullet\mathbf{v}'/c)|\mathbf{x}-\mathbf{x}'|}\right]\left[\frac{\partial t'}{\partial t}\right] \quad (39)$$

When taking the derivative of the first factor in brackets, we will again ignore all terms that vary as $1/|\mathbf{x}-\mathbf{x}'|^2$ and retain only the terms that vary as $1/|\mathbf{x}-\mathbf{x}'|$. With this in mind, we obtain

$$\frac{\partial\mathbf{A}}{\partial t}=\frac{q}{c}\left[\frac{\mathbf{a}'}{(1-\hat{\mathbf{n}}\bullet\mathbf{v}'/c)|\mathbf{x}-\mathbf{x}'|}+\frac{\mathbf{v}'\hat{\mathbf{n}}\bullet\mathbf{a}'/c}{(1-\hat{\mathbf{n}}\bullet\mathbf{v}'/c)^2|\mathbf{x}-\mathbf{x}'|}+\ldots\right]\times\left[\frac{1}{1-\hat{\mathbf{n}}\bullet\mathbf{v}'/c}\right] \quad (40)$$

where, as in Eq. (38), the dots stand for all the extra terms that we have ignored.

---

*More precisely, it varies as $1/[|\mathbf{x}-\mathbf{x}'|]=1/|\mathbf{x}-[\mathbf{x}']|$, because $\mathbf{x}'$ must be evaluated at the retarded time. Throughout the remainder of this section, this retardation for $\mathbf{x}'$ is to be understood, even when it is not explicitly indicated.

Combining Eqs. (38) and (40), we find those terms in the electric field that vary as $1/|\mathbf{x}-\mathbf{x}'|$:

$$\mathbf{E} = -\nabla\Phi - \frac{1}{c}\frac{\partial \mathbf{A}}{\partial t} \tag{41}$$

$$= \frac{q}{c^2}\left[\frac{\hat{\mathbf{n}}\hat{\mathbf{n}}\bullet\mathbf{a}'}{(1-\hat{\mathbf{n}}\bullet\mathbf{v}'/c)^3|\mathbf{x}-\mathbf{x}'|} - \frac{\mathbf{a}'}{(1-\hat{\mathbf{n}}\bullet\mathbf{v}'/c)^2|\mathbf{x}-\mathbf{x}'|} - \frac{\mathbf{v}'\hat{\mathbf{n}}\bullet\mathbf{a}'/c}{(1-\hat{\mathbf{n}}\bullet\mathbf{v}'/c)^3|\mathbf{x}-\mathbf{x}'|}\right]+\ldots \tag{42}$$

It is easy to check that this expression is identically equal to

$$\mathbf{E} = \frac{q}{c^2}\left[\frac{\hat{\mathbf{n}}\times\{(\hat{\mathbf{n}}-\mathbf{v}'/c)\times\mathbf{a}'\}}{(1-\hat{\mathbf{n}}\bullet\mathbf{v}'/c)^3|\mathbf{x}-\mathbf{x}'|}\right]+\ldots \tag{43}$$

■ ***Exercise 3.*** Check that the left sides of Eqs. (42) and (43) are equal. ■

For the calculation of the magnetic field, we need $\nabla\times\mathbf{A}$. As before, we will retain only the terms that vary as $1/|\mathbf{x}-\mathbf{x}'|$; these arise from the dependence of **A** on t', not from the explicit dependence on **x**:

$$\mathbf{B} = \nabla\times\mathbf{A} = \nabla t'\times\frac{\partial\mathbf{A}}{\partial t'}+\ldots$$

$$= \frac{q}{c^2}\left[\frac{-\hat{\mathbf{n}}\times\mathbf{a}'(1-\hat{\mathbf{n}}\bullet\mathbf{v}'/c)-(\hat{\mathbf{n}}\times\mathbf{v}')(\hat{\mathbf{n}}\bullet\mathbf{a}'/c)}{(1-\hat{\mathbf{n}}\bullet\mathbf{v}'/c)^3|\mathbf{x}-\mathbf{x}'|}\right]+\ldots \tag{44}$$

The electric and magnetic fields given by Eqs. (42) and (44) are the *radiation fields* of the point charge. These fields are directly proportional to the magnitude of the (retarded) acceleration $\mathbf{a}'$ and inversely proportional to the distance. The electric and magnetic fields are both perpendicular to the vector $[\hat{\mathbf{n}}]$, whose direction is outward from the retarded position of the point charge. Furthermore, if we multiply Eq. (42) by $[\hat{\mathbf{n}}]\times$, we find

$$\mathbf{B} = [\hat{\mathbf{n}}]\times\mathbf{E} \tag{45}$$

Thus **B** is perpendicular to **E** and has the same magnitude, exactly as expected for electromagnetic waves in vacuum (see Fig. 13.6). Obviously, the radiation fields are spherical waves spreading outward from the retarded position of the point charge. The retardation in our equations implies that a wave pulse generated by some given acceleration pulse takes a time $|\mathbf{x}-\mathbf{x}'|/c$ to travel from the retarded position $|\mathbf{x}'$ to $\mathbf{x}|$; that is, the wave pulse travels at the speed of light.

Figure 13·6

The instantaneous position of the point charge is P, and the retarded position is $\mathbf{x}'$. The unit vector $[\hat{\mathbf{n}}]$ is directed outward from the retarded position. The radiation fields $\mathbf{E}$ and $\mathbf{B}$ are perpendicular to each other and to the unit vector $[\hat{\mathbf{n}}]$.

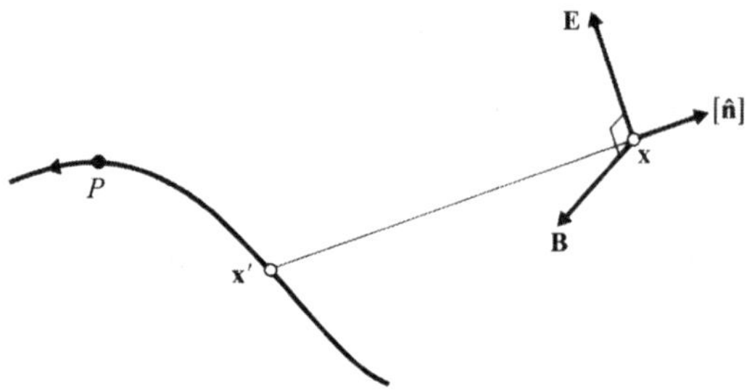

In the above calculations of the derivatives of the potential, we have omitted all terms that vary as $1/|\mathbf{x}-\mathbf{x}'|^2$. To distinguish them from the radiation fields, these terms are called the *near fields*. These terms depend on the velocity of the point charge but not on its acceleration. Evidently, this means that these fields must be the same as for a point charge that moves with constant velocity, equal to the velocity $[\mathbf{v}']$ at the retarded time, $[\mathbf{v}'] = \mathbf{v}'(t - |\mathbf{x}-\mathbf{x}'|/c)$. The near fields therefore consist of the Coulomb field of the moving charge with its relativistic corrections and the quasi-static magnetic field with its relativistic corrections. These fields are exactly the same as we obtained in Section 8.2 by a Lorentz transformation, provided that the velocity and the position of the point charge in Eqs. (8.46) and (8.47) are interpreted as the retarded velocity $[\mathbf{v}']$ and the extrapolated position $\mathbf{x}_{\text{extp}} = \mathbf{x}' + [\mathbf{v}']|\mathbf{x}-\mathbf{x}'|/c$, that is, the position the point charge would have reached at time t if it had moved with constant velocity $\mathbf{v}'$ from its retarded position (see Fig. 13.7). Thus,

Figure 13·7

The extrapolated position $\mathbf{x}_{\text{extp}}$ of the point charge is obtained from the retarded position $\mathbf{x}'$ by a displacement in the direction of the retarded velocity $[\mathbf{v}']$. Note that $\mathbf{x}_{\text{extp}}$ is different for different choices of the point $\mathbf{x}$, because the magnitude of the time retardation $|\mathbf{x}-\mathbf{x}'|/c$ is different. The near fields contributed at the point $\mathbf{x}$ are those of a uniformly moving charge instantaneously located at $\mathbf{x}_{\text{extp}}$.

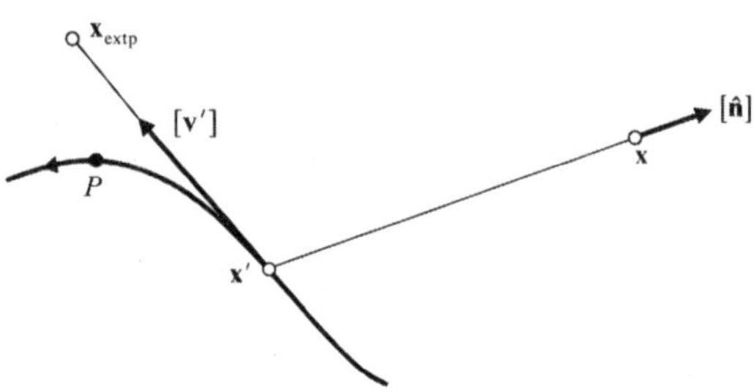

if we had retained all these terms in our present calculation, we would have lost much labor and gained nothing new.

13·4 The Radiation Pattern

The radiation fields carry energy away from the point charge. The energy flux is given by the Poynting vector

$$\mathbf{S} = \frac{c}{4\pi}\mathbf{E}\times\mathbf{B} \tag{46}$$

In view of Eq. (45), this is the same as

$$\mathbf{S} = \frac{c}{4\pi}E^2[\hat{\mathbf{n}}] \tag{47}$$

We will find it convenient to deal separately with the nonrelativistic case and the relativistic case.

EXAMPLE: *radiation field for non-relativistic motion*

Nonrelativistic motion $(v' \ll c)$. With the approximation $v' \ll c$, our formula (43) for the radiation field reduces to

$$\mathbf{E} = \frac{q}{c^2}\left[\frac{\hat{\mathbf{n}}\times(\hat{\mathbf{n}}\times\mathbf{a}')}{|\mathbf{x}-\mathbf{x}'|}\right] \tag{48}$$

If θ is the angle between $[\hat{\mathbf{n}}]$ and $[\mathbf{a}']$ (see Fig. 13.8), the corresponding magnitude of the electric field is

$$E = \frac{q}{c^2}\left[\frac{a'\sin\theta}{|\mathbf{x}-\mathbf{x}'|}\right] \tag{49}$$

The magnitude of the magnetic field is the same as that of the electric field,

$$B = \frac{q}{c^2}\left[\frac{a'\sin\theta}{|\mathbf{x}-\mathbf{x}'|}\right] \tag{50}$$

Accordingly, the energy flux equals

Figure 13·8

Directions of the electric and magnetic radiation fields for an accelerated point charge with nonrelativistic motion. The electric field is in the plane of $[\mathbf{a}']$ and $[\hat{\mathbf{n}}]$ (evaluated at the retarded time), and the magnetic field is perpendicular to this plane.

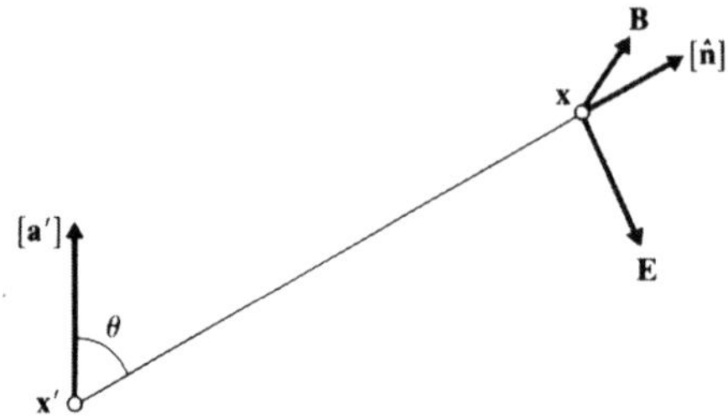

$$\mathbf{S} = \frac{c}{4\pi} E^2 [\hat{\mathbf{n}}] = \frac{q^2}{4\pi c^3} \left[a'^2 \frac{\sin^2\theta}{|\mathbf{x}-\mathbf{x}'|^2} \hat{\mathbf{n}} \right] \tag{51}$$

and the instantaneous power radiated into a solid angle $d\Omega$ with an associated area $|\mathbf{x}-\mathbf{x}'|^2\, d\Omega$ is

$$dP = S \times |\mathbf{x}-\mathbf{x}'|^2\, d\Omega \tag{52}$$

and the power per unit solid angle is

$$\frac{dP}{d\Omega} = |\mathbf{x}-\mathbf{x}'|^2\, S = \frac{q^2}{4\pi c^3} [a'^2 \sin^2\theta] \tag{53}$$

Figure 13.9 shows this radiation pattern, called the dipole radiation pattern. The power per unit solid angle is maximum at right angles to the direction of the acceleration ($\theta = \pi/2$), and it is zero along the direction of the acceleration ($\theta = 0$ and $\theta = \pi$).

The total power radiated into all angles is the integral of $dP/d\Omega$ over all solid angles:

$$P = \int_0^{2\pi}\int_0^{\pi} \frac{dP}{d\Omega} \sin\theta\, d\theta\, d\phi = \int_0^{2\pi}\int_0^{\pi} \frac{q^2}{4\pi c^3} [a'^2 \sin^2\theta] \sin\theta\, d\theta\, d\phi \tag{54}$$

or

$$P = \frac{2}{3}\frac{q^2}{c^3}[a'^2] \tag{55}$$

Figure 13·9

Polar plot of $dP/d\Omega$ for an accelerated point charge with nonrelativistic motion. The length of the line from the origin to the curve represents the magnitude of $dP/d\Omega$.

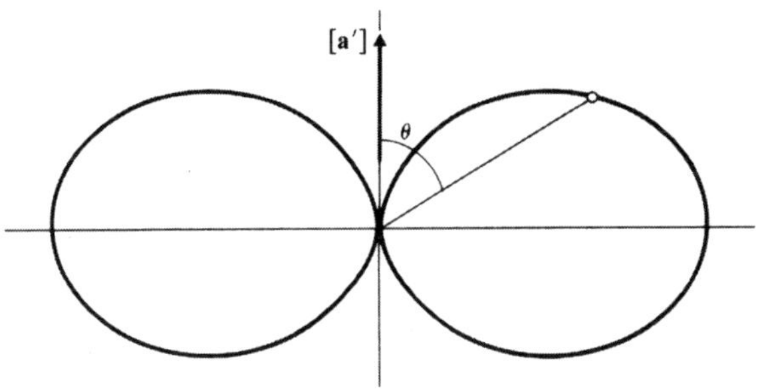

This is *Larmor's formula* for the power radiated by an accelerated charge.

EXAMPLE: *radiation field for highly relativistic motion*

Relativistic motion $(v' \simeq c)$. For a charge with relativistic motion, the radiation pattern depends on the direction of the acceleration and on the direction of the velocity. We will discuss only two special cases: acceleration parallel to velocity and acceleration perpendicular to velocity. If the acceleration is parallel to the velocity, then $\mathbf{v}' \times \mathbf{a}' = 0$ and Eq. (43) yields

$$\mathbf{E} = \frac{q}{c^2}\left[\frac{\hat{\mathbf{n}}\times(\hat{\mathbf{n}}\times\mathbf{a}')}{(1-\hat{\mathbf{n}}\cdot\mathbf{v}'/c)^3\,|\mathbf{x}-\mathbf{x}'|}\right] \tag{56}$$

This differs from the nonrelativistic formula only by the extra factor $1/(1-\hat{\mathbf{n}}\cdot\mathbf{v}'/c)^3 = 1/(1-v'\cos\theta/c)^3$. The power per unit solid angle is

$$\frac{dP}{d\Omega} = \frac{q^2}{4\pi c^3}\left[\frac{a'^2\sin^2\theta}{(1-v'\cos\theta/c)^6}\right] \tag{57}$$

For v' comparable with c, the factor $1/(1 - v'\cos\theta/c)^6$ in this formula is large in the forward direction ($\theta \simeq 0$), leading to a strong enhancement of the radiation in the forward direction. Figure 13.10 shows a plot of the radiation pattern. Comparing this with Fig. 13.9, we see that the relativistic pattern is a distorted version of the nonrelativistic pattern, with a strong elongation in the forward direction. We can understand this by remembering our formula for the aberration of light [Eq. (12.51)]. Any radiation emitted in the rest frame of the charge at, say, right angles to the direction of the velocity will appear at a smaller angle in the laboratory frame; this results in a beaming of the radiation in the forward direction.

Figure 13·10

Polar plot of $dP/d\Omega$ for an accelerated point charge with relativistic motion, with $v' = 0.3c$. The acceleration is parallel to the velocity.

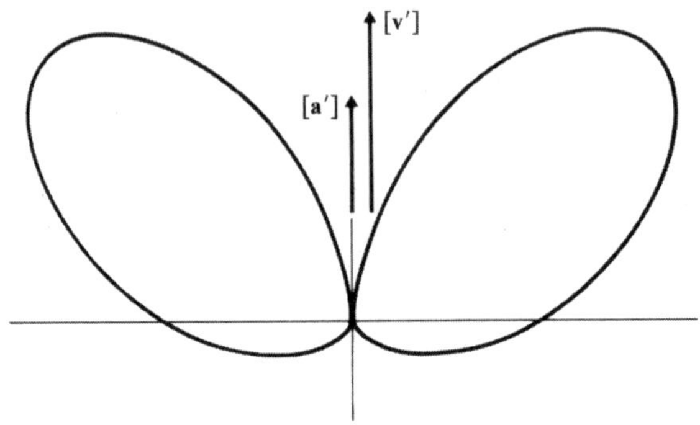

For extreme relativistic motion ($v' \to c$), the forward beam becomes very narrow and very sharp. We can arrive at a simple estimate of the angle of the peak by using a small-angle approximation in Eq. (57):

$$\frac{dP}{d\Omega} \simeq \frac{q^2}{4\pi c^3}\left[\frac{a'^2\theta^2}{(1 - v'/c + v'\theta^2/2c)^6}\right]$$

$$\simeq \frac{q^2}{4\pi c^3}\left[\frac{a'^2\theta^2}{(1 - v'/c + \theta^2/2)^6}\right] \tag{58}$$

To find the maximum of this expression, we demand

$$\frac{d}{d\theta}\frac{\theta^2}{(1 - v'/c + \theta^2/2)^6} = 0$$

from which we find that the angle between the peak and the line of motion is

$$\theta \simeq \sqrt{\frac{2}{5}\left(1 - \frac{v'}{c}\right)} \tag{59}$$

For $v'/c \simeq 1$, we have

$$\gamma = \frac{1}{\sqrt{1 - v'^2/c^2}} = \frac{1}{\sqrt{1 + v'/c}\sqrt{1 - v'/c}} \simeq \frac{1}{\sqrt{2(1 - v'/c)}}$$

so

$$\theta \simeq \frac{1}{\sqrt{5}\gamma} \qquad (60)$$

EXAMPLE: *radiation emitted by highly relativistic electron*

Since γ is proportional to the energy of the particle ($p^0 = mc^2\gamma$), Eq. (60) shows that the angle of the peak is inversely proportional to the energy the angle is very small if the energy is large. For instance, electrons of an energy p^0 = 200 MeV have $\gamma = p^0/(m_e c^2)$ = 390 and $\theta \simeq 1.1\times10^{-3}$ radian.

For the extreme relativistic motion, the power per unit solid angle at the peak in the radiation pattern is

$$\left.\frac{dP}{d\Omega}\right|_{\max} = \frac{q^2}{4\pi c^3}\left[\frac{a'^2/5\gamma^2}{(1/2\gamma^2 + 1/10\gamma^2)^6}\right] \qquad (61)$$

$$= \frac{2\times10^5}{6^6}\frac{q^2}{4\pi c^3}[a'^2\gamma^{10}] \qquad (62)$$

This is proportional to the tenth power of the particle's energy!

The total power is the integral of $dP/d\Omega$ over all angles. This integral can be evaluated exactly, but for $v' \to c$, it receives almost its entire contribution from the sharp beam in the radiation pattern, and we can therefore use the approximation (58):

$$P \simeq \int_0^{2\pi}\int_0^{\infty}\frac{q^2}{4\pi c^3}\left[\frac{a'^2\theta^2}{(1 - v'/c + \theta^2/2)^6}\right]\theta\, d\theta\, d\phi \qquad (63)$$

The upper limit for the integration over θ is irrelevant, provided that it is much larger than the angle of the peak [Eq. (60)]; we have made the convenient choice $\theta = \infty$ for this limit. The integration is elementary and yields

$$P \simeq \frac{1}{20}\frac{q^2}{c^3}\left[\frac{a'^2}{(1 - v'/c)^4}\right] \simeq \frac{4}{5}\frac{q^2}{c^3}[a'^2\gamma^8] \qquad (64)$$

which is proportional to the eighth power of the energy.

Incidentally: When comparing the formulas for the radiated power in this section with the formulas in some other textbooks,* it is necessary to keep in mind the distinction between the power arriving at a fixed receiver and the power released by the (moving) particle. For a particle in relativistic motion, these powers are different. To see how this difference arises, imagine that the particle is receding from the receiver at high speed. If the particle is to generate a given amount of power at the

*For instance, J. D. Jackson, *Classical Electrodynamics*.

receiver, it must not only provide that power but must also pump extra power into the increasing volume of space, filled with electromagnetic fields, between the particle and the receiver. Thus, for a receding particle, the power received is smaller than the power generated. Conversely, for an approaching particle, the power received is larger than the power generated. All our equations for power represent power received at some fixed location. This is the quantity of interest in most experimental measurements.

If the acceleration is perpendicular to the velocity, the formula for the radiation pattern is somewhat more complicated than the formula for the parallel case discussed above, and we will not bother to write it. Figure 13.11 shows a plot of the radiation pattern. Again, the relativistic pattern is a distorted version of the nonrelativistic pattern, with a strong elongation in the forward direction. For extreme relativistic motion, the width of the strong forward beam is $\theta \simeq 1/\gamma$, approximately the same as the width of the beam for the parallel case [see Eq. (60)]. The maximum power per unit solid angle and the total power are roughly the same as in Eqs. (62) and (64), with some minor changes in the numerical factors.

The acceleration is at right angles to the velocity for a particle moving in a circular orbit in a magnetic field. If the particle is relativistic, its radiation pattern will be as shown in Fig. 13.11. The radiation is polarized in the plane of the orbit, because this is the plane of the acceleration. An observer at some fixed position in the plane of the orbit will receive a strong but short pulse of radiation each time the forward beam sweeps over his position.

The radiation produced by particles spiraling in a magnetic field is called *synchrotron radiation*. Electrons in synchrotron accelerators emit this kind of radiation. For extreme relativistic electrons in a circular orbit, the energy loss by radiation is quite large; this sets practical limitations on the maximum electron energy attainable in circular accelerators,

Figure 13·11

Polar plot of $dP/d\Omega$ for an accelerated point charge with relativistic motion, with $v' = 0.3c$. The acceleration is perpendicular to the velocity.

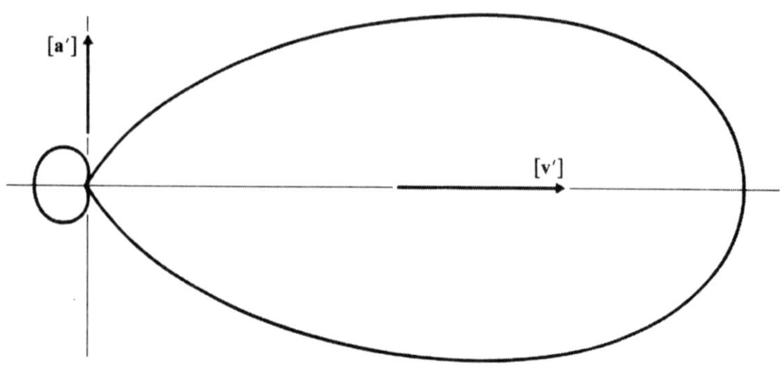

about 10 GeV. To attain higher energies, one must rely on linear accelerators.

Synchrotron radiation plays a crucial role in astrophysics it is the mechanism that generates most of the radio emissions from nebulas and from galaxies. Very energetic electrons orbiting in magnetic fields emit not only at radio frequencies but even at visible frequencies and at X-ray frequencies. For instance, synchrotron radiation by the Crab nebula spans a frequency range from $\omega \simeq 10^8$ to 10^{16} /s.

EXAMPLE: *frequency of synchrotron radiation*

The following calculation is an estimate of the frequency of the synchrotron radiation emitted by an electron in extreme relativistic motion. For such an electron, the radiation is concentrated in a beam of angular width $\Delta\theta \simeq 1/\gamma$ [see Eq. (60)]. An observer in the plane of the orbit receives a sharp pulse of radiation each time this beam sweeps over his position. If the frequency of the orbital motion of the electron is ω_0, then the time the electron takes to complete the emission of the pulse is

$$\Delta t' = \frac{\Delta\theta}{\omega_0} \simeq \frac{1}{\gamma\omega_0} \tag{65}$$

Since the electron is moving toward the observer during this emission (see Fig 13.12), the observer will receive the pulse in a shorter time interval

$$\Delta t = \Delta t'\left(1 - \frac{v'}{c}\right) \tag{66}$$

(This relation becomes obvious if we multiply both sides by c,

$$c\Delta t = c\Delta t' - v'\Delta t' \tag{67}$$

Figure 13·12

The beam of the radiation pattern sweeps over the observer while the electron is moving toward the observer.

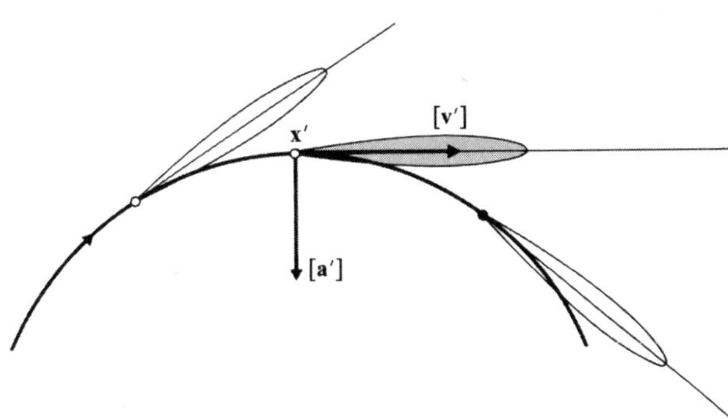

It then says that the length $c\Delta t$ of the pulse in space equals the length $c\Delta t'$ that the pulse would have if the electron were not moving toward the observer minus the distance $v'\Delta t'$ the electron moves during emission.) With the usual approximation $1 - v'/c \simeq 1/2\gamma^2$, Eq. (66) becomes

$$\Delta t \simeq \frac{1}{2\gamma^3\omega_0} \tag{68}$$

A well-known result of Fourier analysis tells us that if a wave pulse lasts a time Δt, then it will contain a spectrum of frequencies ranging from zero up to $\omega_c \simeq 1/\Delta t$. Hence the maximum frequency present in our pulse is roughly

$$\omega_c \simeq \gamma^3\omega_0 \tag{69}$$

This shows that for synchrotron radiation emitted by extreme relativistic electrons, the frequency of the radiation can be much larger than the orbital frequency. For example, electrons of energy 10^6 MeV (and $\gamma \simeq 2\times10^6$) in a magnetic field of 10^{-4} gauss in the Crab nebula have an orbital frequency

$$\omega_0 = \frac{eB}{m_e c\gamma} = 9\times10^{-4}/\text{s} \tag{70}$$

but their synchrotron radiation has frequencies up to

$$\omega_c \simeq \gamma^3\omega_0 \simeq 7\times10^{15}/\text{s} \tag{71}$$

Thus such electrons emit visible and ultraviolet light.

Figure 13.13 shows the spectral distribution (power per unit frequency interval) of synchrotron radiation. This spectral curve has been

Figure 13·13

Spectral distribution of the power emitted in synchrotron radiation.

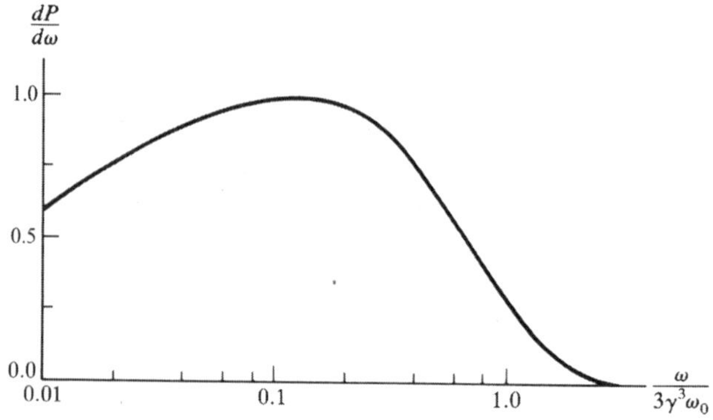

obtained from detailed theoretical calculations. Note that the spectrum extends up to a frequency of about $\gamma^3\omega_0$, in agreement with the rough estimate given in Eq. (69).

13·5 Radiation in a Dielectric Medium; the Cherenkov Effect

If the radiating distribution of charge and current is embedded in a dielectric medium, such as a plastic or gas, then the formulas of the preceding sections require some modifications. Let us assume that the medium is linear, homogenous, and isotropic, with a dielectric constant ε, a permittivity constant μ, and an index of refraction $n = \sqrt{\mu\varepsilon}$. For simplicity, we assume that ε and μ are constant not only throughout the medium, but also are constant over the relevant range of frequencies. Expressing the electric and magnetic fields in terms of the potential and the vector potential, with $\mathbf{E} = -\nabla\Phi - (1/c)\partial\mathbf{A}/\partial t$ and $\mathbf{B} = \nabla \times \mathbf{A}$, we find that the Maxwell equations (12.80) and (12.81) become

$$\frac{n^2}{c^2}\frac{\partial^2\Phi}{\partial t^2} - \nabla^2\Phi = \frac{4\pi}{\varepsilon}\rho \tag{72}$$

$$\frac{n^2}{c^2}\frac{\partial^2\mathbf{A}}{\partial t^2} - \nabla^2\mathbf{A} = \frac{4\pi\mu}{c}\mathbf{j} \tag{73}$$

Here it is understood that ρ and $\mathbf{j}$ refer to the free charges. Furthermore, we have imposed the "modified" Lorentz condition

$$\nabla\cdot\mathbf{A} + \frac{\mu\varepsilon}{c}\frac{\partial\Phi}{\partial t} = 0$$

By analogy with Eq. (3), we immediately see that the solutions to these equations are

$$\Phi(t,\mathbf{x}) = \frac{1}{\varepsilon}\int \frac{\rho(t - |\mathbf{x}-\mathbf{x}'|n/c, \mathbf{x}')}{|\mathbf{x}-\mathbf{x}'|}dV' \tag{74}$$

$$\mathbf{A}(t,\mathbf{x}) = \frac{\mu}{c}\int \frac{\mathbf{j}(t - |\mathbf{x}-\mathbf{x}'|n/c, \mathbf{x}')}{|\mathbf{x}-\mathbf{x}'|}dV' \tag{75}$$

Note that these equations differ from the corresponding equations in vacuum only by the overall factors ε and μ and by the modification of the retarded time $t' = t - |\mathbf{x} - \mathbf{x}'|\, n/c$, which is now calculated with the speed of light c/n appropriate to the medium.* These equations can therefore be given an interpretation similar to that for Eq. (3): each small element of charge or current generates a disturbance of potential or vector potential, which spreads outward in all directions at the speed c/n.

We can apply Eqs. (74) and (75) to calculate the electromagnetic fields of a charged particle moving in the dielectric medium. If the speed of the particle is less than the speed of light in the medium ($v < c/n$), then the calculation can be carried out in much the same way as in Section 13.3. Since the only difference is the presence of some factors of ε, , and n in the equations, the results are qualitatively similar to those of Section 13.3. However, if the speed of the particle exceeds the speed of light in the medium ($v > c/n$), then the calculation becomes rather more complicated, and the final results are drastically different. A particle of this kind will emit radiation not only when it is in accelerated motion but also when it is in uniform motion. The emission of radiation by a particle in uniform motion with a speed exceeding the speed of light in the medium is called the *Cherenkov effect*. In essence, the Cherenkov radiation is a shock wave, analogous to the sonic boom generated by a supersonic aircraft. The radiation is confined to a "Mach cone" trailing the particle the electromagnetic field is zero ahead of the particle and at all points outside the cone. The angle of the cone can be deduced from Fig. 13.14, which shows the particle at one instant of time and the outer limits of the

Figure 13·14

At one instant of time, the particle is at point P. The arcs of circles indicate the outer limits of the electromagnetic disturbances emitted by the particle at earlier times, while it was at the positions marked by open dots.

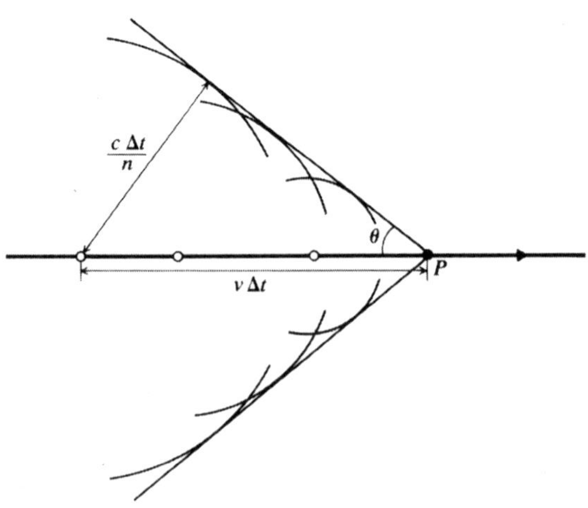

---

*This assumes that the index of refraction is a constant, independent of frequency.

electromagnetic disturbances that emerged from the particle at several earlier times. The disturbance initiated at a time Δt earlier has spread out to a distance $c\ \Delta t/n$; meanwhile, the particle has moved a distance $v\ \Delta t$. Hence the half-angle of the cone is given by

$$\sin\theta = \frac{c}{nv} \tag{76}$$

The electromagnetic field at any given point within the cone can be calculated from the general equations (74) and (75) for the potentials, but it is necessary to take into account that there are always *two* distinct retarded positions of the particle that contribute to the field at any given instant that is, there are always two solutions $\mathbf{x}'$ of Eq. (25). We can understand this in terms of the information-collector shell described in Section 13.1. This shell is centered on the point $\mathbf{x}$, and it collapses toward this point at speed c/n so as to arrive at this point at time t. Figure 13.15 shows the shell at two instants of time, t_1 and t_2, both earlier than t. The particle enters the shell at t_1 and leaves it at t_2, when the shell has shrunk to a smaller radius; thus the particle crosses the shell *twice*. This twofold crossing is not possible for a particle moving at a speed less than c/n; such a particle can never catch up with the shell and therefore cannot enter the shell—it can cross the shell only once, while leaving. Because of these complications, we will not attempt to calculate the Cherenkov radiation fields. Suffice it to say that the results of the calculation confirm our intuitive expectation that the electric and magnetic fields are very strong along the surface of the Mach cone, where the potentials suddenly increase from zero to a finite value. A detector at a fixed posi-

Figure 13·15

Information-collector shell at two instants of time, t_1 and t_2.

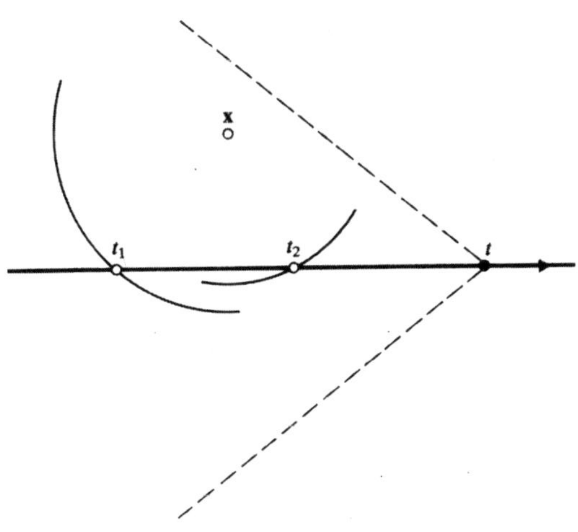

tion to one side of the path of the particle will register a strong pulse of radiation when the cone sweeps over it. Since the direction of propagation of the radiation lies at right angles to the cone, a detector with directional sensitivity can be used to determine the angle of the cone, from which the speed of the particle can be calculated. The Cherenkov effect is commonly exploited in high-energy physics for such determinations of the speeds of fast particles. In the Cherenkov counters used for this purpose, the dielectric medium consists of a transparent plastic or a gas under high pressure.

References

1. O. D. Jefimenko, *Electricity and Magnetism* (Appleton-Century-Crofts, New York, 1966), Section 15-7; *Am. J. Phys.* **58**, 505 (1990).

2. D. J. Griffiths and M. A. Heald, *Am. J. Phys.* **59**, 2 (1991).

Problems

1. Use the Liénard-Wiechert potentials to deduce the electric and magnetic fields of a point charge moving with uniform velocity, and show that the result agrees with the fields obtained in Section 8.2.

2. A charged particle approaches the origin moving inward along the $+x$ axis, stops at the origin, and then moves away along the $+x$ axis. The particle moves with uniform acceleration according to the equation $x = \frac{1}{2}at^2$. Assume that the motion is nonrelativistic. Find the electric and magnetic radiation fields at the point $y = b$ on the y axis at the time $t = 0$.

3. A point charge q moves around a circle of radius R_0 in the x-y plane such that its position vector is

$$\mathbf{r} = \hat{\mathbf{x}}R_0 \cos \omega t + \hat{\mathbf{y}}R_0 \sin \omega t$$

where ω is constant. Find the electric and magnetic radiation fields and near fields produced by the point charge at $x = y = z = 0$ at time $t = 0$. Give explicit expressions for E_x, E_y, E_z, B_x, B_y, B_z at this time. The motion is *relativistic*.

4. A free electron is exposed to a plane electromagnetic wave of frequency ω, amplitude E_0.

(a) Find the average power radiated by the electron while accelerated by the electric field of the wave; assume the motion is nonrelativistic.

(b) In the calculation of part (a), we neglected the effect on the electron of the *magnetic* field of the wave. Show that this is a good approximation if $E_0 \ll m_e c\omega / e$.

5. (a) The Stanford Linac accelerates electrons to a final energy of 40 GeV. The acceleration is achieved along a straight path 3000 m long. Assume that the energy gained by an electron per unit path length is constant. What is the radiated power at the beginning of the acceleration process? At the end?
   (b) In a circular accelerator, the radiated power would be much larger. For comparison, suppose that an electron of 40 GeV is in a circular orbit of radius 30 m. What is the radiated power in this case?

6. A circularly polarized wave, with an electric field as given by Eq. (12.24), is incident on a free electron initially located in the x-y plane, at the origin.
   (a) What is the acceleration of the electron? Assume the motion is nonrelativistic.
   (b) Find the radiation fields **E** and **B** emitted by the electron at a large distance R along the x axis. What is the polarization of the radiation?
   (c) Find the radiation fields emitted by the electron at a large distance R along the z axis. What is the polarization of the radiation?

7. A plane electromagnetic wave propagating in the z direction, with wavelength λ, amplitude E_0, and polarization in the x direction is incident on a pair of free electrons with z coordinates $z=0$ and $z=\lambda/2$. Find the magnitude of the time-average power radiated by these electrons jointly, as a function of the polar angles θ and ϕ. Assume that the motion is nonrelativistic and that the radiation is examined at a distance much larger than λ. Ignore magnetic forces and interactions of the electrons.

8. Suppose that in the collision of an electron with a proton, the distance of closest approach is 8×10^{-11} cm. According to classical physics, calculate the (retarded) electric radiation field at a distance of 3 cm from this point in a direction at right angles to the instantaneous acceleration, and compare it with the electrostatic field of the electron at the same distance.

9. A charged particle has a velocity $\mathbf{v}'$ and acceleration $\mathbf{a}'$. Find the directions (relative to $\mathbf{v}'$, $\mathbf{a}'$) in which no radiation is emitted.

10. A nonrelativistic proton makes a head-on collision with a fixed nucleus of charge Ze. The force acting on the proton is the Coulomb force. The velocity of approach is v_0 when at large distance; the proton approaches along a radius, is reflected, and returns to large distance. Find the total energy radiated during the collision, according to the classical formula (55).

11. Consider an electron in a circular orbit in a hydrogen atom. Assume that the electron moves according to classical mechanics and classical electrodynamics so the radiated power is given by Eq. (55). As the electron loses energy, it gradually spirals toward the nucleus. Assume that at each instant, the electron moves along a circular orbit so its orbital energy as a function of radius is $-e^2/2r$. Calculate the instantaneous rate of decrease of the radius. Calculate how long the electron takes to spiral from the Bohr radius to the center.

12. An electron is in a circular orbit of radius

$$r=\frac{\hbar^2}{m_e e^2}n^2$$

in a hydrogen atom. Assume that n is a large integer. Calculate the power radiated according to classical electrodynamics. Calculate how long it takes the electron to radiate an amount of energy

$$\Delta E = \frac{1}{2} m_e c^2 \left(\frac{e^2}{\hbar c} \right)^2 \left[\frac{1}{(n-1)^2} - \frac{1}{n^2} \right]$$

Find a numerical result for $n = 6$ and compare with the quantum-mechanical result, which is $\approx 10^{-6}$ s.

13. *Bremsstrahlung* is the radiation emitted when electrons are decelerated upon impact on a target, usually consisting of a chunk of metal. Most of this radiation is emitted during collisions between the electrons and the nuclei in the target. Calculate, according to classical electrodynamics, the total energy radiated by a nonrelativistic electron incident with impact parameter b on a nucleus of charge Ze. Assume that the motion of the electron is so fast that it is a good approximation to treat the trajectory as a (infinite) straight line at each point of which the electron has an acceleration due to the Coulomb force (as in Problem 2.10, regard the deviation from straight-line motion produced by this acceleration as a small quantity to be ignored in the first approximation). Evaluate numerically for an electron of 20 keV incident on a molybdenum nucleus at an impact parameter 5×10^{-11} cm.

14. According to Eq. (55), a nonrelativistic charged particle radiates energy at the rate $2q^2a'^2/3c^3$. On account of this energy loss, what is the force required to keep the particle in uniformly accelerated rectilinear motion? In uniform circular motion?

15. Suppose that a magnetic monopole of magnetic charge g is subjected to an acceleration a'.

(a) What are the formulas analogous to Eqs. (48), (50), and (55) for the electric field, the magnetic field, and the power radiated? Assume that the motion is nonrelativistic.

(b) What is the rate of radiation for a magnetic monopole of magnetic charge $c\hbar/2e$ and mass 1×10^{-8} g in free fall in the gravitational field near the surface of the Earth?

16. Integrate Eq. (57) over all solid angles, and show that the resulting exact formula for the power agrees with Eq. (64) if the motion is extremely relativistic $(\gamma \gg 1)$.

17. Figure 13.16 shows a segment of the orbit of a particle in uniform circular motion, with the acceleration perpendicular to the velocity.

(a) In terms of the angles θ and ϕ defined in this figure, show that the radiated power per unit solid angle is

$$\frac{dP}{d\Omega} = \frac{q^2}{4\pi c^3} \left[\frac{a'^2}{(1 - v'\cos\theta/c)^4} \left\{ 1 - \frac{\sin^2\theta\cos^2\phi}{\gamma^2(1 - v'\cos\theta/c)^2} \right\} \right]$$

(b) This radiation pattern has a maximum at $\theta = 0$. Find the value of $dP/d|$ at this maximum.

(c) Show that in the plane of the circular orbit ($\phi = 0$), this radiation pattern has zeros at $\theta = \pm\cos^{-1}(v'/c)$.

18. If a particle is to deliver an amount of energy ($dP/d\Omega$) $\times$ dt to a detector in a time interval dt, it must radiate an equal amount of energy ($dP'/d\Omega$) $\times$ dt' in a retarded time interval dt'. Hence the power $dP/d\Omega$ received by the detector and the power $dP/d\Omega$ radiated by the particle are not equal; they differ by a factor dt/dt',

Figure 13·16

Segment of a circular orbit in the z-x plane.

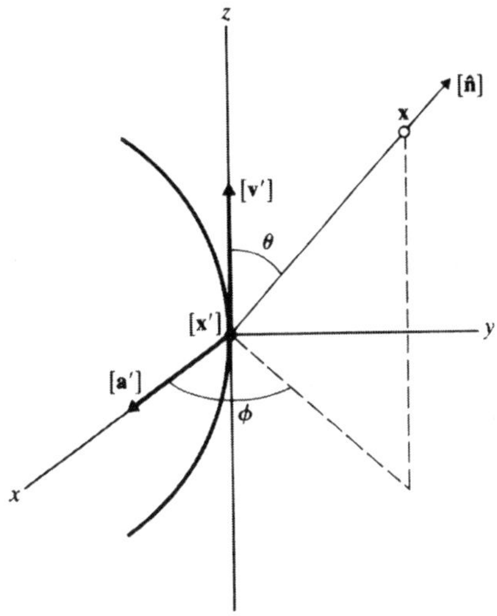

$$\frac{dP'}{d\Omega} = \frac{dP}{d\Omega}\frac{dt}{dt'} = \frac{dP}{d\Omega}\left[1 - \frac{\hat{\mathbf{n}}\cdot\mathbf{v}'}{c}\right]$$

From this, show that the power radiated by a particle whose acceleration is parallel to the velocity is

$$\frac{dP'}{d\Omega} = \frac{q^2}{4\pi c^3}\left[\frac{a'^2 \sin^2\theta}{(1 - v'\cos\theta / c)^5}\right]$$

Integrate this over all solid angles, and show that if the motion is extremely relativistic ($\gamma \gg 1$),

$$P' = \frac{2}{3}\frac{q^2}{c^3}[a'^2\gamma^6]$$

Does it make sense that this radiated power is *smaller* than the received power [Eq. (64)]?

19. Derive Eqs. (74) and (75), and use the conservation equation for electric charge to verify that they satisfy the "modified" Lorentz condition.

20. Starting with Eqs. (74) and (75), deduce formulas analogous to Eqs. (43) and (44) for the electric and magnetic radiation fields of a point charge q moving in a dielectric medium at a speed $v < c/n$. Verify that $\mathbf{B} = \sqrt{\mu\varepsilon}[\hat{\mathbf{n}}]\times\mathbf{E}$.

21. The index of refraction of Lucite is 1.5. Plot the angle of emission of Cherenkov radiation as a function of the kinetic energy of the charged particle for the range 0.5 mc^2 to 5.0 mc^2.

22. (a) What is the minimum (kinetic) energy that an electron must have to emit visible Cherenkov radiation in Lucite, with an index of refraction of 1.50?

(b) What is the energy of an electron that is observed to emit Cherenkov radiation at an angle of 45° relative to its direction of motion?

(c) What is the maximum angle at which an electron of extreme energy can emit Cherenkov radiation in Lucite?

23. Instead of calculating the electric and magnetic fields via the retarded potentials, we can calculate them from the following explicit solution of the Maxwell equations, which expresses **E** and **B** directly in terms of the retarded charge and current densities:

$$\mathbf{E}(\mathbf{x},t) = \int \left([\rho(\mathbf{x}')] \frac{\mathbf{x}-\mathbf{x}'}{|\mathbf{x}-\mathbf{x}'|^3} + \left[\frac{\partial \rho}{\partial t}\right] \frac{\mathbf{x}-\mathbf{x}'}{|\mathbf{x}-\mathbf{x}'|^2 c} - \left[\frac{\partial \mathbf{j}}{\partial t}\right] \frac{1}{|\mathbf{x}-\mathbf{x}'| c^2} \right) dV'$$

$$\mathbf{B}(\mathbf{x},t) = \frac{1}{c} \int \left([\mathbf{j}(\mathbf{x}')] \times \frac{\mathbf{x}-\mathbf{x}'}{|\mathbf{x}-\mathbf{x}'|^3} + \left[\frac{\partial \mathbf{j}}{\partial t}\right] \frac{\mathbf{x}-\mathbf{x}'}{|\mathbf{x}-\mathbf{x}'|^2 c} \right) dV'$$

Here, as usual, the bracket indicates that the charges and currents are evaluated at the retarded time $t-|\mathbf{x}-\mathbf{x}'|/c$. This solution was obtained by Jefimenko, and it can be regarded as a generalization of the Coulomb and Biot-Savart laws for time-dependent charge and current distributions.[1, 2] Verify that these expressions for $\mathbf{E}(\mathbf{x},t)$ and $\mathbf{B}(\mathbf{x},t)$ satisfy the Maxwell equations. [*Hint*: The space derivatives of the charge and current densities can be reduced to time derivatives, e.g., $\nabla\rho(\mathbf{x}', t-|\mathbf{x}-\mathbf{x}'|/c) = \nabla(-|\mathbf{x}-\mathbf{x}'|/c)\partial\rho/\partial t = \nabla'(|\mathbf{x}-\mathbf{x}'|/c)\partial\rho/\partial t$; use Gauss' theorem or integration by parts to transform terms and bring about cancellations.]

CHAPTER

14

Radiation by a System of Charges

The electromagnetic wave emitted by, say, a radio or radar antenna is a coherent superposition of the individual radiation fields of a large number of electrons surging back and forth along the antenna. In principle, we could calculate the radiation field of such an antenna by performing a summation of the individual radiation fields of all the accelerated point charges on the antenna. For this purpose, we would need to keep track of the detailed motions of all the point charges. This is an impossible task, and it is much more convenient to describe the system of point charges in terms of the current density j_μ. In this chapter, we will develop a formalism for calculating the electric and magnetic radiation fields directly from the current density. Our procedure will involve two steps: first use Eq. (13.3) to write the four-vector potential A generated by the given current density j_μ, and then calculate the **E** and **B** fields from this potential. As we will see, any time dependence of the current will lead to the emission of radiation.

The formalism of this chapter is useful not only for finding the radiation emitted by an antenna but also for finding the radiation emitted by an atom or by a nucleus. We cannot calculate the radiation field of an atom or a nucleus from the detailed motion of the electrons, protons, or neutrons because the positions and velocities of these subatomic particles are not well defined classical mechanics fails within the atom. However, it turns out that although quantum mechanics does not give us the detailed motions of the subatomic particles, it does give us an expression for the current density, and from this we can calculate the radiation.

14·1 The Radiation Fields

If a system of charges has a given current density $j_\mu(t,\mathbf{x}')$, then Eq. (13.3) tells us that the corresponding four-vector potential at a given point $\mathbf{x}$ [see Fig. 14.1(a)] must be

$$A^\mu(t,\mathbf{x}) = \frac{1}{c}\int \frac{j^\mu(t-|\mathbf{x}-\mathbf{x}'|/c,\mathbf{x}')}{|\mathbf{x}-\mathbf{x}'|}dV' \qquad (1)$$

In the calculation of the electric and magnetic radiation fields, we can assume that $r \equiv |\mathbf{x}|$ is much larger than the size of the system. For any point $\mathbf{x}'$ in the region of integration, we can then approximate

$$|\mathbf{x}-\mathbf{x}'| \simeq r - \mathbf{x}'\cdot\hat{\mathbf{r}} \tag{2}$$

where $\hat{\mathbf{r}}$ is the unit vector $\mathbf{x}/|\mathbf{x}|$. This amounts to treating $\mathbf{x}$ and $\mathbf{x}-\mathbf{x}'$ as roughly parallel [see Fig. 14.1(b)]. Furthermore, in the integral (1), we need retain only those terms that vary as $1/r$; the terms that vary as $1/r^2$, $1/r^3$, . . . , will not affect the radiation fields. Consequently, $|\mathbf{x}-\mathbf{x}'| \simeq r$ is a sufficiently good approximation in the *denominator* of the integrand in Eq. (1). With these approximations, the four-vector potential becomes

$$A^{\mu}(t,\mathbf{x}) = \frac{1}{cr}\int j^{\mu}(t - r/c + \mathbf{x}'\cdot\hat{\mathbf{r}}/c, \mathbf{x}')\,dV' \tag{3}$$

The magnetic field can now be calculated by taking the curl of $\mathbf{A}$:

$$\mathbf{B} = \nabla\times\mathbf{A} = \frac{1}{c}\int\left[\frac{\nabla\times\mathbf{j}}{r} + \nabla\frac{1}{r}\times\mathbf{j}\right]dV' \tag{4}$$

The second term in the integrand can be ignored because it varies as $1/r^2$. To evaluate $\nabla\times\mathbf{j}$, we observe that space derivatives of $\mathbf{j}$ can be expressed in terms of time derivatives according to the chain rule

$$\frac{\partial}{\partial x}\mathbf{j}(t - r/c + \mathbf{x}'\cdot\hat{\mathbf{r}}/c, \mathbf{x}') = \frac{\partial\mathbf{j}}{\partial t}\frac{1}{c}\frac{\partial}{\partial x}(-r + \mathbf{x}'\cdot\hat{\mathbf{r}})$$

and corresponding rules for $\partial/\partial y$ and $\partial/\partial z$. This leads to

Figure 14·1

(a) A system of charges is contained in a finite volume surrounding the origin.

(b) If $|\mathbf{x}|$ is very large compared with $|\mathbf{x}'|$, then $\mathbf{x}$ and $\mathbf{x}-\mathbf{x}'$ are nearly parallel, and the difference in the lengths of these vectors is $\mathbf{x}'\cdot\hat{\mathbf{r}}$.

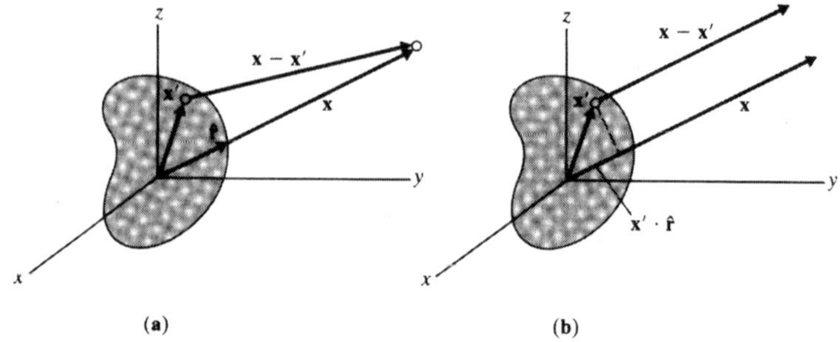

$$\nabla \times \mathbf{j}(t - r/c + \mathbf{x}'\cdot\hat{\mathbf{r}}/c, \mathbf{x}') = \frac{1}{c}\nabla(-r + \mathbf{x}'\cdot\hat{\mathbf{r}}) \times \frac{\partial \mathbf{j}}{\partial t}$$
$$= \frac{1}{c}[-\hat{\mathbf{r}} + \nabla(\mathbf{x}'\cdot\hat{\mathbf{r}})] \times \frac{\partial \mathbf{j}}{\partial t} \tag{5}$$

Furthermore, $\nabla(\mathbf{x}'\cdot\hat{\mathbf{r}})$ can be neglected because it gives rise to a $1/r^2$ term in the field. The magnetic radiation field is then simply

$$\mathbf{B} = -\frac{1}{c^2}\int \frac{\hat{\mathbf{r}}}{r} \times \frac{\partial \mathbf{j}}{\partial t} dV' = -\frac{1}{c}\hat{\mathbf{r}} \times \frac{\partial \mathbf{A}}{\partial t} \tag{6}$$

The electric field can also be calculated in much the same way:

$$\mathbf{E} = -\frac{1}{c}\frac{\partial \mathbf{A}}{\partial t} - \nabla A^0 \simeq -\frac{1}{c}\frac{\partial \mathbf{A}}{\partial t} + \frac{1}{c}\hat{\mathbf{r}}\frac{\partial A^0}{\partial t} \tag{7}$$

■ ***Exercise 1.*** Derive Eq. (7). ■

Since A^μ satisfies the Lorentz condition $\partial_\mu A^\mu = 0$, or

$$\frac{1}{c}\frac{\partial A^0}{\partial t} + \nabla\cdot\mathbf{A} = 0$$

we can replace Eq. (7) by

$$\mathbf{E} \simeq -\frac{1}{c}\frac{\partial \mathbf{A}}{\partial t} - \hat{\mathbf{r}}\nabla\cdot\mathbf{A} \simeq -\frac{1}{c}\frac{\partial \mathbf{A}}{\partial t} + \frac{1}{c}\hat{\mathbf{r}}\,\hat{\mathbf{r}}\cdot\frac{\partial \mathbf{A}}{\partial t} \tag{8}$$

The last step in this derivation again involved expressing space derivatives in terms of time derivatives, as in Eq. (5). We can combine the two terms in Eq. (8) in a double cross product:

$$\mathbf{E} \simeq \frac{1}{c}\hat{\mathbf{r}} \times \left(\hat{\mathbf{r}} \times \frac{\partial \mathbf{A}}{\partial t}\right) \tag{9}$$

and comparison with Eq. (6) then shows that the electric radiation field is related to the magnetic radiation field by

$$\mathbf{E} = -\hat{\mathbf{r}} \times \mathbf{B} \tag{10}$$

The electric and magnetic fields are therefore perpendicular to each other and of the same magnitude, exactly as expected for radiation (see Section 13.3). In what follows, we will find it convenient to calculate **B** from Eq. (6) and then use Eq. (10) to find **E**.

Before we look at an example, let us summarize the assumptions that went into our derivation of the radiation fields:

(*i*) We have made the approximation (2). This assumes that r is much larger than the size of the system,

$$r \gg a \tag{11}$$

where a is the (largest) dimension of the system.

(*ii*) We have neglected $1/r^2$ terms in **E** and **B**. These neglected terms are typically of the order of [see Eq. (4)]

$$\simeq \frac{1}{c}\int \frac{1}{r^2}\mathbf{j}\,dV' \tag{12}$$

The terms we have kept are of the order of

$$\simeq \frac{1}{c^2}\int \frac{1}{r}\frac{\partial \mathbf{j}}{\partial t}\,dV' \tag{13}$$

If we estimate $|\partial \mathbf{j}/\partial t| \simeq |\mathbf{j}/T|$, where T is the period of the system, or the characteristic time needed for an appreciable change to occur in the current, then we see that (12) is smaller than (13) by a factor cT/r. Since cT will be roughly the wavelength of the emitted radiation, we can also express this factor as λ/r, and the condition that this be small is

$$r \gg \lambda \tag{14}$$

14·2 The Half-Wave Antenna

A very simple antenna consists of two pieces of wire into which a high-frequency generator feeds an oscillating current (see Fig. 14.2). The oscillating charge distribution on the antenna will then emit radio waves. In practical applications, the length of such a simple antenna is usually chosen to be about one-half of the wavelength of the radio wave. The current along the antenna then has the form of a standing wave,

$$I = I_0 \cos\frac{2\pi z}{\lambda}\cos\omega t \tag{15}$$

where ω is the frequency of the generator and $\lambda = 2\pi c/\omega$. This current has a maximum at the midpoint of the antenna, and it is zero at the endpoints $z = \pm\lambda/4$. The standing wave comes about in the following way:

Figure 14·2

A half-wave antenna lying along the z axis.

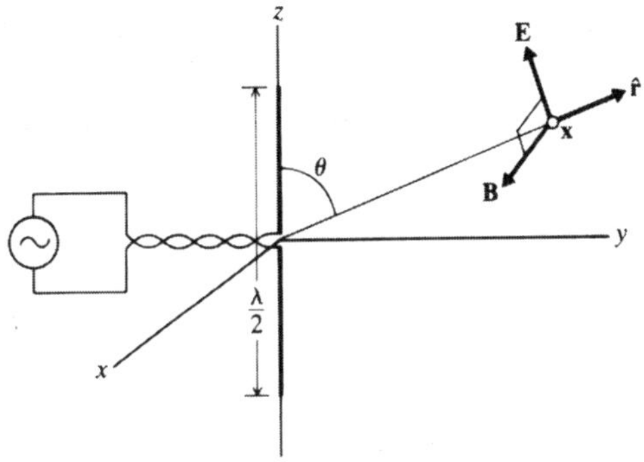

The pulses of current fed into the midpoint of the antenna by the generator travel along the antenna, reflect at the endpoints, and return to the midpoint. The speed of propagation of the pulses along the antenna is approximately the speed of light; hence the wavelength of the standing wave must be this speed multiplied by the period of oscillation, that is, $c \times 2\pi/\omega$ as in Eq. (15). One important advantage of the half-wave antenna is that the factor $\cos 2\pi z/\lambda$ is everywhere positive, and hence all charges on the antenna contribute to the radiation fields with the same sign (constructive interference).

The description of the current distribution in terms of a simple standing wave is only a rather crude approximation. The exact description of the oscillations of the current distribution on an antenna of a given shape is a very complicated boundary-value problem, a problem further aggravated by the fact that the current radiates electromagnetic waves and thereby suffers a loss of energy and a loss of intensity. In practice, "half-wave" antennas are often constructed with telescoping segments so that, by trial and error, the length can be adjusted to achieve maximum radiated power. The adjusted length of a "half-wave" antenna is somewhat shorter than one-half of a wavelength; we can crudely understand this by noting that the speed of propagation of current pulses along the antenna is somewhat less than the speed of light. Incidentally: The antennas used by radio broadcast stations commonly consist of a vertical wire of length $\frac{1}{4}\lambda$ with the lower end near the ground. Since the ground is a conductor, it effectively acts as a mirror and forms an "image" of the wire. The wire plus its "image" then make up a half-wave antenna. For the sake of simplicity, we will ignore all these complications, and we will proceed with the calculation of the radiation fields for the simple current distribution given by Eq. (15).

According to Eq. (3), the vector potential for this current distribution is

$$A_z = A^3 = \frac{1}{cr}\int_{-\lambda/4}^{\lambda/4} I_0 \cos\frac{2\pi z'}{\lambda}\cos\omega\left(t - \frac{r}{c} + \frac{z'}{c}\cos\theta\right)dz' \tag{16}$$

With the identity

$$\cos\alpha\cos\beta = \tfrac{1}{2}\cos(\alpha+\beta) + \tfrac{1}{2}\cos(\alpha-\beta)$$

this becomes

$$\begin{aligned} A_z = \frac{I_0}{2cr}\int_{-\lambda/4}^{\lambda/4} &\left\{\cos\left[\omega\left(t - \frac{r}{c}\right) + z'\left(\frac{2\pi}{\lambda} + \frac{\omega}{c}\cos\theta\right)\right]\right. \\ &\left. + \cos\left[\omega\left(t - \frac{r}{c}\right) - z'\left(\frac{2\pi}{\lambda} - \frac{\omega}{c}\cos\theta\right)\right]\right\}dz' \end{aligned} \tag{17}$$

If we integrate this and replace λ by $2\pi c/\omega$, we obtain

$$\begin{aligned} A_z &= \frac{I_0}{2r}\left\{\frac{\sin[\omega(t - r/c) + \tfrac{1}{2}\pi + \tfrac{1}{2}\pi\cos\theta] - \sin[\omega(t - r/c) - \tfrac{1}{2}\pi - \tfrac{1}{2}\pi\cos\theta]}{\omega(1+\cos\theta)}\right. \\ &\qquad \left. - \frac{\sin[\omega(t - r/c) - \tfrac{1}{2}\pi + \tfrac{1}{2}\pi\cos\theta] - \sin[\omega(t - r/c) + \tfrac{1}{2}\pi - \tfrac{1}{2}\pi\cos\theta]}{\omega(1-\cos\theta)}\right\} \\ &= \frac{I_0}{2r}\left[\frac{2\cos\omega(t - r/c)\sin(\tfrac{1}{2}\pi + \tfrac{1}{2}\pi\cos\theta)}{\omega(1+\cos\theta)}\right. \\ &\qquad \left. + \frac{2\cos\omega(t - r/c)\sin(\tfrac{1}{2}\pi - \tfrac{1}{2}\pi\cos\theta)}{\omega(1-\cos\theta)}\right] \end{aligned} \tag{18}$$

Since sin $(\tfrac{1}{2}\pi \pm \tfrac{1}{2}\pi\cos\theta)$ = cos $(\tfrac{1}{2}\pi\cos\theta)$, we see that the numerators of both terms in the brackets contain the same factors. Upon adding these terms, we find

$$A_z = \frac{2I_0}{r\omega}\cos\omega\left(t - \frac{r}{c}\right)\frac{\cos(\tfrac{1}{2}\pi\cos\theta)}{\sin^2\theta} \tag{19}$$

This is the only nonzero component of **A**.

We can now calculate the magnetic field by means of Eq. (6). Since $\partial\mathbf{A}/\partial t$ is in the direction of $\hat{\mathbf{z}}$, the magnetic field is in the direction of $\hat{\mathbf{r}}\times\hat{\mathbf{z}}$ (see Fig. 14.2). The magnitude of the magnetic field is

$$B = -\sin\theta\frac{1}{c}\frac{\partial A_z}{\partial t} = \frac{2I_0}{cr}\sin\omega\left(t - \frac{r}{c}\right)\frac{\cos(\tfrac{1}{2}\pi\cos\theta)}{\sin\theta} \tag{20}$$

The electric field is, of course, perpendicular to the magnetic field, and it has the same magnitude.

The Poynting vector is in the direction of $\hat{\mathbf{r}}$; it has a magnitude

$$S = \frac{c}{4\pi} B^2 = \frac{I_0^2}{\pi c r^2} \sin^2 \omega \left(t - \frac{r}{c} \right) \frac{\cos^2(\frac{1}{2}\pi \cos\theta)}{\sin^2\theta} \tag{21}$$

The time-average Poynting vector has a magnitude

$$\overline{S} = \frac{I_0^2}{2\pi c r^2} \frac{\cos^2(\frac{1}{2}\pi \cos\theta)}{\sin^2\theta} \tag{22}$$

The time-average power $d\overline{P}/d\Omega$ radiated per unit solid angle equals $\overline{S}r^2$,

$$\frac{d\overline{P}}{d\Omega} = \overline{S}r^2 = \frac{I_0^2}{2\pi c} \frac{\cos^2(\frac{1}{2}\pi \cos\theta)}{\sin^2\theta} \tag{23}$$

Figure 14.3 shows the polar plot of this radiation pattern; for comparison, Fig. 14.3 includes a plot of the radiation pattern for an electric dipole (see below). The half-wave antenna tends to beam the radiation more strongly in the horizontal plane than the dipole.

The total radiated power is the integral of $d\overline{P}/d\Omega$ over all solid angles,

$$\overline{P} = \int \frac{d\overline{P}}{d\Omega} d\Omega = \frac{I_0^2}{2\pi c} \int_0^{2\pi} \int_0^{\pi} \frac{\cos^2(\frac{1}{2}\pi \cos\theta)}{\sin^2\theta} d\theta d\phi$$

Figure 14·3

Polar plots of $d\overline{P}/d\Omega$ for a half-wave antenna and for an electric dipole.

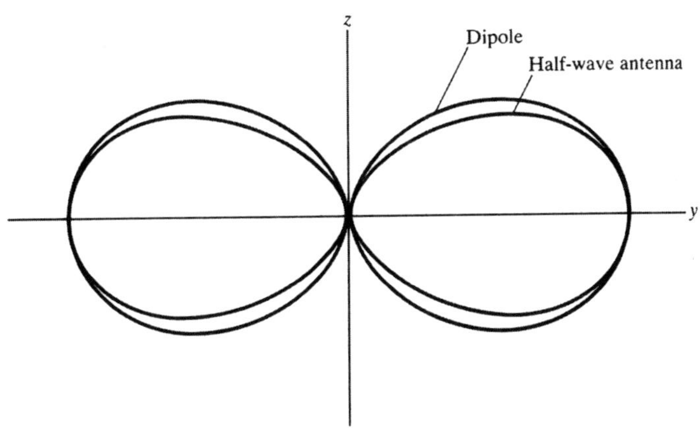

The integral over θ cannot be expressed in terms of elementary functions and must be done numerically. This leads to

$$\overline{P} = 1.22\frac{I_0^2}{c} \tag{24}$$

The radiation from an antenna of some other shape, with a given current distribution, can be calculated by the same methods. The radiation from an antenna array, consisting of, say, several half-wave antennas, can be calculated by superposing the fields of the individual antennas.

14·3 The Multipole Expansion; Dipole Radiation

If the size of the radiating system is small compared with the wavelength of the radiation, it is often convenient to evaluate Eq. (3) by means of a power-series expansion for the current. This method is unsatisfactory for an antenna, such as the half-wave antenna, whose size is comparable with the wavelength the power series then converges slowly or not at all. But the method is very successful for an atom or a nucleus whose size is much smaller than the wavelength—one or two terms of the power series are then often sufficient to obtain an excellent approximation for the radiation fields. Of course, atoms and nuclei are quantum systems, in which classical mechanics is not valid. However, it is possible to calculate the emission of radiation by a semiclassical approximation, in which the motion of the particles is treated quantum-mechanically but the radiation fields are treated classically. This effectively means that we can use the classical expressions for the radiation rates of oscillating dipoles, quadrupoles, etc., but the relevant dipole moments, quadrupole moments, etc. must be calculated quantum-mechanically.

The assumptions (*i*) and (*ii*) [or Eqs. (11) and (14)] were all that we needed in the derivation of our expressions (6) and (10) for the radiation fields. But in order to deal with the radiation fields generated by a charge and current distribution of small size, we will make an extra assumption:

(*iii*) We will suppose that the dependence of the current on the term $\mathbf{x}'\cdot\hat{\mathbf{r}}/c$ can be expressed as a Taylor series:

$$\mathbf{j}(t - r/c + \mathbf{x}'\cdot\hat{\mathbf{r}}/c, \mathbf{x}') = \mathbf{j}(t - r/c, \mathbf{x}') + \frac{1}{c}\mathbf{x}'\cdot\hat{\mathbf{r}}\frac{\partial \mathbf{j}(t - r/c, \mathbf{x}')}{\partial t} + \dots \tag{25}$$

The second term differs from the first by a factor of $\simeq a/cT$; the Taylor series will give a good approximation provided that this factor is small:

$$a \ll cT \tag{26}$$

We can also write this as

$$a \ll \lambda \tag{27}$$

Note that a/T is comparable with, or perhaps larger than, the typical velocity v of the particles in the system. Hence Eq. (26) implies that

$$v \ll c \tag{28}$$

which means that the condition (27) is valid only for nonrelativistic systems.

If we combine the conditions (14) and (27), we obtain

$$a \ll \lambda \ll r \tag{29}$$

This obviously implies (11) and therefore summarizes all the assumptions we need.

Inserting the Taylor series into Eq. (3), we obtain the following expansion for **A**:

$$\mathbf{A}(t,\mathbf{x}) = \frac{1}{cr}\int \mathbf{j}(t - r/c, \mathbf{x}')\,dV' + \frac{1}{c^2 r}\int \mathbf{x}'\cdot\hat{\mathbf{r}}\frac{\partial \mathbf{j}(t - r/c, \mathbf{x}')}{\partial t}dV' + \ldots \tag{30}$$

This is called the *multipole expansion* for radiation. The first term gives electric dipole radiation; the second, magnetic dipole and electric quadrupole radiation; and the higher-order terms indicated by +... in Eq. (30) give magnetic quadrupole radiation, electric octopole radiation, and so on.

We will now concentrate on the first term of Eq. (30). We start with the following useful identity:

$$\int \mathbf{j}\,dV' = -\int \mathbf{x}'(\nabla'\cdot\mathbf{j})\,dV' \tag{31}$$

■ ***Exercise 2.*** Derive this identity. [*Hint:* Integrate the right side by parts.] ■

But current conservation implies

$$\nabla'\cdot\mathbf{j} = -\frac{\partial \rho}{\partial t} \tag{32}$$

so

$$\mathbf{A} = \frac{1}{cr}\int \mathbf{j}\,dV' = \frac{1}{cr}\int \mathbf{x}'\frac{\partial \rho}{\partial t}dV' = \frac{1}{cr}\frac{d}{dt}\int \mathbf{x}'\rho\,dV' \tag{33}$$

If we define the dipole moment in the usual way as

$$\mathbf{p} = \int \mathbf{x}'\rho \, dV' \tag{34}$$

then

$$\mathbf{A}(t,\mathbf{x}) = \frac{1}{cr}\left[\frac{d\mathbf{p}}{dt}\right]_{t-r/c} \tag{35}$$

The right side is evaluated at the time $t - r/c$ because this is the time that appears with the current in Eq. (30).

The magnetic radiation field corresponding to Eq. (35) is

$$\mathbf{B}(t,\mathbf{x}) = -\frac{1}{c}\hat{\mathbf{r}}\times\frac{\partial \mathbf{A}}{\partial t} = -\frac{1}{c^2 r}\hat{\mathbf{r}}\times\left[\frac{d^2\mathbf{p}}{dt^2}\right]_{t-r/c} \tag{36}$$

In concise notation, we then have the following expressions for the radiation fields of a dipole:

$$\mathbf{B} = -\frac{1}{c^2 r}\hat{\mathbf{r}}\times[\ddot{\mathbf{p}}] \tag{37}$$

$$\mathbf{E} = -\hat{\mathbf{r}}\times\mathbf{B} \tag{38}$$

where the dots denote time derivatives and the bracket [ ] indicates time retardation.

The Poynting vector is

$$\mathbf{S} = \frac{c}{4\pi}\mathbf{E}\times\mathbf{B} = \frac{1}{4\pi c^3}\frac{\hat{\mathbf{r}}}{r^2}\left|[\ddot{\mathbf{p}}]\right|^2 \sin^2\theta \tag{39}$$

where θ is the angle between $[\ddot{\mathbf{p}}]$ and $\hat{\mathbf{r}}$ (see Fig. 14.4). The power $dP/d\Omega$ emitted per unit solid angle is Sr^2, so

Figure 14·4

Directions of $[\ddot{\mathbf{p}}]$ and $\hat{\mathbf{r}}$ for a radiating electric dipole.

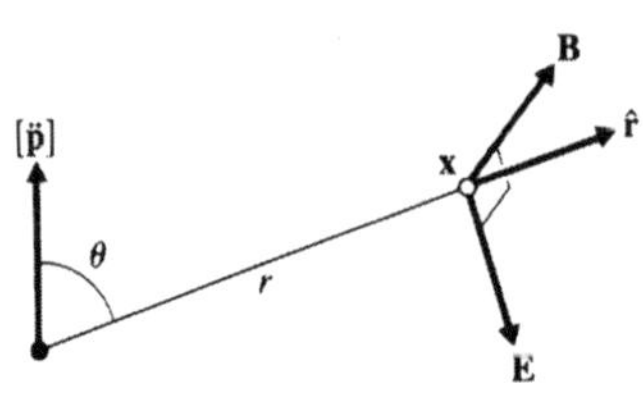

$$\frac{dP}{d\Omega} = \frac{1}{4\pi c^3} [\![\ddot{\mathbf{p}}]\!]^2 \sin^2\theta \tag{40}$$

and the total power emitted in all directions is

$$P = \frac{2}{3c^3} [\![\ddot{\mathbf{p}}]\!]^2 \tag{41}$$

■ ***Exercise 3.*** Integrate Eq. (40) over all angles and derive Eq. (41). ■

Note that if the "system" of charges consists of a single point charge, then $\mathbf{p} = q\mathbf{x}'$ and we recover the familiar result [compare Eq. (13.55)]

$$P = \frac{2q^2}{3c^3} [\![\ddot{\mathbf{x}}']\!]^2 \tag{42}$$

EXAMPLE: *radiation by a harmonic oscillator*

For the special case of an oscillating dipole consisting of, say, a charged particle moving with simple harmonic motion along a straight line with a dipole moment

$$\mathbf{p} = \mathbf{p}_0 \cos\omega t \tag{43}$$

Eqs. (37), (38), and (40) give

$$\mathbf{B} = \frac{1}{c^2 r} \hat{\mathbf{r}} \times \mathbf{p}_0 \omega^2 \cos\omega(t - r/c) \tag{44}$$

$$\mathbf{E} = -\hat{\mathbf{r}} \times \mathbf{B} \tag{45}$$

and

$$\frac{dP}{d\Omega} = \frac{1}{4\pi c^3} p_0^2 \omega^4 \sin^2\theta \cos^2\omega(t - r/c) \tag{46}$$

The electric field lines at successive instants of time are shown in Fig. 14.5. The magnetic field lines are circles around the axis of the dipole; they are perpendicular to the electric field lines. The angular distribution of radiation is proportional to $\sin^2\theta$ (see Fig. 14.3); this radiation pattern is the same as that of an accelerated point charge (see Fig. 13.8).

The instantaneous emitted power is

$$P = \frac{2}{3c^3} p_0^2 \omega^4 \cos^2\omega(t - r/c)$$

This oscillates between zero and a maximum value. The time-average emitted power is

Figure 14·5

Pattern of electric field lines surrounding a radiating electric dipole. The diagrams (a)–(f) span one half cycle, at intervals of 1/12 cycle. [After R. H. Good, Intermediate Physics Demonstrations (diskette for personal computer, 1984).]

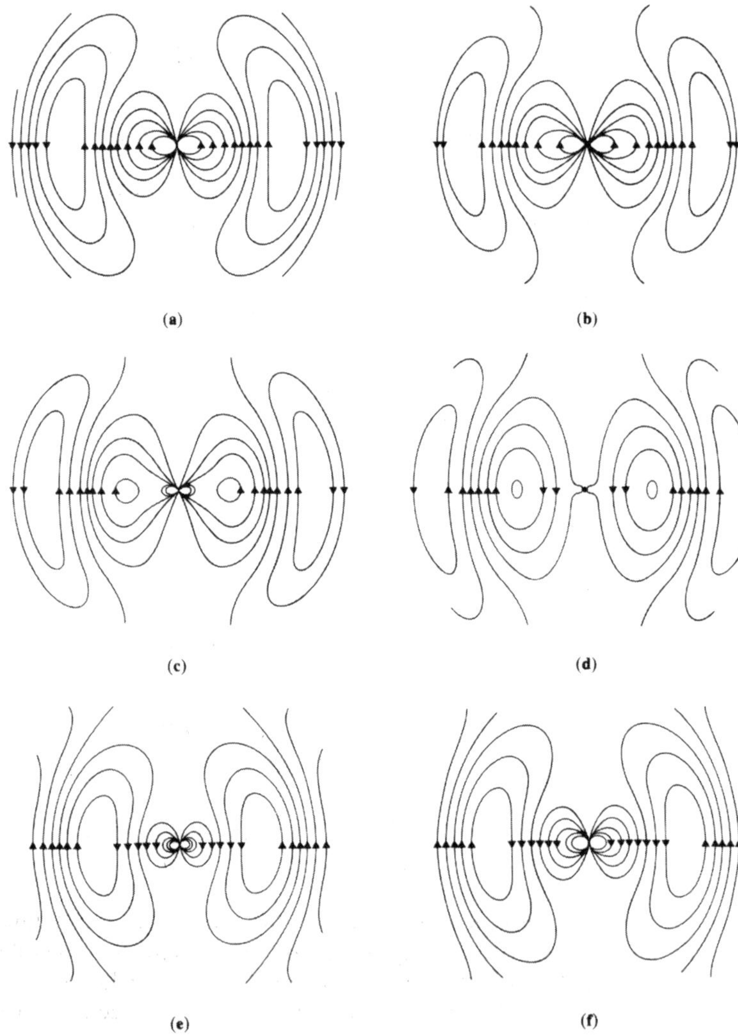

$$\overline{P} = \frac{1}{3c^3}\omega^4 p_0^2 \tag{47}$$

It is important to keep in mind that a charge in simple harmonic motion has not only an oscillating dipole moment but also a quadrupole moment.

Hence the charge emits not only dipole radiation but also quadrupole radiation. However, we will see in Section 14.5 that the intensity of the latter radiation differs from that of the former by a factor of roughly (size of system)$^2/\lambda^2$, or a^2/λ^2, which we assume to be small. For a first approximation, we can therefore neglect quadrupole radiation.

EXAMPLE: *electric dipole radiation by atoms and nuclei*

An atom is a beautiful example of a radiating system of charges. The size of the system is $a \simeq 0.1$ nm, and the wavelength of the emitted light is typically $\lambda \simeq 500$ nm; hence $a \ll \lambda$ is very well satisfied, and the dipole approximation will be very good. However, we must take into account the quantum-mechanical properties of the system. When this is done, it turns out that the formula (47) is still correct, but p_0 must be calculated according to the rule

$$\mathbf{p}_0 = \left| -2\int \psi_b * e\mathbf{x}'\psi_a \, dV' \right| \tag{48}$$

where ψ_a and ψ_b are, respectively, the quantum-mechanical wavefunctions describing the initial state and the final state of the system. The integral appearing in (48) is called the *transition dipole moment.*

It is useful to define the *lifetime* τ of the initial state as

$$\tau = \frac{\hbar\omega}{\overline{P}} \tag{49}$$

This is the time required for the atom to emit an amount of $\hbar\omega$ of energy. Since an atom radiating at the frequency ω always emits exactly the energy $\hbar\omega$ (one photon), the quantity τ does indeed tell us how long it takes for an atom to complete the emission process. The inverse of τ is called the *transition rate*; by Eq. (47), we have

$$\frac{1}{\tau} = \frac{1}{\hbar\omega}\frac{1}{3c^3}\omega^4 p_0^2 \tag{50}$$

This formula for the transition rate is correct, even though the arguments we have used to arrive at it are somewhat questionable (it makes little sense to talk of radiated power if the system emits only *one* photon!).

For a rough estimate of the typical transition rate for an atomic system, we can guess that p_0 is of the order of magnitude of the atomic size multiplied by the electronic charge, so

$$\frac{1}{\tau} \simeq \frac{1}{\hbar\omega}\frac{1}{3c^3}\omega^4 e^2 a^2 \tag{51}$$

With a typical optical frequency, Eq. (51) gives a transition rate of the order of 10^7 s^{-1}, or a lifetime of 10^{-7} s.

Equation (51) can also be applied to the gamma rays emitted by a nucleus. The transition rate is much larger than in the atomic case, because the higher frequency of the gamma rays more than compensates for the smaller size of the nucleus.

14·4 Magnetic Dipole Radiation

Next we will look at the second integral appearing in Eq. (30). Let us ignore the time derivative for a start; then we must evaluate

$$\int (\hat{\mathbf{r}}\cdot\mathbf{x}')\mathbf{j}\, dV' \tag{52}$$

Using the identity

$$(\mathbf{x}'\times\mathbf{j})\times\hat{\mathbf{r}} = (\hat{\mathbf{r}}\cdot\mathbf{x}')\mathbf{j} - (\hat{\mathbf{r}}\cdot\mathbf{j})\mathbf{x}' \tag{53}$$

we can write (52) as

$$\frac{1}{2}\int [(\hat{\mathbf{r}}\cdot\mathbf{x}')\mathbf{j} + (\hat{\mathbf{r}}\cdot\mathbf{j})\mathbf{x}']dV' + \frac{1}{2}\int (\mathbf{x}'\times\mathbf{j})\times\hat{\mathbf{r}}\, dV' \tag{54}$$

The first integral in (54) gives electric quadrupole radiation; we will save this for the next section. The second integral gives the following contribution to **A**:

$$\mathbf{A} = \frac{1}{2c^2 r}\frac{d}{dt}\int (\mathbf{x}'\times\mathbf{j})\times\hat{\mathbf{r}}\, dV'$$

This can also be written as

$$\mathbf{A} = -\frac{1}{cr}\hat{\mathbf{r}}\times\left[\frac{d\mathbf{m}}{dt}\right] \tag{55}$$

where

$$\mathbf{m} = \frac{1}{2c}\int (\mathbf{x}'\times\mathbf{j})\, dV' \tag{56}$$

is the magnetic dipole moment and the bracket stands for retardation.

The magnetic and electric radiation fields produced by the magnetic dipole are then readily found to be

$$\mathbf{B} = \frac{1}{c^2 r}\hat{\mathbf{r}}\times(\hat{\mathbf{r}}\times[\ddot{\mathbf{m}}]) \tag{57}$$

$$\mathbf{E} = \frac{1}{c^2 r}\hat{\mathbf{r}} \times [\ddot{\mathbf{m}}] \tag{58}$$

■ ***Exercise 4.*** Derive Eqs. (57) and (58). ■

The power emitted per unit solid angle and the total power are, respectively,

$$\frac{dP}{d\Omega} = \frac{1}{4\pi c^3}\left|[\ddot{\mathbf{m}}]\right|^2 \sin^2\theta \tag{59}$$

and

$$P = \frac{2}{3c^3}\left|[\ddot{\mathbf{m}}]\right|^2 \tag{60}$$

where θ is the angle between $[\ddot{\mathbf{m}}]$ and $\hat{\mathbf{r}}$.

An oscillating magnetic dipole can be constructed by, say, taking a circular ring of wire and making an alternating current flow around it (see Fig. 14.6).

■ ***Exercise 5.*** Find the magnetic moment for such an arrangement. Show that the electric dipole and quadrupole moments are zero. ■

For such an oscillating magnetic dipole, the pattern of field lines has the same general appearance as shown in Fig. 14.5, but the lines of **E** are now lines of **B**, and conversely; also, all the directions of *one* of these sets of lines (**E** or **B**) must be reversed. The angular distribution of radiation is obviously the same as for the electric dipole. The time average of the total emitted power is

$$\overline{P} = \frac{1}{3c^3}\omega^4 m_0^2 \tag{61}$$

where m_0 is the maximum value of the oscillating magnetic moment.

Figure 14·6

A magnetic dipole antenna lying in the *x*-*y* plane.

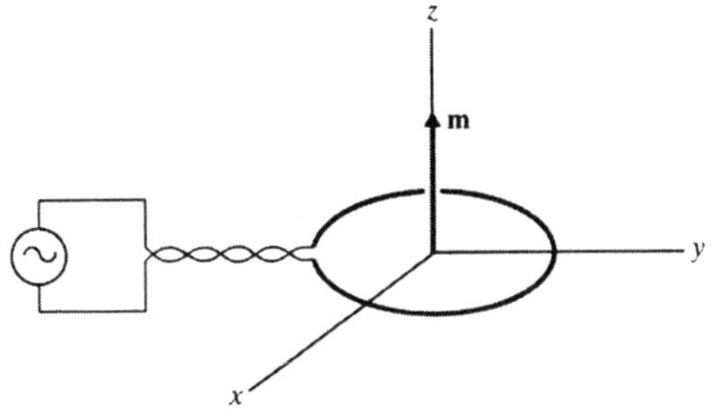

EXAMPLE: *magnetic dipole radiation by atoms and nuclei*

We can apply Eq. (61) to atomic and nuclear systems and obtain a transition rate

$$\frac{1}{\tau} = \frac{1}{\hbar\omega}\frac{1}{3c^3}\omega^4 m_0^2 \tag{62}$$

where, of course, m_0 must be evaluated according to quantum mechanics. For an estimate, it will suffice to remember that the magnetic moment due to orbital motion is related to the angular momentum by

$$\mathbf{m} = \frac{e}{2Mc}\mathbf{L} \tag{63}$$

where M is the mass of the moving particle. Typically, the angular momentum will be of the order $L \simeq \hbar$, so

$$\frac{1}{\tau} \simeq \frac{1}{\hbar\omega}\frac{1}{3c^3}\omega^4\left(\frac{e\hbar}{2Mc}\right)^2 \tag{64}$$

This transition rate is smaller than the electric dipole transition rate (51) by a factor

$$\approx\left(\frac{\hbar}{Mca}\right)^2 \tag{65}$$

For atomic systems, this factor is quite small, of the order of 10^{-5}. Hence a magnetic dipole transition is usually much less likely than an electric dipole transition (of course, if the latter is forbidden by some selection rule, then the former will be important). For nuclear systems, the factor (65) is not quite as small; furthermore, if the magnetic moment (63) is due to the spin of protons or neutrons, it will be somewhat larger than we have estimated. Magnetic dipole transitions are much more likely for nuclei than for atoms.

14·5 Electric Quadrupole Radiation

Finally, we come to the first integral appearing in Eq. (54). We can transform this term some more by another clever identity:

$$\frac{1}{2}\int [(\hat{\mathbf{r}}\cdot\mathbf{x}')\mathbf{j} + (\hat{\mathbf{r}}\cdot\mathbf{j})\mathbf{x}']dV' = -\frac{1}{2}\int \mathbf{x}'(\hat{\mathbf{r}}\cdot\mathbf{x}')\nabla'\cdot\mathbf{j}\,dV' \tag{66}$$

■ ***Exercise 6.*** Prove this identity. [*Hint*: Integrate the right side by parts.] ■

Since $\nabla'\bullet\mathbf{j} = -\partial\rho/\partial t$, we can also write Eq. (66) as

$$\frac{1}{2}\int[(\hat{\mathbf{r}}\bullet\mathbf{x}')\mathbf{j}+(\hat{\mathbf{r}}\bullet\mathbf{j})\mathbf{x}']dV' = \frac{1}{2}\frac{d}{dt}\int\mathbf{x}'(\hat{\mathbf{r}}\bullet\mathbf{x}')\rho\,dV' \tag{67}$$

This gives the following contribution to **A**:

$$\mathbf{A} = \frac{1}{2c^2r}\frac{d^2}{dt^2}\int\mathbf{x}'(\hat{\mathbf{r}}\bullet\mathbf{x}')\rho\,dV' \tag{68}$$

We can add to this expression for **A** an extra term

$$-\frac{1}{6c^2r}\hat{\mathbf{r}}\frac{d^2}{dt^2}\int r'^2\rho\,dV' \tag{69}$$

This extra term can be included in (68) because it does not affect the magnetic field in any way. To see this, we have only to remember that **B** is calculated according to

$$\mathbf{B} = -\frac{1}{c}\hat{\mathbf{r}}\times\frac{\partial\mathbf{A}}{\partial t} \tag{70}$$

so if **A** includes (69), this will only give an extra term proportional to $\hat{\mathbf{r}}\times\hat{\mathbf{r}} = 0$ in (70).

This new expression for **A** is then*

$$\mathbf{A} = \frac{1}{6c^2r}\frac{d^2}{dt^2}\int[3\mathbf{x}'(\hat{\mathbf{r}}\bullet\mathbf{x}') - \hat{\mathbf{r}}r'^2]\rho\,dV' \tag{71}$$

We will write this as

$$\mathbf{A} = \frac{1}{6c^2r}\frac{d^2}{dt^2}[\mathbf{D}] \tag{72}$$

where

$$[\mathbf{D}] \equiv \int[3\mathbf{x}'(\hat{\mathbf{r}}\bullet\mathbf{x}') - \hat{\mathbf{r}}r'^2]\rho\,dV' \tag{73}$$

The vector **D** is related to the quadrupole moment of the system:

$$D^k = Q^{kl}n^l \tag{74}$$

where n^l is the l component of $\hat{\mathbf{r}}$ and where

---

*The change from (68) to (71) is a *gauge transformation* of **A**.

$$Q^{kl} = \int (3x'^k x'^l - \delta^{kl} r'^2)\rho\, dV' \tag{75}$$

is the quadrupole-moment tensor.

■ ***Exercise 7.*** Verify Eq. (74). ■

The magnetic field given by Eq. (70) is

$$\mathbf{B} = -\frac{1}{6c^3 r}\hat{\mathbf{r}} \times [\dddot{\mathbf{D}}] \tag{76}$$

and the angular distribution of the emitted power is

$$\frac{dP}{d\Omega} = \frac{1}{144\pi c^5}\left|\hat{\mathbf{r}} \times [\dddot{\mathbf{D}}]\right|^2 \tag{77}$$

This angular distribution is quite messy. But it has one noteworthy simple feature: From $\dddot{D}^k = \dddot{Q}^{kl} n^l$, it follows that whenever $\hat{\mathbf{r}}$ is an eigenvector of $\dddot{Q}^{kl}$, then $\dddot{\mathbf{D}}$ is proportional to $\hat{\mathbf{r}}$ and therefore (77) is exactly zero. Since $\dddot{Q}^{kl}$ is a symmetric matrix, it has three orthogonal eigenvectors; this means that there are at least three perpendicular directions (and their opposite directions) with zero intensity. (If two of the eigenvectors belong to the same eigenvalue, then the intensity is also zero everywhere in the plane containing these two eigenvectors.)

We can calculate the total emitted power as follows:

$$P = \frac{1}{144\pi c^5}\int [\hat{\mathbf{r}} \times \dddot{\mathbf{D}}]\cdot[\hat{\mathbf{r}} \times \dddot{\mathbf{D}}]\,d\Omega = \frac{1}{144\pi c^5}\int \hat{\mathbf{r}}\cdot[\dddot{\mathbf{D}} \times (\hat{\mathbf{r}} \times \dddot{\mathbf{D}})]\,d\Omega$$

$$= \frac{1}{144\pi c^5}\int \hat{\mathbf{r}}\cdot[\hat{\mathbf{r}}(\dddot{\mathbf{D}}\cdot\dddot{\mathbf{D}}) - (\dddot{\mathbf{D}}\cdot\hat{\mathbf{r}})\dddot{\mathbf{D}}]\,d\Omega = \frac{1}{144\pi c^5}\int [(\dddot{\mathbf{D}})^2 - (\dddot{\mathbf{D}}\cdot\hat{\mathbf{r}})^2]\,d\Omega$$

$$= \frac{1}{144\pi c^5}\int [\dddot{Q}^{kl} n^l \dddot{Q}^{kr} n^r - \dddot{Q}^{kl} n^l n^k \dddot{Q}^{rs} n^s n^r]\,d\Omega \tag{78}$$

The Q's appearing in (78) are constants as far as the angular integration is concerned. Hence what we need are the values of the integrals $\int n^l n^r d\Omega$ and $\int n^k n^l n^r n^s\, d\Omega$. Obviously, the former is zero unless $l = r$ (for $l \neq r$, the integrand is odd under reversal of direction); hence

$$\int n^l n^r\, d\Omega = b\delta^{lr} \tag{79}$$

where b is some constant. To find this constant, we take $l = r$ and sum over l in Eq. (79):

$$\int n^l n^l \, d\Omega = b\delta^{ll} \tag{80}$$

But n^l is the component of the unit vector $\hat{\mathbf{r}}$, so $n^l n^l = 1$. Furthermore, $\delta^{ll} = 3$, so Eq. (80) reduces to

$$\int d\Omega = 3b \tag{81}$$

which implies $b = 4\pi/3$ and gives

$$\int n^l n^r \, d\Omega = \frac{4\pi}{3}\delta^{lr} \tag{82}$$

In a similar way, we obtain

$$\int n^k n^l n^r n^s \, d\Omega = \frac{4\pi}{15}(\delta^{kl}\delta^{rs} + \delta^{kr}\delta^{ls} + \delta^{ks}\delta^{lr}) \tag{83}$$

■ ***Exercise 8.*** Derive Eq. (83) [*Hint*: The indices must be equal in pairs for a nonzero result; furthermore, the right side must be symmetric under exchange of any of the indices. Hence

$$\int n^k n^l n^r n^s \, d\Omega = b(\delta^{kl}\delta^{rs} + \delta^{kr}\delta^{ls} + \delta^{ks}\delta^{lr}) \qquad] \tag{84}$$

■

Inserting Eqs. (82) and (83) into Eq. (78), we find that the result for the total power is

$$P = \frac{1}{180c^5}[\dddot{Q}^{kr}\dddot{Q}^{kr}] \tag{85}$$

■ ***Exercise 9.*** Derive this result. ■

A simple oscillating quadrupole radiator consists of two charges $+q$ at equal and opposite distances from an origin (see Fig. 14.7). This system has no electric dipole moment and no magnetic dipole moment. It does have an electric quadrupole moment and other higher-order multipole moments. We will ignore these higher moments. With the z axis as in Fig. 14.7, Eq. (75) gives

$$Q^{11} = Q^{22} = -2qz^2 \tag{86}$$

$$Q^{33} = 4qz^2 \tag{87}$$

Figure 14·7

A system of charges with an electric quadrupole moment.

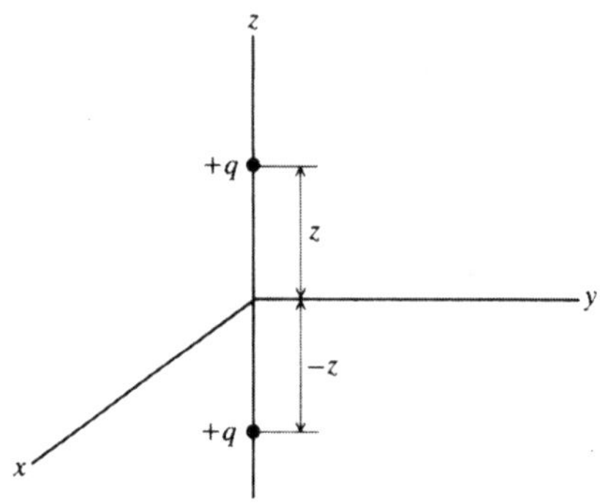

$$Q^{12} = Q^{13} = Q^{23} = 0 \tag{88}$$

If z oscillates according to

$$z = a \sin \omega t$$

then

$$\left[\frac{d^3 z^2}{dt^3}\right] = \frac{d^3}{dt^3} a^2 \sin^2 \omega(t - r/c) = -4\omega^3 a^2 \sin 2\omega(t - r/c) \tag{89}$$

■ ***Exercise 10.*** Check this equation. ■

Equation (89) implies that $\dddot{Q}$ oscillates with a frequency 2ω; hence the emitted quadrupole radiation will have a frequency 2ω. This is reasonable because after one-half period of the motion, the particles will have exchanged their positions, and the quadrupole moment will be the same as at the initial instant. In general, if the position of the charges oscillates harmonically with frequency ω, then the emitted dipole radiation will have frequency ω, the quadrupole radiation will have frequency 2ω, the octopole radiation will have frequency 3ω *and also* ω, and so on.

The vector $\dddot{\mathbf{D}}$ is

$$(\dddot{D}^1, \dddot{D}^2, \dddot{D}^3) = (\dddot{Q}^{1l} n^l, \dddot{Q}^{2l} n^l, \dddot{Q}^{3l} n^l) = (\dddot{Q}^{11} n^1, \dddot{Q}^{22} n^2, \dddot{Q}^{33} n^3)$$

$$= -2q \frac{d^3 z^2}{dt^3}(n^1, n^2, -2n^3) \tag{90}$$

and the angular intensity distribution is

$$\frac{dP}{d\Omega} = \frac{1}{144\pi c^5}\left|\hat{\mathbf{r}} \times [\dddot{\mathbf{D}}]\right|^2 = \frac{1}{144\pi c^5}\left(\left|[\dddot{\mathbf{D}}]\right|^2 - \left|[\dddot{\mathbf{D}}]\cdot\hat{\mathbf{r}}\right|^2\right)$$

$$= \frac{1}{144\pi c^5}\left(2q\left[\frac{d^3z^2}{dt^3}\right]\right)^2 \left\{(n^1)^2 + (n^2)^2 + 4(n^3)^2 - [(n^1)^2 + (n^2)^2 - 2(n^3)^2]^2\right\} \tag{91}$$

Introducing polar coordinates, with

$$n^1 = \sin\theta\cos\phi \tag{92}$$

$$n^2 = \sin\theta\sin\phi \tag{93}$$

$$n^3 = \cos\theta \tag{94}$$

we readily find that Eq. (91) reduces to

$$\frac{dP}{d\Omega} = \frac{9}{144\pi c^5}[8q\omega^3 a^2 \sin^2 2\omega(t - r/c)]^2 \sin^2\theta\cos^2\theta \tag{95}$$

A plot of the function $\sin^2\theta\cos^2\theta$ is shown in Fig. 14.8. The intensity is zero along the z axis, and it is also zero everywhere in the x-y plane.

Figure 14·8

Polar plot of $dP/d\Omega$ for an electric quadrupole.

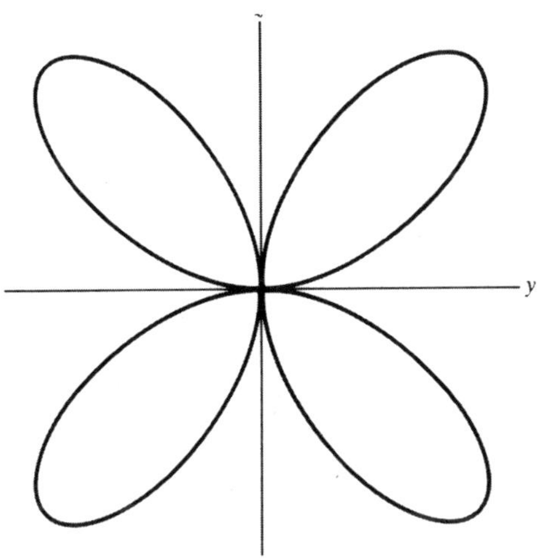

The total radiated power is

$$P = \frac{1}{180c^5}[\dddot{Q}^{kl}\dddot{Q}^{kl}] = \frac{6}{180c^5}[8q\omega^3 a^2 \sin^2 2\omega(t - r/c)]^2 \qquad (96)$$

■ ***Exercise 11.*** Supply the steps leading from Eq. (91) to (95) and then to (96). ■

To obtain the time-average power, we must replace $\sin^2 2\omega(t - r/c)$ by . Note that Eq. (96) varies as ω^6, so the power increases sharply with frequency.

EXAMPLE: *quadrupole radiation by atoms and nuclei*

We can estimate the transition rate for quadrupole emission of atomic and nuclear systems by taking, in Eq. (96), $q = e$ and $a \simeq$ (size of system); then

$$\frac{1}{\tau} \simeq \frac{1}{\hbar\omega}\frac{1}{c^5}\omega^6(ea^2)^2 \qquad (97)$$

This is smaller than the electric dipole transition rate by a factor

$$\simeq 3\frac{\omega^2 a^2}{c^2} \simeq 120\frac{a^2}{\lambda^2} \qquad (98)$$

For an atom, $a \simeq 1$ nm and $\lambda \simeq 500$ nm, so this factor is of the order of about 10^{-4}, which makes electric quadrupole transitions rather unlikely; but for a nucleus, such transitions are almost as likely as magnetic dipole transitions.

Problems

1. If the current on a half-wave antenna is as given by Eq. (15), what is the electric charge per unit length on the antenna? If we were to truncate the antenna, so its length was $L < \lambda/2$, would Eq. (15) still be an acceptable description of the current in the interval $-L/2 \le z \le L/2$?

2. Equation (24) is superficially analogous to the equation for the power dissipated in a resistive AC circuit with given resistance and a given current $I_{rms} = I_0/\sqrt{2}$. On the basis of this analogy, what is the effective resistance of the antenna? Convert your answer to ohms.

3. A full-wave antenna consists of two aligned half-wave antennas placed end to end. The currents in the two antennas are in phase, that is, the currents are always in the same direction. The current is represented mathematically by $I = I_0 \sin(2\pi|z|/\lambda) \times\cos\omega t$, with $-\lambda/2 \le z \le \lambda/2$. Show that the time-average power radiated per unit solid angle is

$$\frac{d\bar{P}}{d\Omega} = \frac{2I_0^2}{\pi c}\frac{\cos^4(\tfrac{1}{2}\pi\cos\theta)}{\sin^2\theta}$$

Integrate this numerically over all solid angles, and show that $\bar{P} = 3.35 I_0^2/c$.

4. Consider an out-of-phase, full-wave antenna consisting of two half-wave antennas placed end to end, with their currents in *opposite* directions. The current in this antenna is represented mathematically by $I = I_0 \sin(2\pi z/\lambda) \cos\omega t$. Find the time-average power radiated per unit solid angle, and compare it with the result of the preceding problem.

5. An antenna array consists of two vertical half-wave antennas separated by a horizontal distance of one-half wavelength. Calculate the time-average power radiated per unit solid angle if the two antennas oscillate in phase. In what direction is the power per unit solid angle maximum?

6. Repeat the preceding problem if the two antennas oscillate 180° out of phase.

7. An antenna array consists of two vertically aligned half-wave antennas separated by a vertical center-to-center distance of one wavelength. The antennas oscillate in phase. Find the time-average power per unit solid angle radiated by this antenna array, and prepare a polar plot similar to Fig. 14.3.

8. A half-wave antenna is placed parallel to a very large, conducting sheet, at a distance $\lambda/4$ from the sheet. Assume that the conducting sheet fills the z-x plane and that the antenna is oriented parallel to the z axis, with its center on the y axis. Calculate time-average power radiated per unit solid angle.

9. Assume that an atom emits light of wavelength 550 nm. Estimate the transition rate for each of the following kinds of radiation:
   (a) Electric dipole radiation.
   (b) Magnetic dipole radiation.
   (c) Electric quadrupole radiation.

10. Assume that a nucleus with a size of approximately 7×10^{-13} cm emits a 1-MeV gamma ray. Estimate the transition rate for each of the following kinds of radiation:
   (a) Electric dipole radiation.
   (b) Magnetic dipole radiation, due to the magnetic moment of the proton, $2.79e\hbar/(2m_p c)$.
   (c) Electric quadrupole radiation.

11. The Earth has a magnetic dipole moment of 8.1×10^{25} erg/gauss, and the direction of this dipole makes an angle of 11° with the axis of rotation. What is the power radiated by this dipole?

12. A neutron star has a magnetic dipole moment of magnitude m. The neutron star rotates at angular velocity ω about an axis inclined at 45° with respect to the magnetic axis.
   (a) What is the power radiated by the magnetic dipole? Consider a neutron star of radius 10 km rotating at $\omega = 100$ radians/s with a magnetic field of 10^{12} gauss at its (magnetic) equator. Give a numerical value for the emitted power.
   (b) Taking the z axis as the axis of rotation, sketch the radiation pattern (polar plot) in the y-z plane at the instant when the dipole moment is in this plane. Sketch the radiation pattern in the x-z plane at the same instant.
   (c) Describe the polarization of the radiation emitted along the z axis and along the x axis.

13. Figure 14.9 shows a circular loop of wire of radius R in the x-y plane. The loop carries an oscillating current $I = I_0 \sin \omega t$.

(a) At some arbitrary point with polar coordinates r, θ (where $r \gg R$), find the instantaneous Poynting vector associated with magnetic dipole radiation; take retardation into account.

(b) In words or in a diagram, describe the direction of the **E** and **B** fields at a few typical points.

14. A sphere has mass M and charge Q uniformly distributed over its volume. The sphere spins quickly about its axis, with spin angular momentum S. The sphere is placed in a uniform magnetic field $\mathbf{B}_0$ and precesses about the magnetic field, with an angle θ between the directions of the magnetic field and the spin. What is the rate of radiation of electromagnetic energy?

15. Two electrons, initially at rest at large distances, are attracted by a (bare) nucleus of charge Ze, and they fall toward the nucleus from *opposite* directions. Calculate the energy radiated by this system in the form of quadrupole radiation as the electrons fall from $r = \infty$ to $r = r_0$. Assume that classical mechanics is applicable, and neglect the electron-electron repulsion.

16. A nonrelativistic proton makes a head-on collision with a fixed nucleus of charge Ze. The force acting on the proton is the Coulomb force. The velocity of approach is v_0 when at large distance; the proton approaches along a radius, is reflected, and returns to large distance. Find the total energy radiated in the form of *quadrupole radiation* during the collision, according to classical electrodynamics. Compare with the result obtained in Problem 13.10, which gives the dipole radiation for this collision.

17. System A consists of a particle of charge q whose position vector is given by $\mathbf{r} = \hat{\mathbf{x}} \cos \omega t + \hat{\mathbf{y}} \sin \omega t$. System B consists of two particles of charge q whose position vectors are, respectively, $\mathbf{r} = \hat{\mathbf{x}} \cos \omega t$ and $\mathbf{r} = \hat{\mathbf{y}} \sin \omega t$. Do both systems radiate at the same rate? Prove it if they do; find the (approximate) difference in the radiation rates if they do not.

18. A single particle of charge q moves along the z axis. Its position is given by $z = b \sin \omega t$. What is the power radiated at frequency ω? At frequency 2ω?

Figure 14·9

A circular loop of wire in the x-y plane.

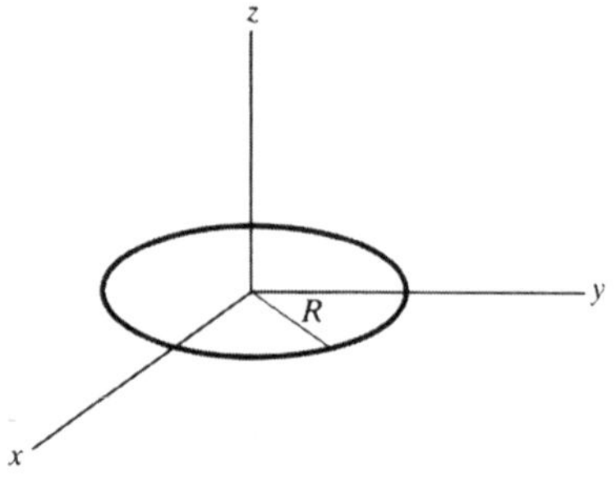

19. A rod of length L has charge Q uniformly distributed along its length. The rod rotates at angular velocity ω about an axis perpendicular to the rod, through its center. Does this rod radiate dipole radiation? Magnetic dipole radiation? Quadrupole radiation? What is the approximate rate of radiation of electromagnetic energy of each type? What is the frequency of the radiation?

20. Two identical particles of mass m and charge q are initially moving at equal speeds v_0 in opposite directions. Calculate the electromagnetic energy radiated by these particles during a head-on collision. Assume that the motion is nonrelativistic and that the only force on the particles is their mutual Coulomb force; assume that the initial and the final separations are very large. [*Hint:* This is a case of electric quadrupole radiation.]

21. An even number N of charged particles of charge q are uniformly spaced on the circumference of a circle that rotates at a frequency ω. Find the total power radiated (consider dipole, magnetic dipole, and quadrupole radiation). What happens to your result if you take the limit $N \to \infty$, $q \to 0$ such that Nq = constant?

22. For the transition from the $n = 2$, $l = 1$, $m = 0$ state to the $n = 1$ state of the hydrogen atom, the value of the integral in Eq. (48) is $\hat{\mathbf{z}}ea_0 \times 0.745$, where $a_0 = 0.0529$ nm is the Bohr radius. The transition frequency is $\omega = 1.549 \times 10^{16}$/s. What are the transition rate and the lifetime for this transition?

23. If a system of charges has a time-dependent electric dipole moment $\mathbf{p}$, a magnetic dipole moment $\mathbf{m}$, and an electric quadrupole moment Q^{kl}, then the magnetic radiation field will be

$$\mathbf{B}(t,\mathbf{x}) = \left[-\frac{1}{rc^2}\hat{\mathbf{r}}\times[\ddot{\mathbf{p}}] + \frac{1}{rc^2}\hat{\mathbf{r}}\times(\hat{\mathbf{r}}\times[\ddot{\mathbf{m}}]) - \frac{1}{6rc^3}\hat{\mathbf{r}}\times[\dddot{\mathbf{D}}] \right]$$

The energy flux in the direction $\hat{\mathbf{r}}$ is proportional to $|\mathbf{B}|^2$, and this contains interference terms among the different multipoles. Show that these interference terms do *not* contribute to the total power radiated by the system—that is, show that the integral over the energy flux radiated in all directions is simply the sum of the usual expressions,

$$\overline{P} = \overline{P}_{\text{electric dipole}} + \overline{P}_{\text{magnetic dipole}} + \overline{P}_{\text{electric quadrupole}}$$

[*Hint*: Remember that $\mathbf{D}$ contains a unit vector (see Eq. (74)). Express all unit vectors $\hat{\mathbf{r}}$ in terms of their components n^s, and express cross products in terms of ε^{klm}; then integrate over all directions using Eqs. (82) and (83).]

CHAPTER

Resonant Cavities and Waveguides 15

An organ pipe is an example of an acoustic resonant cavity. If we stimulate an organ pipe mechanically, the air column in its interior vibrates. If the frequency of stimulation coincides with one of the eigenfrequencies of the pipe, a standing acoustic wave of large amplitude is excited—the pipe resonates.

In this chapter, we will see that any closed container made of a conducting material can serve as an electromagnetic resonant cavity. Electromagnetic waves can be excited in such a cavity by means of a small antenna inserted into the interior. If the frequency coincides with one of the eigenfrequencies of the cavity, a standing electromagnetic wave of large amplitude is excited. If the antenna is removed, the electromagnetic fields continue to oscillate—in a cavity made of a perfect conductor, the oscillations would go on forever. In real cavities, the induced electric currents in the cavity walls are subject to ohmic losses, so the electromagnetic energy is gradually dissipated into heat; this means that electromagnetic oscillations are damped by the "friction" of the current in the walls.

We will also examine the propagation of electromagnetic waves in waveguides, that is, long tubes made of a conducting material. The mathematical treatment of waves in such waveguides is quite similar to that of waves in cavities, but instead of standing waves we have to deal with traveling waves.

15·1 Eigenvalue Problem for a Cavity

The problem of electromagnetic cavity oscillations is mathematically much more complicated than the corresponding acoustic problem. We must solve for two *vector* fields **E** and **B**; these fields are coupled by Maxwell's equations, and both must satisfy their own boundary conditions.

We will assume that the interior of the cavity is empty,* so the Maxwell equations (8.61)–(8.64) are

---

*It would be very easy to take into account the presence of a dielectric material of some given dielectric constant.

$$\nabla \cdot \mathbf{E} = 0 \tag{1}$$

$$\nabla \cdot \mathbf{B} = 0 \tag{2}$$

$$\nabla \times \mathbf{B} = \frac{1}{c} \frac{\partial \mathbf{E}}{\partial t} \tag{3}$$

$$\nabla \times \mathbf{E} = -\frac{1}{c} \frac{\partial \mathbf{B}}{\partial t} \tag{4}$$

In the usual way (see Section 12.1), we can obtain from (1)–(4) the following wave equations for **E** and **B**:

$$\nabla^2 \mathbf{E} - \frac{1}{c^2} \frac{\partial^2 \mathbf{E}}{\partial t^2} = 0 \tag{5}$$

$$\nabla^2 \mathbf{B} - \frac{1}{c^2} \frac{\partial^2 \mathbf{B}}{\partial t^2} = 0 \tag{6}$$

Let us look for solutions that oscillate harmonically in time. In complex notation, the fields have the form

$$\mathbf{E}(t, \mathbf{x}) = \mathbf{E}(\mathbf{x}) \mathrm{e}^{-i\omega t} \tag{7}$$

$$\mathbf{B}(t, \mathbf{x}) = \mathbf{B}(\mathbf{x}) \mathrm{e}^{-i\omega t} \tag{8}$$

where it is understood that the real part is to be taken as the physical field. Our wave equations for **E** and **B** then become

$$\nabla^2 \mathbf{E}(\mathbf{x}) = -\frac{\omega^2}{c^2} \mathbf{E}(\mathbf{x}) \tag{9}$$

$$\nabla^2 \mathbf{B}(\mathbf{x}) = -\frac{\omega^2}{c^2} \mathbf{B}(\mathbf{x}) \tag{10}$$

These equations are typical *eigenvalue* equations. Thus the x component of Eq. (9) is

$$\nabla^2 E_x = -\frac{\omega^2}{c^2} E_x$$

which says that E_x is an eigenvector of the operator ∇^2 with eigenvalue $-\omega^2 / c^2$.

Our problem is then to solve the eigenvalue equations (9) and (10) with some suitable boundary conditions. However, there is a complication: The eigenvalue equations (9), (10) do not carry all of the information that is contained in Maxwell's equations. In seeking a solution for the six equations (9) and (10), we must keep in mind that the components of **E** and **B** are not all independent of one another. In general, this makes for a rather messy situation, but we will see in the next section that in the special case of cylindrical cavities and waveguides, these complications can be bypassed.

The boundary conditions at the walls of the cavity were obtained in Sections 4.3 and 11.4: the component of **E** parallel to the surface and the component of **B** perpendicular to the surface must be continuous. For the sake of simplicity, we will assume that the walls are perfect conductors, so all fields are exactly zero inside the wall. Continuity then requires

$$\mathbf{E}_{\parallel} = 0 \tag{11}$$

$$\mathbf{B}_{\perp} = 0 \tag{12}$$

at the surface.

These are the *only* boundary conditions that we have to impose if the walls are *perfect conductors*. The usual condition on the continuity of $\mathbf{E}_{\perp}$ and $\mathbf{B}_{\parallel}$ does not apply in this case. To see this, consider the "pillbox" of Fig. 15.1. According to Gauss' law.

$$\int_S \mathbf{E} \cdot d\mathbf{S} = 4\pi \int \rho \, dV \tag{13}$$

The surface of the conductor can carry a surface charge density σ, so $\int \rho \, dV = \sigma A$. Furthermore, since $\mathbf{E} = 0$ inside the conductor, only the area just outside the conductor contributes to $\int \mathbf{E} \cdot d\mathbf{S}$; the value of this integral is then simply $E_{\perp} A$. Thus Eq. (13) implies

$$E_{\perp} = 4\pi\sigma \tag{14}$$

This is *not* a boundary condition on $E_{\perp}$ but, rather, a condition on σ: the surface charge density always adjusts itself in such a way that Eq. (14) is satisfied.

Figure 15·1

A closed Gaussian surface of base area A intersects a perfectly conducting wall. The charge enclosed in this surface is σA.

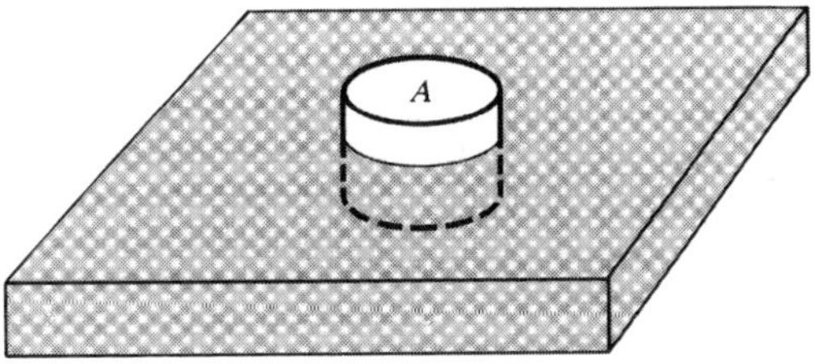

A similar argument, based on Fig. 15.2 and Ampère's law, can be used to show that

$$B_{\|} = \frac{4\pi}{c}\kappa \tag{15}$$

where κ is the surface current density, or the amount of current flowing along the conducting surface intercepted by a unit length lying in the surface parallel to $\mathbf{B}_{\|}$ (see Fig. 15.2). The direction of the vector $\boldsymbol{\kappa}$ is perpendicular to the direction of $\mathbf{B}_{\|}$, according to the usual right-hand rule.

■ ***Exercise 1.*** Derive Eq. (15). [*Hint:* The loop of Fig. 15.2 is oriented perpendicular to the surface current.] ■

We can use Eqs. (14) and (15) to calculate the charges and currents on the walls of the cavity *after* we have found the fields inside the cavity.

Although our boundary conditions (11) and (12) are based on the unrealistic assumption of infinite conductivity, they are good approximations for finite conductivity, provided the frequency of the electromagnetic waves is high. From Section 12.8 we know that waves of high frequency penetrate a distance of only about one skin depth δ into the conductor; and if the frequency is high, this skin depth is very short, and the wave penetrates very little. In contrast, if the frequency is low, the skin depth is large, and the wave penetrates deep into the conductor. This changes the character of the wave. For instance, in a waveguide, high-frequency waves are essentially radio waves that bounce along the inside of the waveguide on a zigzag path, and their speed is close to the speed of light. But low-frequency waves are more like current pulses flowing within the conducting walls of the waveguide, and their speed is considerably less than the speed of light. The propagation of such current pulses is governed by a different equation (the "telegraphic equation"), and we will not deal with this here.

Figure 15·2

The short sides of a closed path penetrate the conducting wall. One of the long sides, of length l, is outside the conductor; the other is inside. The path is perpendicular to the surface current and links an amount of current $l \times \kappa$.

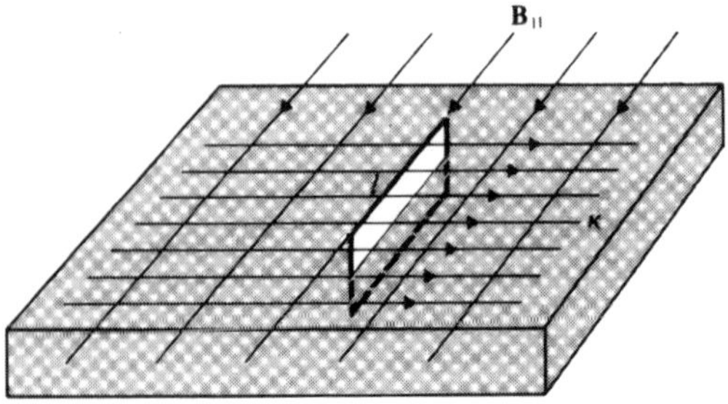

15·2 Cylindrical Cavities

Let us now consider the special case of cylindrical cavities. The cross section of the cylinder can be arbitrary (for example, a rectangular cross section or a circular cross section), but according to the general definition of *cylinder*, the cross section must remain the same everywhere along the length of the cylinder. We will take the z axis along the length of the cylinder (see Fig. 15.3).

Again, we assume that the time dependence of all fields is $e^{-i\omega t}$, because we are interested in oscillatory solutions. Furthermore, we assume that the z dependence of all fields is $e^{\pm ikz}$ (where k is a constant), so

$$\mathbf{E}(t,\mathbf{x}) = \mathbf{E}(x,y)e^{-i\omega t \pm ikz} \tag{16}$$

$$\mathbf{B}(t,\mathbf{x}) = \mathbf{B}(x,y)e^{-i\omega t \pm ikz} \tag{17}$$

This assumption about the z dependence involves no loss of generality because, by Fourier's theorem, any arbitrary function of z can be written as a superposition of the exponential functions.

The expressions (16) and (17) represent traveling waves. Of course, what we really want for our cavity oscillations are *standing waves*. Nevertheless, it is more convenient to perform the calculations with the simple exponential functions appearing in (16) and (17) and afterward construct the standing waves by superposition of two traveling waves.

With (16) and (17), the Maxwell equations (3) and (4) become

Figure 15·3

A cylinder of arbitrary cross section.

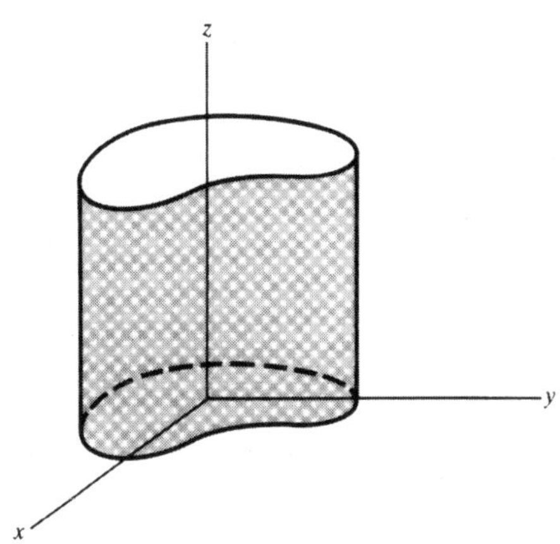

$$\frac{\partial B_z}{\partial y} \mp ikB_y = -\frac{i\omega}{c} E_x \tag{18}$$

$$\pm ikB_x - \frac{\partial B_z}{\partial x} = -\frac{i\omega}{c} E_y \tag{19}$$

$$\frac{\partial B_y}{\partial x} - \frac{\partial B_x}{\partial y} = -\frac{i\omega}{c} E_z \tag{20}$$

$$\frac{\partial E_z}{\partial y} \mp ikE_y = \frac{i\omega}{c} B_x \tag{21}$$

$$\pm ikE_x - \frac{\partial E_z}{\partial x} = \frac{i\omega}{c} B_y \tag{22}$$

$$\frac{\partial E_y}{\partial x} - \frac{\partial E_x}{\partial y} = \frac{i\omega}{c} B_z \tag{23}$$

■ ***Exercise 2.*** Derive Eqs. (18)–(23). ■

We can regard Eqs. (18), (19), (21), and (22) as four equations for the four unknowns E_x, E_y, B_x, and B_y, everything else in these equations being assumed known. We can then solve for these field components in terms of E_z and B_z. For instance,

$$E_x = \frac{1}{\omega^2/c^2 - k^2}\left(\pm ik\frac{\partial E_z}{\partial x} + \frac{i\omega}{c}\frac{\partial B_z}{\partial y}\right) \tag{24}$$

Since $\pm ikE_z = \partial E_z/\partial z$, we can also write Eq. (24) as

$$E_x = \frac{1}{\omega^2/c^2 - k^2}\left(\frac{\partial^2 E_z}{\partial x \partial z} + \frac{i\omega}{c}\frac{\partial B_z}{\partial y}\right) \tag{25}$$

The corresponding results for the other field components are

$$E_y = \frac{1}{\omega^2/c^2 - k^2}\left(\frac{\partial^2 E_z}{\partial y \partial z} - \frac{i\omega}{c}\frac{\partial B_z}{\partial x}\right) \tag{26}$$

$$B_x = \frac{1}{\omega^2/c^2 - k^2}\left(\frac{\partial^2 B_z}{\partial x \partial z} - \frac{i\omega}{c}\frac{\partial E_z}{\partial y}\right) \tag{27}$$

$$B_y = \frac{1}{\omega^2/c^2 - k^2}\left(\frac{\partial^2 B_z}{\partial y \partial z} + \frac{i\omega}{c}\frac{\partial E_z}{\partial x}\right) \tag{28}$$

■ ***Exercise 3.*** Derive Eqs. (26)–(28) from Eqs. (18), (19), (21), and (22). ■

Note that Eqs. (25)–(28) are independent of the sign of k and, hence, will hold equally well for e^{+ikz}, e^{-ikz}, and *any superposition* of these. Consequently these equations are also valid for standing waves with a z dependence cos kz or sin kz.

The importance of Eqs. (25)–(28) is that these equations show that if E_z and B_z are known, then all the other field components are uniquely determined. This means we can use the following general method for finding the cavity oscillations: Begin by solving the eigenvalue equations

$$\nabla^2 E_z = -\frac{\omega^2}{c^2} E_z \tag{29}$$

$$\nabla^2 B_z = -\frac{\omega^2}{c^2} B_z \tag{30}$$

for E_z and B_z, using the boundary conditions (11) and (12). Then calculate E_x, E_y, B_x, and B_y from E_z and B_z by means of Eqs. (25)–(28).

Before we look at some examples, we note that there are two obvious possibilities. It could happen that $E_z = 0$, but $B_z \neq 0$; then, we need to worry only about Eq. (30). Such a mode of oscillation is called *transverse electric* (TE) because the electric field has no longitudinal component. Alternatively, we could have $B_z = 0$ and $E_z \neq 0$; then, we need only Eq. (29). This mode is called *transverse magnetic* (TM). It will obviously be convenient to treat these two modes separately; the general mode of oscillation can afterward be found as a superposition of TE and TM modes. We summarize the characteristics of these modes as follows:

Mode	E_z	B_z	**Eigenvalue Equation**	**Boundary Condition**
TE	0	≠0	$\nabla^2 B_z = -\frac{\omega^2}{c^2} B_z$	$\mathbf{B}_\perp = 0$ at surface
TM	≠0	0	$\nabla^2 E_z = -\frac{\omega^2}{c^2} E_z$	$\mathbf{E}_\parallel = 0$ at surface

15·3 Rectangular Cavities

Figure 15.4 shows a rectangular cavity of dimensions $a \times b \times d$, with $a < b < d$. We will look for TM modes of this cavity. According to the preceding discussion, this involves solving the wave equation (29) for E_z. It is easy to check that

Figure 15·4

A rectangular cavity of dimensions $a \times b \times d$, with $a < b < d$.

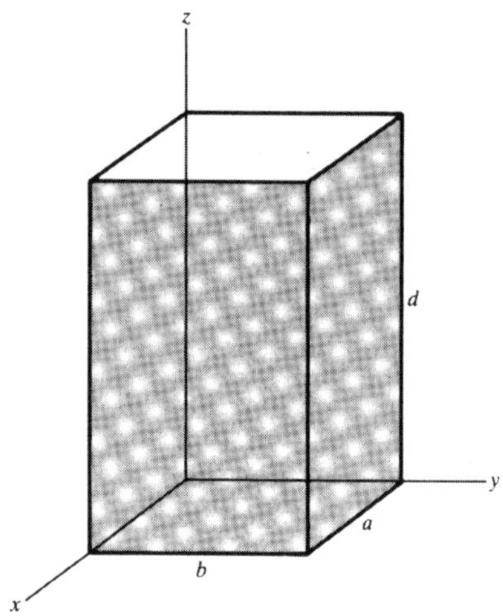

$$E_z(x,y,z) = E_0 \sin\left(\frac{m\pi x}{a}\right)\sin\left(\frac{n\pi y}{b}\right)\begin{Bmatrix}\sin kz \\ \text{or} \\ \cos kz\end{Bmatrix} \tag{31}$$

is a solution of Eq. (29) with an eigenvalue

$$-\frac{\omega^2}{c^2} = -\left(\frac{m\pi}{a}\right)^2 - \left(\frac{n\pi}{b}\right)^2 - k^2 \tag{32}$$

■ ***Exercise 4.*** Check that (31) is a solution of the wave equation, and prove that if *m, n* are integers, then (31) satisfies the boundary condition $E_{\|} = 0$ along the sides of the cavity (by *sides* we mean the four planes $x = 0$, $x = a$, $y = 0$, and $y = b$). ■

EXAMPLE: *electric and magnetic fields for TM mode of rectangular cavity*

At the top ($z = d$) and bottom ($z = 0$) of the cavity, the boundary condition $E_{\|} = 0$ requires that both $E_x = 0$ and $E_y = 0$. If we use cos kz in Eq. (31) and calculate E_x and E_y according to Eqs. (25) and (26), we find

$$E_x(x,y,z) = \frac{E_0}{\omega^2/c^2 - k^2}\left(-k\frac{m\pi}{a}\right)\cos\left(\frac{m\pi x}{a}\right)\sin\left(\frac{n\pi y}{b}\right)\sin kz \tag{33}$$

$$E_y(x,y,z) = \frac{E_0}{\omega^2/c^2 - k^2}\left(-k\frac{n\pi}{b}\right)\sin\left(\frac{m\pi x}{a}\right)\cos\left(\frac{n\pi y}{b}\right)\sin kz \tag{34}$$

It is clear that the boundary condition at $z = d$ demands that

$$k = \frac{l\pi}{d} \tag{35}$$

with l an integer. It is also clear that if we use sin kz in Eq. (31), then we cannot meet the boundary condition at $z = 0$.

The eigenfunction E_z is therefore

$$E_z(x, y, z) = E_0 \sin\left(\frac{m\pi x}{a}\right)\sin\left(\frac{n\pi y}{b}\right)\cos\left(\frac{l\pi z}{d}\right) \tag{36}$$

with an eigenfrequency

$$\frac{\omega^2}{c^2} = \left(\frac{m\pi}{a}\right)^2 + \left(\frac{n\pi}{b}\right)^2 + \left(\frac{l\pi}{d}\right)^2 \tag{37}$$

To obtain the time dependence of the field, we must multiply by $e^{-i\omega t}$ and take the real part. Assuming that E_0 is real, this results in

$$E_z(t, x, y, z) = E_0 \sin\left(\frac{m\pi x}{a}\right)\sin\left(\frac{n\pi y}{b}\right)\cos\left(\frac{l\pi z}{d}\right)\cos\omega t \tag{38}$$

The time-dependent E_x and E_y fields can be obtained likewise from Eqs. (33) and (34); the time dependence is again $\cos\omega t$.

■ ***Exercise 5.*** Show that for this TM mode, Eqs. (27) and (28) yield

$$B_x(x, y, z) = \frac{E_0}{\omega^2/c^2 - k^2}\left(-\frac{i\omega}{c}\right)\frac{n\pi}{b}\sin\left(\frac{m\pi x}{a}\right)\cos\left(\frac{n\pi y}{b}\right)\cos\left(\frac{l\pi z}{d}\right) \tag{39}$$

$$B_y(x, y, z) = \frac{E_0}{\omega^2/c^2 - k^2}\frac{i\omega}{c}\frac{m\pi}{a}\cos\left(\frac{m\pi x}{a}\right)\sin\left(\frac{n\pi y}{b}\right)\cos\left(\frac{l\pi z}{d}\right) \tag{40}$$

■

■ ***Exercise 6.*** Show that the time dependence of the real part of the B_x and B_y fields is sin ωt. Hence the electric and magnetic fields are a quarter of a cycle out of phase; **E** reaches its maximum value when **B** is zero, and conversely. ■

The lowest values for l, n, m that give a nontrivial field E_z are $l = 0$, $n = 1$, $m = 1$. The corresponding mode of oscillation is designated TM_{011} (the indices are listed in the same order as that of the *decreasing* dimensions of the cavity; l is listed first because the d dimension was assumed to be the largest, and so on). The field lines of **E** and **B** for this mode are displayed in Fig. 15.5. It must be kept in mind that the

Figure 15·5

Pattern of field lines for the TM_{011} mode.

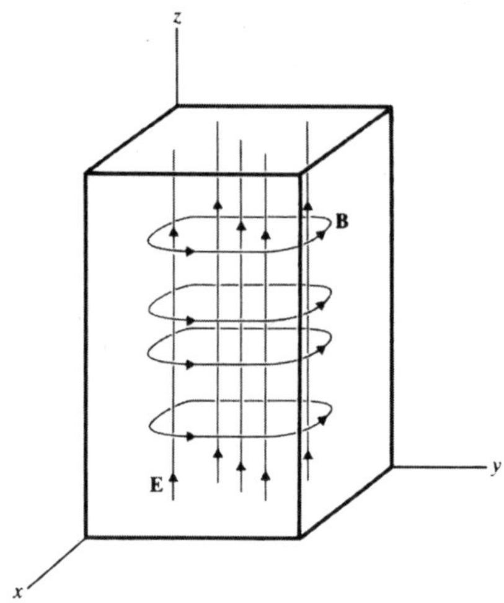

maximum electric and magnetic fields (pictured in Fig. 15.5) occur not simultaneously but with a time lag of a quarter of a cycle (see Exercise 6). The eigenfrequency of this mode is

$$\omega_{011} = \pi c \sqrt{\frac{1}{a^2} + \frac{1}{b^2}} \tag{41}$$

Figure 15.6 displays the field lines for the TM_{111} mode; this mode has an eigenfrequency

$$\omega_{111} = \pi c \sqrt{\frac{1}{a^2} + \frac{1}{b^2} + \frac{1}{d^2}} \tag{42}$$

Note that if, in Fig. 15.5, we had placed the z axis where the x axis is, then the mode shown would have been a TE mode. It is true in general that any mode that has the **E** and **B** fields entirely perpendicular to each other can be regarded as TM or TE, depending on the choice of axes.

■ ***Exercise 7.*** Is it possible to transform the TM_{111} mode (shown in Fig. 15.6) into a TE mode by a choice of coordinate axes? ■

Figure 15·6

Pattern of field lines for the TM_{111} mode.

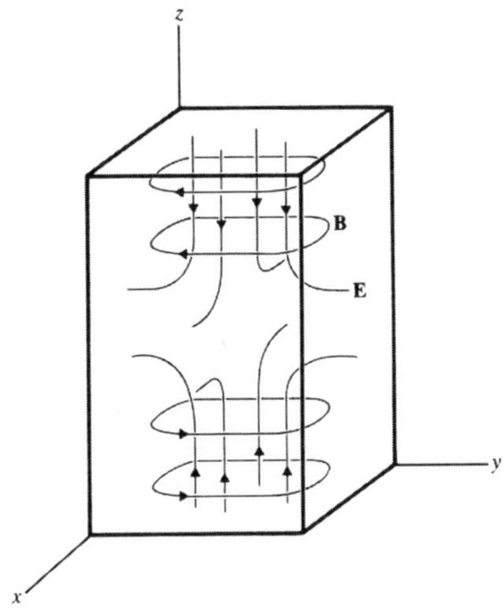

15·4 Energy Dissipation in a Cavity

We now take a brief look at energy dissipation in resonant cavities. The solution we obtained for the perfectly conducting cavity does not include any damping—according to Eq. (38), the field oscillates harmonically with no decrease in amplitude. Real cavities, made of materials of finite conductivity, will gradually lose electromagnetic energy because the walls absorb some of this energy. To get a *rough* estimate of the amount of electromagnetic energy lost per cycle, we can argue as follows: In the interior of the cavity, the energy density is

$$T^{00} = \frac{1}{8\pi}(E^2 + B^2) \tag{43}$$

and the total energy is the volume integral of Eq. (43). For our rough estimate, we need not bother to use the exact energy density. We simply take some typical (average) value of T^{00} and calculate the total energy by multiplying this T^{00} by the volume of the cavity:

$$(\text{energy in cavity}) \simeq T^{00} V \tag{44}$$

If the walls have finite conductivity, then the fields will penetrate the walls for a distance of about one skin depth δ [see Eq. (12.174)]. Within this thickness δ, the fields, and hence the energy density, are comparable with the fields inside the cavity; deeper in the metal, the fields are quite small. The electromagnetic energy in the skin-depth layer is then approximately

$$T^{00} \times (\text{volume of skin}) \simeq T^{00} \times A\delta \tag{45}$$

where A is the interior area of the cavity.

The quantity (45) is roughly the energy absorbed by the walls per cycle. This is easy to understand if we recall that the wavelength of the radiation in the metal is approximately δ;* hence, in one period, all the energy originally in the skin-depth layer moves out of this layer and deeper into the metal, where it is absorbed.

It will be convenient to express the loss rate in terms of the fraction of the energy lost per cycle. Taking the ratio of Eqs. (44) and (45) we find

$$\frac{(\text{energy})}{(\text{energy lost per cycle})} \simeq \frac{V}{A\delta} \tag{46}$$

The quantity

$$Q \equiv \frac{(\text{energy})}{(\text{energy lost per cycle})} \tag{47}$$

is called the Q of the cavity.† Equation (46) shows that the Q is roughly equal to the ratio of cavity volume to "skin volume." A more precise calculation modifies Eq. (46) by an extra dimensionless factor that takes into account the shape of the cavity.

EXAMPLE: *Q of a resonant cavity*

The Q of the cavity tells us how the energy will gradually decrease if no extra energy is fed into the cavity to make good the losses. We can rewrite Eq. (47) as

$$\frac{dU}{dt} = -\frac{\omega}{2\pi Q} U \tag{48}$$

where U stands for the total energy in the cavity. Integration of Eq. (48) gives us

$$U = U_0 e^{-\omega t/2\pi Q} \tag{49}$$

which shows that the energy disappears exponentially with a damping time $2\pi Q/\omega$. Qualitatively, we can say that after a time comparable with, or slightly larger than, $2\pi Q/\omega$, the free oscillations of the cavity will have pretty much disappeared. Note that the damping time, or "ring-

---

*$2\pi\delta$, to be exact. See Eq. (12.172).

†*Warning:* Q is sometimes defined in terms of energy lost per *radian* of the oscillation, which differs from our definition (47) by 2π.

down" time, of the cavity is simply $Q \times$ (period of cavity). For a good cavity (made of a good conductor), Q can be as much as 10^4, so the cavity oscillates on its own for 10^4 or so cycles whenever it is started. Superconducting cavities with Q as large as 10^8 have been built for use in accelerators.

15·5 Waveguides

A *waveguide* is a pipe used to carry radio-frequency electromagnetic waves. One of the simplest waveguides is the *coaxial cable*, which consists of a cylindrical outer shell with a concentric, solid cylinder (or a wire) in its interior (see Fig. 15.7). The region between the cylinders is usually filled with dielectric, but we will assume it is empty; we will also assume that the material of the walls is a perfect conductor. In general, the modes of wave propagation in a cylindrical geometry are quite complicated. However, there exists one very simple mode of propagation for the coaxial cable: the TEM mode. In this mode, both the electric field and magnetic field are purely transverse ($E_z = B_z = 0$). It then follows from Eqs. (18), (19), (21) and (22) that

$$k = \frac{\omega}{c} \tag{50}$$

and that

$$B_y = \pm E_x \qquad B_x = \mp E_y \tag{51}$$

■ ***Exercise 8.*** Verify Eqs. (50) and (51). ■

Figure 15·7

Coaxial cable consisting of a wire surrounded by a cylindrical shell.

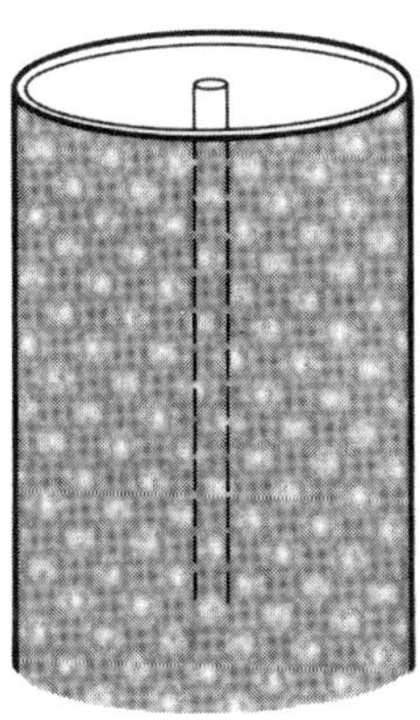

The upper sign in Eq. (51) corresponds to propagation in the $+z$ direction; the lower sign, to propagation in the $-z$ direction. This can be seen from the z and t dependence of the fields, which is $e^{-i\omega t \pm ikz}$, respectively [see Eqs. (16)–(17)].

Equation (50) shows that the relation between k and ω for a TEM wave is exactly the same as for a free electromagnetic wave. Equation (51) shows that **B** is perpendicular to **E** and of the same magnitude—again, just as for a free wave. It then remains only to find out how E_x and E_y vary with x and y. For this purpose, we use Eqs. (1) and (23):

$$\frac{\partial E_x}{\partial x} + \frac{\partial E_y}{\partial y} = 0 \tag{52}$$

$$\frac{\partial E_y}{\partial x} - \frac{\partial E_x}{\partial y} = 0 \tag{53}$$

These Equations for E_x and E_y are exactly the familiar equations for an *electrostatic* field in two dimensions. The x-y dependence of the electric field is therefore exactly the same as that of the fields in a cylindrical capacitor. This electric field can be conveniently expressed in cylindrical coordinates:

$$E_\rho \propto \frac{1}{\rho}$$

According to Eq. (51), the magnetic field is then

$$B_\phi \propto \pm\frac{1}{\rho}$$

Including the z and t dependence of these fields and taking the real part, we find

$$E_\rho(t,\rho,\phi,z) = E_0\frac{a}{\rho}\cos(\omega t \pm kz)$$

$$B_\phi(t,\rho,\phi,z) = \pm E_0\frac{a}{\rho}\cos(\omega t \pm kz) \tag{54}$$

where E_0 is the amplitude of the electric field at the radius a, say, the radius of the inner conductor.

Our conclusions on the behavior of TEM fields can be generalized to other than circular cylindrical geometries. It is obvious that Eqs. (50)–(53) do not depend on the assumption of rotational symmetry.

For the transmission of large amounts of power at high frequencies, hollow waveguides are best. In a coaxial cable, there is too much loss in the central conductor, which has to carry large current densities, and in the dielectric needed to support the central conductor in its place. Hollow waveguides are usually rectangular in cross section and are often silver-plated on the interior to increase the conductivity.

■ ***Exercise 9.*** Prove that it is not possible to have a TEM mode in a hollow rectangular waveguide or in any kind of hollow waveguide. [*Hint:* Does the boundary condition on **E** permit a nontrivial solution of Eqs. (52) and (53)?] ■

EXAMPLE: *TE mode in rectangular waveguide*

Both TE and TM modes can, of course, exist in a rectangular waveguide. Figure 15.8 shows a rectangular waveguide with a cross section $a \times b$. Let us consider a TE mode of this waveguide. We must then solve the wave equation

$$\nabla^2 B_z = -\frac{\omega^2}{c^2} B_z \tag{55}$$

For a wave traveling in the $+z$ direction, we need

$$B_z = B_z(x, y)e^{-i\omega t + ikz} \tag{56}$$

so

$$\left(\frac{\partial^2}{\partial x^2} + \frac{\partial^2}{\partial y^2}\right) B_z(x, y) = \left(-\frac{\omega^2}{c^2} + k^2\right) B_z(x, y) \tag{57}$$

Figure 15·8

Rectangular waveguide of cross section $a \times b$.

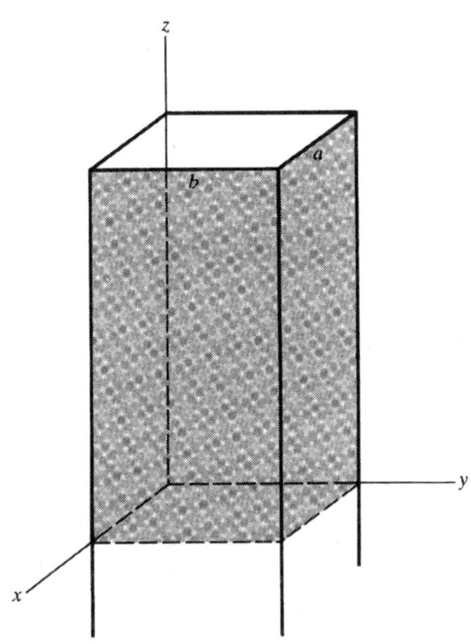

This is again an eigenvalue problem: B_z is an eigenfunction of $\partial^2/\partial x^2 + \partial^2/\partial y^2$ with eigenvalue $-\omega^2/c^2 + k^2$. Note that, whereas in the case of a resonant cavity, $-\omega^2/c^2$ was the eigenvalue, now $-\omega^2/c^2$ is an adjustable parameter whose value is fixed by the frequency of the generator that is feeding the power into the waveguide; obviously, we can run this generator at whatever frequency we choose.* Hence the eigenvalue obtained from Eq. (57) determines the value of k.

The boundary condition on **B** is $B_\perp = 0$ at the walls. The solution of Eq. (57) satisfying this boundary condition is

$$B_z(x,y) = B_0 \cos\left(\frac{m\pi x}{a}\right)\cos\left(\frac{n\pi y}{b}\right) \tag{58}$$

where m, n are integers (at least one of m, n must be nonzero).

■ ***Exercise 10.*** Check that (58) is a solution of the wave equation and that B_x and B_y satisfy the boundary condition for $B_\perp$. ■

The eigenvalue associated with Eq. (58) is $-(m\pi/a)^2 - (n\pi/b)^2$. Thus

$$-\frac{\omega^2}{c^2} + k^2 = -\left(\frac{m\pi}{a}\right)^2 - \left(\frac{n\pi}{b}\right)^2 \tag{59}$$

which we can solve for k:

$$k = \sqrt{\frac{\omega^2}{c^2} - \left(\frac{m\pi}{a}\right)^2 - \left(\frac{n\pi}{b}\right)^2} \tag{60}$$

This expression shows that if ω is too small, then k becomes imaginary. We then do not get propagation but, rather, exponential attenuation of the fields. The critical *cutoff frequency,* below which the waveguide ceases to transmit, is given by

$$\omega_{mn} = \pi c\sqrt{\left(\frac{m}{a}\right)^2 + \left(\frac{n}{b}\right)^2} \tag{61}$$

This is the cutoff frequency for the TE_{mn} mode. Although propagation in this mode is impossible if ω is below ω_{mn}, it may still be possible to send a wave through the guide if another mode, of lower cutoff frequency, is available. However, if the frequency ω is smaller than *all* the cutoff frequencies of the waveguide, then all transmission comes to an end. For

*However, it turns out that if the frequency is too low, then there will be no wave propagation (see below).

example, if $a > b$, then the lowest cutoff frequency is that of the TE_{10} mode,

$$\omega_{10} = \frac{\pi c}{a} \tag{62}$$

and this is the minimum frequency required for transmission of any sort.

■ ***Exercise 11.*** Calculate all the electric and magnetic field components for a wave in the TE_{mn} mode. Include the z and t dependence. ■

The field lines for the TE_{10} mode are shown in Fig. 15.9. In contrast to the cavity fields, **E** and **B** are now *in phase*, and Fig. 15.9 shows them at one instant of time. The whole pattern of field lines moves along the waveguide with the phase velocity of the wave.

This brings us to the question of wave velocity of our guided waves. The phase velocity is larger than c,

$$u_{\mathrm{p}} = \frac{\omega}{k} = c\frac{\omega}{\sqrt{\omega^2 - (m\pi c/a)^2 - (n\pi c/b)^2}} > c \tag{63}$$

However, the group velocity is smaller than c,

Figure 15·9

Pattern of field lines for the TE_{10} mode of a rectangular waveguide.

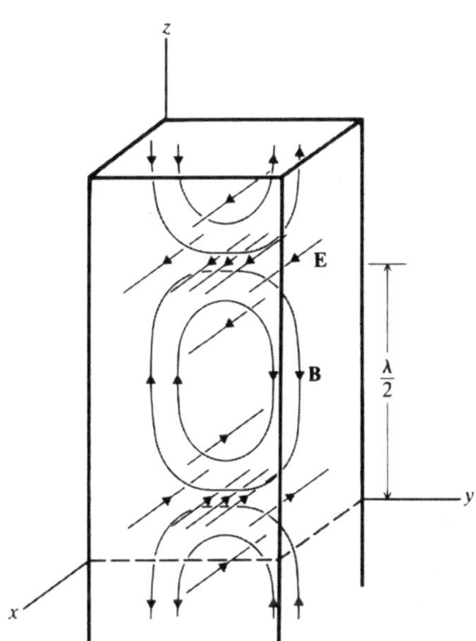

$$u_g = \frac{1}{dk/d\omega} = c\frac{\sqrt{\omega^2 - (m\pi c/a)^2 - (n\pi c/b)^2}}{\omega} < c \qquad (64)$$

The group velocity is the velocity at which signals—and energy—move along the waveguide.

To gain a better feeling for how the difference between the velocities (63) and (64) comes about, consider a free plane wave approaching a mirror at an angle (see Fig. 15.10). The intersection point of a wave crest with the mirror moves along the mirror with velocity $c/\sin\theta$; however, energy moves parallel to the mirror with a velocity $c\sin\theta$. The phase and group velocities along the mirror are therefore

$$u_p = \frac{c}{\sin\theta} \qquad (65)$$

$$u_g = c\sin\theta \qquad (66)$$

The reason why this is relevant to propagation in a waveguide is that the guided wave can be regarded as a plane wave that is reflected back and forth between the walls of the guide. Figure 15.11 shows the wavefronts of a TE_{10} wave reflected back and forth in this manner (the electric field vector is perpendicular to the plane of the figure). The wave fronts moving toward the upper right are assumed to be 180° out of phase with the wave fronts moving toward the lower right with which they intersect. Hence the electric field cancels at the walls, and the boundary condition is satisfied.

EXAMPLE: *phase and group velocities, geometrical argument*

We can now use a simple geometrical argument based on Fig. 15.11 to establish that Eqs. (63) and (64) are equivalent to Eqs. (65) and (66). From this figure, we find that the wavelength of the guided wave (the repeat distance measured *along* the waveguide) can be expressed as

Figure 15·10

Wave crests approaching a mirror.

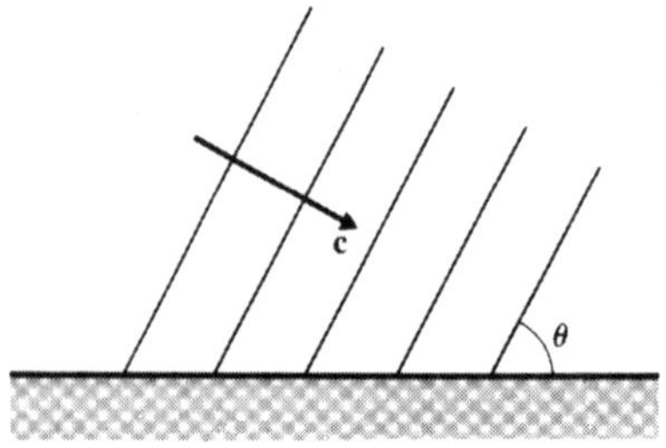

Figure 15·11

Wave fronts of a wave reflected back and forth between the parallel walls of a waveguide.

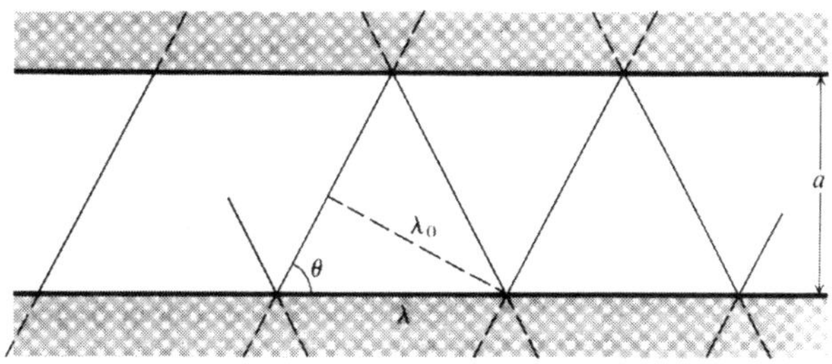

$$\lambda = \frac{2a}{\tan\theta} \tag{67}$$

and that the wavelength of the plane wave (the repeat distance measured perpendicularly to the wave fronts) can be expressed as

$$\lambda_0 = 2a\cos\theta \tag{68}$$

■ ***Exercise 12.*** Check this. ■

Since the wavelength of the plane wave equals $2\pi c/\omega$, Eq. (68) gives us

$$\cos\theta = \frac{\pi c}{\omega a} \tag{69}$$

Hence the phase and group velocities become

$$u_{\mathrm{p}} = \frac{c}{\sqrt{1-\cos^2\theta}} = \frac{c}{\sqrt{1-(\pi c/\omega a)^2}} \tag{70}$$

$$u_{\mathrm{g}} = c\sqrt{1-\cos^2\theta} = c\sqrt{1-(\pi c/\omega a)^2} \tag{71}$$

in agreement with Eqs. (63) and (64).

15·6 Dielectric Waveguides

A *dielectric waveguide* consists of a rod of dielectric material used for the transmission of electromagnetic waves. Thin rods of this kind, called optic fibers, are used to transmit modulated light signals in modern telephone systems. Essentially, the light wave is confined within the dielectric by total internal reflection—the wave is confined without the need of any conducting walls. Very thin, optic fibers are much easier to manufacture than hollow, metallic pipes, but as we mentioned above, the dielectric tends to absorb the wave to some extent.

The mathematical treatment of the propagation of an electromagnetic wave is rather more difficult for a dielectric waveguide than for a hollow waveguide. We know from Section 12.6 that the internally reflected wave penetrates for some distance beyond the surface of the dielectric—the wave decays exponentially with distance from the dielectric. Thus it is necessary to solve Maxwell's equations both inside and outside the dielectric waveguide and to join these solutions by means of the appropriate boundary conditions. This is a fairly complicated mathematical problem. Furthermore, the dielectric employed in optic fibers often has a radial dependence of the index of refraction, which leads to extra complications.

Here, we will deal only with a very simple case of a dielectric waveguide: a homogeneous dielectric slab of uniform thickness, such as a very large plate of glass. This simple case exhibits most of the typical features of wave propagation in a dielectric waveguide but is fairly easy to handle mathematically. Figure 15.12 shows the slab; its thickness is b, and its midplane coincides with the x-z plane. We can regard this slab as a rectangular waveguide of infinite width in the x direction. For a wave propagating in the z direction in this slab, we can then assume there is no x dependence. The partial differential equation for the x-y dependence of the wave [the eigenvalue equation; see Eq. (76)] therefore reduces to an ordinary differential equation involving only the y dependence; this equation is easily solved.

Figure 15·12

Very large dielectric slab of width b. The slab is parallel to the x-z plane.

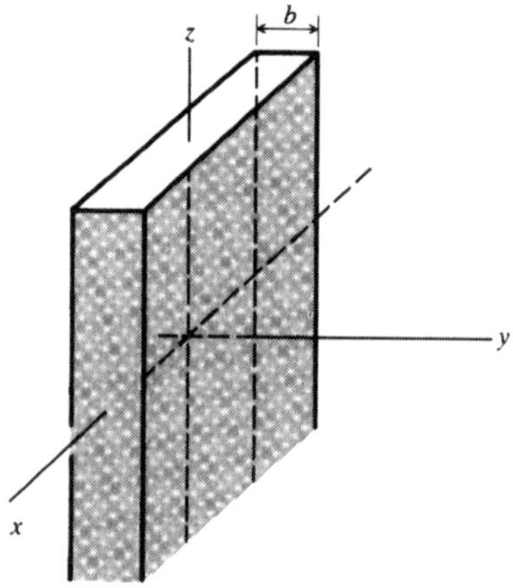

In a dielectric medium we must modify Eqs. (25)–(30) by including suitable factors of ε and μ. The modified equations are

$$E_x = \frac{1}{\mu\varepsilon\omega^2/c^2 - k^2}\left(\frac{\partial^2 E_z}{\partial x \partial z} + \frac{i\omega}{c}\frac{\partial B_z}{\partial y}\right) \tag{72}$$

$$E_y = \frac{1}{\mu\varepsilon\omega^2/c^2 - k^2}\left(\frac{\partial^2 E_z}{\partial y \partial z} - \frac{i\omega}{c}\frac{\partial B_z}{\partial x}\right) \tag{73}$$

$$B_x = \frac{1}{\mu\varepsilon\omega^2/c^2 - k^2}\left(\frac{\partial^2 B_z}{\partial x \partial z} - \frac{i\omega}{c}\mu\varepsilon\frac{\partial E_z}{\partial y}\right) \tag{74}$$

$$B_y = \frac{1}{\mu\varepsilon\omega^2/c^2 - k^2}\left(\frac{\partial^2 B_z}{\partial y \partial z} + \frac{i\omega}{c}\mu\varepsilon\frac{\partial E_z}{\partial x}\right) \tag{75}$$

$$\nabla^2 E_z = -\frac{\mu\varepsilon\omega^2}{c^2}E_z \tag{76}$$

$$\nabla^2 B_z = -\frac{\mu\varepsilon\omega^2}{c^2}B_z \tag{77}$$

EXAMPLE: *TM wave in dielectric waveguide*

Let us look for a TM wave propagating in the z direction. The wave depends on y but not on x. Thus

$$E_z = E_z(y)\mathrm{e}^{-i\omega t + ikz} \tag{78}$$

and Eq. (76) implies

$$\frac{\partial^2}{\partial y^2}E_z(y) + \left(\frac{\mu\varepsilon\omega^2}{c^2} - k^2\right)E_z(y) = 0 \tag{79}$$

The corresponding equation outside the slab is

$$\frac{\partial^2}{\partial y^2}E_z(y) + \left(\frac{\omega^2}{c^2} - k^2\right)E_z(y) = 0 \tag{80}$$

The solution of Eq. (79) is $\cos\gamma y$ or $\sin\gamma y$, with

$$-\gamma^2 + \left(\frac{\mu\varepsilon\omega^2}{c^2} - k^2\right) = 0 \tag{81}$$

The cosine solution is symmetric under reversal of the y coordinate, and the sine solution is antisymmetric. Here we will adopt the cosine solution, and we will leave the sine solution for Problem 21:

$$E_z(y) = E_0 \cos \gamma y \qquad \text{for } |y| \le b/2 \tag{82}$$

The solution of Eq. (80) is

$$E_z(y) = \begin{cases} A\mathrm{e}^{-\beta y} & \text{for } y \ge b/2 \\ A\mathrm{e}^{+\beta y} & \text{for } y \le -b/2 \end{cases} \tag{83}$$

where

$$\beta^2 + \left(\frac{\omega^2}{c^2} - k^2 \right) = 0 \tag{84}$$

Here we have adopted the solution that decreases exponentially with distance from the surface of the slab. This decrease of the fields with distance is essential to prevent the escape of the wave from the waveguide.

The general boundary conditions at the dielectric surfaces $y = \pm b/2$ are given by Eqs. (12.121)–(12.124). Since there is no perpendicular **B** field at the surface, the relevant boundary conditions are the first three: continuity of E_z, continuity of D_y (or εE_y), and continuity of H_x (or B_x/μ). With Eqs. (82) and (83), the first of these conditions becomes

$$E_0 \cos \frac{\gamma b}{2} = A\mathrm{e}^{-\beta b/2} \tag{85}$$

For the second boundary condition, we need to evaluate E_y, which we can do by means of Eq. (73). This leads to

$$\frac{\varepsilon E_0}{\mu\varepsilon\omega^2/c^2 - k^2} \left(-ik\gamma \sin \frac{\gamma b}{2} \right) = \frac{A}{\omega^2/c^2 - k^2} (-ik\beta \mathrm{e}^{-\beta b/2}) \tag{86}$$

The third boundary condition turns out to coincide with Eq. (86) and can therefore be ignored.

■ ***Exercise 13.*** Check the third boundary condition. ■

Taking the ratio of Eqs. (85) and (86), we find

$$\frac{\varepsilon\gamma}{\mu\varepsilon\omega^2/c^2 - k^2} \tan \frac{\gamma b}{2} = \frac{\beta}{\omega^2/c^2 - k^2}$$

or

$$\frac{\varepsilon}{\gamma}\tan\frac{\gamma b}{2} = -\frac{1}{\beta} \tag{87}$$

If we combine Eqs. (81) and (84), we can express β in terms of γ:

$$\gamma^2 + \beta^2 = \frac{\omega^2(\mu\varepsilon - 1)}{c^2} \tag{88}$$

or

$$\beta = \sqrt{\frac{\omega^2(\mu\varepsilon - 1)}{c^2} - \gamma^2} \tag{89}$$

With this, Eq. (87) becomes

$$\tan\frac{\gamma b}{2} = -\frac{\gamma/\varepsilon}{\sqrt{\omega^2(\mu\varepsilon - 1)/c^2 - \gamma^2}} \tag{90}$$

or, in terms of the dimensionless quantity $\gamma b/2$,

$$\tan\frac{\gamma b}{2} = -\frac{\gamma b/2\varepsilon}{\sqrt{\omega^2 b^2(\mu\varepsilon - 1)/4c^2 - (\gamma b/2)^2}} \tag{91}$$

This equation determines the permitted values of γ for the transversely symmetric TM modes of our dielectric waveguide, and it therefore determines the permitted values of k.

The solutions of Eq. (90) must be obtained numerically or graphically. For our purposes, a graphical solution will be adequate. This is obtained by plotting the right side and the left side of Eq. (91) on the same graph, with $\gamma b/2$ as abscissa (see Fig. 15.13); the intersections of the two plots give the solutions. Note that the number of intersections, or the number of permitted modes, depends on the magnitude of ω. For the value of the frequency assumed in Fig. 15.13, there are two permitted modes. By sketching analogous plots for lower values of the frequency, we find that there is no permitted mode unless

$$\omega > \frac{\pi c/b}{\sqrt{\mu\varepsilon - 1}} \tag{92}$$

This gives the lowest frequency required for propagation of a TM mode in our dielectric waveguide; that is, it gives the cutoff frequency.

Figure 15·13

Plots of the right side and left side of Eq. (91) as a function of $\gamma b/2$. The value of the frequency assumed for the plot of the right side is $\omega = (4\pi c/b)\sqrt{\mu\varepsilon - 1}$ and $\varepsilon = 1.1$. At this frequency, there are two intersections, that is, two permitted modes.

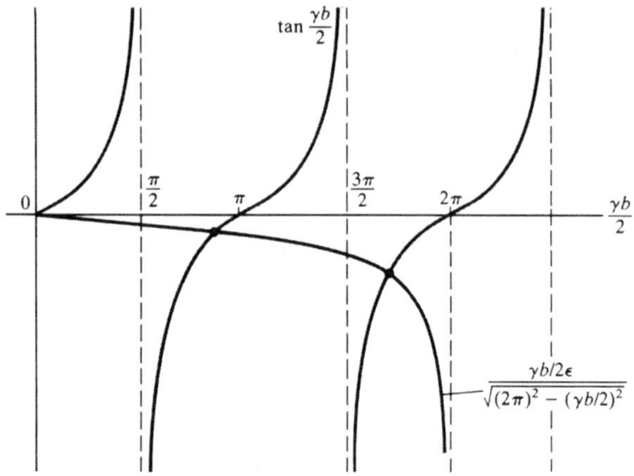

15·7 The Laser as a Cavity

A laser can be regarded as a resonant cavity in which electromagnetic fields oscillate at optical frequencies, of the order of 10^{15} Hz. The laser cavity is defined by two parallel mirrors, and the normal modes of this cavity are simply the standing waves trapped between these mirrors. Intense standing light waves are excited in the laser cavity by stimulated emission from atoms or molecules. Figure 15.14 shows a schematic diagram of a He-Ne laser, with a mixture of neon and helium atoms in a discharge tube. The neon atoms provide the stimulated emission that feeds the standing wave, while the helium atoms serve to "pump" the neon atoms into excited states, by helium-neon collisions. The ends of the discharge tube are closed off by "Brewster windows" consisting of glass plates inclined at the Brewster angle [see Eq. (12.141)]. A light wave striking such a glass plate at this angle is totally transmitted if its plane of polarization is parallel to the plane of the page (plane of incidence), but it is partially reflected if its plane of polarization is perpendicular to the plane of the page. This means that the parallel-polarized wave travels back and forth within the laser without loss, whereas the perpendicular-polarized wave is partially reflected and quickly eliminated from the laser beam. Thus the Brewster windows favor the amplification of the parallel-polarized wave. The parallel mirrors defining the laser cavity are placed just beyond the Brewster windows. One of these parallel mirrors is partially silvered, and the amplified polarized wave

Figure 15·14

Schematic diagram of a He-Ne laser. The two parallel mirrors M and M' define the ends of the laser cavity. The discharge tube containing the He-Ne mixture is closed off by Brewster windows at its ends; these windows permit total transmission of waves with polarization in the plane of the page.

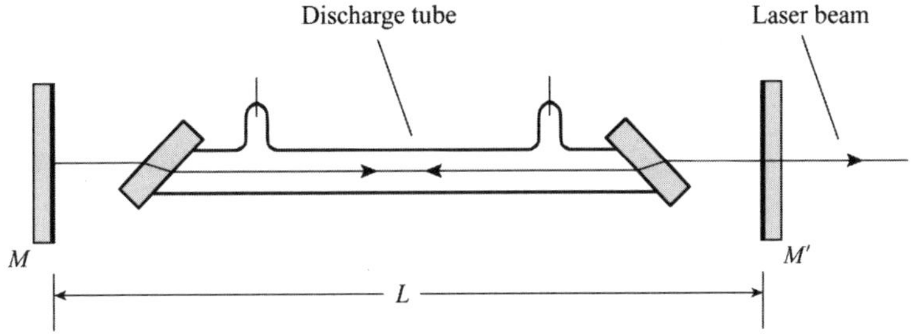

ultimately emerges through this mirror, forming the external beam of the laser.

EXAMPLE: *normal modes in laser cavity*

Suppose that the distance between the parallel mirrors M and M' is L. The wavelength and the frequency of the normal modes of the laser cavity are determined by the familiar condition for a standing wave: the length L must equal an integral number of half-wavelengths,

$$L = m \times \lambda / 2 \qquad m = 1, 2, 3, \ldots \tag{93}$$

To make this more precise, we have to take into account that the path of the light wave is angled upward or downward within the glass plates of the Brewster windows and that the index of refraction in the glass plates and in the discharge tube differs from 1. But this does not affect the validity of Eq. (93), provided we interpret the length L as the optical path length, that is, the integral of length multiplied by the index of refraction. For a length of, say, 50 cm and a wavelength of 600 nm, Eq. (93) tells us that the number m is very large, $m \simeq 10^6$. Consequently, the fractional difference in wavelength between one mode and the next is small, about one part in 10^6. The width of the atomic spectral lines in a laser is larger than that—the intrinsic width of the spectral line of an isolated atom is quite small, but in the lasing medium the atoms are subjected to perturbations, for instance, collisions and random thermal motions (with concomitant random Doppler shifts) if the medium is a gas, or local inhomogeneities if the medium is a crystal. These perturbations alter the wavelengths of different atoms in different ways, so the spectral line emitted by the ensemble of atoms spreads over a range of wavelengths (and frequencies) centered on the wavelength (and frequency) of an undisturbed atom. This broadened range of wavelengths typically spans several dozen of the normal modes of the laser cavity, and the laser therefore typically operates in several modes at once.

Moreover, in practice, the mirrors at the ends of the laser cavity are usually slightly spherical, rather than plane. This prevents the light wave from drifting away from the axis of the laser. However, such spherical mirrors permit off-axis normal modes, with standing waves that include a small admixture of transverse wave speeds. The wavelengths of the off-axis modes are of the same order of magnitude as those of the on-axis modes specified by Eq. (93), and the laser will typically operate in several off-axis modes as well as several on-axis modes. The presence of such off-axis modes makes the external laser beam slightly divergent, with a typical angular spread of 1 milliradian.

The pumping of atoms into excited states and the amplification of the wave by stimulated emission are quantum phenomena, which we cannot describe by classical mechanics and electrodynamics. However, we can describe some aspects of the growth of wave amplitude in classical terms; and by examining the energy balance between the light wave and the oscillating electrons in the atoms, we can obtain a simple criterion for attaining the threshold for initiating laser action.

For such a semiclassical description of the operation of the laser, we need to know the electric susceptibility of the laser medium. Under the assumptions that only two atomic states participate directly in the lasing action—a state of low energy E_1 and a state of a higher energy E_2—and that the frequency of the standing light wave in the laser cavity is near the transition frequency for these energy levels, the susceptibility derived from quantum mechanics is given by the simple formula that we obtained in Section 12.5 from the naive classical model for electrons bound to their equilibrium position by elastic forces. This classical model for the behavior of electrons gave us a susceptibility involving a sum over all the resonant frequencies of the electron [see Eq. (12.105)]. But if the frequency is near one of these resonant frequencies, say, $\omega \simeq \omega_0$, then the corresponding term in the sum is very large, and we can ignore all the other terms, so Eq. (12.105) reduces to

$$\chi \simeq \frac{Ne^2}{m_e} \frac{f_0}{\omega_0^2 - \omega^2 - i\Gamma\omega} \tag{94}$$

Here, as in Section 12.5, ω is the frequency of the light wave. But according to quantum mechanics, the other quantities in Eq. (94) must be reinterpreted as follows: The resonance frequency ω_0 is the quantum-mechanical transition frequency, proportional to the energy difference between the two states, $\omega = (E_2 - E_1)/\hbar$. The damping constant Γ is the width of the spectral line, that is, the range of frequencies over which the spectral line spreads. And N is the difference between the number of atoms per unit volume in the lower and the upper states, $N = N_1 - N_2$. Atoms in the upper states contribute to the susceptibility with an opposite sign to those in the lower state; because of this sign change, atoms in the upper state are said to act as "negative oscillators." The operation of the laser hinges on the presence of a *population inversion*, that is, $N_2 > N_1$ and therefore $N < 0$. This sign change implies that in the laser medium, the electric susceptibility has the opposite sign from that expected in the classical case.

To examine the energy balance between the wave and the oscillating electrons, consider the power absorbed by the wave from one electron, with velocity dx/dt. According to the classical model, this power is

$$-(\text{force} \times \text{velocity}) = -(-eE)\frac{dx}{dt} \tag{95}$$

If we multiply this by the number of electrons per unit volume, we find that the rate of change of the energy density of the wave can be expressed in terms of the polarization *P*, or the dipole moment per unit volume; and this, in turn, can be expressed in terms of the susceptibility, as follows:

$$\frac{dT^{00}}{dt} = -E\frac{d}{dt}(-eNx) = -E\frac{d}{dt}P = -E\frac{d}{dt}\chi E \tag{96}$$

Since the time dependence of the electric field is $e^{-i\omega t}$, we obtain

$$\frac{dT^{00}}{dt} = E \times (i\omega\chi E) \tag{97}$$

In the evaluation of the right side, we must remember that both of the factors E and $i\omega\chi E$ are complex and that the physically relevant portion of their product is $\text{Re}\, E \times \text{Re}(i\omega\chi E)$. If we write χ as a sum of real and imaginary parts, $\chi = \chi' + i\chi''$, we obtain

$$\frac{dT^{00}}{dt} = E_0 \cos\omega t \times (\omega\chi' E_0 \sin\omega t - \omega\chi'' E_0 \cos\omega t) \tag{98}$$

The time-average rate of change of the energy density is then

$$\frac{d\overline{T}^{00}}{dt} = -\frac{1}{2}E_0^2\omega\chi'' \tag{99}$$

According to Eq. (94), the imaginary part of χ is

$$\chi'' = \frac{Ne^2 f_0 \Gamma \omega / m_e}{(\omega_0^2 - \omega^2)^2 + \Gamma^2 \omega^2} \tag{100}$$

Thus, to achieve an energy gain ($d\overline{T}^{00}/dt > 0$), we need $\chi'' < 0$ in Eq. (99), and we need $N < 0$ in Eq. (100); that is, we need a population inversion.

Furthermore, if the wave amplitude is to increase with time, the rate of energy gain (99) must be larger than the rate of energy loss within the cavity. The energy losses of the laser cavity can be characterized by a Q value, as in Section 15.4. This Q value represents losses due to imperfect reflections (including losses through the half-silvered mirror from which the external beam emerges) and losses by scattering and absorption in the medium. The rate of decrease of the energy density due to such losses can be written as in Eq. (48):

$$\left.\frac{d\overline{T}^{00}}{dt}\right|_{\text{loss}} = -\frac{\omega}{2\pi Q}\overline{T}^{00} \simeq -\frac{\omega}{2\pi Q} \times \frac{E_0^2}{8\pi} \tag{101}$$

If we demand that the rate of gain given by Eq. (99) exceed the rate of loss, we obtain the threshold condition

$$-\chi'' > \frac{1}{2\pi Q} \times \frac{1}{4\pi} \tag{102}$$

EXAMPLE: *threshold condition for laser*

If we evaluate this for a frequency $\omega \simeq \omega_0$, at or near the resonance frequency of the atom, the threshold condition becomes

$$-N > \frac{m_e c}{4\pi e^2} \frac{\Gamma}{f_0} \frac{1}{\lambda_0 Q} \tag{103}$$

Here $\lambda_0 = 2\pi c / \omega_0$ is the wavelength that corresponds to this resonance frequency. For a given value of Q, Eq. (103) tells us the value of $|N|$ required to initiate lasing; that is, it tells us the required magnitude of the population inversion (see Problem 23 for typical numbers for a He-Ne laser). The width of the resonance maximum of χ'' is $\simeq \Gamma$, and this is usually larger than the separation between the frequencies of the normal modes of the laser cavity. Hence there will be several normal modes with frequencies near ω_0 for which the threshold condition (103) is satisfied, and all of these will be amplified intensely.

As long as the condition (103) is satisfied, the amplitudes of the standing waves will continue to grow. However, when the amplitude becomes too large, our expression for the susceptibility is modified by a quantum-mechanical correction—effectively, the damping constant Γ increases with the amplitude. This increases the right side of Eq. (103)

and brings the laser below threshold. The amplitude then stops growing, that is, the amplitude saturates at some critical value.

Problems

1. From the expressions given in Section 15.3 for the electric and magnetic fields in a rectangular cavity oscillating in the TM_{011} mode, find the energy density and find the total energy in the cavity.

2. Sketch the electric and magnetic field lines (top view, front view, and side view) of a rectangular cavity oscillating in the TM_{112} mode.

3. Find the components of the electric and magnetic fields, including the time dependence, for a rectangular cavity of dimensions $a \times b \times d$ oscillating in a TE mode. Sketch the field lines for the TE_{110} mode.

4. Consider a rectangular cavity of dimensions $a \times b \times d$ oscillating in the TM_{011} mode. Find the charge density everywhere on the walls of the cavity. Find the surface current density.

5. A rectangular cavity with copper walls measures 3.5 cm $\times$ 7.5 cm $\times$ 9.0 cm. Find the lowest frequency for a TM mode in this cavity. Estimate the damping time for this mode. The conductivity of copper is 5.22×10^{17} (statohms·cm)$^{-1}$.

6. Suppose that a rectangular cavity is filled with a dielectric of dielectric constant ε and permittivity constant μ. How does this modify Eqs. (32)–(40)?

7. A coaxial waveguide consists of a solid, conducting cylinder of radius a surrounded by a conducting cylindrical shell of radius b (see Fig. 15.7). Find the power carried by this waveguide in the TEM mode, given that the frequency of the wave is ω and the amplitude of the electric field at the surface of the inner conductor is E_0.

8. A rectangular waveguide of cross section $a \times b$ is propagating a wave in the TE_{11} mode. Sketch the field lines of the electric and magnetic fields.

9. A common rectangular waveguide has a cross section of 7.21 cm $\times$ 3.40 cm.
(a) What is the cutoff frequency for the TE_{10} mode?
(b) Suppose you feed a TE_{10} wave of frequency 1.8×10^9 Hz into this waveguide. Within what distance along the waveguide will the fields be attenuated by a factor e?

10. Calculate the time-average power carried by a rectangular waveguide of cross section $a \times b$ in a TE_{mn} mode with the magnetic field given by Eq. (58). What is the value of the power when ω is much larger than the cut-off frequency?

11. (a) Show that the phase and group velocities for the rectangular waveguide [see Eqs. (63) and (64)] satisfy the identity $u_p u_g = c^2$.
(b) Show that this identity is valid in general for any waveguide. [*Hint:* Show that this identity obtains whenever the squares of the frequency and the wave number obey a linear relationship of the form $\omega^2/c^2 - k^2 = \text{constant}$.]

12. Consider a rectangular waveguide excited in a TE_{mn} mode with the magnetic field given by Eq. (58). What is the time-average Poynting vector? What is the time-average energy density? Verify that the time-average Poynting vector integrated over the cross section of the waveguide equals the time-average energy density integrated over the cross section and multiplied by the group velocity,

$$\int \overline{S}_z \, dx\, dy = u_g \int \overline{T}^{00} \, dx\, dy$$

13. A rectangular waveguide of cross section $a \times b$ is propagating a wave in the TM_{11} mode. Calculate the electric and magnetic fields. Sketch the field lines of the electric and magnetic fields.

14. Suppose that the z component of the magnetic field in a rectangular waveguide is

$$B_z = B_0 \cos\left(\frac{\pi x}{a}\right) \cos\left(\frac{\pi y}{b}\right) e^{-i\omega t + ikz}$$

The lateral surfaces of the waveguide are the planes $x = 0$, $y = 0$, $x = a$, $y = b$. Find the instantaneous surface current density (all components) and surface charge density at each point on the lateral surfaces of the waveguide.

15. Suppose that a wave is propagating in a waveguide at a frequency above the cutoff frequency. Show that if the waveguide has walls of finite conductivity, the attenuation distance, or the distance within which the intensity of the wave decreases by a factor of e, is $Q \times$ (wavelength). Show that for a rectangular waveguide, this is approximately

$$\frac{ab}{2(a+b)\delta} \times (\text{wavelength})$$

16. The accompanying table gives the dimensions of some commercially available waveguides made of copper. For each of these waveguides, calculate the cutoff frequency, and estimate the attenuation distance for a wave of twice the cutoff frequency (see the preceding problem). The conductivity of copper is 5.22×10^{17} (statohms·cm)$^{-1}$.

Cross sections of waveguides:

3.48 cm	× 1.58 cm
4.75	× 2.21
7.21	× 3.40

17. A waveguide consists of two thin, parallel wires separated by a distance d. An electromagnetic wave propagates along these wires in a TEM mode. Find the electric and magnetic fields if the frequency is ω. What are the wavelength, phase velocity, and group velocity of the wave?

18. Two very large, parallel, perfectly conducting plates are separated by a distance b. One plate coincides with the plane $y = 0$, the other with the plane $y = b$.

(a) Find the electric and magnetic fields for a TEM wave of frequency ω propagating in the z direction between the plates. Sketch the fields. What are the wavelength, phase velocity, and group velocity of the wave?

(b) If the plates are not perfect conductors but have some finite conductivity σ, then the amplitude of the wave will decrease exponentially as the wave travels along the space between the plates. Estimate what distance in the z direction the wave can travel before it is attenuated by a factor of e. Express your answer in terms of ω, σ, b; assume that the thickness of the plates is much larger than the skin depth.

19. Consider the pair of parallel conducting plates described in the preceding problem. Find the electric and magnetic fields of a TE_{0n} wave of frequency ω propagating in the z direction between these plates. Assume that the electric field of the wave is in the x direction, parallel to the plates.

20. Prove that a TEM wave cannot propagate in a dielectric waveguide.

21. Assume that the transverse dependence of the electric field of a TM wave in the dielectric slab described in Section 15.6 is $E_z(y) = E_0 \sin \gamma y$. Under this assumption, derive an equation analogous to Eq. (90).

22. Consider TE waves propagating in the dielectric slab described in Section 15.6. Find an equation analogous to Eq. (90) for such a wave. What is the lowest frequency permitted for TE waves?

23. Consider a He-Ne laser with a laser cavity of length L = 50 cm, with one totally reflective end mirror and one end mirror that reflects 99%. The stimulated emission of the neon atoms has $\lambda_0 = 632.8\ \text{nm}$, $\Gamma = 9.4\times10^9/\text{s}$, and $f_0 = 8\times10^{-3}$.

(a) What is the value of m for the normal mode at the resonance maximum? What range of normal modes is within the width of this resonance maximum, from $\omega = \omega_0 - \Gamma/2$ to $\omega = \omega_0 + \Gamma/2$?

(b) What is the Q for the laser cavity? [*Hint*: In one cycle of oscillation, 1% of the wave energy approaching the 99% mirror from within a distance of one wavelength will escape.]

(c) What is the threshold value for $|N|$?

CHAPTER

16

Scattering and Diffraction

When an electromagnetic wave strikes a free electron or a loosely bound electron, it is *scattered*: the wave shakes the electron and causes the emission of radiation; this radiation spreads outward as a spherical wave and carries away some of the energy of the original incident wave. A familiar example of this phenomenon is seen in the blue color of the sky, which is due to the scattering of sunlight by the electrons in molecules of air.

In this chapter, we will find that the scattering produced by a single electron is quite easy to calculate. However, if we want to calculate the scattering produced by a grain of dust, by a droplet of water, or by some other macroscopic obstacle, we must add the coherent effects of many electrons. This is a formidable mathematical problem, best handled by means of the macroscopic Maxwell equations rather than by the microscopic equations of motion of the individual electrons. We are then faced with a boundary-value problem: the obstacle is a medium of some given dielectric constant and conductivity, and we seek the solution of the wave equation for a given incident wave. For an obstacle of some complicated shape, the exact solution of this problem is quite intractable. We will not deal with any exact solution, but only with an approximate solution furnished by the Kirchhoff integral. Of course, the most common application of this integral is in *diffraction*, or the bending of light around obstacles, which can be regarded as a special case of scattering. But the Kirchhoff integral also furnishes an approximate solution of more general cases of scattering.

16·1 Thomson Scattering

Suppose that a plane electromagnetic wave is incident on a single, free electron. If the wave is propagating in the z direction and is polarized in say, the x direction, then the electric field is given by

$$\mathbf{E}_{\text{in}} = E_0 \hat{\mathbf{x}} e^{-i(\omega t - kz)} \tag{1}$$

Here, we have used complex notation, leaving it understood that the physical electric field is the real part of the above formula. Under the influence of this electric field, an electron at $z = 0$ will acquire an acceleration in the x direction,

$$a = \frac{d^2x}{dt^2} = -\frac{e}{m_e} E_0 e^{-i\omega t} \tag{2}$$

According to Eq. (13.53), the power radiated by the electron per unit solid angle is proportional to the square of the acceleration. Since the acceleration a is complex, we must take the real part before we proceed to the square:

$$\frac{dP}{d\Omega} = \frac{e^2}{4\pi c^3} \mathrm{Re}(a) \cdot \mathrm{Re}(a) \sin^2 \Theta \tag{3}$$

where Θ is the angle between the direction of polarization and the direction of radiation (see Fig. 16.1). For our purposes, the quantity of interest is not the instantaneous power but the time-average power,

$$\frac{d\overline{P}}{d\Omega} = \frac{e^2}{4\pi c^3} \overline{\mathrm{Re}(a) \cdot \mathrm{Re}(a)} \sin^2 \Theta \tag{4}$$

The time average $\overline{\mathrm{Re}(a) \cdot \mathrm{Re}(a)}$ can be conveniently evaluated by means

Figure 16·1

The direction of polarization of the incident wave and the direction of acceleration of the electron are parallel to the x axis. The intensity of the scattered radiation is a function of the angle Θ.

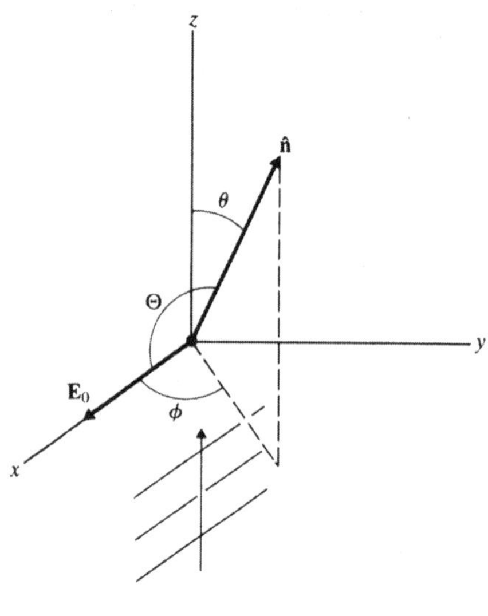

of the following formula, which is valid for any function that has a time-dependence of the form $e^{\pm i\omega t}$:

$$\overline{\text{Re}(a)\cdot\text{Re}(a)} = \frac{1}{2}aa^* \tag{5}$$

■ ***Exercise 1.*** Prove Eq. (5). More generally, prove that for two complex quantities a and b with the same time dependence $e^{-i\omega t}$, the time average is

$$\overline{\text{Re}(a)\cdot\text{Re}(b)} = \frac{1}{2}\text{Re}(ab^*) \quad ■$$

With the acceleration $-(e/m_e)E_0e^{-i\omega t}$, Eq. (4) then leads to the time-average power

$$\frac{d\bar{P}}{d\Omega} = \frac{e^2}{8\pi c^3}\left(\frac{e}{m_e}E_0\right)^2 \sin^2\Theta \tag{6}$$

or

$$\frac{d\bar{P}}{d\Omega} = \frac{c}{8\pi}r_e^2E_0^2\sin^2\Theta \tag{7}$$

Here r_e is the so-called classical electron radius,

$$r_e = \frac{e^2}{m_ec^2} = 2.82\times10^{-13}\ \text{cm} \tag{8}$$

We define the *differential-scattering cross section* as the ratio of $d\bar{P}/d\Omega$ to the time-average incident energy flux,

$$\frac{d\sigma}{d\Omega} \equiv \frac{d\bar{P}/d\Omega}{(\text{incident flux})} \tag{9}$$

The quantity $d\sigma/d\Omega$ has the dimensions of an area; it is a measure of how much power is scattered into the solid angle $d\Omega$:

$$\frac{d\bar{P}}{d\Omega} = (\text{incident flux})\times\frac{d\sigma}{d\Omega} \tag{10}$$

Note that $d\sigma/d\Omega$ is the scattered power per unit flux; this is perhaps the easiest way to remember the definition of the cross section.

In our case, the scattered power is given by Eq. (7), and the time-average incident flux is given by the Poynting vector $\bar{S} = cE_0^2/8\pi$. Thus the differential cross section is

$$\frac{d\sigma}{d\Omega}=\frac{(c/8\pi)r_e^2E_0^2\sin^2\Theta}{(c/8\pi)E_0^2}=r_e^2\sin^2\Theta \tag{11}$$

In terms of the usual polar angles (see Fig. 16.1),

$$\sin^2\Theta=1-\cos^2\Theta=1-(\sin\theta\cos\phi)^2 \tag{12}$$

so $d\sigma/d\Omega$ is a function of θ and ϕ:

$$\frac{d\sigma}{d\Omega}=r_e^2(1-\sin^2\theta\cos^2\phi) \tag{13}$$

The total scattering cross section, or the total scattered power per unit flux, is the integral of $d\sigma/d\Omega$ over all solid angles,

$$\sigma=\int\frac{d\sigma}{d\Omega}d\Omega=\int_0^{2\pi}\int_0^{\pi}r_e^2(1-\sin^2\theta\cos^2\phi)\sin\theta\,d\theta d\phi=\frac{8\pi}{3}r_e^2 \tag{14}$$

This is called the *Thomson cross section.* Its numerical value is

$$\frac{8\pi}{3}r_e^2=6.65\times10^{-25}\text{ cm}^2 \tag{15}$$

The Thomson cross section is independent of the frequency of the incident wave. However, for very high frequencies $(\omega\geq m_e c^2/\hbar)$, our classical calculation fails and quantum effects come into play. The cross section then acquires a frequency dependence (Klein-Nishina cross section).

Next, we want to calculate the scattering by an electron bound in an atom. We will again rely on the simple model of Section 12.5: the binding force on the electron is represented by an elastic restoring force $-m_e\omega_0^2x$, and energy losses of the electron are represented by an effective frictional force $-m_e\Gamma dx/dt$. The equation of motion of the bound electron driven by the electric field (1) is then

$$m_e\frac{d^2x}{dt^2}+m_e\Gamma\frac{dx}{dt}+m_e\omega_0^2x=-eE_0e^{-i\omega t} \tag{16}$$

Under steady-state conditions, this equation leads to an acceleration

$$a=\frac{d^2x}{dt^2}=-\frac{e}{m_e}\frac{\omega^2E_0e^{-i\omega t}}{\omega^2-\omega_0^2+i\Gamma\omega} \tag{17}$$

■ ***Exercise 2.*** Derive Eq. (17). ■

This differs from the acceleration of a free electron [see Eq. (2)] by the extra factor

$$\frac{\omega^2}{\omega^2 - \omega_0^2 + i\Gamma\omega} \tag{18}$$

Therefore, the cross section will differ from the cross section of a free electron by an extra factor

$$\left|\frac{\omega^2}{\omega^2 - \omega_0^2 + i\Gamma\omega}\right|^2 = \frac{\omega^4}{(\omega^2 - \omega_0^2)^2 + \Gamma^2\omega^2} \tag{19}$$

and the total scattering cross section for the bound electron is

$$\sigma = \frac{8\pi}{3} r_e^2 \frac{\omega^4}{(\omega^2 - \omega_0^2)^2 + \Gamma^2\omega^2} \tag{20}$$

If $\omega \gg \omega_0$ and $\omega \gg \Gamma$, this result reduces to $\sigma \simeq (8\pi/3)r_e^2$, the free-electron cross section. When subjected to a high-frequency wave, the bound electron radiates like a free electron because the steady-state binding force is negligible compared with the external driving force. If $\omega \ll \omega_0$, Eq. (20) reduces to

$$\sigma \simeq \frac{8\pi}{3} r_e^2 \left(\frac{\omega}{\omega_0}\right)^4 \tag{21}$$

Thus, at low frequencies, the cross section increases as the fourth power of the frequency; this dependence on frequency is called *Rayleigh's law.*

EXAMPLE: *blue color of sky*

The blue color of the sky is a consequence of Rayleigh's law. The scattering cross section for a molecule of air is roughly given by Eq. (21), with some suitable correction factor to account for the number of vibrating electrons in the molecule. Most of the resonant frequencies for N_2, O_2, and H_2O molecules are in the ultraviolet, so the assumption $\omega \ll \omega_0$ is reasonably well satisfied for visible light. Hence light at the blue end of the spectrum ($\lambda \simeq 400$ nm) is scattered much more than light at the red end of the spectrum ($\lambda \simeq 700$ nm); the ratio of the scattering cross sections is roughly $(\frac{7}{4})^4 \simeq 9$. The blue light is scattered out of the beam of sunlight and fills the entire sky. But in this simple explanation of the blue color of the sky, we have overlooked a subtle question. A volume of air contains many molecules, and we must ask how the contributions of the individual molecules are combined. In the above explanation, we have implicitly assumed that the individual molecules contribute incoherently to the total scattering—each molecule produces a scattered intensity according to Eq. (7), and the net scattered intensity is the sum of these intensities. The validity of this procedure would be immediately evident for a very low-density gas or plasma in which the scatterers are separated by large distances, that is, distances much larger than a wavelength. Under these conditions, the accelerations and the radiations of the scatterers would differ in phase by large and random amounts—the scatterers would act as incoherent sources, and the net intensity of the scattered radiation would be the sum of the individual intensities. But in air at normal density, electrons in adjacent molecules move and radiate in phase—they act as coherent sources, and the net amplitude of the scattered radiation is the sum of the individual *amplitudes*. The net intensity is then proportional to the square of the net amplitude.

If air were a perfectly uniform dielectric, such coherent radiation would result in zero net amplitude for the scattered wave because the radiations from different volumes of air would cancel by destructive interference. For instance, Fig. 16.2 shows two thin, equal volumes of air or some other dielectric material, each a thin slab facing the direction of incidence; the slabs are separated by a distance of exactly one-half wavelength. Consider the scattered radiation emitted at right angles to the direction of incidence, radiation that ultimately reaches an observer far away on the y axis. The incident wave at the first slab is always exactly 180° out of phase with the incident wave at the second slab. Hence the scattered radiations contributed by any pair of molecules located at corresponding points in this pair of slabs will suffer destructive interference at the position of the observer—the net scattered amplitude is zero! But this perfect cancellation of the scattering hinges on the perfect uniformity of the dielectric material. In air, perfect uniformity fails because the density is subject to small random fluctuations. On the average, the two small volumes in Fig. 16.2 contain an equal number N of molecules, but at any given instant, the actual numbers of molecules are $N + \delta N_1$ and $N + \delta N_2$. The amplitude of the scattered radiation produced by each volume

Figure 16·2

Two thin slabs of air, separated by a distance of one-half wavelength.

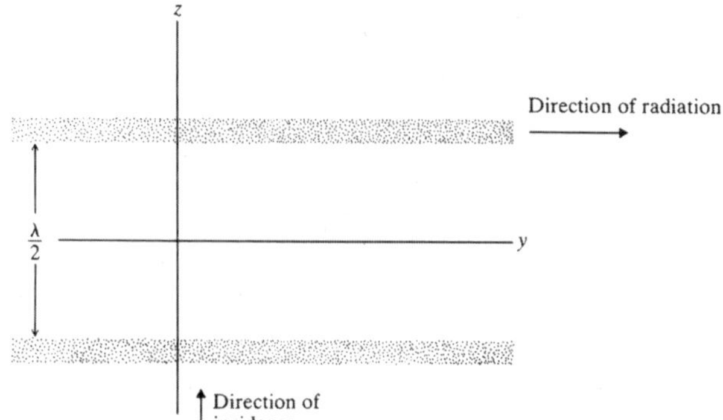

is proportional to the number of molecules; since the radiations differ in phase by 180°, the net amplitude is proportional to the difference between the number of molecules,

$$(\text{amplitude}) \propto (N+\delta N_1)-(N+\delta N_2)=\delta N_1-\delta N_2 \tag{22}$$

The intensity is proportional to the square of this difference,

$$(\text{intensity}) \propto (\delta N_1-\delta N_2)^2=(\delta N_1)^2+(\delta N_2)^2-2\delta N_1\delta N_2 \tag{23}$$

Statistical mechanics tells us that, on the average, $(\delta N_1)^2=(\delta N_2)^2=N$, and that $\delta N_1\delta N_2=0$, because the two fluctuations are uncorrelated, and they are positive and negative at random. The net intensity is then proportional to 2*N*; that is, it is proportional to the net number of molecules in the two volumes. This means that the net contribution to the scattering is exactly what it would be if the scatterers were incoherent, and thus the density fluctuations make the coherent scattering in air seem incoherent. This justifies our use of the cross section for the description of the scattering of light in the sky. The discovery of the crucial role that density fluctuations play in the scattering of light in gases was one of Einstein's early achievements in physics.

16·2 The Forward Scattering Amplitude; the Optical Theorem

According to Eq. (13.49), the amplitude of the electric radiation field produced by an electron accelerated by an incident wave is

$$E_{\text{rad}}=\frac{e}{c^2}\sin\Theta\frac{a(t-r/c)}{r} \tag{24}$$

where Θ is, again, the angle between the direction of polarization of the incident wave and the direction of radiation (see Fig. 16.1). For a bound electron, the acceleration $a(t)$ is given by Eq. (17), and we obtain

$$E_{\text{rad}}=-\frac{e^2}{m_e c^2}\frac{\omega^2 E_0 \sin\Theta}{\omega^2-\omega_0^2+i\Gamma\omega}\frac{e^{-i(\omega t-kr)}}{r} \tag{25}$$

The factor multiplying the outgoing spherical wave $E_0 e^{-i(\omega t-kr)}/r$ is called the *scattering amplitude*,

$$A = -\frac{e^2}{m_e c^2} \frac{\omega^2 \sin\Theta}{\omega^2 - \omega_0^2 + i\Gamma\omega} \tag{26}$$

Evaluating this in the forward direction (or the direction of propagation of the incident wave), we obtain the *forward scattering amplitude*:

$$A_{\text{forward}} = -\frac{e^2}{m_e c^2} \frac{\omega^2}{\omega^2 - \omega_0^2 + i\Gamma\omega} \tag{27}$$

Let us examine the imaginary part of this function,

$$\text{Im}(A_{\text{forward}}) = \frac{e^2}{m_e c^2} \frac{\Gamma\omega^3}{(\omega^2 - \omega_0^2)^2 + \Gamma^2\omega^2} \tag{28}$$

It turns out that this expression is related to the total cross section, that is, the total rate at which the electron removes energy from the incident wave. To calculate this rate, we note that under steady-state conditions, the power absorbed by the electron from the incident wave must match the power lost by the electron to friction,

$$P_{\text{absorbed}} = -\frac{dx}{dt} F_{\text{friction}} = \frac{dx}{dt}\left(m_e \Gamma \frac{dx}{dt} \right) = m_e \Gamma \frac{dx}{dt} \frac{dx}{dt} \tag{29}$$

Using the formula (5), we find that the time average of this absorbed power is

$$\overline{P}_{\text{absorbed}} = \tfrac{1}{2} m_e \Gamma \left(\frac{dx}{dt} \frac{dx^*}{dt} \right) = \tfrac{1}{2} m_e \Gamma \left(\frac{e}{m_e} \right)^2 \frac{\omega^2 E_0^2}{(\omega^2 - \omega_0^2)^2 + \Gamma^2\omega^2} \tag{30}$$

The total cross section is defined as the ratio of $\overline{P}_{\text{absorbed}}$ to the time-average incident energy flux,

$$\begin{aligned} \sigma_{\text{tot}} &= \frac{\overline{P}_{\text{absorbed}}}{(\text{incident flux})} = \frac{\overline{P}_{\text{absorbed}}}{cE_0^2/8\pi} \\ &= \frac{4\pi}{c} \frac{e^2}{m_e} \frac{\Gamma\omega^2}{(\omega^2 - \omega_0^2)^2 + \Gamma^2\omega^2} \end{aligned} \tag{31}$$

Comparing this with Eq. (28), we immediately see that

$$\sigma_{\text{tot}} = \frac{4\pi}{k} \text{Im}(A_{\text{forward}}) \tag{32}$$

This equality is called the *optical theorem.* Although we have derived this theorem only for the special case of an electromagnetic wave scattered by an electron with elastic binding, a generalized version of the theorem is true for any kind of wave scattered by any kind of obstacle. The optical theorem finds important applications in quantum theory in the computation of cross sections.

Note that the total cross section includes the scattering cross section. This is so because the frictional force $-m_e\Gamma dx/dt$ includes the radiation-reaction force, that is, the recoil force that the electron experiences while emitting radiation. The radiation-reaction force removes energy from the electron at a rate equal to the radiated power; thus it amounts to an effective frictional force. It can be shown that under steady-state conditions, the radiation-reaction force for an electron with harmonic motion is $-\frac{2}{3}(e^2/c^3)\omega^2 dx/dt$, which is proportional to the velocity and can be included in the usual frictional term in Eq. (16).*

16·3 The Kirchhoff Integral

If a light wave is incident on a large, opaque plate perforated by a few apertures of arbitrary shape, the wave will exhibit diffraction: it will spill into the edges of the shadow zone behind the plate and produce a pattern of bright and dark fringes. We can regard the wave amplitude beyond the plate as a superposition of the incident wave plus a scattered wave generated by the coherent radiation of the electrons in the atoms of the plate. Thus the absence of light in the dark shadow zone is interpreted as a cancellation of the incident wave and the scattered wave—the waves interfere destructively. Likewise, at the dark fringes, the incident and the scattered waves interfere destructively, whereas at the bright fringes, they interfere constructively.

For the calculations of this and the following sections, we will assume that the incident wave is a plane wave,

$$\mathbf{E}_{\text{in}} = E_0 \hat{\mathbf{x}} e^{-i(\omega t - kz)}$$

*If the motion is not harmonic, the radiation-reaction force is $\frac{2}{3}(e^2/c^3)d^3x/dt^3$.

Many of the older optics textbooks assume that the incident wave is a spherical wave, because in the days before the invention of lasers, the only available sources of coherent light were pointlike sources (pinholes), which generate spherical waves. Nowadays, diffraction experiments are almost exclusively performed with laser light sources, which generate nearly perfect plane waves.* Our choice of a plane wave is therefore in accord with modern experimental practice. Furthermore, we will assume that the wave is normally incident on the opaque plate; but this restriction is not essential, and most of our formulas can be easily generalized for oblique incidence.

We can derive an approximate expression for the electric field at a point beyond the plate by the following argument: Imagine that we close all the holes in the plate with precisely-fitting opaque flat plugs, or shields. Then the entire region beyond the plate will be shadowed—that is, the electric field will be zero. As we saw above, this electric field is the superposition of the field of the incident wave plus the fields radiated by all the electrons; thus

$$0 = \mathbf{E}_{\text{in}} + \mathbf{E}_{\text{rad}} \tag{33}$$

But the electric field $\mathbf{E}_{\text{rad}}$ consists of two contributions: the electric field radiated by the electrons of the perforated plate and the electric field radiated by the electrons of the plugs. This gives us the identity

$$0 = \mathbf{E}_{\text{in}} + \mathbf{E}_{\text{rad,plate}} + \mathbf{E}_{\text{rad,plug}} \tag{34}$$

Parenthetically, we remark that this identity is not quite exact. An unperforated plate, made of a single piece of material, is not exactly equivalent to a perforated plate plus a close-fitting but separate plug. The motion of the electrons on the perforated plate and on the plug is subject to the constraint that the electrons cannot cross the boundary separating the perforated plate and the plug; the motion on the unperforated plate is not subject to this constraint. This difference in boundary conditions leads to small deviations from Eq. (34), but these deviations are important only in regions fairly near the plate-plug boundary. At large distances, Eq. (34) is an excellent approximation.

Now imagine that we open all the apertures in the plate, discarding the plugs. The diffraction field beyond the perforated plate will then be

$$\mathbf{E}_{\text{plate}} = \mathbf{E}_{\text{in}} + \mathbf{E}_{\text{rad,plate}} \tag{35}$$

which, in view of Eq. (34), equals

---

*The plane wave from a laser can be converted into a spherical wave by passage through a lens. Spherical waves are important in holography, but we will not deal with that topic here.

$$\mathbf{E}_{\text{plate}} = -\mathbf{E}_{\text{rad,plugs}} \tag{36}$$

This tells us that we can determine the diffraction field beyond the perforated plate by simply calculating the radiation contributed by the electrons of the plugs. Except for the minus sign, Eq. (36) is mathematically equivalent to the familiar Huygens-Fresnel principle of optics, which asserts that the electric field beyond the plate can be calculated by pretending that each point of the aperture is the source of a spherical wave. The minus sign in Eq. (36) indicates that the Huygens-Fresnel principle contains a sign mistake, but that does not affect the final result for the intensity.

Next we want to express the right side of Eq. (36) explicitly as an integral over the electrons in the plugs. As stated above, we assume that the wave is normally incident on the plate (see Fig. 16.3). If the number of electrons per unit area on the plate and on the flat plugs is η, then the number in a small area is $\eta dx'dy'$, and the electric radiation field contributed by this area is, according to Eq. (25),

$$dE_{\text{plate}} = -dE_{\text{rad,plugs}} = \frac{e^2}{m_e c^2} \frac{\omega^2 E_0 \sin\Theta}{\omega^2 - \omega_0^2 + i\Gamma\omega} \frac{\mathrm{e}^{-i(\omega t - kr)}}{r} \eta dx'dy' \tag{37}$$

Here it is expedient to collect all the factors that depend on the atomic properties of the material and to introduce an abbreviation for the product of these factors:

Figure 16·3

An opaque plate with some apertures. A wave is incident from below. If the point of observation is sufficiently distant, the apertures and the corresponding flat plugs will subtend only a small angle.

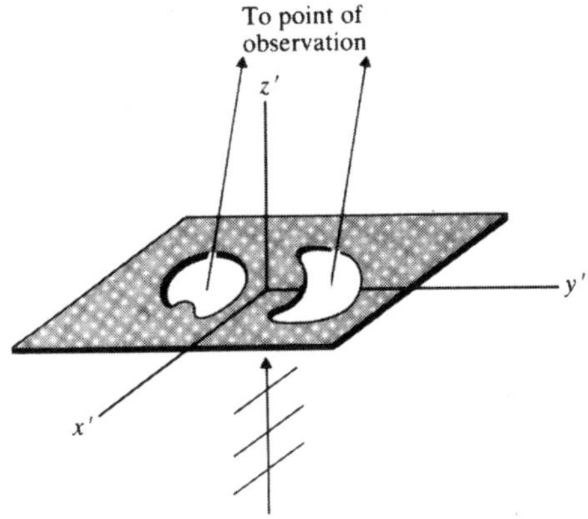

$$C = \frac{e^2}{m_e c^2} \frac{\omega^2}{\omega^2 - \omega_0^2 + i\Gamma\omega} \eta \tag{38}$$

Then

$$dE_{\text{plate}} = CE_0 \sin\Theta \frac{e^{-i(\omega t - kr)}}{r} dx'dy' \tag{39}$$

For the integration of the electric field, we ought to take into account that $d\mathbf{E}_{\text{plate}}$ is a vector. However, in most practical diffraction problems, the angle subtended by the region of integration as seen from the point of observation is small (see Fig. 16.3); the contributions of the electric field vectors are then approximately parallel, and it suffices to integrate the magnitudes dE_{plate} given by Eq. (39). It is consistent with this approximation to take sin Θ as constant over the region of integration. Hence

$$E_{\text{plate}} = CE_0 \sin\Theta\, e^{-i\omega t} \iint_{\text{aperture}} \frac{e^{ikr}}{r} dx'dy' \tag{40}$$

Finally, we have to adjust the constant C so as to satisfy the requirement that the plate be opaque. Since C is related to the atomic properties of the plate [see Eq. (38)], this really amounts to an adjustment of these atomic constants, but in what follows, we will bypass the discussion of the microscopic properties of the material. Mathematically, the requirement that the plate be opaque is just Eq. (33),

$$-\mathbf{E}_{\text{rad}} = \mathbf{E}_{\text{in}} \tag{41}$$

The right side of this equation is simply $E_0 \hat{\mathbf{x}} e^{-i(\omega t - kx)}$. The left side is an integral of the form (40), with the region of integration covering the entire plate. However, before we write an expression for this integral, we must make two minor modifications to Eq. (40). Since we are now going to integrate over the entire plate, we cannot treat sin Θ as constant; instead, we must keep it in the integrand. Furthermore, we cannot treat all the contributing electric field vectors as parallel; instead, we must take the component of the electric field along the direction of polarization of the incident wave (x direction), this being the only component that survives the integration. In Fig. 16.1, the direction of the radiated field is in the plane of $\mathbf{E}_0$ and $\hat{\mathbf{n}}$, at right angles to $\hat{\mathbf{n}}$; hence the component of $\mathbf{E}_{\text{plate}}$ along the direction of $\mathbf{E}_0$ is $E_{\text{plate}} \times \sin\Theta$.

With these modifications taken into account, Eq. (41) becomes

$$CE_0 e^{-i\omega t} \iint_{\text{aperture}} \sin^2\Theta \frac{e^{ikr}}{r} dx'dy' = E_0 e^{-i(\omega t - kz)} \tag{42}$$

The integration is best done in cylindrical coordinates with origin at the point of observation (see Fig. 16.4). The area element on the plate is $\rho\, d\phi\, d\rho$, which, in view of $r^2 = \rho^2 + z^2$, is the same as $r\, d\phi\, dr$. The polar angles θ and ϕ at the point of observation and the point of radiation are the same; hence Eq. (12) remains valid,

$$\sin^2 \Theta = 1 - (\sin\theta \cos\phi)^2 \qquad (43)$$

The integral on the left side of Eq. (42) is then

$$\begin{aligned}\int_z^{\infty}\int_0^{2\pi} [1-(\sin^2\theta\cos^2\phi)]\frac{e^{ikr}}{r} r\, d\phi dr &= 2\pi\int_z^{\infty} (1-\tfrac{1}{2}\sin^2\theta)e^{ikr}\, dr \\ &= 2\pi\int_z^{\infty} \tfrac{1}{2}(1+\cos^2\theta)e^{ikr}\, dr \\ &= \pi\int_z^{\infty} e^{ikr}\, dr + \pi\int_z^{\infty} \frac{z^2}{r^2}e^{ikr}\, dr \qquad (44)\end{aligned}$$

The integral

$$\int_z^{\infty} e^{ikr}\, dr = \frac{1}{ik}(e^{ik\infty} - e^{-ikz}) \qquad (45)$$

is indeterminate at its upper limit. This ambiguity can be blamed on our unrealistic assumptions about the incident plane wave: we have assumed that the plane wave has infinite extent in the x and y directions. If we replace this unrealistic plane wave by a realistic plane wave of large but

Figure 16·4

Polar coordinates for the integration of Eq. (42). The origin is at the point of observation.

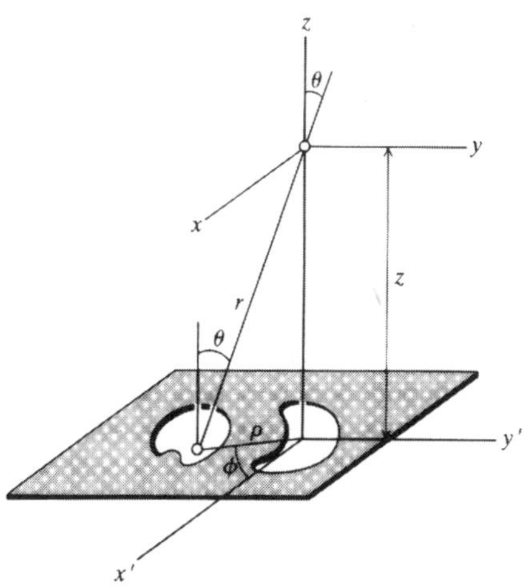

finite extent, the amplitude E_0 at the upper limit of integration will be zero, and the term $e^{-ik\infty}$ will be eliminated. The other integral in Eq. (44) can be handled by integration by parts:

$$\int_z^\infty \frac{z^2}{r^2} e^{ikr} dr = \left[\frac{1}{ik}\frac{z^2}{r^2}e^{ikr}\right]_z^\infty - \int_z^\infty \frac{-2z^2}{r^3}\frac{e^{ikr}}{ik} dr$$

$$= -\frac{1}{ik}e^{ikz} + \int_z^\infty \frac{2z^2}{r^3}\frac{e^{ikr}}{ik} dr \qquad (46)$$

Successive integrations by parts yield a power series in $1/(kz)$. If z is much larger than a wavelength, it is a good approximation to neglect all terms in this power series except the first term, $-e^{ikz}/ik$. Putting all of this together, we find that Eq. (42) reads

$$CE_0 e^{-i\omega t}\left(\frac{-2\pi e^{ikz}}{ik}\right) = E_0 e^{-i(\omega t - kz)} \qquad (47)$$

from which

$$C = -\frac{ik}{2\pi} \qquad (48)$$

Our final version of the diffraction integral is then

$$E_{\text{plate}} = -\frac{ik}{2\pi} E_0 \sin\Theta e^{-i\omega t} \iint_{\text{aperture}} \frac{e^{ikr}}{r} dx'dy' \qquad (49)$$

This is the *Kirchhoff integral* for diffraction. The restrictions on the validity of this integral must be kept in mind: the angle Θ is supposed to be about the same for all the apertures, and the distance r is supposed to be much larger than a wavelength.*

Before we proceed to the detailed solution of several examples of diffraction problems, let us briefly return to the identities derived at

---

*The angular factor sin Θ in Eq. (49), called the *obliquity factor,* differs from angular factors proposed in some other textbooks on optics or electrodynamics. The differences among the proposed factors reflect differences in the approximation methods and differences in the assumptions about the characteristics of the plate. The usual approximation employed in optics (see, for example, A. Sommerfeld, *Optics*) ignores the polarization and pretends that the electric field is a scalar. The vectorial approximation sometimes employed in electrodynamics (see, for example, J. D. Jackson, *Electrodynamics)* pretends that the plate is a perfect conductor of zero thickness. Apart from different obliquity factors, all these approximations lead to the same diffraction integral, as in Eq. (49). For most practical applications of diffraction, the obliquity factor can be ignored because the point of observation is near the forward direction as seen from the aperture, so $\sin\Theta \simeq 1$.

the beginning of this section and discuss another of their consequences. We know that the diffraction field produced by the plate equals the negative of the radiation field of the plug [see Eq. (36)],

$$\mathbf{E}_{\text{plate}} = -\mathbf{E}_{\text{rad,plug}} \tag{50}$$

If we remove the perforated plate and instead place the plugs in the path of the incident wave, then these plugs will produce a diffraction field. The plugs can be regarded as a new plate *complementary* to the original plate (see Fig. 16.5) By an argument similar to that which led to Eq. (50), the diffraction field produced by the plugs equals the negative of the radiation field of the original plate,

$$\mathbf{E}_{\text{plug}} = -\mathbf{E}_{\text{rad,plate}} \tag{51}$$

Substituting these expressions into Eq. (34), we obtain

$$\mathbf{E}_{\text{plate}} + \mathbf{E}_{\text{plug}} = \mathbf{E}_{\text{in}} \tag{52}$$

This relation between the diffraction fields of a plate and its complement is called *Babinet's principle.* It tells us the diffraction pattern produced by the plugs in terms of the diffraction pattern produced by the plate, and conversely; thus we can calculate one pattern from the other (for an example of this, see the next section). But the most remarkable application of Babinet's principle is the following: Suppose that the incident wave has a limited extent in the transverse direction; then at any point outside the beam of the incident wave, the electric field $\mathbf{E}_{\text{in}}$ is zero. Hence

$$\mathbf{E}_{\text{plate}} = -\mathbf{E}_{\text{plug}} \tag{53}$$

EXAMPLE: *diffraction by plate and by complementary plate*

This means that at such points, the intensities produced by the plate and by the plugs are identical—that is, *complementary plates produce the same diffraction patterns.* For instance, a plate with an arrangement of star-shaped apertures [see Fig. 16.6(*a*)] and the complementary "plate" with an arrangement of stars suspended in midair [see Fig. 16.6(*b*)] produce the same diffraction patterns. Figures 16.7(*a*) and 16.7(*b*) show

Figure 16·5

(*a*) An opaque plate. (*b*) The complementary plate.

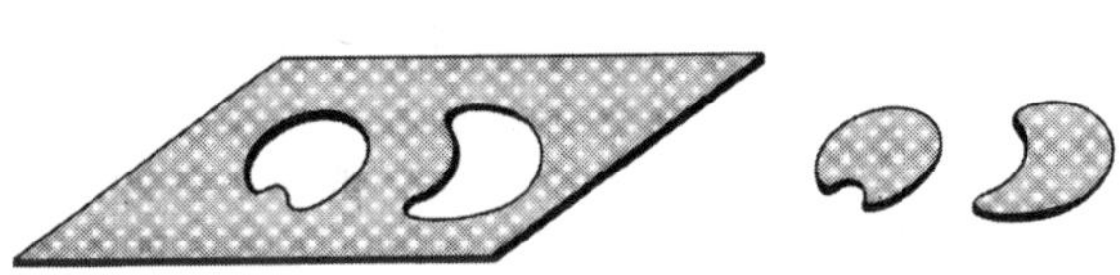

(**a**) (**b**)

Figure 16·6

(*a*) Opaque plate, with star-shaped apertures. (*b*) Complementary plate, with opaque stars.

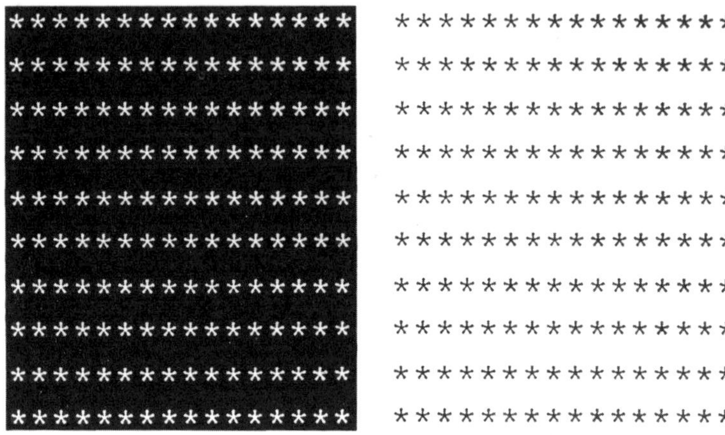

(a) (b)

Figure 16·7

Photographs of diffraction patterns of complementary plates. (*a*) Diffraction pattern of the plate shown in Fig. 16.6(*a*). (*b*) Diffraction pattern of the plate shown in Fig. 16.6(*b*).

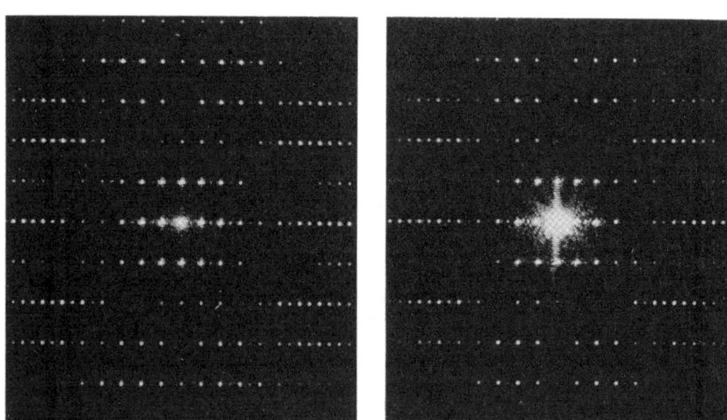

(a) (b)

photographs of these two diffraction patterns. As predicted, they are identical, except in the central region, which is in the path of the incident beam.

16·4 Fraunhofer Diffraction

We will now apply the Kirchhoff integral to some examples of diffraction. Figure 16.8 shows the coordinates x', y' of a point in the aperture of a plate and the coordinates x, y, z of the point of observation. The origin $x' = y' = 0$ is chosen at the center of the aperture or at some convenient point within or near the aperture. The distance between a point in the aperture and the point of observation is

$$r = \sqrt{(x - x')^2 + (y - y')^2 + z^2} \tag{54}$$

The integrand in Eq. (49) depends on this distance. To ease the labor of integration, we will approximate r as follows:

$$r = \sqrt{r_0^2 - 2xx' - 2yy' + \ldots} \simeq r_0 - \frac{x}{r_0}x' - \frac{y}{r_0}y' + \ldots \tag{55}$$

where

$$r_0 = \sqrt{x^2 + y^2 + z^2} \tag{56}$$

is the distance from the point $x' = y' = 0$ to the point of observation. The approximation (55) characterizes *Fraunhofer diffraction.* This approximation is valid if the distance r_0 is much larger than the size of the aperture $(r_0 \gg x',\ r_0 \gg y')$.

With the approximation (55), the Kirchhoff integral becomes

Figure 16·8

All the rays from the aperture to a very distant point of observation are nearly parallel. The path difference between the two rays shown in the z'-y' plane is $y' \sin\theta_y = y'y/r_0$.

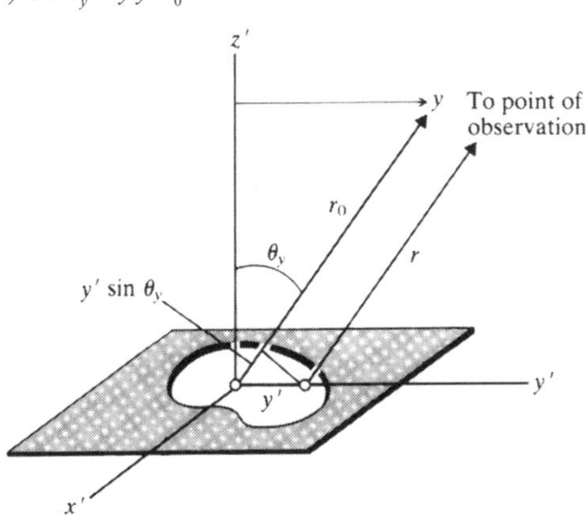

$$E_{\text{plate}} = -\frac{ik}{2\pi} E_0 \sin\Theta \frac{e^{-i(\omega t - kr_0)}}{r_0} \iint e^{-ik(xx'/r_0 + yy'/r_0)}\, dx'dy' \tag{57}$$

Note that in the denominator of the integrand, we have approximated $r \simeq r_0$; this is adequate because, in contrast to the numerator, the denominator is not very sensitive to small changes.

Geometrically, the Fraunhofer approximation corresponds to the assumption that the point of observation is so far away that the rays from the aperture to this point are nearly parallel (see Fig. 16.8). The path difference between the ray originating at $x' = y' = 0$ and another ray originating at some arbitrary point in the aperture is then (see Fig. 16.8)

$$\begin{aligned}(\text{path difference}) &= x'\sin\theta_x + y'\sin\theta_y \\ &= \frac{x'x}{r_0} + \frac{y'y}{r_0}\end{aligned} \tag{58}$$

and the phase difference is

$$(\text{phase difference}) = k \times (\text{path difference}) = k\left(\frac{x'x}{r_0} + \frac{y'y}{r_0}\right) \tag{59}$$

This phase difference accounts for the phase factor in the integrand of Eq. (57).

EXAMPLE: *diffraction by rectangular aperture*

The first example we will examine involves a rectangular aperture. For the sake of simplicity, we will suppose that x and y are small compared with r_0, so $\sin\Theta \simeq 1$. Figure 16.9 shows the dimensions a and b of the rectangular aperture. The integral (57) is then

Figure 16·9

An opaque plate with a rectangular aperture.

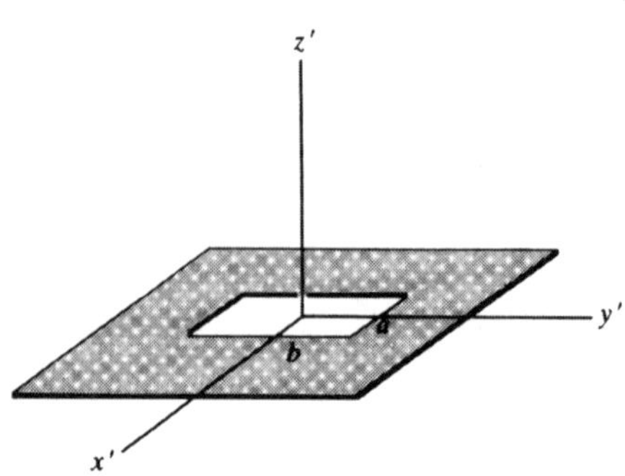

$$E_{\text{plate}} = -\frac{ik}{2\pi} E_0 \frac{e^{-i(\omega t - kr_0)}}{r_0} \int_{-b/2}^{b/2} \int_{-a/2}^{a/2} e^{-ik(xx'/r_0 + yy'/r_0)} \, dx' dy' \tag{60}$$

$$= -\frac{ik}{2\pi} E_0 \frac{e^{-i(\omega t - kr_0)}}{r_0} \frac{2r_0}{kx} \frac{2r_0}{ky} \sin\left(\frac{kax}{2r_0}\right) \sin\left(\frac{kby}{2r_0}\right) \tag{61}$$

The intensity, or the time-average Poynting vector, is

$$\bar{S} = \frac{c}{4\pi} \frac{1}{2} E_{\text{plate}} E^*_{\text{plate}} = \left(\frac{ck^2a^2b^2E_0^2}{32\pi^3 r_0^2}\right) \frac{\sin^2(kax/2r_0)}{(kax/2r_0)^2} \frac{\sin^2(kby/2r_0)}{(kby/2r_0)^2} \tag{62}$$

Figure 16.10 displays a plot of this intensity as a function of x, for y = constant. The intensity has a central maximum at $x = y = 0$, and it has a series of minima and secondary maxima. The minima occur at the locations

$$\frac{kax}{2r_0} = \pm\pi, \pm 2\pi, \ldots \qquad \frac{kby}{2r_0} = \pm\pi, \pm 2\pi, \ldots \tag{63}$$

In terms of the angles $\theta_x = \sin^{-1}(x/r_0)$ and $\theta_y = \sin^{-1}(y/r_0)$, these locations can be expressed in the familiar form

$$a \sin\theta_x = \pm\lambda, \pm 2\lambda, \ldots \qquad b \sin\theta_y = \pm\lambda, \pm 2\lambda, \ldots \tag{64}$$

Figure 16·10

Intensity as a function of x for the diffraction pattern of a rectangular aperture.

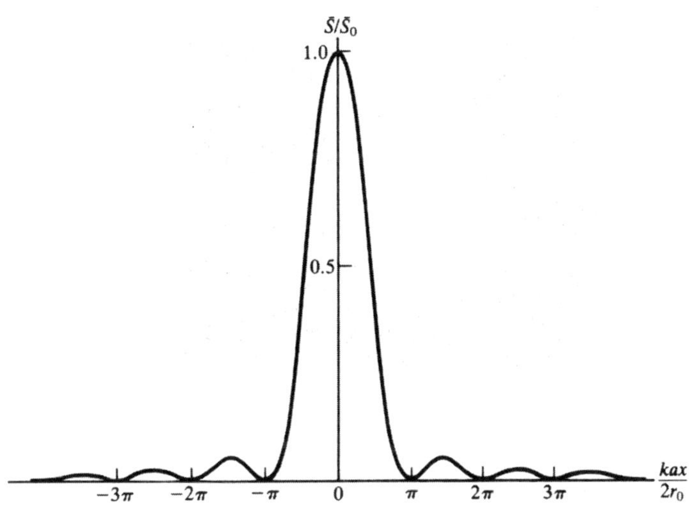

Figure. 16.11 is a photograph of the diffraction pattern of a rectangular aperture.

An interesting property of our result for Fraunhofer diffraction by a rectangular aperture is that the amplitude (61) equals—within an overall constant of proportionality—the Fourier transform of the step function $H(x,y)$ whose "support"* is the aperture:

$$H(x,y)=\begin{cases}1 & \text{if } |x|\le a/2, |y|\le b/2\\ 0 & \text{if } |x|> a/2, |y|> b/2\end{cases} \tag{65}$$

This is a general property of Fraunhofer diffraction, and it provides us with an optical-analog method for the determination of the Fourier transform of any two-dimensional step function with some arbitrary and complicated boundary. We need only cut a plate with an aperture with a boundary of the same shape and illuminate it with a plane wave; the diffraction pattern will show us the Fourier transform or, rather, the square of the Fourier transform. For a precise application of this method, we must take the square root of the observed intensity to find the amplitude.

EXAMPLE: *diffraction by rectangular plate*

An opaque rectangular plug (see Fig. 16.12) is complementary to our plate with a rectangular aperture. Babinet's principle permits us to write directly the diffraction amplitude for this plug:

$$\begin{aligned}E_{\text{plug}} &= E_{\text{in}} - E_{\text{plate}}\\ &= E_0 e^{-i\omega t}\left[e^{ikz} + \frac{e^{ikr_0}}{r_0}\frac{ik}{2\pi}\frac{2r_0}{kx}\frac{2r_0}{ky}\sin\left(\frac{kax}{2r_0}\right)\sin\left(\frac{kby}{2r_0}\right)\right]\end{aligned} \tag{66}$$

Figure 16·11

Photograph of the diffraction pattern of a rectangular aperture with $a/b = 7/8$.

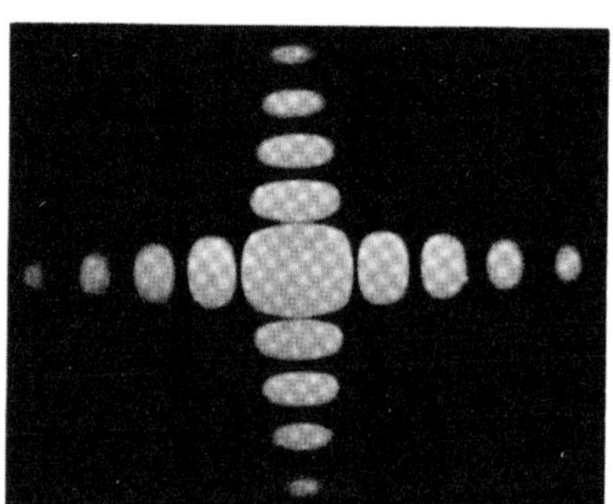

*The *support* of a function is the set of those points for which the function is nonzero.

Figure 16·12

An opaque rectangle, complementary to the opaque plate of Fig. 16.9.

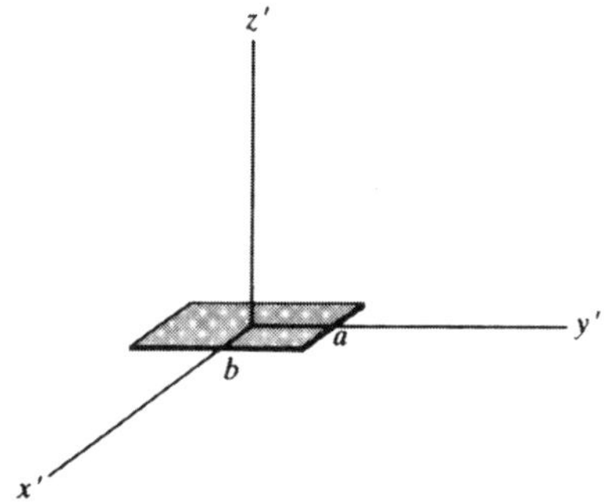

The corresponding intensity is

$$\bar{S}=\frac{c}{4\pi}\frac{1}{2}E_{\text{plug}}E^*_{\text{plug}}=\frac{cE_0^2}{8\pi}\left[1+\frac{k^2a^2b^2}{4\pi^2r_0^2}\frac{\sin^2(kax/2r_0)}{(kax/2r_0)^2}\frac{\sin^2(kby/2r_0)}{(kby/2r_0)^2}\right.$$

$$\left.+\frac{k}{\pi}\frac{ab}{r_0}\sin k(z-r_0)\frac{\sin(kax/2r_0)}{kax/2r_0}\frac{\sin(kby/2r_0)}{kby/2r_0}\right] \tag{67}$$

Figure 16.13 shows a plot of this intensity as a function of x for a constant value of y. Note that there is no central dark region—the plate

Figure 16·13

Intensity as a function of x for the diffraction pattern of an opaque rectangle, with $ka = 150$, $kb = 150$, $kr_0 = 2000$.

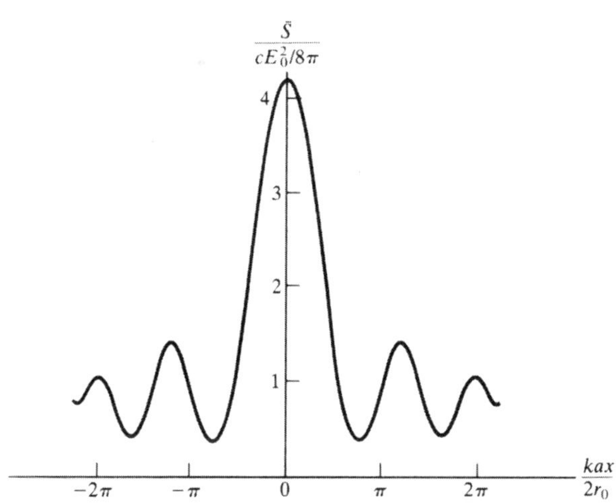

throws no shadow. The reason for this absence of a shadow is that in the context of the Fraunhofer approximation, our point of observation is so far away that the light has completely diffracted into the shadow zone. For the mathematical investigation of the shadow thrown by an obstacle of finite size, we would have to retain at least one more term in the power-series expansion (55); this would lead us into the Fresnel approximation (see the next section).

EXAMPLE: *diffraction by circular aperture*

Another worthwhile example is the circular aperture. Figure 16.14 shows an aperture of radius a. If we replace all the coordinates by cylindrical coordinates ρ', ϕ', we find

$$xx' + yy' = \rho\rho'(\cos\phi\cos\phi' + \sin\phi\sin\phi') = \rho\rho'\cos(\phi - \phi') \qquad (68)$$

The integral (57) then becomes, with $\sin\Theta \simeq 1$,

$$E_{\text{plate}} = -\frac{ik}{2\pi}E_0\frac{e^{-i(\omega t - kr_0)}}{r_0}\int_0^a\int_0^{2\pi} e^{-ik\rho\rho'\cos(\phi-\phi')/r_0}\,\rho' d\phi' d\rho' \qquad (69)$$

Since ϕ' is integrated over the entire circle, we can replace $\cos(\phi - \phi')$ by $\cos\phi'$ without changing the value of the integral. With the abbreviations

$$u = \frac{k}{r_0}\rho a \qquad v = \frac{k}{r_0}\rho\rho' \qquad (70)$$

our integral becomes

$$E_{\text{plate}} = -\frac{ik}{2\pi}E_0\frac{e^{-i(\omega t - kr_0)}}{r_0}\frac{a^2}{u^2}\int_0^u\int_0^{2\pi} e^{-iv\cos\phi'}\,v\,d\phi' dv \qquad (71)$$

Figure 16·14

An opaque plate with a circular aperture.

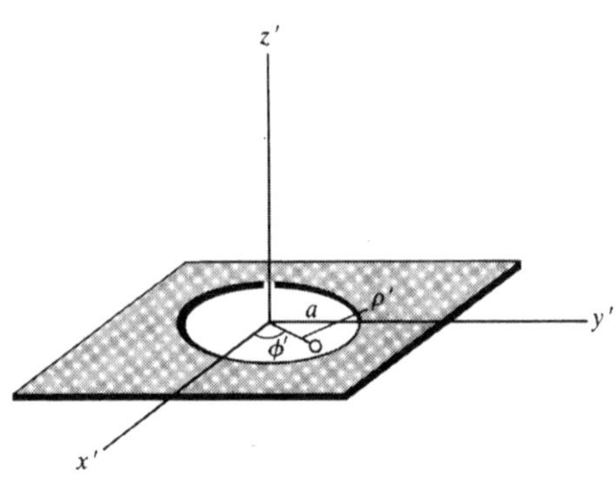

To evaluate this, we expand $e^{-iv\cos\phi'}$ in a power series,

$$e^{-iv\cos\phi'} = 1 - iv\cos\phi' - \frac{1}{2!}v^2\cos^2\phi' + \frac{i}{3!}v^3\cos^3\phi' + \ldots \tag{72}$$

and integrate term by term with respect to ϕ' and v. This leads to

$$E_{\text{plate}} = -\frac{ik}{2\pi}E_0\frac{e^{-i(\omega t - kr_0)}}{r_0}\frac{2\pi a^2}{u^2}\left[\frac{u^2}{2} - \frac{u^4}{2^4} + \frac{u^6}{3\cdot 2^7} + \ldots\right] \tag{73}$$

■ ***Exercise 3.*** Integrate the right side of Eq. (71) term by term and verify Eq. (73). ■

The series appearing in brackets in Eq. (73) can be written in the concise form

$$\frac{u^2}{2}\sum_{l=0}^{\infty}\frac{(-1)^l}{l!(l+1)!}\left(\frac{u}{2}\right)^{2l} \tag{74}$$

Comparison with Eq. (3.64) shows this series is simply $uJ_1(u)$, where J_1 is the Bessel function of order 1. The diffraction field of the circular aperture is therefore

$$E_{\text{plate}} = -ikE_0\frac{e^{-i(\omega t - kr_0)}}{r_0}a^2\frac{J_1(u)}{u} \tag{75}$$

and the intensity is

$$\overline{S} = \frac{c}{4\pi}\frac{1}{2}E_{\text{plate}}E^*_{\text{plate}} = \frac{cE_0^2}{8\pi}\frac{k^2a^4}{r_0^2}\left[\frac{J_1(u)}{u}\right]^2 \tag{76}$$

Figure 16.15 shows a plot of this intensity. As in the case of the rectangular aperture, there is a strong central maximum and there are weaker secondary maxima. The locations of the minima are given by the roots of the Bessel function $J_1(u)$:

$$u = 3.83, 7.02, 10.17, \ldots$$

or

$$\rho = 3.83\frac{r_0}{ka},\ 7.02\frac{r_0}{ka},\ 10.17\frac{r_0}{ka}, \ldots$$

In optics, this is usually expressed in terms of the diameter $D = 2a$ of the aperture, the wavelength $\lambda = 2\pi/k$, and the angle $\theta = \sin^{-1}(\rho/r_0)$:

$$\sin\theta = \frac{\rho}{r_0} = 1.22\frac{\lambda}{D}, 2.23\frac{\lambda}{D}, 3.24\frac{\lambda}{D}, \ldots \tag{77}$$

Figure 16·15

Intensity as a function of ρ for the diffraction pattern of a circular aperture.

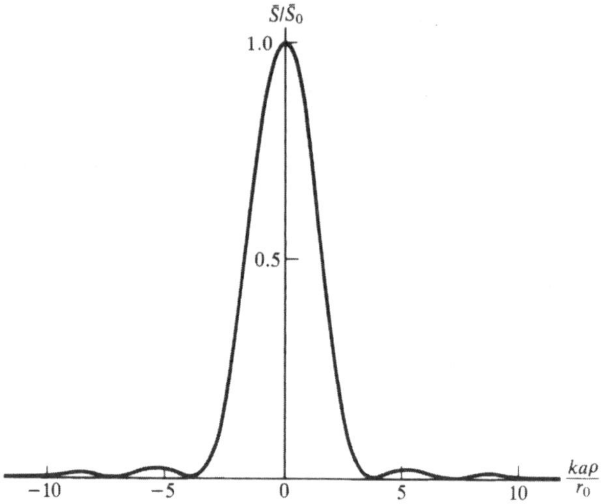

Figure 16·16

Photograph of the diffraction pattern of a circular aperture.

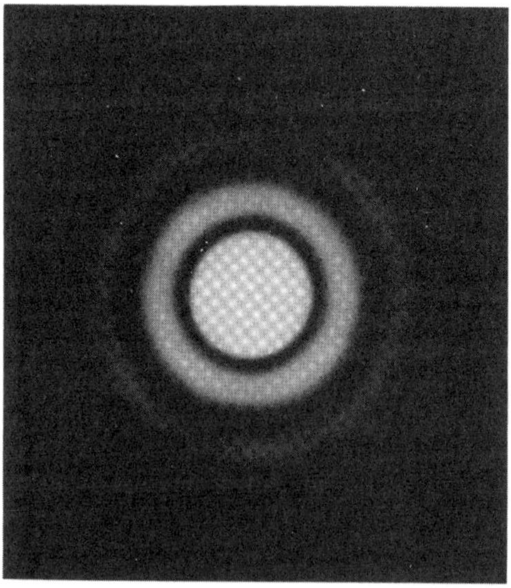

Figure 16.16 is a photograph of the diffraction pattern of a circular aperture.

16·5 Fresnel Diffraction

If the distance r_0 from the aperture to the point of observation is not extremely large, the approximation we have used in Eq. (55) is inadequate. We must then retain higher-order terms in the power-series expansion of r, for instance, terms of order x'^2 and y'^2:

$$r = \sqrt{r_0^2 - 2xx' - 2yy' + x'^2 + y'^2}$$

$$\simeq r_0 - \frac{x}{r_0}x' - \frac{y}{r_0}y' + \frac{1}{2r_0}(x'^2 + y'^2) - \frac{1}{2r_0}\left(\frac{x}{r_0}x' + \frac{y}{r_0}y'\right)^2 + \ldots \qquad (78)$$

The approximation (78) characterizes *Fresnel diffraction.* The ranges of validity of Fraunhofer and Fresnel diffraction are not sharply separated; as a rough rule, we can say that Fraunhofer diffraction is applicable when $r_0 \gg a^2/\lambda$, and Fresnel diffraction is applicable when $r_0 \sim a^2/\lambda$, where a is the characteristic size of the aperture (the condition $r_0 \gg a^2/\lambda$ implies that the phases contributed by the second-order terms in (78) are small—not just small compared with the first-order terms, but small in themselves). But we must keep in mind that sometimes the Fraunhofer approximation is completely invalid, even when r_0 is large. For instance, the diffraction pattern of a straight edge (one-half of an infinite opaque plate; see below) cannot be calculated from the Fraunhofer approximation, because the Kirchhoff integral diverges, regardless of the value of r_0.

With the Fresnel approximation (78), the integrand of the Kirchhoff integral becomes

$$\frac{1}{r_0}\exp\left\{ik\left[r_0 - \frac{x}{r_0}x' - \frac{y}{r_0}y' + \frac{1}{2r_0}(x'^2 + y'^2) - \frac{1}{2r_0}\left(\frac{x}{r_0}x' + \frac{y}{r_0}y'\right)^2\right]\right\} \qquad (79)$$

where, as in the preceding section, we have approximated $r \simeq r_0$ in the denominator. Since the argument of the exponential function has a rather messy dependence on x' and y', we find it convenient to simplify this argument by a trick: for any given point of observation, we shift the origin of coordinates in the plane of the plate so that this origin is exactly opposite the point of observation (see Fig. 16.17); this makes $x = y = 0$ and leaves us with only

$$\frac{1}{r_0}\exp\left\{ik\left[r_0 + \frac{1}{2r_0}(x'^2 + y'^2)\right]\right\} \qquad (80)$$

Figure 16·17

Choice of the origin of coordinates for Fresnel diffraction. The origin of coordinates has been shifted so that it lies exactly opposite to (or below) the point of observation. The coordinates of the point of observation are now $x = 0, y = 0, z = r_0$.

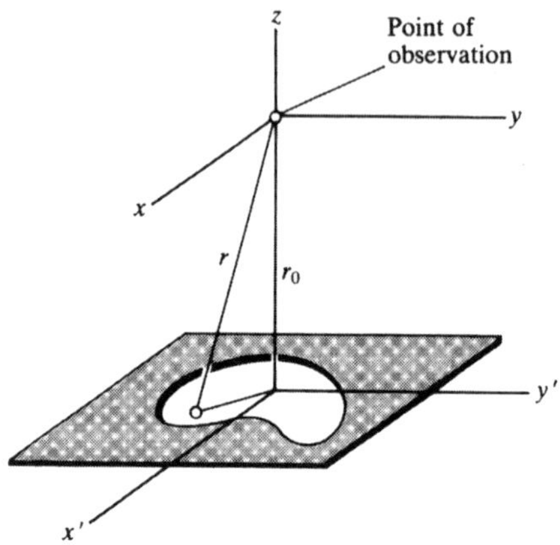

Of course, since we have chosen the origin to suit the point of observation, our limits of integration will now depend on the position of the point of observation, but we will find this dependence quite easy to deal with. With Eq. (80), the Kirchhoff integral reduces to

$$E_{\text{plate}} = -\frac{ik}{2\pi} E_0 \sin\Theta \frac{e^{-i(\omega t - kr_0)}}{r_0} \iint e^{ik(x_0'^2 + y'^2)/2r_0}\, dx'dy' \qquad (81)$$

EXAMPLE: *rectangular aperture, Fresnel approximation*

For a first example of the Fresnel approximation, we return to the case of the rectangular aperture illustrated in Fig. 16.9. As in other calculations, we will suppose that $\sin\Theta \simeq 1$. If the point of observation has coordinates x, y with respect to the center of the rectangle, then this center has coordinates $-x$, $-y$ with respect to the point of observation. Therefore, with our shifted origin of coordinates, the limits of integration for x' are $-x - a/2$, $-x + a/2$; and the limits of integration for y' are $-y - b/2, -y + b/2$. Introducing the abbreviations

$$w_1 = \sqrt{\frac{k}{\pi r_0}}\left(-x - \frac{a}{2}\right) \qquad w_2 = \sqrt{\frac{k}{\pi r_0}}\left(-x + \frac{a}{2}\right) \qquad (82)$$

$$v_1 = \sqrt{\frac{k}{\pi r_0}}\left(-y - \frac{b}{2}\right) \qquad v_2 = \sqrt{\frac{k}{\pi r_0}}\left(-y + \frac{b}{2}\right) \qquad (83)$$

we readily obtain from Eq. (81)

$$E_{\text{plate}} = -\frac{i}{2}E_0 \mathrm{e}^{-i(\omega t - kr_0)}[F(w_2) - F(w_1)][F(v_2) - F(v_1)] \tag{84}$$

where $F(w)$ is the integral

$$F(w) = \int_0^w \mathrm{e}^{i\pi\tau^2/2}\, d\tau \tag{85}$$

This integral is called *Fresnel's integral*; it is usually expressed in terms of real and imaginary parts:

$$F(w) = \mathscr{C}(w) + i\,\mathscr{S}(w) \tag{86}$$

with

$$\mathscr{C}(w) = \int_0^w \cos(\pi\tau^2/2)\, d\tau \tag{87}$$

$$\mathscr{S}(w) = \int_0^w \sin(\pi\tau^2/2)\, d\tau \tag{88}$$

These integrals cannot be evaluated in terms of elementary functions; they must be evaluated numerically.

Figure 16.18 gives a graphical representation of $F(w)$ in the complex plane. The real and the imaginary parts of $F(w)$ are regarded as

Figure 16·18

The Cornu spiral. The numbers marked along the curve give the values of the parameter w.

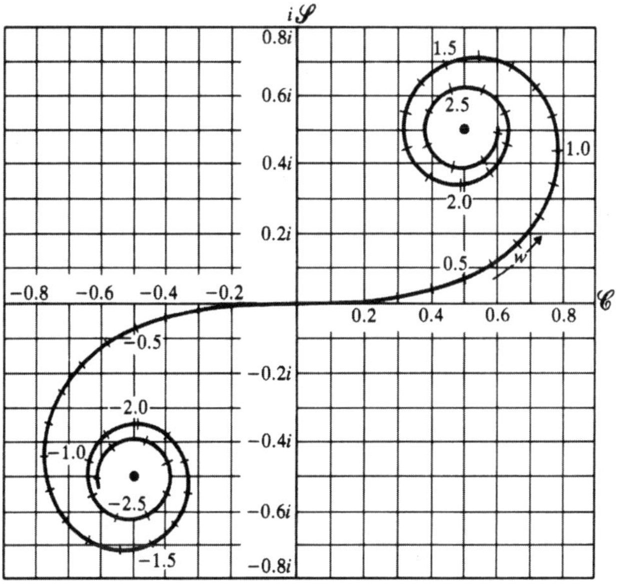

horizontal and as vertical coordinates in this plane, and w is regarded as a parameter. As a function of w, the horizontal and vertical coordinates trace out a curve, called the *Cornu spiral*. The spiral is symmetric with respect to the origin, $F(-w)=-F(w)$. For $w \to \pm\infty$, the spiral winds around the points $\pm(1+i)/2$. Note that the quantity $F(w_2) - F(w_1)$ appearing in Eq. (84) is represented graphically by the vector from the point w_1 to the point w_2 on the Cornu spiral.

The intensity in the diffraction pattern is

$$\begin{aligned}\bar{S} &= \frac{c}{4\pi}\frac{1}{2}E_{\text{plate}}E^*_{\text{plate}} \\ &= \frac{cE_0^2}{32\pi}\left\{[\mathscr{C}(w_2)-\mathscr{C}(w_1)]^2+[\mathscr{S}(w_2)-\mathscr{S}(w_1)]^2\right\} \\ &= \frac{cE_0^2}{32\pi}\left\{[\mathscr{C}(w_2)-\mathscr{C}(w_1)]^2+[\mathscr{S}(w_2)-\mathscr{S}(w_1)]^2\right\} \\ &\quad\times\left\{[\mathscr{C}(v_2)-\mathscr{C}(v_1)]^2+[\mathscr{S}(v_2)-\mathscr{S}(v_1)]^2\right\}\end{aligned} \tag{89}$$

Figure 16.19 shows a plot of this intensity pattern as a function of x for constant y. This differs somewhat from the Fraunhofer pattern but tends toward the Fraunhofer pattern when $r_0 \to \infty$.

■ ***Exercise 4.*** Prove that when $r_0 \to \infty$, the intensity formula (89) approaches the Fraunhofer formula (62). ■

Figure 16·19

Intensity as a function of x for the diffraction pattern of a rectangular aperture, with $a\sqrt{k/\pi r_0} = 1.80$. The right half of this plot gives the Fresnel approximation; the left half gives the Fraunhofer approximation, for comparison.

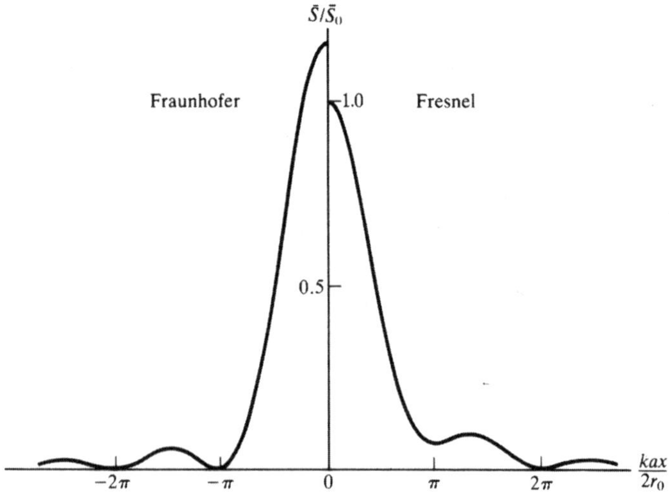

EXAMPLE: *straight edge, Fresnel approximation*

The diffraction pattern produced by a straight edge (one-half of an infinite opaque plate; see Fig. 16.20) follows from Eq. (89) if we take the limit $w_1 \to -\infty$, $v_1 \to -\infty$, $v_2 \to \infty$ and set $w_2 = -\sqrt{k/\pi r_0}\,x$. This leads to

$$\bar{S} = \frac{cE_0^2}{16\pi}\left\{\left[\mathscr{C}(w_2)+\frac{1}{2}\right]^2 + \left[\mathscr{S}(w_2)+\frac{1}{2}\right]^2\right\} \tag{90}$$

Figure 16.21 displays a plot of the intensity as a function of the distance x measured from the edge of the geometrical shadow. In the shadow zone ($x > 0$), the intensity gradually decreases toward zero. In the illuminated zone ($x < 0$), the intensity exhibits a series of diffraction fringes. At the edge of the shadow ($x = 0$), the intensity is exactly ¼ of the intensity of the incident wave.

EXAMPLE: *circular aperture, Fresnel approximation*

Next we reexamine the case of the circular aperture. For the sake of simplicity, we assume that the point of observation is on the axis of the aperture. In polar coordinates, the integral (81) is then

$$E_{\text{plate}} = -\frac{ik}{2\pi}E_0\frac{e^{-i(\omega t - kr_0)}}{r_0}\int_0^a\int_0^{2\pi} e^{ik\rho'^2/2r_0}\,\rho' d\phi' d\rho' \tag{91}$$

The integration is trivial and yields

$$E_{\text{plate}} = -E_0 e^{-i(\omega t - kr_0)}[e^{ika^2/2r_0} - 1] \tag{92}$$

The corresponding intensity is

$$\bar{S} = \frac{cE_0^2}{2\pi}\sin^2\left(\frac{ka^2}{4r_0}\right) \tag{93}$$

Figure 16·20

An opaque plate occupies the x'-y' plane, from $x' = 0$ to $x' = \infty$. The edge of the plate coincides with the y' axis.

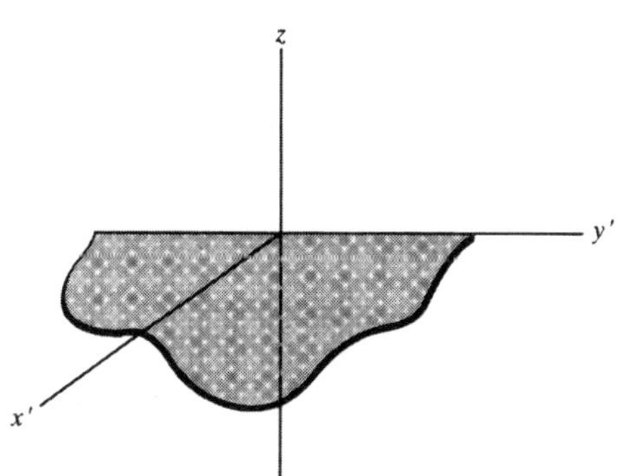

Figure 16·21

Intensity as a function of x for the diffraction pattern of a straight edge. The shadow zone corresponds to $x > 0$ (see Fig. 16.20).

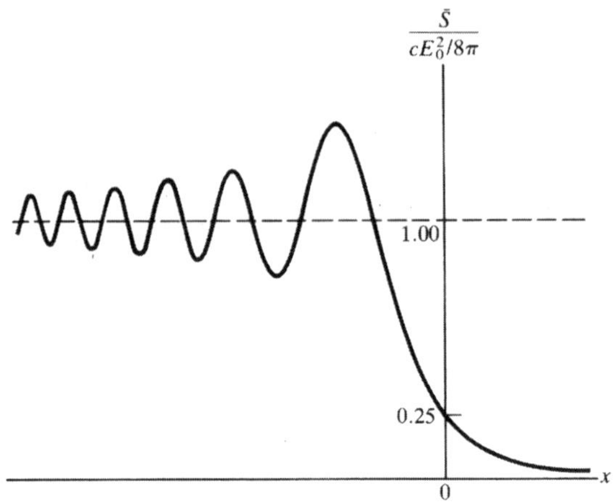

As a function of r_0, this intensity oscillates between $cE_0^2/2\pi$ and zero (see Fig. 16.22)—there are points on the axis of the hole at which there is no light!

■ ***Exercise 5.*** Use Babinet's principle and Eq. (92) to show that the intensity on the axis of an opaque disk is $cE_0^2/8\pi$, the same as the intensity of the incident wave. Thus the axis of a disk is not dark. The bright spot in the center of the shadow of a disk is called *Poisson's spot.* ■

Figure 16·22

Intensity on the axis of a circular aperture as a function of the distance r_0.

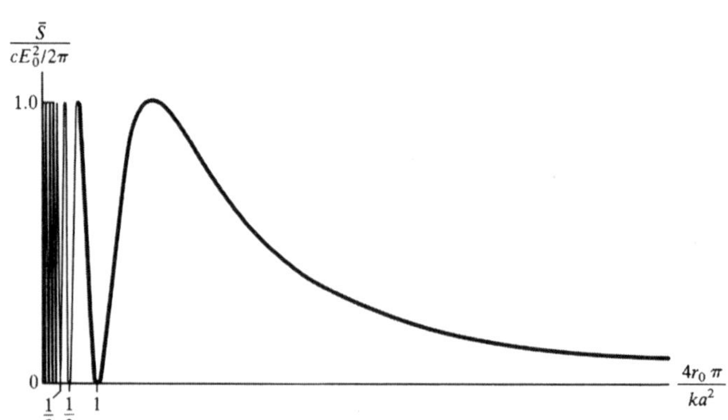

EXAMPLE: *zone plate*

At the maxima, Eq. (93) yields an intensity four times as large as that of the incident light; this is a substantial enhancement of intensity. Much larger enhancements can be produced by a *zone plate* consisting of alternate opaque and transparent concentric rings. The sizes of these rings are chosen so that the opaque rings block out those radial intervals in the integral (91) that give rise to destructive interference. The integrand has a positive imaginary part in the interval $0 < \rho' < \sqrt{2\pi r_0 / k}$; a negative imaginary part in the interval $\sqrt{2\pi r_0 / k} < \rho' < \sqrt{4\pi r_0 / k}$; again, a positive imaginary part in the next interval; and so on (see Fig. 16.23). Thus the integrand has a positive phase (between 0° and 180°) in the first interval; a negative phase (between 0° and –180°) in the second interval; again, a positive phase in the third interval; and so on. If we manufacture a plate with opaque rings that block out all the radial intervals with negative phases, the diffraction field of this plate will have a very strong maximum at a point on the axis where the waves from adjacent transparent rings come together with positive phases. The required radii defining the boundaries of the opaque rings are

$$\rho_l' = \sqrt{\frac{2\pi l r_0}{k}} = \sqrt{l r_0 \lambda} \qquad l = 1, 2, 3, \ldots \tag{94}$$

Figure 16.24 shows such a zone plate. This zone plate, with radii given by Eq. (94), yields a strong bright spot at $r_0 = \rho_1'^2 / \lambda$.

■ ***Exercise 6.*** Show that this zone plate also yields bright spots at, roughly,

$$r_0 = \rho_1'^2/3\lambda,\ \rho_1'^2/5\lambda,\ \rho_1'^2/7\lambda, \ldots \tag{95}$$

[*Hint:* For $r_0 = \rho_1'^2/3\lambda$, the phase in each transparent radial interval spans a range of $3 \times 180°$.] ■

Figure 16·23

The function $\sin(k\rho'^2 / 2r_0)$. This function changes sign at the points $\rho_l' = \sqrt{2\pi l r_0 / k}$.

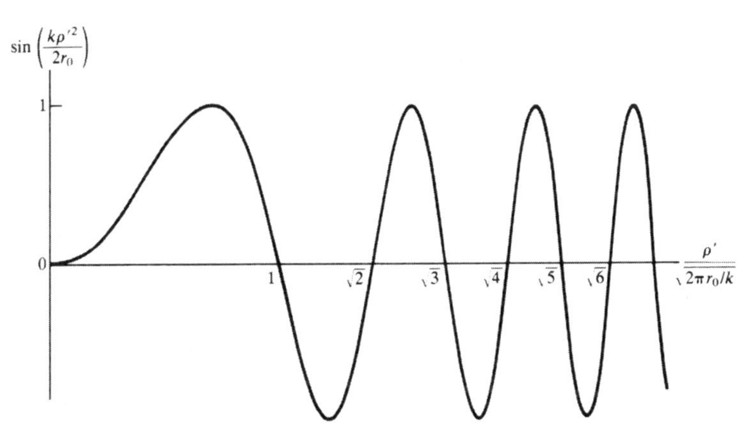

Figure 16·24

A zone plate. The radius of the inner circle is $\rho_1' = 0.8$ cm; when illuminated with light of wavelength 500 nm, this zone plate yields a bright spot at a distance $r_0 = (0.8\text{ cm})^2/(500\times10^{-7}\text{ cm}) = 1.3\times10^4$ cm. Reduction of this zone plate to $1/10$ of its size would yield a bright spot at 1.3×10^2 cm.

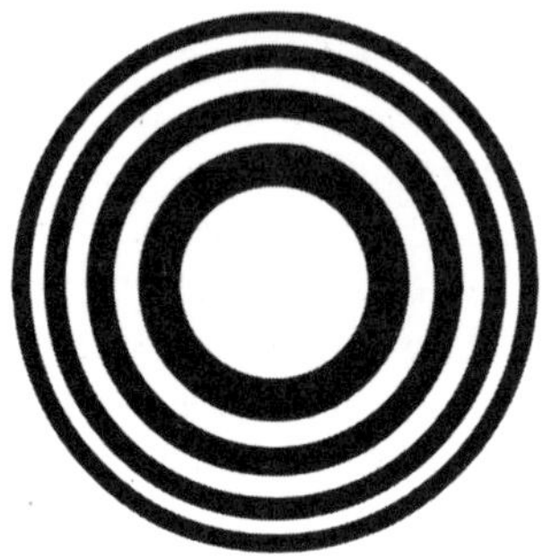

EXAMPLE: *diffraction by a lens*

Our final example involves a rather different problem. We want to calculate the intensity produced by a lens when a plane wave is incident on it. A lens can be thought of as a circular aperture; thus we can again use the integral (91). However, the lens differs from an empty aperture in that it gives a phase shift to the wave passing through. For instance, suppose the lens is plano-convex with a radius of curvature R (see Fig. 16.25). The thickness of the lens as a function of the radial coordinate ρ' is then approximately (for $\rho' \ll R$)

$$h \simeq h_0 - \frac{1}{2}\frac{\rho'^2}{R} \tag{96}$$

Figure 16·25

A thin, plano-convex lens. The curved surface has a radius R. The thickness of the lens at the center is h_0.

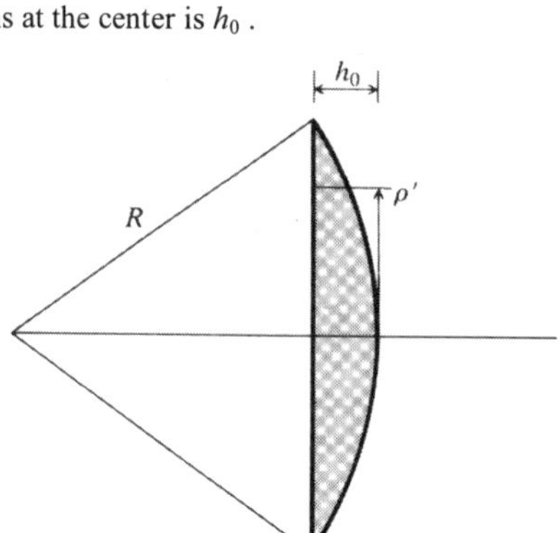

If the index of refraction is n, the phase shift that a wave acquires when it passes through the lens at the radial distance ρ' is nkh, where k is the wave vector in empty space, $k = \omega/c$. The extra phase shift, as compared with the phase shift for an empty aperture, is therefore

$$\delta(\rho') = (n-1)kh = (n-1)kh_0 - (n-1)\frac{k\rho'^2}{2R} \tag{97}$$

The constant phase shift $(n-1)kh_0$ in Eq. (97) is of no consequence and can be ignored.

With the inclusion of the extra phase shift, Eq. (91) is altered to

$$E_{\text{lens}} = -\frac{ik}{2\pi}E_0\frac{e^{-i(\omega t - kr_0)}}{r_0}\int_0^a\int_0^{2\pi} e^{ik\rho'^2/2r_0}e^{-i(n-1)k\rho'^2/2R}\,\rho' d\phi' d\rho' \tag{98}$$

This, of course, is valid only along the axis of the lens. The integration is again trivial:

$$E_{\text{lens}} = -E_0 e^{-i(\omega t - kr_0)}\frac{\exp\{ika^2[1/r_0 - (n-1)/R]/2\} - 1}{1-(n-1)r_0/R} \tag{99}$$

The intensity is

$$\overline{S} = \frac{cE_0^2}{2\pi}\frac{\sin^2\{ka^2[1/r_0 - (n-1)/R]/4\}}{[1-(n-1)r_0/R]^2} \tag{100}$$

This has a maximum at the point

$$r_0 = \frac{R}{n-1} \tag{101}$$

where the denominator vanishes. Obviously, this must be the focal point of the lens, and the distance $R/(n-1)$ must be the focal distance. Our result for the focal distance agrees with the result derived from geometrical optics (lens maker's formula).

According to geometrical optics, the intensity at the focal point is infinite. But our formula (100) shows that the intensity remains finite at $r_0 = R/(n-1)$, and it has the value

$$\overline{S} = \frac{cE_0^2}{2\pi}\left(\frac{ka^2}{4r_0}\right)^2 = \frac{cE_0^2}{8\pi}\times\left(\frac{\pi a^2}{\lambda r_0}\right)^2 \tag{102}$$

On the right side of this equation, the first factor represents the intensity of the incident plane wave, and the second factor represents the intensity enhancement produced by the lens at its focal point. According to Eq. (100) the width of the intensity peak in the longitudinal direction (measured from maximum to nearest minimum) is given by $\delta(ka^2/4r_0) = \pm\pi$, or approximately $(ka^2/4r_0^2)\delta r_0 = \pm\pi$, which yields

$$\delta r_0 \simeq 2\frac{r_0^2\lambda}{a^2} \qquad (103)$$

The width of the intensity peak in the transverse direction cannot be obtained from our formula (99), because this applies only along the axis of the lens. The transverse dimension is determined by diffraction effects and is given by Eq. (77),

$$\delta\rho = r_0 \sin\theta = \frac{1.22}{2}\frac{r_0\lambda}{a} \qquad (104)$$

Comparing the longitudinal and transverse dimensions, we see that, apart from a numerical factor of the order of magnitude of 1, they differ by a factor a/r_0. This is a geometrical effect: a ray from the edge of the lens that misses the focus by a longitudinal distance δr_0 will miss the focus by a smaller transverse distance $\delta r_0 \times (a/r_0)$, where the extra factor is due to the slope of the ray. The size of the focal spot given by Eqs. (103) and (104) is quite small, roughly of the order of magnitude of a wavelength; and, correspondingly, the intensity enhancement given by Eq. (10) is very large. In practice, the focusing by lenses is often limited by aberration or by the angular spread within the incident beam.

Note that at the focal point, the denominator in Eq. (99) changes sign. This leads to the well-known phase reversal at the focal point: the electric field just in front of the focal point is 180° out of phase with the electric field just behind the focal point.

Problems

1. The scattered blue light in the sky is partially polarized. You can detect this with Polaroid sunglasses. In what direction, relative to the Sun, must you look to observe maximum polarization? What is the plane of this maximum polarization of the light?

2. Equation (66) gives the Fraunhofer approximation for the amplitude of the light diffracted by a rectangular plate. Deduce the differential scattering cross section; express your answer in polar coordinates. If you integrate this expression for the differential cross section over all solid angles, will you obtain the correct total scattering cross section? Explain.

3. According to Eq. (66), what is the forward scattering amplitude of a rectangular plate of dimension $a \times b$? Use the optical theorem to deduce the total scattering cross section of the plate, and compare it with the geometric cross section $a \times b$.

4. By Babinet's principle, an opaque perforated plate S and the complementary opaque plug S' produce the same diffraction pattern (with the same intensity) in those directions that are outside the incident beam. Show that if the complementary plug S' is not opaque but, rather, is made of transparent material that shifts the phase of the transmitted light by 180° (half-wave phase-retarding material), then this plug S' produces the same diffraction pattern as the perforated plate S but with an intensity four times as large.

5. Find the Fraunhofer diffraction pattern produced by a square aperture measuring $a \times a$ whose center is blocked off by a smaller square insert measuring $b \times b$ [see Fig. 16.26)]. What is the (angular) position of the first minimum in the direction perpendicular to a side if $b = 0.5a$? If $b = 0.9\,a$?

6. Calculate the Fraunhofer diffraction pattern produced by a square hole measuring $a \times a$ if the direction of incidence of the wave is tilted by an angle α toward one side of the square. What is the angular position of the central maximum? What are the angular widths of this maximum in the two orthogonal directions perpendicular to the sides of the square?

7. A circular aperture of diameter D is being illuminated by two very distant, pointlike light sources—such as two stars—with a small angular separation. The light incident on the aperture then consists of two approximately plane waves of slightly different directions of propagation. The waves will produce two overlapping diffraction patterns of the form (76) beyond the aperture. According to *Rayleigh's criterion,* the two patterns are regarded as barely resolved (barely distinguishable) if the maximum of one falls on the first minimum of the other. Show that this requires an angular separation of $1.22\lambda/D$ between the light sources. Evaluate this angular separation for the Space Telescope (λ = 550 nm, D = 2.4 m) and for the Arecibo radio telescope (λ = 4 cm, D = 300 m). Which of these achieves the better angular resolution?

Figure 16·26

Square aperture of dimension $a \times a$ with its center blocked off by an opaque square of dimension $b \times b$.

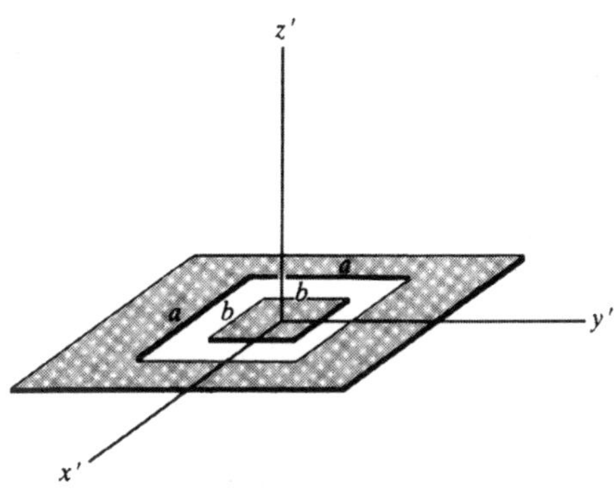

8. How would the pattern of dots in Fig. 16.7(a) change if each of the stars in the plate shown in Fig. 16.6(a) were made half as large? What if the separation between the stars were increased by a factor of 2? Explain qualitatively.

9. A radio telescope consists of 20 small antennas arranged on a straight line with a distance of 60 m between adjacent antennas. If the wavelength of the radio waves is 1 m, what is the angular resolution of the telescope? That is, what is the minimum angular separation that two point radio sources in the sky must have if they are to show up as separate in the telescope?

10. The colors of the rainbow are due to refraction of sunlight by raindrops. For sharp color separation, the angular spread produced by diffraction of light by the raindrops must be significantly smaller than the width of the rainbow, which is roughly ½°. If this condition is not satisfied, the colors overlap and merge—the result is a *white rainbow.* Estimate the width of the main diffraction peak produced by a raindrop of diameter 0.4 mm; repeat for a raindrop of diameter 0.04 mm.

11. A rectangular plate of glass of index of refraction n has height a, width b, and thickness d. Calculate the Fraunhofer diffraction pattern generated by this plate. Assume that the glass does not reflect any of the incident light; that is, the glass only produces a phase shift. Furthermore, assume that the plate is very thin ($d \ll \lambda$) but that $n \gg 1$, so nd is of the order of magnitude of λ.

12. Find the Fraunhofer diffraction pattern for two adjacent square holes in an opaque plate. The holes measure $a \times a$, they have the same orientation, and their center-to-center distance is $d > a$.

13. Find the Fraunhofer diffraction pattern for an opaque plate with two very long slits of width a intersecting at right angles (a cross). Roughly sketch the pattern of bright zones in the x-y plane.

14. Consider the point $x = 16\lambda$, $y = 0$, $r_0 = 120\lambda$ in the diffraction field of a rectangular aperture with $a = 12\lambda$, $b = 6\lambda$. Use the graphical method based on the Cornu spiral (Fig. 16.18) for an approximate evaluation of the intensity (89) at this point.

15. According to Eq. (89), what is the intensity of light in the forward direction ($x = 0$, $y = 0$) beyond a square aperture measuring $20\lambda \times 20\lambda$? Use the Cornu spiral to plot the intensity as a function of r_0, for $r_0 = 100\lambda$ to 600λ.

16. Consider a very long, opaque strip of width a illuminated with a plane wave. Use the Fresnel approximation to find the intensity as a function of distance along a line perpendicular to the strip and passing through the center of the strip.

17. A strip zone plate consists of very long, parallel adjacent bands, alternately opaque and transparent. What must be the widths of successive bands if they are to produce constructive interference of light of wavelength λ along a parallel line at a distance r_0 beyond the centerline of the strip plate?

18. Use Eq. (84) in conjunction with Babinet's principle to obtain the Fresnel diffraction pattern for an opaque rectangle. What is the intensity in the forward direction ($x = 0$, $y = 0$)?

19. A semi-infinite plate of glass fills the lower half ($y < 0$) of the x-y plate. The glass has index of refraction n and thickness d. Use the Fresnel approximation to calculate the diffraction pattern. Assume that the glass does not reflect any of the incident

light; that is, the glass only produces a phase shift. Furthermore, assume that the plate is very thin ($d \ll \lambda$) but that $n \gg 1$, so nd is of the order of magnitude of λ.

20. Find the Fresnel diffraction pattern for a very long slit of width a in an opaque plate illuminated with a plane wave. What is the intensity in the forward direction ($x = 0$, $y = 0$)?

21. Calculate the Fresnel diffraction pattern for a very long, opaque strip of width a illuminated with a plane wave. Do this calculation in two ways and compare the results:

(a) Use Eq. (81) with appropriate limits to integrate over the region exterior to the strip.

(b) Use Eq. (84) (for $b = \infty$) in conjunction with Babinet's principle.

22. Show that for $r_0 \gg ka^2$, Eq. (93) gives Eq. (76) at $u = 0$.

23. (a) Consider an exactly plane wave of light of wavelength 600 nm incident on a convex lens of radius 4 cm and focal length 20 cm. What is the (transverse) size of the focal spot, and what is the intensity enhancement at the focus?

(b) Consider, instead, incident light from a laser, with an intrinsic angular spread of 1 milliradian. What is the resulting size of the focal spot, and what is the intensity enhancement at the focus?

APPENDIX

Fundamental Constants 1

The SI values in this table are from P. J. Mohr and B. N. Taylor, "CODATA Recommended Values of the Fundamental Physical Constants: 2002," *Rev. Mod. Phys.* **76**, no. 4 (2004) (also accessible online at http://physics.nist.gov/constants). This compilation contains a more extensive list of best available values of fundamental constants. The digits in parentheses are the one-standard-deviation uncertainties in the last digits of the listed value. Whenever two alternative formulas are given for a quantity (e.g., Bohr magneton, $e\hbar/2m_e$ or $e\hbar/2m_e c$), the first is in SI units and the second in CGS electrostatic units.

Quantity	Symbol	Value (and uncertainty)	SI Units	CGS Units
Speed of light in vacuum	c	299792458	$m\ s^{-1}$	10^2 cm s^{-1}
Permeability of vacuum	μ_0	$4\pi = 12.566370614...$	10^{-7} N A^{-2}	-
Permittivity of vacuum	ε_0	$1/\mu_0 c^2 = 8.854187817...$	10^{-12} F m^{-1}	-
Newtonian constant of gravitation	G	6.6742(10)	10^{-11} m^3 kg^{-1} s^{-2}	10^{-8} cm^3 gm^{-1} s^{-2}
Planck constant	h	6.6260693(11)	10^{-34} J s	10^{-27} erg s
in electron volts		4.13566743(35)	10^{-15} eV s	-
$h/2\pi$	$\hbar$	1.05457168(18)	10^{-34} J s	10^{-27} erg s
in electron volts		6.58211915(56)	10^{-16} eV s	-
Planck mass, $(\hbar c/G)^{1/2}$	m_P	2.17645(16)	10^{-8} kg	10^{-5} gm
Planck length, $(\hbar G/c^3)^{1/2}$	l_P	1.61624(12)	10^{-35} m	10^{-33} cm
Elementary charge	e	1.60217653(14)	10^{-19} C	-
		4.80320440(42)	-	10^{-10} esu
Magnetic flux quantum, $h/2e$ or $hc/2e$	Φ_0	2.06783372(18)	10^{-15} Wb	10^{-7} gauss cm^2
Conductance quantum	$2e^2/h$	7.748091733(26)	10^{-5} Ω^{-1}	-
		6.963637570(23)	-	10^7 s^{-1} cm
Bohr magneton, $e\hbar/2m_e$ or $e\hbar/2m_e c$	μ_B	9.27400949(80)	10^{-24} J T^{-1}	10^{-21} erg $gauss^{-1}$
Nuclear magneton, $e\hbar/2m_p$ or $e\hbar/2m_p c$	μ_N	5.05078343(43)	10^{-27} J T^{-1}	10^{-24} erg $gauss^{-1}$
Fine-structure constant, $e^2/4\pi\varepsilon_0\hbar c$ or $e^2/\hbar c$	α	7.297352568(24)	10^{-3}	10^{-3}
Inverse fine-structure constant	α^{-1}	137.03599911(46)	1	1
Rydberg constant, $\alpha^2 m_e c/2h$	R_∞	10973731.568525(73)	m^{-1}	10^{-2} cm^{-1}
$R_\infty hc$ in eV		13.6056923(12)	eV	-
Bohr radius, $\hbar/m_e c\alpha$	a_0	0.5291772108(18)	10^{-10} m	10^{-8} cm

Quantity	Symbol	Value (and uncertainty)	SI Units	CGS Units
Electron mass	m_e	9.1093826(16)	10^{-31} kg	10^{-28} g
in u		5.4857990945(24)	10^{-4} u	10^{-4} u
in electron volts		0.510998918(44)	MeV/c^2	-
Electron-muon mass ratio	m_e/m_μ	4.83633167(13)	10^{-3}	10^{-3}
Electron-proton mass ratio	m_e/m_p	5.4461702173(25)	10^{-4}	10^{-4}
Electron charge to mass ratio	$-e/m_e$	−1.75882012(15)	10^{11} C kg^{-1}	-
		−5.27281007(45)		10^{17} esu g^{-1}
Compton wavelength, h/m_ec	λ_C	2.426310238(16)	10^{-12} m	10^{-10} cm
$\lambda_C/2\pi = \alpha a_0$	$\bar{\lambda}_C$	3.861592678(26)	10^{-13} m	10^{-11} cm
Classical electron radius, $\alpha^2 a_0$	r_e	2.817940325(28)	10^{-15} m	10^{-13} cm
Thomson cross section, $(8\pi/3)r_e^2$	σ_e	0.665245873(13)	10^{-28} m^2	10^{-24} cm^2
Electron magnetic moment	μ_e	−9.28476412(80)	10^{-24} J T^{-1}	10^{-21} erg gauss^{-1}
in Bohr magnetons	μ_e/μ_B	−1.0011596521859(38)	1	1
in nuclear magnetons	μ_e/μ_N	−1838.28197107(85)	1	1
Electron magnetic-moment anomaly, $\lvert\mu_e\rvert/\mu_B - 1$	a_e	1.1596521859(38)	10^{-3}	10^{-3}
Electron g-factor, $2(1 + a_e)$	g_e	2.0023193043718(75)	1	1
Muon mass	m_μ	1.88353140(33)	10^{-28} kg	10^{-25} g
in u		0.1134289264(30)	u	u
in electron-volts		105.6583692(94)	MeV/c^2	-
Muon-electron mass ratio	m_μ/m_e	206.7682838(54)	1	1
Muon magnetic moment	μ_μ	-4.49044799(40)	10^{-26} J T^{-1}	10^{-23} erg gauss^{-1}
in Bohr magnetons	μ_μ/μ_B	−4.84197045(13)	10^{-3}	10^{-3}
in nuclear magnetons	μ_μ/μ_N	−8.89059698(23)	1	1
Muon g-factor	g_μ	2.0023318396(12)	1	1
Tau mass	m_τ	3.16777(52)	10^{-27} kg	10^{-24} gm
in u		1.90768(31)	u	u
in electron-volts		1776.99(29)	MeV/c^2	-
Tau-electron mass ratio	m_τ/m_e	3477.48(57)	1	1
Proton mass	m_p	1.67262171(29)	10^{-27} kg	10^{-24} g
in u		1.00727646688(13)	u	u
in electron-volts		938.272029(80)	MeV/c^2	-

Quantity	Symbol	Value (and uncertainty)	SI Units	CGS Units
Proton-electron mass ratio	m_p / m_e	1836.15267261(85)	1	1
Proton charge to mass ratio	e / m_p	9.57883376(82)	10^7 C kg^{-1}	-
		2.87166212(25)	-	10^{14} esu g^{-1}
Proton Compton wavelength, $h/m_p c$	$\lambda_{C,p}$	1.3214098555(88)	10^{-15} m	10^{-13} cm
$\lambda_{C,p}/2\pi$	$\bar{\lambda}_{C,p}$	2.103089104(14)	10^{-16} m	10^{-14} cm
Proton rms charge radius	R_p	0.8750(68)	10^{-15} m	10^{-13} cm
Proton magnetic Moment	μ_p	1.41060671(12)	10^{-26} J T^{-1}	10^{-23} erg gauss^{-1}
in Bohr magnetons	μ_p / μ_B	1.521032206(15)	10^{-3}	10^{-3}
in nuclear magnetons	μ_p / μ_N	2.792847351(28)	1	1
Proton *g*-factor	g_p	5.585694701(56)	1 1	
Neutron mass	m_n	1.67492728(29)	10^{-27} kg	10^{-24} g
in u		1.00866491560(55)	u	u
in electron-volts		939.565360(81)	MeV/c^2	-
Neutron-electron mass ratio	m_n / m_e	1838.6836598(13)	1	1
Neutron-proton mass ratio	m_n / m_p	1.00137841870(58)	1	1
Neutron Compton wavelength, $h / m_n c$	$\lambda_{C,n}$	1.3195909067(88)	10^{-15} m	10^{-13} cm
$\lambda_{C,n}/2\pi$	$\bar{\lambda}_{C,n}$	0.2100194157(14)	10^{-15} m	10^{-13} cm
Neutron magnetic moment	μ_n	−0.96623645(24)	10^{-26} J T^{-1}	10^{-23} erg gauss^{-1}
in Bohr magnetons	μ_n / μ_B	−1.04187563(25)	10^{-3}	10^{-3}
in nuclear magnetons	μ_n / μ_N	−1.91304273(45)	1	1
Avogadro constant	N_A	6.0221415(10)	10^{23} mol^{-1}	10^{23} mol^{-1}
Atomic mass unit	u	1.66053886(28)	10^{-27} kg	10^{-24} g
in electron-volts		931.494043(80)	MeV/c^2	-
Faraday constant, $N_A e$	F	96485.3383(83)	C mol^{-1}	-
		2.89255767(25)	-	10^{14} esu mol^{-1}
Molar gas constant	R	8.314472(15)	J mol^{-1} K^{-1}	10^7 erg mol^{-1} K^{-1}
Boltzmann constant, R/N_A	k	1.3806505(24)	10^{-23} J K^{-1}	10^{-16} erg K^{-1}
in electron-volts		8.617343(15)	10^{-5} eV K^{-1}	-
Molar volume (ideal gas), RT/p at $T = 273.15$ K, $p = 101325$ Pa	V_m	22.413996(39)	L/mol	L/mol
Loschmidt constant, N_A / V_m	n_0	2.6867773(47)	10^{25} m^{-3}	10^{19} cm^{-3}
Stefan-Boltzmann constant, $\pi^2 k^4 / 60\hbar^3 c^2$	σ	5.670400(40)	10^{-8} W m^{-2} K^{-4}	10^{-5}erg s^{-1}cm^{-2} K^{-4}
First radiation constant, $2\pi hc^2$	c_1	3.74177138(64)	10^{-16} W m^2	10^{-5} erg s^{-1} cm^2
Second radiation constant, hc / k	c_2	1.4387752(25)	10^{-2} m K	1 cm K
Wien displacement law constant, $b = \lambda_{max} T$	b	2.8977685(51)	10^{-3} m K	10^{-1} cm K

APPENDIX

2

Correspondence Between CGS and SI Units

In the SI system, the base units used for electrodynamics are the meter, the kilogram, the second, and the ampere; hence this system is sometimes called MKSA. The ampere (A) is defined as the amount of current that produces a specified magnetic force per unit length between two long, straight parallel wires separated by a specified distance. The unit of electric charge is the coulomb (C), which is regarded as a derived unit:

$$1 \text{ coulomb} = 1 \text{ ampere} \times 1 \text{ second}$$

The equations of electrodynamics in the CGS electrostatic, or Gaussian, system and the SI system are different because the latter system treats the vacuum as though it were a material medium of permittivity ε_0 and permeability μ_0. Hence, factors of ε_0 and μ_0 appear in many equations in the SI system. Besides, sundry extra factors of 4π are inserted into some equations for the sake of simplifying a few formulas heavily used by engineers.

Because of these differences, the CGS and SI units for electric and magnetic quantities do not carry the same dimensions, and they cannot be regarded as multiples or submultiples of each other. However, for any given amount of a physical quantity—say the electric charge of one electron—there is a correspondence between the CGS numerical value of this charge and the SI numerical value of this charge.

Table A.1 lists these correspondences between the numerical values of physical quantities expressed in CGS and SI units. For instance, according to this table, 1 coulomb of charge corresponds to 2.998×10^9 esu. Hence, the charge of an electron, -4.80×10^{-10} esu in CGS units, can be expressed in SI units by dividing by 2.998×10^9,

$$4.80\times10^{-10} \text{ esu} \Leftrightarrow \frac{4.80\times10^{-10}}{2.998\times10^{9}} \text{ coulomb} = 1.60\times10^{-19} \text{ coulomb}$$

Note that this formula has been written as an equivalence rather than an equality, since the CGS and SI units differ in their dimensions.

Table A · 1 Numerical Correspondences*

Physical Quantity	Amount in SI Units	Amount in CGS Units
Charge	1 coulomb (C)	2.998×10^9 esu
Charge density	1 coulomb/m^3 (C/m^3)	2.998×10^3 esu/cm^3
Electric E field	1 volt/m (V/m)	$\frac{1}{2.998}\times10^{-4}$ statvolt/cm
Potential	1 volt (V)	$\frac{1}{299.8}$ statvolt
Dipole moment	1 coulomb·m (C·m)	2.998×10^{11} esu·cm
Polarization	1 coulomb/m^2 (C/m^2)	2.998×10^5 esu/cm^2
Electric D field	1 coulomb/m^2 (C/m^2)	$2.998\times4\pi\times10^5$ statvolts/cm
Capacitance	1 farad (F)	$(2.998)^2\times10^{11}$ cm
Magnetic B field	1 tesla (T)	10^4 gauss
Current	1 ampere (A)	2.998×10^9 statamperes
Current density	1 ampere/m^2 (A/m^2)	2.998×10^5 statamperes/cm^2
Conductivity	1 ohm^{-1}·m^{-1} (Ω^{-1}·m^{-1})	$(2.998)^2\times10^9$ s^{-1}
Resistivity	1 ohm·m (Ω·m)	$\frac{1}{(2.998)^2}\times10^{-9}$ s
Resistance	1 ohm (Ω)	$\frac{1}{(2.998)^2}\times10^{-11}$ s·cm^{-1}
Magnetic moment	1 ampere·m^2 (A·m^2)	10^3 gauss·cm^3
Magnetization	1 ampere/m (A/m)	10^{-3} gauss
Magnetic H field	1 ampere-turn/m (A/m)	$4\pi\times10^{-3}$ gauss†
Magnetic flux	1 weber (Wb)	10^8 gauss·cm^2
Inductance	1 henry (H)	$\frac{1}{(2.998)^2}\times10^{-11}$ s^2 cm^{-1}

*Based on a table in Jackson, *Classical Electrodynamics*. The numerical factor 2.998 appearing in this table is the same factor that appears in the speed of light; its exact value is 2.997 924 58.

†For historical reasons, the CGS unit for H is often called the *oersted*; but this coincides with the gauss.

Any equation of electrodynamics in the CGS language (used in this book) can be translated into a corresponding equation in the SI language by making suitable replacements for all the electromagnetic quantities appearing in the equation.

Table A.2 lists the replacements that accomplish this translation from one language to another. For instance, with the replacements for *q, E*, **B**, and *c* listed in the table, the equation for the Lorentz force

$$\mathbf{F} = q\mathbf{E} + q\frac{\mathbf{v}}{c}\times\mathbf{B} \qquad \text{(CGS)}$$

becomes

$$\mathbf{F} = \frac{q}{\sqrt{4\pi\varepsilon_0}}\sqrt{4\pi\varepsilon_0}\,\mathbf{E} + \frac{q}{\sqrt{4\pi\varepsilon_0}}\mathbf{v}\sqrt{\mu_0\varepsilon_0}\times\sqrt{\frac{4\pi}{\mu_0}}\mathbf{B}$$

that is,

$$\mathbf{F} = q\mathbf{E} + q\mathbf{v}\times\mathbf{B} \qquad \text{(SI)}$$

Note that the table can also be used to translate an equation from SI language into CGS language. For instance, the SI equation for the divergence of **D**,

$$\nabla\bullet\mathbf{D} = \rho \qquad \text{(SI)}$$

becomes

$$\nabla\bullet\left(\sqrt{\frac{\varepsilon_0}{4\pi}}\mathbf{D}\right) = \sqrt{4\pi\varepsilon_0}\,\rho$$

Canceling the common factor $\sqrt{\varepsilon_0}$ we find

$$\nabla\bullet\mathbf{D} = 4\pi\rho \qquad \text{(CGS)}$$

which is, indeed, the correct CGS equation.

Table A · 2 Dictionary for CGS–SI translations

Quantity	CGS Expression	SI Expression
Speed of light	c	$1/\sqrt{\mu_0\varepsilon_0}$
Charge	q	$q/\sqrt{4\pi\varepsilon_0}$
Charge Density	ρ	$\rho/\sqrt{4\pi\varepsilon_0}$
Electric E field	$\mathbf{E}$	$\sqrt{4\pi\varepsilon_0}\,\mathbf{E}$
Potential	Φ	$\sqrt{4\pi\varepsilon_0}\,\Phi$
Dipole moment	$\mathbf{p}$	$\mathbf{p}/\sqrt{4\pi\varepsilon_0}$
Polarization	$\mathbf{P}$	$\mathbf{P}/\sqrt{4\pi\varepsilon_0}$
Electric D field	$\mathbf{D}$	$\sqrt{4\pi/\varepsilon_0}\,\mathbf{D}$
Dielectric constant	ε	$\varepsilon/\varepsilon_0$
Capacitance	C	$C/4\pi\varepsilon_0$
Magnetic B field	$\mathbf{B}$	$\sqrt{4\pi/\mu_0}\,\mathbf{B}$
Current	I	$I/\sqrt{4\pi\varepsilon_0}$
Current Density	$\mathbf{j}$	$\mathbf{j}/\sqrt{4\pi\varepsilon_0}$
Conductivity	σ	$\sigma/4\pi\varepsilon_0$
Resistivity	ρ	$4\pi\varepsilon_0\rho$
Resistance	R	$4\pi\varepsilon_0 R$
Magnetic moment	$\mathbf{m}$	$\sqrt{\mu_0/4\pi}\,\mathbf{m}$
Magnetization	$\mathbf{M}$	$\sqrt{\mu_0/4\pi}\,\mathbf{M}$
Magnetic H field	$\mathbf{H}$	$\sqrt{4\pi\mu_0}\,\mathbf{H}$
Permeability	μ	μ/μ_0
Inductance	L	$4\pi\varepsilon_0 L$

Bibliography

The first section of this bibliography recommends some other textbooks. The second part lists further readings for each chapter. Apart from the books in this bibliography, the *American Journal of Physics* is a great resource, with many excellent articles discussing delicate points in the study of electricity and magnetism. The contents of the journal can be searched online at http://scitation.aip.org/ajp

Recommended Books

Among the graduate-level books, the most comprehensive is

Jackson, J. D., *Classical Electrodynamics*, 3rd ed. (Wiley, New York, 1999). This well-written book provides a judicious balance of physical insights and mathematical techniques, including a nice introduction to Green's functions. Unfortunately, with nearly 900 pages, the third edition suffers from overweight; the first edition (1962) was leaner and more sharply focused, and it demanded less stamina from the reader.

Panofsky, W. K. H., and Phillips, M., *Classical Electricity and Magnetism* (Addison-Wesley, Reading, MA, 1962) is another excellent graduate-level book.

Eyges, L., *The Classical Electromagnetic Field* (Dover, New York, 1980) is a very readable book which contains a critical examination of the conventional macroscopic treatment of electricity and magnetism and shows that the electric **D** field and the magnetic **H** field are superfluous.

Landau, L. D., and Lifshitz, E. M., *Classical Theory of Fields* (Wiley, New York, 1962) provides a concise and clear introduction to electrodynamics as a relativistic field theory, based on the Lagrangian approach.

Among the books at the undergraduate level, some of the most noteworthy are

Booker, H. G., *Energy in Electromagnetism* (Institution of Electrical Engineers, London, 1982) provides a stimulating, fresh view of electromagnetism, with concentration on the theoretical role of electric and magnetic energy.

Carter, G. W., *The Electromagnetic Field*, 2nd ed. (Elsevier, New York, 1967) emphasizes some practical applications of interest to engineers but is nevertheless well balanced.

Cook, D. M., *The Theory of the Electromagnetic Field* (Prentice-Hall, Englewood Cliffs, NJ, 1975) includes computer algorithms for the evaluation of integrals and the solution of the differential equations encountered in electromagnetism (for instance, an algorithm for tracing field lines).

Griffiths, D. J., *Introduction to Electrodynamics*, 3rd ed. (Prentice-Hall, Upper Saddle River, NJ, 1999).

Lorrain, P., and Corson, D., *Electromagnetic Fields and Waves*, 3rd ed. (Freeman, San Francisco, 1970)

Reitz, J. R., Milford, F. J., and Christy, R. W., *Foundations of Electromagnetic Theory*, 3rd ed. (Addison-Wesley, Reading, MA, 1979).

Wangsness, R. K., *Electromagnetic Fields* (Wiley, New York, 1979).

These seven books all suffer from the defect that relativity is treated only tangentially or as an afterthought. In contrast,

Schwartz, M., *Principles of Electrodynamics* (McGraw-Hill, New York, 1972) recognizes the intimate connection between relativity and electrodynamics and exploits this to derive the Maxwell equations.

At somewhat lower mathematical level, the following books emphasize a grasp of physical principles:

Becker, R., *Electromagnetic Fields and Interactions* (Dover, New York, 1982) is a classic that has been around a long time. The first edition, written by A. Föppl, made its appearance nearly a hundred years ago, and was the book from which Einstein learned Maxwell's theory. It went through many revisions at the hands of M. Abraham and R. Becker. The latest edition, like the earlier ones, is distinguished by its lucid organization and well-balanced coverage.

Joos, G., *Theoretical Physics* (Hafner Press, New York, 1950) includes a concise but thorough section on electrodynamics.

Feynman, R. P., Leighton, R. B., and Sands, M., *The Feynman Lectures on Physics*, vols. 1 and 2 (Addison-Wesley, Reading, MA, 1964) contains a wealth of scintillating discussions providing sharp insights into the basic physics underlying complicated phenomena, such as the propagation of light in a material medium, electromagnetic oscillations of a capacitor, phase shifts produced by the vector potential, and so on.

Duffin, W. J., *Electricity and Magnetism* (Wiley, New York, 1973) emphasizes the empirical basics of electromagnetism and presents some clever and simple derivations of standard results. The commentaries at the ends of the chapters draw attention to oft-neglected conundrums; for instance, how can the electric field lines in a current-carrying wire follow the tortuous meanderings of a typical network?

Purcell, E. M., *Electricity and Magnetism* (McGraw-Hill, New York, 1965) gives careful and clear discussions of the basic principles of electromagnetism, and includes an especially nice analysis of the electric fields of moving charges and how the theory of relativity, applied to a moving charge on a wire, leads to the conclusion that there must exist magnetic fields.

The historical development of electric and magnetism is discussed in great detail in

Whittaker, E. A., *A History of the Theories of Aether and Electricity* (Harper & Row, New York, 1960). Although Whittaker provides good coverage for the eighteenth and nineteenth centuries, his view of twentieth-century developments is distorted by his attempt to give all the credit for the theory of relativity to Lorentz rather than to Einstein.

Further Readings

Chapter 1—Vector Calculus

Schey, H. M., *div, grad, curl, and all that* 3rd ed. (Norton, New York, 2005) is an "informal text on vector calculus" which provides a concise and very readable overview of the differential and integral operations with vectors.

Lengthier, more complete, and more rigorous treatments of vector calculus will be found in

Boas, M. L., *Mathematical Methods in the Physical Sciences* (Wiley, New York, 1983).

Brand, L., *Vector Analysis* (Wiley, New York, 1957), which includes chapters on the applications of vector calculus in particle mechanics, fluid mechanics, and electrodynamics.
Margenau, H., and Murphy, G. M., *The Mathematics of Physics and Chemistry* (Van Nostrand, Princeton, 1943), Chapter 4.
Wills, A. P., *Vector Analysis with an Introduction to Tensor Analysis* (Dover, New York, 1958).
Wrede, R. C., *Introduction to Vector and Tensor Analysis* (Wiley, New York, 1963).

The general definitions of vectors and tensors of any rank by means of their transformation properties under coordinate transformations are spelled out clearly in Margenau and Murphy, in Wrede, and also in

Bergmann, P. G., *Introduction to the Theory of Relativity* (Dover, New York, 1976).
Sokolnikoff, I. S., *Tensor Analysis* (Wiley, New York, 1951).

A nice introduction to vector analysis in the language of differential forms and the exterior calculus of Cartan is contained in

Misner, C. W., Thorne, K. S., and Wheeler, J. A., *Gravitation* (Freeman, San Francisco, 1973), Chapter 2–4.

Differential forms permit a very neat and concise formulation of Maxwell's equations, but the sleights of hand involved in this formulation tend to obscure the physics.

Chapter 2—Electrostatics

Many examples of problems in electrostatics involving a diversity of charge distributions are treated in the older books, such as

Webster, A. G., *The Theory of Electricity and Magnetism* (Macmillan, London, 1897).
Jeans, J., *The Mathematical Theory of Electricity and Magnetism* (University Press, Cambridge, 1960).
MacMillan, W. D., *The Theory of the Potential* (Dover, New York, 1958).
Smythe, W. R., *Static and Dynamic Electricity* (McGraw-Hill, New York, 1939).

Chapter 3—The Boundary-Value Problem

Thorough treatments of boundary-value problems, with examples, will be found in Jackson, *op. cit.,* and Panofsky and Philips, *op. cit.*

Besides the method of images, there are two other methods of solution of the Laplace equation that rely on magical tricks: the method of inversion and the method of complex transformation (in two dimensions). Panofsky and Philips and Jeans discuss these methods and give examples.

More information on the boundary-value problem and on the special functions that arise in the solution of the Laplace equation–such as the Bessel functions briefly mentioned in Section 3.3–will be found in books on mathematical physics; for instance

Boas, *op. cit.*
Margenau and Murpy, *op. cit.*
Morse, P. M., and Feshbach, H., *Methods of Theoretical Physics* (McGraw-Hill, New York, 1953), vol. 2.
Sommerfeld, A., *Partial Differential Equations in Physics* (Academic Press, New York, 1964).
Whittaker, E. T., and Watson, G. N., *A Course of Modern Analysis* (University Press, Cambridge, 1958).

General theorems regarding the existence and uniqueness of solutions are presented in

Courant, R., and Hilbert, D., *Methods of Mathematical Physics* (Interscience, New York, 1966), vol. 2.

Tables of values of special functions and lists of identities are available in

Jahnke, E., and Emde, F., *Tables of Functions* (Dover, New York, 1945).
Abramowitz, M., and Steigman, I. A., *Handbook of Mathematical Functions* (National Bureau of Standards, Washington, D.C., 1964).

Chapter 4—Dielectrics

A careful examination of the averaging procedures required to define the macroscopic electric and magnetic fields will be found in

Eyges, *op, cit.,* Chapter 6.
Jackson, *op. cit*., Section 6.7.
Rosenfeld, L., *Theory of Electrons* (Dover, New York, 1965), Chapter 2.

The definition of the **D** field depends on just what charges are included in the polarization. It is possible to include the "free" charges in an extra term in the polarization and thereby achieve a formal simplification of the Maxwell equations. This is shown in Booker, *op. cit*., Chapter 2.

Chapter 5— Electric Energy

For a clear and thorough discussion of energy in the electric field and the exchange of energy and momentum between field and charges, see Booker, *op. cit.*, Chapters 3 and 4.

Chapter 6— The Theory of Special Relativity

An English translation of Einstein's first paper on relativity, "Zur Elektrodynamik bewegter Körper" [*Ann. d. Physik* **17**, 891 (1905)], is available in a reprint volume:

Lorentz, H. A., Einstein, A., Minkowski, H., and Weyl, H., *The Principle of Relativity* (Dover, New York). Einstein's paper is amazingly complete; it contains the principle of relativity, the definition of simultaneity, the derivation of the Lorentz transformations, the transformations of the electric and magnetic fields, and even the equation of motion and the expression for the energy of a relativistic particle (but the expression Einstein obtains for the transverse mass is erroneous).

Good introductory books on relativity are:

French, A. P., *Special Relativity* (Norton, New York, 1968), which contains clear and thorough discussions of the traditional experiments test of relativity.
Ohanian, H. C., *Special Relativity: A Modern Introduction* (Physics Curriculum and Instruction, Lakeville, MN, 2001). This includes not only the classical experimental tests but also the most recent tests, such as the new measurements of the one-way speed of light. It also gives a simple discussion of how relativity implies an intimate connection between the electric and magnetic fields of moving and accelerating charges.

Taylor, E. F., and Wheeler, J. A., *Spacetime Physics*, 2nd ed. (Freeman, New York, 1992). This book emphasizes that the theory of relativity is a theory of the geometry of spacetime. It contains an abundance of intriguing puzzles and paradoxes.

More advanced treatises on relativity are

Anderson, J. L., *Principles of Relativity Physics* (Academic Press, New York, 1967), which includes a careful discussion of the spacetime structures underlying Newtonian and relativistic physics.
Bergmann, P. G., *Introduction to the Theory of Relativity* (Dover, New York, 1976). The first part of this book is an introduction to special relativity, with emphasis on the formalism of tensors.
Møller, C., *The Theory of Relativity* (Clarendon Press, Oxford, England, 1952). This very complete book includes a general discussion of relativistic energy, angular momentum, and center of mass for arbitrary closed systems.
Pauli, W., *Theory of Relativity* (Pergamon Press, London, 1958). This is a classic written when Pauli was a student of only 21. At the time of its first publication in 1921, it was a complete review of all the progress in relativity; by now, some parts are outdated, but the mathematical material remains valid and informative.
Rindler, W., *Essential Relativity*, 2nd ed. (Springer-Verlag, New York, 1977). The first six chapters deal with special relativity, presenting a clear and sharp critical examination of the conceptual foundation of the theory. There are also exceptionally nice qualitative treatments of length contraction and time dilation.
Synge, J. L., *Relativity: The Special Theory* (North-Holland, Amsterdam, 1965) discusses systems of particles and collisions, and contains a simple definition of the energy-momentum tensor.

Chapter 7—Four-Vectors and Tensors

A simple introduction to the energy-momentum vector of a particle is given in Taylor and Wheeler, *op. cit.,* Chapter 2.

More on the energy-momentum vector and other four vectors will be found in

Jackson, J. D., *Classical Electrodynamics*, *op. cit*, Chapter 12.
Panofsky and Philips, *op. cit.*, Chapter 17.
Aharoni, J., *The Special Theory of Relativity* (Dover, New York, 1965), Chapter 5.
Landau and Lifshitz, *op. cit.*, Chapter 2.
Hagedorn, R., *Relativistic Kinematics and Precession Polarization* (Benjamin, New York, 1963).

Chapter 8—Relativistic Electrodynamics

The derivation in this chapter of the complete set of Maxwell equations from Coulomb's law was inspired by Schwartz, *op. cit.*, Chapter 3.

For more rigorous derivation of Maxwell's equations from Coulomb's law, relativity, and suitable auxiliary assumptions, see

Frisch, D. H., and Wilets, L., *Am. J. Phys.* **24**, 574 (1956).
Tessman, J. R., *Am. J. Phys.* **34**, 1048 (1966).
Elliott, R. S., *IEEE Spectrum,* March 1966, p. 140.
Kobe, D. H., *Am. J. Phys.* **54**, 631 (1986).

The most rigorous of these is the paper by Frisch and Wilets, which shows that the complete set of Maxwell's equations can be deduced from Coulomb's law under the assumption that the fields depend only on the velocity and the acceleration (but not on any higher derivatives of the position) of the charges that act as sources.

The conventional approach is to begin with the three-dimensional form of Maxwell's equations and from these deduce the four-dimensional form. This is the approach taken by

Jackson, *op. cit.*
Panofsky and Philips, *op. cit.*
Marion, J. B., *Classical Electromagnetic Radiation* (Academic Press, New York, 1965).

General discussions of electrodynamics as a field theory will be found in

Aharoni, *op. cit.*, Chapters 3 and 4.
Landau and Lifshitz, *op. cit.*, Chapters 3 and 4.
Soper, D. E., *Classical Field Theory* (Wiley, New York, 1975).
Davis, W. R., *Classical Fields, Particles, and the Theory of Relativity* (Gordon and Breach, New York, 1970).

A very lucid introduction to the Lagrangian formulation for fields is contained in the last chapter of

Goldstein, H., *Classical Mechanics*, 3rd ed. (Addison-Wesley, Reading, MA, 2002).

Thorough discussions of energy flow in electromagnetic fields and energy and momentum transfer between electromagnetic fields and matter, with good illustrative examples, will be found in

Adler, R. B., Chu, L. J., and Fano, R. M., *Electromagnetic Energy Transmission and Radiation* (Wiley, New York, 1966).
Booker, *op. cit.*

Chapter 9—Magnetostatics

A few nice examples of exact solutions of the current flow in conductors and the energy flow in resistors and electron beams are treated in

Magid, L. M., *Electromagnetic Fields, Energy, and Waves* (Wiley, New York, 1972).

Some general theorems concerning DC Circuits (for example, Thevenin's and Norton's theorems) and some further theorems concerning magnetostatic fields are stated in Duffin, *op. cit.*, Chapters 6, 7, and 8.

Practical applications of magnetic forces on currents in electric motors are discussed in

Hague, B., *The Principles of Electromagnetism Applied to Electric Machines* (Dover, New York, 1962).
Slemon, G. R., *Magnetoelectric Devices* (Wiley, New York, 1966).

Chapter 10—Electromagnetic Induction

The application of Faraday's law is sometimes quite tricky. Feynman, *op.cit.*, gives some instructive examples.

The calculation in Section 10.1 of the induced electric field in a Betatron can be applied to a cylindrical solenoid, provided that the current flow is purely around the cylinder. However, in a solenoid with a helical winding, the current flow has a component parallel to the axis of the cylinder. A careful calculation, taking into account this component of the current flow and its contribution to the magnetic field, shows that the induced electric field in a helical solenoid has a surprisingly large component parallel to the axis; see Magid, *op. cit.*

Practical engineering applications of induction in electric generators are covered in Slemon, *op. cit.*

The motion of particles in a magnetic field is discussed in

Jackson, *op. cit.*, Chapter 12.
Fermi, E., *Nuclear Physics* (University of Chicago, 1950), Chapter 10.

The example of displacement current illustrated in Fig. 10.18 assumes that the path of integration is far from the fringing fields of the capacitor. The subtle complications that arise when the path is within the fringing fields are examined by

French, A. P., and Tessman, J. R., *Am. J. Phys.* **31**, 201 (1963).

Chapter 11—Magnetic Dipoles and Magnetic Materials

A critical examination of the role of the **H** field is found in Eyges, *op. cit.*, Chapter 9.

Diverse phenomena in ferromagnets are discussed in

Landau, L. D., and Lifshitz, E. M., *Electrodynamics of Continuous Media* (Addison-Wesley, Reading, MA, 1960), Chapter 5.

The physical properties of ferromagnetic materials are described in encyclopedic detail in

Bozorth, R. M., *Ferromagnetism* (Van Nostrand, New York, 1951).

Detailed calculations of the magnetic fields produced by current-carrying wires placed near masses of iron and the magnetic forces exerted on such wires are presented in Hague, *op. cit.*

Chapter 12—Electromagnetic Waves

A neat and concise introduction to electromagnetic waves, including an interesting chapter on the reflection of radio waves by the ionosphere, is found in

White, F. W. G., *Electromagnetic Waves* (Methuen, London, New York, 1950).

The rather complicated mathematics for treating cylindrical and spherical waves is developed in

Stratton, J. A., *Electromagnetic Theory* (McGraw-Hill, New York, 1941).

A more detailed discussion of the group velocity in a dispersive medium, including the Kramers-Kronig dispersion relations for the real and imaginary parts of the index of refraction, is given in Jackson, *op. cit.*, Chapter 7, and Panofsky and Philips, *op. cit.*, Chapter 22.

Further discussion of dispersion and of waves in material media, including inhomogeneous and anisotropic media, will be found in Rosenfeld, *op. cit.*, Chapters 5 and 6, and Landau and Lifshitz, *op. cit.*, Chapters 9 and 11.

A good description of how the index of refraction arises from the superposition of the incident wave and the scattered waves radiated by the electrons of a material medium is given by Feynman, *op. cit.*, and by Schwartz, *op. cit.*, Chapter 7.

Waves in a plasma are discussed in

Alfvén, H., *Cosmical Electrodynamics* (Oxford University Press, 1950).
Jackson, *op. cit.*, Chapter 10.

Much detail on the skin effect in several geometries is presented by

King, R. W. P., *Electromagnetic Engineering,* vol. 1 (McGraw-Hill, New York, 1945).

Chapter 13—Radiation by a Point Charge

The complete evaluation of all the terms in the electric and magnetic field of an accelerated point charge is done by Jackson, *op. cit.,* Chapter 14, and Panofsky and Philips, *op. cit.*, Chapter 20.

A simple derivation of the direction of propagation of the radiation emitted in the Cherenkov effect is given by

Motz, H., and Schiff, L. I., *Am. J. Phys.* **21**, 258 (1953)

Radiation-reaction forces are discussed in

Jackson, *op. cit.*, Chapter 17.
Rohrlich, F., *Classical Charged Particles* (Addison-Wesley, Reading, MA, 1965), Chapter 6.

Chapter 14—Radiation by a System of Charges

Detailed treatments of the radiation by systems of charges, including some examples of antennas, are given by Jackson, *op. cit.,* Chapter 9. and Stratton, *op. cit.,* Chapter 8.

More on antennas can be found in

Sommerfeld, *op. cit.*, Chapter 6.
Aharoni, J., *Antennae: An Introduction to their Theory* (Clarendon Press, Oxford, England, 1946).

Chapter 15—Resonant Cavities and Waveguides

Cylindrical cavities of circular cross section are treated by Jackson, *op. cit.,* Chapter 8.

The propagation of waves along a circular conductor (telegraph wire) is handled by Stratton, *op. cit.,* Chapter 9, and by Sommerfeld A., *Electrodynamics* (Academic Press, New York, 1964), Sections 22 and 23.

Detailed treatments of the general properties of wave propagation in waveguides and transmission lines are found in Adler, Chu, and Fano, *op. cit.,* and Magid, *op. cit.*

The following books emphasize applications of waveguides:

Ramo, S., and Whinnery, J. R., *Fields and Waves in Modern Radio* (Wiley, New York, 1953).
Slater, J. C., *Microwave Transmission* (McGraw-Hill, New York, 1942).
Southworth, G. C., *Principles and Applications of Waveguide Transmission* (Van Nostrand, Princeton, 1950).

Dielectric waveguides and their applications are discussed in

Cheo, P. K., *Fiber Optics* (Prentice-Hall, Englewood Cliffs, NJ, 1985).
Ghatak, A. K., and Thyagarjan, K., *Contemporary Optics* (Plenum, New York, 1978).

For an introduction to lasers, see

Sargent, M. III, Scully, M. O., and Lamb, W. E., *Laser Physics* (Addison-Wesley, Reading, MA, 1974).

Chapter 16—Scattering and Diffraction

Diverse aspect of scattering of electromagnetic waves are discussed in Panofsky and Philips, *op. cit.,* Chapter 22, and Landau and Lifshitz, *op. cit.,* Chapter 14.

A very thorough and general discussion of scattering is given in

Newton, R. G., *Scattering Theory of Waves and Particles* (McGraw-Hill, New York 1966), Part I.

More extensive and more rigorous treatments of the Kirchhoff integral and of Babinet's principle will be found in

Born, M., and Wolff, E., *Principles of Optics* (Pergamon Press, Oxford, 1970), Chapters 8 and 11.
Jackson, *op. cit.,* Chapter 9.
Sommerfeld, A., *Optics* (Academic Press, New York, 1964), Chapter 5.
Baker, B. B., and Copson, E. T., *The Mathematical Theory of Huygen's Principle* (Oxford University Press, 1939).
Jones, D. S., *The Theory of Electromagnetism* (Pergamon Press, Oxford, 1964), which gives *exact* solutions of problems in scattering (cylinder, sphere) and diffraction (semi-infinite plane and strip).

Answers to Even-Numbered Problems

Chapter 1

2. $\hat{\mathbf{n}} = \frac{1}{5\sqrt{2}}(-3\hat{\mathbf{x}} + 4\hat{\mathbf{y}} + 5\hat{\mathbf{z}})$

4. $\hat{\mathbf{n}} = \frac{1}{\sqrt{870}}(19\hat{\mathbf{x}} + 5\hat{\mathbf{y}} - 22\hat{\mathbf{z}})$

6. $\frac{14}{11}\hat{\mathbf{x}} - \frac{25}{11}\hat{\mathbf{y}} - \frac{13}{11}\hat{\mathbf{z}}$

8. $\hat{\boldsymbol{\rho}} \times \hat{\boldsymbol{\phi}} = \hat{\mathbf{z}}$, $\hat{\boldsymbol{\phi}} \times \hat{\mathbf{z}} = \hat{\boldsymbol{\rho}}$, $\hat{\boldsymbol{\rho}} \times \hat{\mathbf{z}} = -\hat{\boldsymbol{\phi}}$

14. (a) yes ; $\hat{\mathbf{x}} * \hat{\mathbf{y}} = (\hat{\mathbf{x}} + \hat{\mathbf{y}})/\sqrt{2}, \hat{\mathbf{x}} * \hat{\mathbf{z}} = (\hat{\mathbf{x}} + \hat{\mathbf{z}})/\sqrt{2}, \hat{\mathbf{y}} * \hat{\mathbf{z}} = (\hat{\mathbf{y}} + \hat{\mathbf{z}})/\sqrt{2}$; (b) no

18. $\begin{pmatrix} \cos\phi & \sin\phi & 0 \\ -\cos\theta\sin\phi & \cos\theta\cos\phi & \sin\theta \\ \sin\theta\sin\phi & -\sin\theta\cos\phi & \cos\theta \end{pmatrix}$; the inverse matrix is the transpose

20. $A_\rho = -\frac{3}{\sqrt{5}}, A_\phi = -\frac{4}{\sqrt{5}}, A_z = 3;\ A_r = -\frac{3}{\sqrt{5}}, A_\phi = -\frac{4}{\sqrt{5}}, A_\theta = -3$

26. (a) $\Phi = \frac{1}{n+2} r^{n+2}$; (b) $n = -3$

36. $\frac{8\pi}{3} CR^4$

Chapter 2

2. $\mathbf{E} = \hat{\mathbf{r}}\frac{q}{r^2}(1 + \kappa r)e^{-\kappa r}$; no; yes

4. $F = \frac{L^2}{Q^2}\ln\left[\frac{(d+L)^2}{(d+2L)d}\right]$

6. $2\pi\sigma z\left(\frac{1}{\sqrt{z^2 + R_1^2}} - \frac{1}{\sqrt{z^2 + R_2^2}}\right)\hat{\mathbf{z}}$

8. $\nu = 7.2\times 10^{15}$ Hz

10. 0.039 radian

12. (a) $E_x = 4\pi\rho x$ for $-L < x < L$, $E_x = -4\pi\rho L$ for $x \le -L$, $E_x = 4\pi\rho L$ for $x \ge L$; (b) 0

14. $2\lambda\mathbf{a}/R^2$, where $\mathbf{a}$ is the displacement vector from the axis of the cylinder to the axis of the cavity

16. (a) $\Phi = \frac{Q}{L}\ln\left[\frac{x+L/2+\sqrt{(x+L/2)^2+y^2+z^2}}{x-L/2+\sqrt{(x-L/2)^2+y^2+z^2}}\right]$

18. $\Phi = \frac{Q}{L}\ln\left[\frac{d+L/2+\sqrt{(d+L/2)^2+R^2}}{d-L/2+\sqrt{(d-L/2)^2+R^2}}\right]$

$$E = -\frac{Q}{L}\left[\frac{1}{\sqrt{(d+L/2)^2+R^2}} - \frac{1}{\sqrt{(d-L/2)^2+R^2}}\right]$$

20. (a) $\mathbf{E} = \frac{4\pi k r^{n+1}}{n+3}\hat{\mathbf{r}}$; (b) $\Delta\Phi = \left[\frac{4\pi k}{(n+3)(n+2)}\right]\left(a^{n+2} - b^{n+2}\right)$

22. $\Phi = \frac{q}{\sqrt{r^2+b^2/4+rb\cos\theta}} - \frac{q}{\sqrt{r^2+b^2/4-rb\cos\theta}}$

$$E_r = \frac{q(r+\frac{1}{2}b\cos\theta)}{\left(r^2+b^2/4+rb\cos\theta\right)^{3/2}} - \frac{q(r-\frac{1}{2}b\cos\theta)}{\left(r^2+b^2/4-rb\cos\theta\right)^{3/2}}$$

$$E_\theta = -\frac{qb\sin\theta}{\left(r^2+b^2/4+rb\cos\theta\right)^{3/2}} - \frac{qb\sin\theta}{\left(r^2+b^2/4-rb\cos\theta\right)^{3/2}}$$

24. $\omega = \sqrt{\frac{12qE}{mb}}$

26. (a) $\tau_{\text{on dipole}} = -\frac{pq'\sin\theta}{r^2}$; (b) $\tau_{\text{on charge}} = \frac{pq'\sin\theta}{r^2}$

Chapter 3

4. $\Phi = \frac{q}{\sqrt{r^2-6ar\cos\alpha+9a^2}} - \frac{q/3}{\sqrt{r^2-(2/3)ar\cos\alpha+a^2/9}}$;

$$\sigma = \frac{q}{4\pi a^2}\frac{8}{(10-6\cos\alpha)^{3/2}}$$, where α is angle with x axis

6. $\Phi = 2\lambda\ln\left[\frac{\sqrt{y^2+z^2}}{\sqrt{y^2+(z+2b)^2}}\right]$

8. (a) $\frac{q^2}{4b}$, no ; (b) 0.36 eV

10. $\Phi = \sum_{m,n=1}^{\infty} A_{mn}\sin\left(\frac{n\pi x}{L}\right)\sin\left(\frac{m\pi y}{L}\right)\sinh\left(\gamma_{mn}z\right)$, with

$$A_{mn} = \frac{4V_0}{\pi^2}\frac{[1-(-1)^n][1-(-1)^m]}{mn\sinh(\gamma_{mn}L)}$$, where $\gamma_{mn} = \sqrt{n^2+m^2}\,\frac{\pi}{L}$

12. $\sigma = \frac{1}{2\pi}E_0\cos\phi$

14. $\Phi = -\sum_1^{\infty}\frac{4V_0}{\pi n}\cos\left(\frac{1}{2}n\pi\right)\frac{R^n}{\rho^n}\sin n\phi$

16. $\Phi = \sum_{n\text{ even}}\frac{8V_0}{\pi n}\frac{\rho^n}{R^n}\sin n\phi$

18. $\Phi = \sum_{n \text{ odd}} B_l R^{-l-1} P_l(\cos\theta)$, where $B_l = (2l+1)V_0 R'^{l+1} \dfrac{1}{2^{l+1}} \dfrac{(-1)^{(l+1)/2}}{\{[(l+1)/2)]!\}^2} \dfrac{(l+1)!}{l}$

20. $\Phi = -E_0 r\cos\theta + E_0 \dfrac{R^3}{r^2}\cos\theta + \dfrac{Q}{r}$

22. $E = 4\pi\sigma_0$; $p = 4\pi\sigma_0 R^3$

24. $\Phi = -\dfrac{2}{3}\pi\rho_0 r^2 + A_0 + A_1 r\cos\theta + \dfrac{B_0}{r} + \dfrac{B_1}{r^2}\cos\theta$, where

$$A_0 = \frac{(2/3)\pi\rho_0(a^3/b - b^2) - 2V_0}{a/b - 1}, \quad A_1 = \frac{V_0}{a - b^3/a^2},$$

$$B_0 = \frac{(2/3)\pi\rho_0(b^2 - a^2) + 2V_0}{1/b - 1/a}, \quad B_1 = \frac{V_0}{1/a^2 - a/b^3}$$

26. $Q_{xx} = Q_{yy} = \tfrac{1}{2}b^2 Q$, $Q_{zz} = -b^2 Q$, $Q_{xy} = Q_{yz} = Q_{xz} = 0$;

$$\Phi = \frac{Q}{r} + 0 + \frac{1}{r^5}\frac{b^2 Q}{4}(x^2 + y^2 - 2z^2)$$

28. $\dfrac{q}{2r^7}[5(\mathbf{x}'\cdot\mathbf{x})^3 - 3r'^2 r^2 \mathbf{x}'\cdot\mathbf{x}]$

32. (b) $Q^{ij} = \dfrac{q}{m}(-3I^{ij} + \delta^{ij} I^{kk})$; no ; (c) $Q^{11} = qL^2/12, Q^{22} = qL^2/12, Q^{33} = -qL^2/6$

Chapter 4

4. $\alpha_2 = -\tan^{-1}\left(\dfrac{\varepsilon_2}{\varepsilon_1}\tan\alpha_1\right)$; $E_2 = \sqrt{\left(\dfrac{\varepsilon_1\cos\alpha_1}{\varepsilon_2}\right)^2 + \sin^2\alpha_1}\, E_1$

6. (a) $E = \dfrac{q}{\varepsilon r^2}$, $\Phi = \dfrac{q}{\varepsilon r}$; (b) 0.62 nm, 0.099 eV

8. $\Phi(\mathbf{x}) = \dfrac{q}{|\mathbf{x} - \mathbf{d}|} - \dfrac{\varepsilon - 1}{\varepsilon + 1}\dfrac{q}{|\mathbf{x} + \mathbf{d}|}$ above; $\Phi(\mathbf{x}) = \dfrac{2\varepsilon}{\varepsilon + 1}\dfrac{q}{|\mathbf{x} - \mathbf{d}|}$ below

12. $\Phi = \dfrac{V_0 \ln(1 + bz/a)}{\ln(1 + bd/a)} +$ constant ; $E = -\dfrac{bV_0}{(a + bz)\ln(1 + bd/a)}$;

$$\sigma_P = -\frac{1 - 1/(a + bd)}{4\pi}\frac{bV_0}{\ln(1 + bd/a)}; \quad \sigma_P = \frac{1 - 1/a}{4\pi}\frac{bV_0}{\ln(1 + bd/a)}$$

14. $E_{\text{crit}} = 7.8 \times 10^3$ stavolts/m; E_{crit} increases; E_{crit} decreases

16. $\Phi = -\dfrac{2\varepsilon E_0}{1 + \varepsilon}\rho\cos\phi$, $\mathbf{E} = \dfrac{2\varepsilon E_0}{1 + \varepsilon}\hat{\mathbf{x}}$ inside hole;

$$\Phi = -E_0\rho\cos\phi + \frac{1 - \varepsilon}{1 + \varepsilon}E_0 R^2 \frac{\cos\phi}{\rho},$$

$$\mathbf{E} = E_0\hat{\mathbf{x}} + \frac{1 - \varepsilon}{1 + \varepsilon}E_0 R^2\left(\frac{\cos\phi}{\rho^2}\hat{\boldsymbol{\rho}} + \frac{\sin\phi}{\rho^2}\hat{\boldsymbol{\phi}}\right) \text{ outside;}$$

$$\sigma_P = \frac{2E_0\cos\phi}{4\pi}\frac{1 - \varepsilon}{1 + \varepsilon}$$

18. $\Phi = -\dfrac{2E_0}{\varepsilon + 1}\rho\cos\phi$ inside; $\Phi = -E_0\rho\cos\phi + \dfrac{\varepsilon - 1}{\varepsilon + 1}E_0 R^2 \dfrac{\cos\phi}{\rho}$ outside

20. (a) $E_r = -Ar(3\cos^2\theta - 1)$, $E_\theta = 3Ar\cos\theta\sin\theta$ for $r > R$;

(b) $\Phi = Ar^2 \dfrac{3\cos^2\theta - 1}{2} + \dfrac{2AR^5}{r^3}\dfrac{1-\varepsilon}{2\varepsilon+3}\dfrac{3\cos^2\theta-1}{2}$ for $r > R$;

$\Phi = \dfrac{5Ar^2}{2\varepsilon+3}\dfrac{3\cos^2\theta-1}{2}$ for $r < R$

22. $\Phi = -E_0 r\cos\theta$, $\mathbf{E} = \mathbf{E}_0$ for $r < R_1$;

$\Phi = -\dfrac{E_0}{3}\dfrac{1+2\varepsilon}{\varepsilon} r\cos\theta + \dfrac{E_0 R_1^3}{3}\dfrac{1-\varepsilon}{\varepsilon}\dfrac{\cos\theta}{r^2}$ for $R_1 < r < R_2$;

$\Phi = -E_0 r\cos\theta + E_0(R_2^3 - R_1^3)\dfrac{\varepsilon-1}{3\varepsilon}\dfrac{\cos\theta}{r^2}$ for $r > R_2$

Chapter 5

2. $U = \dfrac{Q^2}{L^2}\left[(2L+d)\ln\left(\dfrac{2L+d}{L+d}\right) - d\ln\left(\dfrac{L+d}{d}\right)\right]$;

$F = -\dfrac{Q^2}{L^2}\left[\ln\left(1+\dfrac{L}{L+d}\right) - \ln\left(1+\dfrac{L}{d}\right)\right]$

4. $U = \dfrac{1}{2}\left(\dfrac{4\pi k}{3-n}\right)^2 \dfrac{R^{5-2n}}{5-2n}$; finite for $n < \dfrac{5}{2}$

6. (a) 1.6×10^{-3} erg ; (b) 1.3×10^{-3} erg ; (c) 0.99×10^{-3} erg

8. (a) $\mathbf{E} = q\dfrac{\hat{\mathbf{x}} - \hat{\mathbf{z}}d}{|\hat{\mathbf{x}} - \hat{\mathbf{z}}d|^3} + Q\dfrac{\mathbf{x}}{|\mathbf{x}|^3}$ for $|\mathbf{x}| > R$; $\mathbf{E} = q\dfrac{\hat{\mathbf{x}} - \hat{\mathbf{z}}d}{|\hat{\mathbf{x}} - \hat{\mathbf{z}}d|^3} + Q\dfrac{\mathbf{x}}{R^3}$ for $|\mathbf{x}| < R$

12. $R^3E_0^2\left[\dfrac{\varepsilon_1}{3}\left(\dfrac{\varepsilon_2-\varepsilon_1}{\varepsilon_2+2\varepsilon_1}\right)^2 + \dfrac{3\varepsilon_2}{2}\left(\dfrac{\varepsilon_1}{\varepsilon_2+2\varepsilon_1}\right)^2\right]$

14. $C = \dfrac{1+\varepsilon}{2}R$

16. (a) $C = \dfrac{3}{2}\varepsilon R$; (b) $E = \dfrac{3}{2}\dfrac{R\Delta\Phi}{r^2}$, $U = \dfrac{3}{4}\varepsilon R(\Delta\Phi)^2$

18. $F = -9.9\times10^2$ dynes

20. (a) $p_{kl} = \dfrac{1}{R}\begin{pmatrix} 1 & 1/2 & 1/3 \\ 1/2 & 1/2 & 1/3 \\ 1/3 & 1/3 & 1/3 \end{pmatrix}$, $c_{kl} = R\begin{pmatrix} 2 & -2 & 0 \\ -2 & 8 & -6 \\ 0 & -6 & 9 \end{pmatrix}$; (b) $U = \dfrac{7}{12}\dfrac{Q^2}{R}$

22. $p_{kl} = \begin{pmatrix} 1/R_2 + 1/\varepsilon R_1 - 1/\varepsilon R_2 & 1/R_2 \\ 1/R_2 & 1/R_2 \end{pmatrix}$

$c_{kl} = \dfrac{\varepsilon}{1/R_1R_2 - 1/R_2^2}\begin{pmatrix} 1/R_2 & -1/R_2 \\ -1/R_2 & 1/R_2 + 1/\varepsilon R_1 - 1/\varepsilon R_2 \end{pmatrix}$

24. (a) $p_{kl} \simeq \begin{pmatrix} 1/R_1 & 1/L \\ 1/L & 1/R_2 \end{pmatrix}$; $c_{kl} \simeq \dfrac{1}{1/R_1R_2 - 1/L^2}\begin{pmatrix} 1/R_2 & -1/L \\ -1/L & 1/R_1 \end{pmatrix}$

(b) $U \simeq \tfrac{1}{2}Q_1^2/R_1 + \tfrac{1}{2}Q_2^2/R_2 + Q_1Q_2/L^2$

26. $p_{kl} \simeq \begin{pmatrix} 1/R & 1/2R & 1/L \\ 1/2R & 1/2R & 1/L \\ 1/L & 1/L & 1/R \end{pmatrix}$

$$c_{kl} \simeq \frac{1}{1/R^3 - 2/RL^2}\begin{pmatrix} 2/R^2 - 4/L^2 & -2/R^2 + 4/L^2 & 0 \\ -2/R^2 + 4/L^2 & 4/R^2 - 4/L^2 & -2/RL \\ 0 & -2/RL & 1/R^2 \end{pmatrix}$$

Chapter 6

2. 2.3×10^{-21} s

4. $y' = \frac{x' + Vt'}{\sqrt{1-V^2/c^2}} \tan \omega \frac{t' + Vx'/c^2}{\sqrt{1-V^2/c^2}}$; no

6. (a) 1.95×10^{-3} eV; (b) 1.35×10^{-7}; (c) no

12. Tucana first, by 1.22 hours

18. (a) $\sin\phi = \sin\theta \frac{1-V^2/c^2}{1-2(V/c)\cos\theta + V^2/c^2}$; (b) $\phi = \theta$

20. 2.0×10^{-10} s

24. 6.0×10^{-12} s

26. $\Delta\tau = \frac{2V_0}{a_0}\sqrt{1-V_0^2/c^2}$; $\Delta\tau = \frac{1}{a_0}\left[V_0\sqrt{1-4V_0^2/c^2} + \frac{1}{2}\sin^{-1} 2V_0/c\right]$; the first

28. $V = 0.92c$; 50.2 years

34. $y' = - \; g\left(\frac{t' + Vx'/c^2}{\sqrt{1-V^2/c^2}}\right)$; no ; no ; no

Chapter 7

2. (b) yes

4. (c) $U = \begin{pmatrix} \sqrt{\frac{\gamma+1}{2}} & -\sqrt{\frac{\gamma-1}{2}} \\ -\sqrt{\frac{\gamma-1}{2}} & \sqrt{\frac{\gamma+1}{2}} \end{pmatrix} e^{i\phi}$, where $\gamma = \frac{1}{\sqrt{1-V^2/c^2}}$ and ϕ is arbitrary

12. $v = (1 - 9.4\times10^{-4})c$

14. 0.51 GeV

16. (a) 165 MeV ; (b) 578 MeV

18. 230 MeV , 198 MeV

20. (b) 3.8×10^{43} tons ; (c) 0.11 tons

26. $-\frac{VI\sin\theta}{c^2\sqrt{1-V^2/c^2}}$

Chapter 8

4. $E = 51$ statvolts/cm, $B = 51$ gauss

6. $\mathbf{E}' = \dfrac{\mathbf{V}/c \times \mathbf{B}}{\sqrt{1-V^2/c^2}}$; no

10. 1.3×10^5 gauss

12. $\mathbf{B} = \dfrac{2\lambda r}{\varepsilon R^2 c}\mathbf{V}\times\hat{\mathbf{r}}$ for $r<R$; $\mathbf{B} = \dfrac{2\lambda}{rc}\mathbf{V}\times\hat{\mathbf{r}}$ for $r>R$

14. (a) $\mathbf{E} = \gamma\mathbf{E}_0$, $\mathbf{B} = \dfrac{\gamma}{c}\mathbf{V}\times\mathbf{E}_0$; (b) $\Phi = -\gamma E_0 z$, $\mathbf{A} = -\dfrac{\gamma V E_0}{c} z\hat{\mathbf{y}}$

(c) $\sigma = \pm\dfrac{\gamma E_0}{4\pi}$, $\mathbf{j} = \pm\dfrac{\gamma V E_0}{4\pi}\hat{\mathbf{y}}$; yes

18. $$T^{\mu\nu} = \begin{pmatrix} (E_0^2+B_0^2)/8\pi & 0 & 0 & E_0B_0/4\pi \\ 0 & (B_0^2-E_0^2)/8\pi & 0 & 0 \\ 0 & 0 & (E_0^2-B_0^2)/8\pi & 0 \\ E_0B_0/4\pi & 0 & 0 & (E_0^2+B_0^2)/8\pi \end{pmatrix}$$

20. false (for motion parallel to **E** the actual Poynting vector is zero)

22. $\mathbf{S} = \dfrac{V}{4\pi}\gamma q^2\left(\gamma\dfrac{y'^2}{|\mathbf{x}'|^6}\hat{\mathbf{x}} - \dfrac{x'y'}{|\mathbf{x}'|^6}\hat{\mathbf{y}}\right)$, where $x' = \dfrac{x-Vt}{\sqrt{1-V^2/c^2}}$, $y' = y$, $z' = z$

24. $F_z = \dfrac{L^2}{8\pi}(E_2^2 - E_1^2)$

Chapter 9

2. $0, (1.4\times10^{-2}\text{ dyne})\hat{\mathbf{y}}$; $(1.4\times10^{-2}\text{ dyne})\hat{\mathbf{y}}$

6. $\mathbf{B} = \dfrac{16}{c}\dfrac{IL^2}{(4z^2+L^2)\sqrt{4z^2+2L^2}}\hat{\mathbf{z}}$

8. $\mathbf{B} = \dfrac{2\pi k}{c}(-\hat{\mathbf{x}}+\hat{\mathbf{y}})$, $\dfrac{2\pi k}{c}(-\hat{\mathbf{x}}-\hat{\mathbf{y}})$, $\dfrac{2\pi k}{c}(\hat{\mathbf{x}}-\hat{\mathbf{y}})$, $\dfrac{2\pi k}{c}(\hat{\mathbf{x}}+\hat{\mathbf{y}})$ for the first, second, third, and fourth quadrants, respectively

10. $\dfrac{2\pi k^2}{c^2}$

14. $B = \dfrac{2Ih}{cR^2}$ in a direction perpendicular to the line from center to center

18. $m = \dfrac{\omega Q}{4c}R^2$

22. (a) 8.1×10^{25} gauss·cm^3 ; (b) 0.24 gauss horizontal, 0.40 gauss vertical

24. (a) $$B_z = \frac{2Q\omega}{cLR^2}\Big[\sqrt{R^2+(z-L/2)^2}(L/2-z) + \sqrt{R^2+(z+L/2)^2}(L/2+z) + (L/2-z)^2 - (L/2+z)^2\Big]$$

(b) $$B_r = \frac{Q\omega}{2c}\left[\frac{R^2}{r^3}\cos\theta + \frac{1}{2r^5}(R^2L^2/2 - R^4)(5\cos^3\theta - 3\cos\theta) + \ldots\right]$$

$$B_\theta = \frac{Q\omega}{2c}\left[\frac{R^2}{2r^3}\sin\theta + \frac{3}{8r^5}\left(R^2L^2/2 - R^4\right)(5\cos^2\theta - 1)\sin\theta + \ldots\right]$$

26. (a) $B_z = \frac{2Q\omega}{cR^2}\left[\frac{2z^2+R^2}{\sqrt{R^2+z^2}} - 2z\right]$

(b) $B_r = \frac{Q\omega}{2c}\left[\frac{R^2}{r^3}\cos\theta - \frac{R^4}{2r^5}(5\cos^3\theta - 3\cos\theta) + ...\right]$

$B_\theta = \frac{Q\omega}{2c}\left[\frac{R^2}{2r^3}\sin\theta - \frac{3R^4}{8r^5}(5\cos^2\theta - 1)\sin\theta + ...\right]$

28. (a) 0.098 gauss; (b) 28 GeV/c

30. $x = a\cos\omega t$, $y = a\sin\omega t$, $z = -\frac{1}{2}\frac{e}{m_e}Et^2$, with $\omega = \frac{eB}{m_e c}$

32. 0.52 Hz

34. rear; front

42. 300 gauss

44. 4.1×10^{-15} statvolt, or 1.2×10^{-12} volt

Chapter 10

2. $\mathcal{E} = \frac{2Ivl}{c^2 r}$; $E' = \frac{2Iv}{c^2 r}$

4. $B = \frac{cRQ}{AN}$

6. (b) $E_z(\rho) \simeq \frac{1}{c^2}\frac{dI}{dt}\ln\frac{L}{\rho}$

8. 13 gauss; 6.1×10^{-7} statvolt/cm

12. $\frac{L}{l} = \frac{4\pi d}{c^2 h}$

16. $\mathcal{E} = \frac{2b}{c^2}\frac{dI}{dt}\ln\left[\frac{(d+h+a)h}{(d+h)(h+a)}\right]$; $M = \frac{2b}{c^2}\ln\left[\frac{(d+h+a)h}{(d+h)(h+a)}\right]$

18. $L = \frac{2N^2}{c^2}(R_1 - R_2)\ln\left(\frac{R_2}{R_1}\right)$

20. (b) $I \simeq 2.8\times10^{15}$ statamps

22. $M = \frac{2\pi^2}{c^2}R'\left[\frac{R'}{R} - \frac{3R'^3}{8R^3} - \frac{15R'^5}{64R^5} + ...\right]$

24. $M_{nk} = \frac{2}{c^2}\begin{pmatrix} \ln 3 & \ln(3/2) \\ \ln(3/2) & \ln 3 \end{pmatrix}$ per unit length

26. $P = \frac{V_0^2\cos(\omega t - \phi)\cos\omega t}{\sqrt{R^2 + (\omega L - 1/\omega C)^2}}$; $\overline{P} = \frac{V_0^2\cos\phi}{2\sqrt{R^2 + (\omega L - 1/\omega C)^2}}$

28. $Z = \frac{1}{i/\omega L + 1/R - i\omega C}$; maximum for $\omega = 0$ and $\omega = \infty$, minimum for $\omega = 1/\sqrt{LC}$

30. (a) 137 MeV/c, 0.15c, 50 km ; (b) latitude $35.3°$; (c) $\alpha_{\text{crit}} = 51.7°$

32. $E_z = \frac{2\omega}{c^2}I_0\sin\omega t\ln\frac{b}{\rho}$; $B_\phi = \frac{2\omega^2}{c^3}I_0\cos\omega t\left(\frac{a^2}{2\rho}\ln\frac{b}{a} + \frac{a^2}{4\rho} - \frac{\rho}{2}\ln\frac{b}{\rho} - \frac{\rho}{4}\right)$

34. (a) $E_{(1)} = \frac{\omega B_0}{2c}\rho\sin\omega t$; (b) $B_{(2)} = -\frac{1}{4}\frac{\omega^2 B_0}{c^2}\rho^2\cos\omega t$

(c) $E_{(2)} = -\frac{1}{16}\frac{\omega^3 B_0}{c^3}\rho^3\sin\omega t$

36. $\simeq 60$ statvolt, or $\simeq 2\times10^4$ volt

Chapter 11

2. 8.025×10^7 radians/s

4. 3.4×10^5 gauss/cm

6. (a) yes; (b) no; (c) 12

8. $\chi = 1.1\times10^{-10}$

10. $F = 2\frac{\mu}{c^2}\frac{I^2}{d}$

12. (a) $\mathbf{B} = \frac{\mathbf{B}_0}{\mu}$, $\mathbf{H} = \frac{\mathbf{B}_0}{\mu}$; (b) $\mathbf{B} = \mathbf{B}_0$, $\mathbf{H} = \mathbf{B}_0$

14. $B = 5500$ gauss, $H = 1.38$ gauss, $M = (B-H)/4\pi \simeq 440$ gauss,

$k = 1.3\times10^{13}$ statamps/cm

16. $B \simeq 11{,}000$ gauss

18. (a) $B_z = 2\pi M\left[\frac{L/2-z}{\sqrt{R^2+(z-L/2)^2}} + \frac{L/2+z}{\sqrt{R^2+(z+L/2)^2}}\right]$

(b) $B_r = 2\pi ML\left[\frac{R^2}{r^3}\cos\theta + \frac{R^4}{4r^5}\left(\frac{L^2}{R^2}-3\right)(5\cos^2\theta - 3\cos\theta) + \ldots\right]$

$B_\theta = 2\pi ML\left[\frac{R^2}{2r^3}\sin\theta + \frac{3R^4}{16r^5}\left(\frac{L^2}{R^2}-3\right)(5\cos^2\theta - 1)\sin\theta + \ldots\right]$

20. With the abbreviation $K = \left[(2\mu+1)(\mu+2) - 2(\mu-1)^2 a^3/b^3\right]^{-1}$,

$\Phi_M = -9\mu B_0 K r\cos\theta$, $B = H = -\nabla\Phi_M$ for $r < a$;

$\Phi_M = -3(\mu+1)B_0 K r\cos\theta + 9(\mu-1)a^3 B_0 K\frac{\cos\theta}{r^2}$, $B/\mu = H = -\nabla\Phi_M$

for $b > r > a$;

$\Phi_M = -B_0 r\cos\theta + (2\mu+1)(\mu-1)(b^3-a^3)B_0 K\frac{\cos\theta}{r^2}$, $B = H = -\nabla\Phi_M$

for $r > b$; factor of $\frac{1}{\mu}\frac{9b^3}{2(b^3-a^3)}$

22. $\Phi_M = -B_0\rho\cos\phi + A\frac{\cos\phi}{\rho}$, $B = H = -\nabla\Phi_M$ for $\rho > b$

$\Phi_M = C\rho\cos\phi + D\frac{\cos\phi}{\rho}$, $B/\mu = H = -\nabla\Phi_M$ for $a < \rho < b$

$\Phi_M = F\rho\cos\phi$, $B = H = -\nabla\Phi_M$ for $\rho < a$

where $A = -[a^2b^2(\mu^2-1) - b^4(\mu^2-1)]B_0 K$, $C = -2b^2(\mu+1)B_0 K$,

$D = -2a^2b^2(\mu-1)B_0 K$, $F = -4b^2\mu B_0 K$, with

$K = [a^2(-\mu^2+2\mu-1) + b^2(\mu^2+2\mu+1)]^{-1}$

Chapter 12

2. $\omega' = \dfrac{\omega}{\sqrt{1-V^2/c^2}}$; $\theta' = \cos^{-1}(-V/c)$

4. 41 arcseconds; 0.11 nm

6. $\mathbf{E} = -E_0(\hat{\mathbf{x}} + i\hat{\mathbf{y}})e^{-i(\omega t+kz)}$; negative helicity; 0

8. 2.9×10^{-5} cm, 2.1×10^{-13} g

12. (c) $\mathbf{H}'_\parallel = \mathbf{H}_\parallel$, $\mathbf{H}'_\perp = \gamma\left(\mathbf{H}_\perp - \dfrac{\mathbf{V}}{c}\times\mathbf{D}\right)$; $\mathbf{D}'_\parallel = \mathbf{D}_\parallel$, $\mathbf{D}'_\perp = \gamma\left(\mathbf{D}_\perp + \dfrac{\mathbf{V}}{c}\times\mathbf{H}\right)$

$\mathbf{M}'_\parallel = \mathbf{M}_\parallel$, $\mathbf{M}'_\perp = \gamma\left(\mathbf{M}_\perp + \dfrac{\mathbf{V}}{c}\times\mathbf{P}\right)$; $\mathbf{P}'_\parallel = \mathbf{P}_\parallel$, $\mathbf{P}'_\perp = \gamma\left(\mathbf{P}_\perp - \dfrac{\mathbf{V}}{c}\times\mathbf{M}\right)$

14. 3.757° , 3.745° , 3.721°

16. (a) $l^* = \dfrac{x}{\tan\{\sin^{-1}[1.33\sin(\tan^{-1}x/0.5\text{ m})]\}}$

(b) $l^* = \dfrac{x}{\tan\{\sin^{-1}[(1/1.33)\sin(\tan^{-1}x/0.5\text{ m})]\}}$

18. yes; yes; requires simultaneous incidence of two coherent waves from upper right and lower right

24. $\theta \geq 89.8898°$

26. (a) rotates in x-z plane

(b) $|E'_{0x}|/E_0 = \sqrt{(1.5\sin\theta)^2 - 1}$, $|E'_{0z}|/E_0 = 1.5\sin\theta$

28. $E''_0 = -\dfrac{2i(n'^2-1)\sin k'd}{e^{ik'd}(n'-1)^2 - e^{-ik'd}(n'+1)^2}E_0$

$E'_0 = -\dfrac{4n'e^{ikd}}{e^{ik'd}(n'-1)^2 - e^{-ik'd}(n'+1)^2}E_0$

30. With the abbreviation $\beta = \sqrt{2\pi\sigma/\mu'\omega}$,

$E''_0 = \sqrt{\dfrac{(\cos\theta - \beta\cos\theta')^2 + \beta^2\cos^2\theta'}{(\cos\theta + \beta\cos\theta')^2 + \beta^2\cos^2\theta'}}\,E_0 e^{i\delta}$

where $\delta = \tan^{-1}\left(\dfrac{-2\beta\cos\theta\cos\theta'}{\cos^2\theta - \beta^2\cos^2\theta'}\right)$

$[\text{reflected fraction}] = \dfrac{(\cos\theta - \beta\cos\theta')^2 + \beta^2\cos^2\theta'}{(\cos\theta + \beta\cos\theta')^2 + \beta^2\cos^2\theta'}$

32. $N > 5.1\times10^8/\text{cm}^3$

Chapter 13

2. $E_x = -\dfrac{qa^3b}{2c^6}\dfrac{1}{\sqrt{1+b^2a^2/c^4} - 1}$, $E_y = -\dfrac{qa^2}{2c^4}$, $B_z = \sqrt{E_x^2 + E_y^2}$

4. (a) $\dfrac{1}{3}\dfrac{e^4E_0^2}{c^3m_e^2}$

6. (a) $\mathbf{a} = -\dfrac{eE_0}{m_e}(\hat{\mathbf{x}}\cos\omega t + \hat{\mathbf{y}}\sin\omega t)$

(b) $\mathbf{E} = -\dfrac{e^2 E_0}{m_e c^2 R}\hat{\mathbf{y}}\cos\omega(t - R/c)$; $\mathbf{B} = \hat{\mathbf{x}}\times\mathbf{E}$; linear polarization

(c) $\mathbf{E} = -\dfrac{e^2 E_0}{m_e c^2 R}[\hat{\mathbf{x}}\cos\omega(t - R/c) + \hat{\mathbf{y}}\sin\omega(t - R/c)]$; $\mathbf{B} = \hat{\mathbf{z}}\times\mathbf{E}$;
circular polarization

8. 7.0×10^{-3} statvolt/cm compared with 5.3×10^{-11} statvolt/cm

10. $E = \dfrac{8}{45}\dfrac{m_p v_0^5}{c^3 Z}$

12. $P = \dfrac{2}{3}\dfrac{e^{14} m_e^2}{c^3 \hbar^8 n^8}$; $t = \dfrac{3}{2}\dfrac{c^3 \hbar^6 n^5}{e^{10} m_e}$; 1.4×10^{-6} s

14. $ma' + \dfrac{2}{3}\dfrac{q^2 a'^2}{c^3 v}$; $\dfrac{mv'^2}{r}$ radial and $\dfrac{2}{3}\dfrac{q^2 v'^3}{c^3 r^2}$ tangential

16. $P = \dfrac{q^2 a'^2}{2v'^3}\left[\dfrac{3 - 2v'/c}{10(1 - v'/c)^4} - \dfrac{3 + 2v'/c}{10(1 + v'/c)^4} - \dfrac{1}{3(1 - v'/c)^3} + \dfrac{1}{3(1 + v'/c)^3}\right]$

20. $\mathbf{E} = \mu\dfrac{q}{c^2}\left[\dfrac{\hat{\mathbf{n}}\times\{(\hat{\mathbf{n}} - \sqrt{\mu\varepsilon}\,\mathbf{v}'/c)\times\mathbf{a}'\}}{(1 - \sqrt{\mu\varepsilon}\,\hat{\mathbf{n}}\cdot\mathbf{v}'/c)^3|\mathbf{x} - \mathbf{x}'|}\right]$

$\mathbf{B} = \mu\sqrt{\mu\varepsilon}\dfrac{q}{c^2}\left[\dfrac{-\hat{\mathbf{n}}\times\mathbf{a}'(1 - \sqrt{\mu\varepsilon}\,\hat{\mathbf{n}}\cdot\mathbf{v}'/c) - \sqrt{\mu\varepsilon}(\hat{\mathbf{n}}\times\mathbf{v}'/c)\hat{\mathbf{n}}\cdot\mathbf{a}'}{(1 - \sqrt{\mu\varepsilon}\,\hat{\mathbf{n}}\cdot\mathbf{v}'/c)^3|\mathbf{x} - \mathbf{x}'|}\right]$

22. (a) 0.17 MeV; (b) 1.0 MeV; (c) 42°

Chapter 14

2. $R = \dfrac{2.44}{c} = 8.13\times10^{-11}$ statohm $\Leftrightarrow$ 73.1 ohm

4. $\dfrac{d\bar{P}}{d\Omega} = \dfrac{I_0^2}{2\pi c}\dfrac{\sin^2(\pi\cos\theta)}{\sin^2\theta}$; this radiation pattern has a nodal plane at $\theta = 90°$,
in contrast to the radiation pattern of the ordinary full wave antenna, which has a maximum at $\theta = 90°$

6. $\dfrac{d\bar{P}}{d\Omega} = \dfrac{2I_0^2}{\pi c}\dfrac{\cos^2(\tfrac{1}{2}\pi\cos\theta)}{\sin^2\theta}\sin^2(\tfrac{1}{2}\pi\sin\theta\cos\phi)$; this radiation pattern has
maxima at $\theta = 90°$, $\phi = 0°$ and $\phi = 180°$

8. $\dfrac{d\bar{P}}{d\Omega} = \dfrac{2I_0^2}{\pi c}\dfrac{\cos^2(\tfrac{1}{2}\pi\cos\theta)}{\sin^2\theta}\sin^2(\tfrac{1}{2}\pi\sin\theta\cos\phi)$

10. (a) 5×10^{15}/s; (b) 8×10^{12}/s; (c) 2×10^{13}/s

12. (a) 1.2×10^{36} erg/s; (c) circular; linear, in *x-y* plane

14. $\bar{P} = \dfrac{1}{3c^3}\left(\dfrac{Q}{2Mc}\right)^6 S^2 B_0^4$

16. $\dfrac{16}{315}\dfrac{m_p v_0^7}{c^5 Z}$

18. $P = \dfrac{2}{3}\dfrac{q^2}{c^3}\left[\omega^2 b\sin\omega(t - r/c)\right]^2$; $P = \dfrac{1}{30}\dfrac{q^2}{c^5}\left[4\omega^3 b^2\sin 2\omega(t - r/c)\right]^2$

20. $\dfrac{2}{7}\left(\dfrac{8}{15}\right)^2 \dfrac{mv_0^7}{c^5}$

22. 1.57×10^8 /s ; $6.37\times10^{-9} s$

Chapter 15

4. At $z=0$ and $z=d$: $\sigma = \pm\dfrac{E_0}{4\pi}\sin(\pi x/a)\sin(\pi y/b)\cos\omega t$

$$\kappa_x = \mp\frac{c^2E_0}{4\omega a}\cos(\pi x/a)\sin(\pi y/b)\sin\omega t$$

$$\kappa_y = \mp\frac{c^2E_0}{4\omega b}\sin(\pi x/a)\cos(\pi y/b)\sin\omega t$$

At $x=0$ and $x=a$: $\sigma=0$, $\kappa_z = \pm\dfrac{c^2E_0}{4\omega a}\sin(\pi y/b)\sin\omega t$

At $y=0$ and $y=b$: $\sigma=0$, $\kappa_z = \pm\dfrac{c^2E_0}{4\omega b}\sin(\pi x/a)\sin\omega t$

6. See Eqs. (72)–(77)

10. $\overline{P} = \dfrac{B_0^2ab\omega}{32\pi}\dfrac{\sqrt{\omega^2/c^2-(m\pi/a)^2-(n\pi/b)^2}}{(m\pi/a)^2+(n\pi/b)^2}$; $\overline{P} \simeq \dfrac{B_0^2ab\omega^2}{32\pi c}\dfrac{1}{(m\pi/a)^2+(n\pi/b)^2}$

12. $$\overline{S}_z = \frac{B_0^2\omega k}{8\pi[(m\pi/a)^2+(n\pi/b)^2]^2}\Big[(n\pi/b)^2\cos^2(m\pi x/a)\sin^2(n\pi y/b) + (m\pi/a)^2\sin^2(m\pi x/a)\cos^2(n\pi y/b)\Big]$$

$$\overline{T}^{00} = \frac{B_0^2}{16\pi[(m\pi/a)^2+(n\pi/b)^2]^2}\times$$
$$\{[(\omega n\pi/cb)^2+(kn\pi/b)^2]\cos^2(m\pi x/a)\sin^2(n\pi y/b)$$
$$+[(\omega m\pi/ca)^2+(km\pi/a)^2]\sin^2(m\pi x/a)\cos^2(n\pi y/b)$$
$$+[(m\pi/a)^2+(n\pi/b)^2]^2\cos^2(m\pi x/a)\cos^2(n\pi y/b)\}$$

14. At $x=0$ and $x=a$: $\kappa_y = \mp\dfrac{cB_0}{4\pi}\cos(\pi y/b)e^{-i\omega t+ikz}$,

$$\kappa_z = \mp\frac{cB_0}{4\pi}\frac{ik}{\omega^2/c^2-k^2}\frac{\pi}{b}\sin(\pi y/b)e^{-i\omega t+ikz}$$

$$\sigma = \mp\frac{B_0}{4\pi}\frac{1}{\omega^2/c^2-k^2}\frac{i\omega\pi}{cb}\sin(\pi y/b)e^{-i\omega t+ikz}$$

At $y=0$ and $y=b$: $\kappa_x = \pm\dfrac{cB_0}{4\pi}\cos(\pi x/a)e^{-i\omega t+ikz}$,

$$\kappa_z = \pm\frac{cB_0}{4\pi}\frac{ik}{\omega^2/c^2-k^2}\frac{\pi}{a}\sin(\pi x/a)e^{-i\omega t+ikz}$$

$$\sigma = \pm\frac{B_0}{4\pi}\frac{1}{\omega^2/c^2-k^2}\frac{i\omega\pi}{ca}\sin(\pi x/a)e^{-i\omega t+ikz}$$

16. 4.30×10^9 Hz, 3.16×10^9 Hz, 2.08×10^9 Hz;

2×10^5 cm, 3×10^5 cm, 5×10^5 cm

18. (a) $E_x=0$, $E_y=E_0\cos(\omega t-\omega z/c)$, $B_x=-E_0\cos(\omega t-\omega z/c)$, $B_y=0$

$\lambda = 2\pi c/\omega, \quad v_p = c, \quad v_g = c$

(b) $Q\lambda$, or $\pi b\sqrt{\pi\sigma/2\omega}$

22. $\tan\dfrac{\gamma b}{2} = -\dfrac{\gamma/\mu}{\sqrt{\omega^2(\mu\varepsilon - 1)/c^2 - \gamma^2}}; \quad \omega = \dfrac{\pi c/b}{\sqrt{\mu\varepsilon - 1}}$

Chapter 16

2. $\dfrac{d\sigma}{d\Omega} = \dfrac{k^2a^2b^2}{4\pi^2}\dfrac{\sin^2(kax/2r_0)}{(kax/2r_0)^2}\dfrac{\sin^2(kby/2r_0)}{(kby/2r_0)^2}$; no, the Fraunhofer approximation breaks down at large angles

6. $\overline{S} = \dfrac{ck^2a^4E_0^2\cos^2\alpha}{32\pi^3r_0^3}\dfrac{\sin^2(kax/2r_0)}{(kax/2r_0)^2}\dfrac{\sin^2(kay\cos\alpha/2r_0)}{(kay\cos\alpha/2r_0)^2}$, where x and y are measured perpendicularly from the line of incidence; central maximum is on the line of incidence; widths are $2\lambda/a$, $2\lambda/a\cos\alpha$

8. The modulation of the intensity of the dark dots (dark zones of missing dots) would expand by a factor of two. The separation between the dots would shrink by a factor of two.

10. 0.19°; 1.9°

12. $\overline{S} = \dfrac{ck^2a^2b^2E_0^2}{8\pi^3r_0^3}\dfrac{\sin^2(kax/2r_0)}{(kax/2r_0)^2}\dfrac{\sin^2(kay/2r_0)}{(kay/2r_0)^2}\cos^2(kdx/2r_0)$

14. $\overline{S} = \dfrac{cE_0^2}{32\pi}\times 0.064$

16. $\overline{S} = \dfrac{cE_0^2}{4\pi}\left\{\left[\tfrac{1}{2} - \mathscr{C}(w)\right]^2 + \left[\tfrac{1}{2} - \mathscr{S}(w)\right]^2\right\}$, where $w = \sqrt{\dfrac{k}{\pi r_0}}\dfrac{a}{2}$

18. $\overline{S} = \dfrac{cE_0^2}{8\pi}\Big|1 + \tfrac{1}{2}i[\mathscr{C}(w_2) + i\mathscr{S}(w_2) - \mathscr{C}(w_1) - i\mathscr{S}(w_1)]\times$

$[\mathscr{C}(v_2) + i\mathscr{S}(v_2) - \mathscr{C}(v_1) - i\mathscr{S}(v_1)]\Big|^2$;

forward: $\overline{S} = \dfrac{cE_0^2}{8\pi}\left|1 + 2i[\mathscr{C}(w) + i\mathscr{S}(w)][\mathscr{C}(v) + i\mathscr{S}(v)]\right|^2$,

where $w = \sqrt{\dfrac{k}{\pi r_0}}\dfrac{a}{2}$, $v = \sqrt{\dfrac{k}{\pi r_0}}\dfrac{b}{2}$

20. $\overline{S} = \dfrac{cE_0^2}{16\pi}\left\{[\mathscr{C}(w_2) - \mathscr{C}(w_1)]^2 + [\mathscr{S}(w_2) - \mathscr{S}(w_1)]^2\right\}$;

forward: $\overline{S} = \dfrac{cE_0^2}{4\pi}\left\{[\mathscr{C}(w)]^2 + [\mathscr{S}(w)]^2\right\}$, where $w = \sqrt{\dfrac{k}{\pi r_0}}\dfrac{a}{2}$

Index